Information Technology and Law Series

Volume 31

More information about this series at http://www.springer.com/series/8857

Simone van der Hof · Ilina Georgieva ·
Bart Schermer · Bert-Jaap Koops
Editors

Sweetie 2.0

Using Artificial Intelligence to Fight Webcam Child Sex Tourism

ASSER PRESS

Springer

Editors
Simone van der Hof
Center for Law and Digital Technologies (eLaw), Leiden Law School
Leiden University
Leiden, The Netherlands

Ilina Georgieva
Faculty Governance and Global Affairs
Institute of Security and Global Affairs (ISGA)
Leiden University
The Hague, The Netherlands

Bart Schermer
Center for Law and Digital Technologies (eLaw), Leiden Law School
Leiden University
Leiden, The Netherlands

Bert-Jaap Koops
Tilburg Institute for Law, Technology, and Society (TILT), Tilburg Law School
Tilburg University
Tilburg, The Netherlands

ISSN 1570-2782 ISSN 2215-1966 (electronic)
Information Technology and Law Series
ISBN 978-94-6265-287-3 ISBN 978-94-6265-288-0 (eBook)
https://doi.org/10.1007/978-94-6265-288-0

Library of Congress Control Number: 2018968089

Published by T.M.C. ASSER PRESS, The Hague, The Netherlands www.asserpress.nl
Produced and distributed for T.M.C. ASSER PRESS by Springer-Verlag Berlin Heidelberg

This T.M.C. ASSER PRESS imprint is published by the registered company Springer-Verlag GmbH, DE part of Springer Nature
The registered company address is: Heidelberger Platz 3, 14197 Berlin, Germany

Series Information

The *Information Technology and Law Series* was an initiative of IT*e*R, the national programme for Information Technology and Law, which was a research programme set up by the Dutch government and The Netherlands Organisation for Scientific Research (NWO) in The Hague. Since 1995 IT*e*R has published all of its research results in its own book series. In 2002 IT*e*R launched the present internationally orientated and English language *Information Technology and Law Series*.

This well-established series deals with the implications of information technology for legal systems and institutions. Manuscripts and related correspondence can be sent to the Series' Editorial Office, which will also gladly provide more information concerning editorial standards and procedures.

Editorial Office

Foreword

Two factors in the developing world, namely rising Internet usage rates and persistent poverty, have fostered the emergence of a new and rapidly growing form of online child sexual exploitation. Webcam Child Sex Tourism (WCST), as it was coined some 5 years ago, was identified as a practice whereby men from wealthier parts of the world pay money to children in developing countries to perform sexually explicit shows in front of a webcam.

At a time when WCST was not really present on the radar of law enforcement as a criminal activity, Terre des Hommes, working with former teenage prostitutes in the Philippines, picked up numerous signals that the phenomenon was spreading like an epidemic in the slums of Manila and Cebu. Recognizing the latter's significance, we realized that focusing exclusively on rescuing victims—our traditional role—was insufficient. A shift in our strategy was required, namely to move the focus of our work from the "supply" side of the problem to the "demand" side. The euphemistically named "clients" or "customers" are perpetrators who recognized years ago the potential of the Internet to live-stream sex shows involving children, evidence of which would disappear the moment the connection was broken, rendering arrest or prosecution virtually impossible.

The demand for live online sexual contact with children has grown exponentially over the past 5 years. Terre des Hommes now maintains the only way to take on this sordid industry is to go after the users themselves by tracking them online.

Therefore, in 2013, Terre des Hommes initiated the Sweetie Project. It aimed to draw attention to the scale of the online sexual exploitation of children and demonstrate that the identification of potential child abusers is relatively simple. By using computer animation technology, a virtual 10-year old Philippine girl was created, which allowed researchers to identify one thousand predators from no less than 71 countries within 10 weeks, using information obtained online in the public domain such as Facebook or Yahoo accounts. This fact alone illustrates that the average online abuser feels largely unthreatened by any exposure of their criminal activity.

"The moment you go online, pretending to be a child from the Philippines, they jump at you, hundreds of them," as one of the researchers on the Sweetie Project stated at the time. According to an estimate by the FBI confirmed by the UN, there are some 750,000 individuals online at any given moment, seeking to engage children in sexual activities. However, at the time of our first research efforts we could only document six successful prosecutions worldwide. Obviously, law enforcement is ill equipped trying to apply twentieth-century-based laws and practices to this twenty-first-century form of serious crimes against children.

At the culmination of the Sweetie Project, the files of the 1,000 perpetrators were handed over to the Dutch desk at Europol for distribution to member states and beyond. Sweetie became world news for months. As far as we know, arrests and convictions have taken place in countries such as Australia, Belgium, Denmark, the Netherlands, Poland, and the UK. The issue is now in the public domain, laws are under review in some countries and law enforcement has acknowledged the seriousness of what is generally now described as "the live streaming of child sex abuse." However, there is growing concern expressed by several police forces that "we cannot arrest our way out of this problem." It is simply too big.

Terre des Hommes advocates a more proactive approach, preventive in nature, to tackle the demand side of WCST. With this in mind, we developed "Sweetie 2.0" software to intercept, identify, and deter individuals who use the internet to sexually abuse children. The updated chatbots enable their administrators to monitor larger parts of the Internet to locate and identify (potential) predators, and to subsequently send them messages that warn of the legal consequences should they proceed. According to research conducted by forensic psychologists, some 25% of men (one in four!) who begin by watching child pornography will then move on to hands-on child sexual abuse. Early intervention, the objective of Sweetie 2.0, could play a significant role in preventing this outcome.

Applying such investigation tools, however, raises serious legal questions. Will law enforcement be allowed to apply these innovative new technologies? What about issues of entrapment or the protection of privacy? Can live-streaming be regarded as a criminal offense in the first place? Not simple questions; difficult to answer and compounded by differences in judicial practice between countries.

It is, therefore, with great pleasure that we welcome this book, the outcome of a joint research initiative by Leiden University's Center for Law and Digital Technologies (eLaw) and Tilburg University's Institute for Law, Technology, and Society (TILT). The team involved took on the study of this new and challenging phenomenon in order to unravel these complicated legal issues in a concise and readable(!) manner, and to provide much needed answers where possible and prudent.

It is no mean feat, for which we at Terre des Hommes are genuinely grateful. We hope you enjoy reading it with the same enthusiasm with which it was conceived and put together.

The Hague, The Netherlands

Hans Guyt
Team Leader, Sweetie Project Terre des Hommes NL

Contents

Editors and Contributors

About the Editors

Simone van der Hof is the Director of the Center for Law and Digital Technologies (eLaw) at Leiden Law School, Programme Director of the Advanced Studies Programme in Law and Digital Technologies, and one of the Directors of the Leiden Law School research profile area Interaction between legal systems. She coordinates and teaches various courses, among which "Regulating Online Child Safety" (Master Youth Law), "Digital Child Rights" (Advanced Master Law and Digital Technologies), "The Rights of the Child in the Digital World" (Advanced Master International Children's Rights). She is a Key Lecturer at the Cyber Security Academy. Simone's particular academic interest is in the field of online privacy, digital child rights, and regulation of online child safety. She was involved in the Sweetie 2.0 project on online webcam child sex abuse, commissioned by children's rights organization Terre des Hommes as well as a project on the levels of protection of personal data of European citizens. She participates in the SCALES project (big data and privacy) and leads the ethics by design work package of the Game Changers project on the development of health games for children.

Ilina Georgieva is a Ph.D. candidate of The Hague Programme for Cyber Norms. In her research, Ilina is focusing on the capacity of networks of intelligence agencies to shape the international community's perception of what is normal in cyberspace. For that purpose, she investigates the networks' normative power by looking into their practice of foreign electronic surveillance. Prior to joining the Institute of Security and Global Affairs, Ilina served as a researcher on the Sweetie Project at eLaw, the Center for Law and Digital Technologies at Leiden University. Her research encompassed a comparative legal study concerning the transborder investigation of Internet sexual crimes against children. Before joining eLaw's team, she worked as an Editor at the Utrecht Journal of International and European Law (October 2013–September 2014). Ilina was also a part of Heidelberg

University's Cluster of Excellence "Asia and Europe in a Global Context" (December 2012–August 2013) and of the Austria Institute for European and Security Policy (summer of 2012) in her capacity as a research assistant. From January 2009 to June 2010, she worked at the Max Planck Institute for Comparative Public Law and International Law in Heidelberg. She also served as a Senior Research Associate and later on as a Counsel for the Public International Law and Policy Group (PILPG) from September 2014 to October 2016.

Bart Schermer is an Associate Professor at eLaw, the Center for Law and Digital Technologies at Leiden University, and a fellow at the E.M. Meijers Institute for Legal Studies. He specializes in privacy, data protection, and criminal law. Apart from his work at the University Bart is Chief Knowledge Officer at Considerati, member of the Cybercrime Expert Group for the Dutch judiciary and member of the Human Rights Committee of the Advisory Council on International Affairs.

Bert-Jaap Koops is Professor of Regulation and Technology at the Tilburg Institute for Law, Technology, and Society (TILT) in the Netherlands. His main research fields are cybercrime, cyber-investigation, privacy, and data protection. He is also interested in topics such as DNA forensics, identity, digital constitutional rights, "code as law", and regulatory implications of human enhancement, genetics, robotics, and neuroscience. With a personal postdoc (1999), VIDI (2003) and VICI (2014) grant, Koops is one of the few Dutch researchers who received all three stages of NWOs (Netherlands Organisation for Scientific Research) personal research-grant scheme. Koops studied mathematics and general and comparative literature at Groningen University, and received his Ph.D. in law at Tilburg University in 1999. From 2005 to 2010, he was a member of De Jonge Akademie, a young-researcher branch of the Royal Netherlands Academy of Arts and Sciences. In 2016/17, he was Distinguished Lorentz Fellow at the Netherlands Institute for Advanced Study (NIAS). He coedited 13 books in English on technology regulation and published many articles and books in English and Dutch on a wide variety of topics.

Hans Guyt was the co-founder of Greenpeace Netherlands in 1978 and the International Campaigns Director for Greenpeace International in the 1980s, and as a former Dutch merchant navy officer joined the Dutch Non-Governmental Organization, Terre des Hommes in 1999 as Director of Programs. This international humanitarian aid and development charity focuses on the welfare of children and operates in some 27 countries in Asia, Africa, Europe, and the Middle East. Terre des Hommes' main theme is child exploitation, dealing with issues such as child prostitution, child trafficking, and child slavery. In 2011, Terre des Hommes and their local partners in the Philippines encountered a new form of child sexual exploitation. Street children trying to keep themselves alive by selling their bodies, explained they used the numerous and cheap Internet facilities in urban centers to establish contact with men in richer parts of the world who would pay for live sex shows. This phenomenon, known in the Philippines as "cybersex", is now widespread and on the increase. In addition to efforts to rehabilitate former young prostitutes and to prevent others from falling victim to the sex industry, Terre des Hommes decided to investigate this "new" demand for sex with children. From the beginning, Hans Guyt has lead the team that brought an innovative and effective approach to monitoring the internet for child predators. Using a specially developed computer image of an 11-year old Filipina girl, the "Sweetie Project" (launched in 2013) set a new standard for early intervention and awareness raising regarding Internet-based child abuse. The progress of the "Sweetie Project" has been watched with interest not only by the general public, but also by the judicial and legislative communities in many countries. It has won several international awards, receives worldwide attention and has brought "live-streaming of child sex abuse", as it is now formally known, out into the open.

Contributors

Stefan Bogaerts Tilburg University, Tilburg, The Netherlands

Ines Bojić Constitutional Court of the Republic of Croatia, European Court of Human Rights, Zagreb, Croatia

Charlotte Conings Stibbe, Brussels, Belgium

Michael Anthony C. Dizon Te Piringa—Faculty of Law, University of Waikato, Hamilton, New Zealand

Rens de Wolf Tracks Inspector, The Hague, The Netherlands

Ilina Georgieva Institute of Security and Global Affairs, Leiden University, Leiden, The Netherlands

Alisdair A. Gillespie Lancaster University, Lancaster, UK

Asaf Harduf Zefat Academic College, Zefat, Israel

Hans Henseler Digital Forensics & E-Discovery, Hogeschool Leiden, Leiden, The Netherlands;
Tracks Inspector, The Hague, The Netherlands

Kaspar Kala Ministry of Social Affairs, University of Tartu School of Law, Tartu, Estonia

Manon Kleijn Tilburg University, Tilburg, The Netherlands

Bert-Jaap Koops Tilburg Institute for Law, Technology, and Society, Tilburg University, Tilburg, The Netherlands

Zvjezdana Kuprešak European Court of Human Rights, Zagreb, Croatia

Gaëlle Marlier Institute of Criminal Law, KU Leuven, Leuven, Belgium

Sofie Royer Institute of Criminal Law, KU Leuven, Leuven, Belgium

Bart Schermer Center for Law and Digital Technologies, Leiden University, Leiden, The Netherlands

Jonathan Unikowski University of California, Berkeley, CA, USA

Gregor Urbas ANU Cybercrime Observatory, Australian National University, Canberra, Australia

Simone van der Hof Center for Law and Digital Technologies, Leiden University, Leiden, The Netherlands

Chapter 1
Legal Aspects of Sweetie 2.0

Bart Schermer, Ilina Georgieva, Simone van der Hof and Bert-Jaap Koops

Contents

B. Schermer (✉) · S. van der Hof
Center for Law and Digital Technologies, Leiden University, Leiden, The Netherlands
e-mail: b.w.schermer@law.leidenuniv.nl

S. van der Hof
e-mail: s.van.der.hof@law.leidenuniv.nl

I. Georgieva
Institute of Security and Global Affairs, Leiden University, Leiden, The Netherlands
e-mail: i.n.georgieva@fgga.leidenuniv.nl

B.-J. Koops
Tilburg Institute for Law, Technology, and Society, Tilburg University, Tilburg, The Netherlands
e-mail: e.j.koops@tilburguniversity.edu

S. van der Hof et al. (eds.), *Sweetie 2.0*, Information Technology and Law Series 31,
https://doi.org/10.1007/978-94-6265-288-0_1

Abstract Webcam sex tourism, the act of engaging children in webcam prostitution, is a growing international problem. Not only does webcam sex tourism provide easy access to child abuse and child abuse images for child abusers, it also a crime that has a comparatively low risk for the offenders. Live webcam performances leave few traces and little evidence that law enforcement can use. Further difficulties arise from the fact that webcam sex tourism often has a cross-border character, which causes jurisdictional conflicts and makes it more difficult to obtain evidence or even launch an investigation. The Dutch children's rights organization Terre des Hommes (TdH) was the first NGO to actively tackle webcam child sex tourism by using a virtual character called 'Sweetie' to identify offenders in chatrooms and online forums. Since then, Sweetie has been further developed into a chatbot, called Sweetie 2.0, to automate interaction with offenders. But using artificial intelligence raises serious legal questions. Sweetie as an investigative tool is so innovative, that it is unclear whether its use is actually covered by existing rules of criminal procedure. However, the question of criminal procedural legality of Sweetie is preceded by a prior substantive criminal law question: is interacting with Sweetie in a sexually charged way a criminal offence in the first place, given that Sweetie is not a person, but a virtual avatar? An answer to this question is important, because if webcam sex tourism with a virtual avatar is not considered criminal, it will be much harder to make the case that Sweetie is an acceptable investigative method. This chapter addresses the application of substantive criminal law and criminal procedure to Sweetie 2.0.

Keywords Criminal Law · Cybercrime · Sexual Offence · Child Pornography · Criminal Procedure · Entrapment

1.1 Introduction

1.1.1 Aims and Challenges

Digital technologies and the Internet may pose risks for the safety and well-being of children. The Internet in particular has created new opportunities for child sexual offenders to find and contact victims, and to sexually exploit them. Sexual predators are active in self-created Internet communities, where they exchange tips and tactics on how to most effectively approach and manipulate children.[1] Perpetrators also use social media platforms and chatrooms to directly engage with victims. This constantly evolving model of sexual exploitation of children is characterised by the exchange of messages with victims via the Internet that typically escalate quickly to sexually-explicit conversations. Once contact has been established, the victim is usually asked or even pressured to undress in front of a webcam, to perform or witness sexual acts, or all of the above. An additional dimension to this problem is the fact that parents or legal guardians may directly be involved in the sexual exploitation of their children. The latter is especially true for families living in developing countries, who prostitute their children as a much needed source of income.

The situation whereby children are engaged in webcam prostitution is generally referred to as webcam child sex tourism. Webcam sex tourism not only causes serious and lasting damage to children,[2] it also challenges the effectiveness of criminal investigations, as live webcam performances leave few traces and little evidence that law enforcement can use. Further difficulties arise from the fact that webcam sex tourism often has a trans-border character, which causes jurisdictional conflicts and makes it more difficult to obtain evidence or even launch an investigation.

The Dutch children's rights organization Terre des Hommes ('TdH') was the first NGO to combat webcam child sex tourism by using a virtual character called 'Sweetie'. Sweetie was used to identify offenders in chatrooms and online forums. The Sweetie avatar, posing as a ten-year old Filipino girl, was operated by an agent of the organisation, whose goal was to gather information on individuals who contacted Sweetie and solicited webcam sex. The gathered information was subsequently handed over to the authorities, who thereupon launched investigations in various countries.[3]

The successful implementation of Sweetie 1.0 inspired the further technological development of Sweetie. This time, a technical team commissioned by TdH engineered an artificial intelligence ('AI') software system, capable of depicting and

---

1 Lovejoy 2007, p. 312.

2 Goldstein 1999, p. 144.

3 Further information on the project known as 'Sweetie 1.0' can be found on www.terredeshommes.nl/en/sweetie-face-webcam-child-sex-tourism [16 June 2016].

acting as Sweetie without human intervention in order to not only identify persistent perpetrators, but also to deter first-time offenders.[4]

However, the creation of the software raises various challenging legal questions on its application in a law enforcement context. The laws that govern the use of special investigative tools and the role of law enforcement need to be considered. It is precisely this background against which the central question of this legal study takes shape. This report aims to illuminate the existing legal framework of criminal laws and procedures in a number of selected countries in order to determine whether said framework allows for investigative methods such as Sweetie to be used by law enforcement agencies in the fight against webcam child sex tourism.

1.1.2 Problem Statement and Research Questions

In this report we address the following problem statement:

> To what extent is it possible for law enforcement agencies to use Sweetie 2.0 for the investigation and prosecution of webcam sex with minors based on the current criminal law framework (in a selected number of countries)?

The answer to this problem statement may yield the conclusion that the (inter)national criminal law framework may currently not be adequate to combat webcam sex using tools like Sweetie. Therefore, we will also consider the following question:

> Which changes to the (inter)national criminal law framework are necessary/desirable in order to facilitate the effective and legitimate use of Sweetie by law enforcement agencies?

To solve this twofold problem statement, we will explore the following research questions:

1. How is webcam sex with minors criminalised in selected jurisdictions?
2. To what extent do existing crime descriptions within substantive criminal law apply to virtual victims (i.e., chatbots like Sweetie 2.0)?

[4] For further information on the second part of the project known as 'Sweetie 2.0', see https://www.terredeshommes.nl/programmas/sweetie-20-webcamseks-met-kinderen-de-wereld-uit [16 June 2016].

3. To what extent does the criminal procedure law framework allow for the (proactive) investigation of webcam sex offences using Sweetie 2.0, taking into account that:
   a. Sweetie 2.0 is an AI that interacts with suspects without direct human control or intervention;
   b. A 'fake identity' is used for the AI.
4. Are there specific limitations in criminal procedure when it comes to entrapment and what are the consequences of this for using Sweetie 2.0?
5. Which forensic requirements apply to the collection of evidence using Sweetie 2.0?

Given the global nature of the issue of webcam sex tourism, we will also discuss issues surrounding international investigations and jurisdiction.

1.1.3 Research Methodology

In this report Sweetie's use will be assessed in the light of the five main legal issues it raises.

In the area of substantive criminal law, these include the application of criminal provisions to virtual victims such as Sweetie, and the criminalisation of preparatory acts and attempts to commit the sexual offences in question. In this context, particular attention will be given to the doctrine of impossible attempts.

In terms of procedural criminal law aspects, we seek to establish whether existing coercive powers can provide a legal basis for the use of Sweetie 2.0. Special investigative powers are usually employed without the suspect's knowledge or consent, which interferes with his/her right to privacy and private life as stipulated in Article 8 ECHR[5] and Article 17 ICCPR.[6] In addition, the right to a fair trial as codified in Article 6 ECHR and Article 14 ICCPR has to be taken into account when assessing the (proactive) use of artificial intelligence agents in criminal investigations and the thereby triggering defence of entrapment.

We will use the following research methodologies to analyse the use of Sweetie:

Desk Research and Literature Study
The basis of the research is desk research and literature study.

Comparative Legal Analysis
After introducing the international instruments that are relevant for this study, the report continues with the comparative legal analysis. The main goal of the

[5] Convention for the Protection of Human Rights and Fundamental Freedoms, CETS no. 194, Rome, 4.XI.1950 (ECHR).

[6] International Covenant on Civil and Political Rights (16 December 1966, entered into force 23 March 1976) 999 UNT 171 (ICCPR).

comparative legal analysis is to compare and contrast the legal approaches to dealing with the issue of webcam child sex tourism and the application of Sweetie 2.0. Furthermore, the comparative legal analysis will highlight any issues when it comes to jurisdictional issues related to criminalisation and cross-border investigation.

Given that webcam sex tourism is a global phenomenon a diverse set of countries (in terms of geographical location and legal systems) was chosen for analysis. The following countries were selected for analysis:

- Argentina
- Australia
- Belgium
- Brazil
- Canada
- Croatia
- England and Wales
- Estonia
- Germany[7]
- Israel
- Netherlands
- Nigeria
- Philippines
- Poland
- Scotland[8]
- South Korea
- Spain
- Turkey
- United States

Selection criteria were: geographic spread, type of legal system, exposure to webcam sex tourism and prior experience with Sweetie 1.0.[9] Practitioners and scholars from the selected jurisdictions were asked to prepare a comprehensive report on their respective criminal system and to evaluate it against the research question(s) of the Sweetie project. The authors of the country reports are acknowledged throughout the present study.

---

[7] The information used to assess the German situation reflects on substantive criminal law issues only. Therefore, Germany is not considered in the analysis of criminal procedure.

[8] Scotland is a part of the United Kingdom, as are England and Wales. However, given the different criminal law system vis-à-vis the rest of the UK, it is listed as a separate jurisdiction.

[9] Some relevant countries (such as Russia or Kenia) did not make the final selection due to the unavailability of a legal expert in the timeframe of the project.

1.1.4 Structure of the Chapter

This study is divided into eight parts, including the present introduction. The report continues with Sects. 1.1.5–1.1.6, which presents Sweetie, the software behind the avatar, its goals and application. Section 1.2 discusses the core legal issues raised by Sweetie in terms of substantive criminal law by establishing a baseline of the international law provisions applicable to the matter, and subsequently turns to a comparative analysis of the legal norms that cover webcam child sex tourism in the studied countries. Section 1.3 does the same with regard to procedural criminal law. Section 1.4 touches upon the standards of digital forensics relevant for the preservation of data and evidence originating from the cyber domain. Section 1.5 elaborates on the jurisdictional issues concerning trans-border investigations of webcam sexual assaults of children. An analysis of the restrictions found in the discussed criminal law systems and suggestions on how to adapt the legal frameworks to the challenge of webcam child sex tourism can be found in Sect. 1.6. Section 1.7 offers a summary and conclusion of our findings.

1.1.5 Sweetie

As a result of the rapid proliferation of devices with cameras, free video chat software (e.g. Skype and Google Hangout), the increase in Internet bandwidth, and the lowering cost of data traffic, people throughout the world now communicate on a daily basis via video. A specific aspect of video chatting is that of a sexual nature: webcam sex.

While webcam sex can take place legally between consenting adults, there are also risks associated with webcam sex, in particular for minors. Risks arise not only because predators actively approach unsuspecting minors, but also because a 'cottage industry' of webcam prostitution of minors has emerged, in particular in developing countries. This relatively new phenomenon of webcam child sex tourism has quickly grown into a hidden, but global problem.

To combat webcam sex tourism and raise awareness for the issue, Terre des Hommes developed the Sweetie programme. Sweetie 1.0 was a virtual 10-year old Filipino girl used to identify and expose pedosexuals engaged in webcam sex tourism. Sweetie was operated by a human agent that engaged in conversation with the suspected webcam sex tourist.

While Sweetie 1.0 was extremely successful, one limitation of its design was the human operator. A human operator can only conduct a number of chat conversation at the same time, while real victims receive up to two hundred sex solicitations an hour. To counter this problem TdH has developed a more advanced version of Sweetie: Sweetie 2.0. The main difference with Sweetie 1.0 is that Sweetie 2.0 is no longer operated by a human but is a now a fully autonomous artificial intelligence that can engage in a meaningful conversation with a suspect.[10] Unlike human operators, the use of this artificial intelligence is in theory infinitely scalable.

1.1.6 Technology

Sweetie is a virtual minor that engages in conversation with a suspect who has a sexual interest in children with the goal of identifying this suspect. Sweetie is comprised of three main technological elements: (1) three-dimensional imagery, (2) a chatbot facility, and (3) an underlying software framework.[11]

3D Imagery
The most striking aspect of the original Sweetie was the use of 3D imagery to create a realistic representation of a virtual girl. The realistic animations of Sweetie were designed to make suspects think that they were dealing with a real minor. For Sweetie 2.0 the animations have been further refined. It is important to note that Sweetie's animations do not show any nudity or images of a sexual nature.

Chatbot Facility
To eliminate the need for human intervention, Sweetie 2.0 employs AI technology. A chatbot character has been built based on the experiences, work instructions and chat logs from the initial Sweetie project. Using results from the past, the conversation model will simulate as realistically as possible a fictitious 10/11-year-old child.

Software Framework
To use the chatbot functionality for various communication platforms a base has been built that interconnects all software components. These components include, but are not limited to:

- Automated chat functionality for the chatrooms and direct chat;
- Functions to drive the generated imagery;
- Management functionality for the chatrooms, characters, chat structure and corresponding question/answer combinations;
- Storage of all chats and related details;
- Processing of identifiable material from the chats for each chat partner;
- Detection functionality to recognize repeating chat partners, indecent proposals/ or explicit materials;
- Dashboard for graphical presentation of all required actions, chat results, as well as statistics for operational, tactical and strategic insight;

[10] This type of artificial intelligence is popularly known as a 'chatbot'.

[11] https://tracksinspector.com/blog/ti-software-sweetie-2-0.html [28 September 2016].

- Reporting module to confront potentially offending chat partners with their own behaviour and chat phrases. This module will also follow up with relevant advice, deterrent warnings and/or possible threat of identification, based on the findings of current academic research for the project.

The chat logs are stored and exchanged data are processed per chat to a profile for each chat partner. This profile can ultimately be used to identify repetitive patterns. All chat reports and extracts of chats are logged in a universally accepted standard which facilitates the exchange of cases. This will take into account generic storage methods used by various national and international (investigation) agencies such as Interpol and Europol in order to simplify matching with other (online) child abuse cases.

1.2 Substantive Criminal Law

Sweetie is an investigative tool that enables law enforcement to engage with sexual predators and interact with them. If law enforcement is to use Sweetie as an investigative method, this means that the actual behaviour under investigation (i.e., interacting with Sweetie) must be deemed criminal behaviour. If this is not the case, then it will be much harder, if not impossible to prove that the suspect committed or attempted a criminal act. This in turn will make it more difficult to justify the use of Sweetie as an investigative method.

In this section we explore whether and how (an attempt) to interact with Sweetie in a sexually oriented way is criminalised in the various jurisdictions under investigation. To this end we first consult the international law instruments to establish a 'baseline' of criminal behaviour and then explore specific substantive law issues in relation to webcam sex in general and webcam sex with an avatar such as Sweetie in particular.

1.2.1 Criminalisation of Abuse of Minors and International Harmonisation

In most, if not all jurisdictions worldwide, sexual abuse of minors is criminalised. Different forms of abuse are criminalised in national criminal law. Apart from criminalisation at the national level, there is also international harmonisation when it comes to the protection of minors and the criminalisation of abuse of minors.

At a global level the protection of minors is codified in different international law instruments. For the purpose of this report we will explore four particularly relevant international law instruments: (1) the UN Convention on the Rights of the Child

(CRC),[12] and (2) the Optional Protocol to the Convention on the Rights of the Child on the sale of children, child prostitution and child pornography (OPSC),[13] (3) the Council of Europe Convention on the Protection of Children against Sexual Exploitation and Sexual Abuse (Lanzarote Convention)[14] and, (4) the Council of Europe Cybercrime Convention.[15,16]

UN Convention on the Rights of the Child

The United Nations Convention on the Rights of the Child (CRC) is an international treaty that sets out the civil, political, economic, social, health and cultural rights of children. At the moment of writing there are 196 countries party to the treaty.[17] The United States is the only member of the UN that has not ratified the document.

The fundamental idea of the CRC is that every child, every human being below the age of eighteen years, is born with fundamental freedoms and the inherent rights of human beings. Moreover, the CRC recalls that children are entitled to special care and assistance because of their vulnerability.[18] According to the preamble children need to grow up 'in a family environment, in an atmosphere of happiness, love and understanding.'[19] Article 20 of the CRC, for example, states that a child temporarily or permanently deprived of his or her family environment shall be entitled to special protection and assistance provided by the state. Furthermore, General Comment 13 to the CRC underlines the importance that every child's life must be free from all forms of violence.[20]

---

[12] Convention on the Rights of the Child (adopted 20 November 1989, entered into force 2 September 1990) 1577 UNTS 3 (CRC or UNCRC), available at: http://www.ohchr.org/en/professionalinterest/pages/crc.aspx [13 June 2016].

[13] Optional Protocol to the Convention on the Rights of the Child on the sale of children, child prostitution and child pornography (adopted on 25 May 2000, entered into force 18 January 2002) A/RES/54/263, available at: http://www.ohchr.org/EN/ProfessionalInterest/Pages/OPSCCRC.aspx [13 June 2016].

[14] Council of Europe Convention on the Protection of Children against Sexual Exploitation and Sexual Abuse, CETS 201, Lanzarote, 25.X.2007 (Lanzarote Convention).

[15] Council of Europe Convention on Cybercrime, CETS No.185, Budapest, 23.XI.2001 (Budapest Convention), available at: http://www.europarl.europa.eu/meetdocs/2014_2019/documents/libe/dv/7_conv_budapest_/7_conv_budapest_en.pdf [13 June 2016].

[16] Both the Lanzarote Convention and the Cybercrime Convention are Council of Europe instruments. In the European Union Directive 2011/93/EU on combating sexual abuse and sexual exploitation of children, and child pornography is also applicable. Given the more global remit of the Council of Europe instruments, we will not discuss Directive 2011/93/EU further in the context of this report.

[17] https://treaties.un.org/Pages/ViewDetails.aspx?src=IND&mtdsg_no=IV-11&chapter=4&lang=en [18 May 2016].

[18] UNCRC, Preamble, paras 4 and 9.

[19] UNCRC, Preamble, para 6.

[20] Committee on the Rights of the Child 2011, *General comment No. 13: The right of the child to freedom from all forms of violence*, CRC/C/GC/13.

While the CRC does not criminalise specific acts against the well-being of children, several articles put a positive obligation on the states to protect children against sexual abuse and exploitation. Article 19 and Article 34 are particularly relevant in this regard:

> **Article 19**
>
> States Parties shall take all appropriate legislative, administrative, social and educational measures to protect the child from all forms of physical or mental violence, injury or abuse, neglect or negligent treatment, maltreatment or exploitation, including sexual abuse, while in the care of parent(s), legal guardian(s) or any other person who has the care of the child.
>
> **Article 34**
>
> States Parties undertake to protect the child from all forms of sexual exploitation and sexual abuse. For these purposes, States Parties shall in particular take all appropriate national, bilateral and multilateral measures to prevent:
>
> (a) The inducement or coercion of a child to engage in any unlawful sexual activity; (b) The exploitative use of children in prostitution or other unlawful sexual practices; (c) The exploitative use of children in pornographic performances and materials.

Optional Protocol to the Convention on the Rights of the Child on the Sale of Children, Child Prostitution and Child Pornography

The Optional Protocol to the Convention on the Rights of the Child on the sale of children, child prostitution and child pornography (OPSC) has been signed by 182 state parties to this date, yet no more than 173 have ratified it.[21] The protocol is intended to achieve the purposes of the articles in the CRC. For example, Article 1 states that parties are to protect the rights and interests of child victims of trafficking, child prostitution, child pornography and child labour. Article 2 (broadly) defines the criminal acts of sale of children, child prostitution and child pornography:

> For the purposes of the present Protocol:
>
> (a) Sale of children means any act or transaction whereby a child is transferred by any person or group of persons to another for remuneration or any other consideration;
> (b) Child prostitution means the use of a child in sexual activities for remuneration or any other form of consideration;
> (c) Child pornography means any representation, by whatever means, of a child engaged in real or simulated explicit sexual activities or any representation of the sexual parts of a child for primarily sexual purposes.

The OPSC obliges states to criminalise these practices. Finally, the protocol sets international standards for mutual assistance in investigations, confiscation of assets and extradition.

---

[21] https://treaties.un.org/Pages/ViewDetails.aspx?src=IND&mtdsg_no=IV-11-c&chapter=4&lang=en [18 May 2016].

Council of Europe Convention on the Protection of Children against Sexual Exploitation and Sexual Abuse (Lanzarote Convention)
The Lanzarote Convention is a Council of Europe Convention aimed at combating sexual exploitation and abuse of minors. The purposes of the Lanzarote Convention are defined in its Article 1:

- Prevent and combat sexual exploitation and sexual abuse of children;
- Protect the rights of child victims of sexual exploitation and sexual abuse;
- Promote national and international co-operation against sexual exploitation and sexual abuse of children.

The treaty obliges parties to pass laws that criminalise any practice that is in conflict with these purposes, for instance child pornography. All Member States of the Council have ratified the treaty.

We will use the Lanzarote Convention as a basis for our discussion of substantive criminal law as it provides the most complete inventory of crimes related to minors.

Council of Europe Cybercrime Convention
Being the first in its kind, the Convention on Cybercrime pursues a common criminal policy aimed at the protection of society against crimes committed via the Internet and other computer networks.[22] The treaty fosters fast and effective international cooperation, harmonisation of domestic criminal law in the area of cybercrime and the provision of domestic criminal procedural law powers necessary for the investigation and prosecution of such crimes. The Convention contains provisions on a wide variety of crimes, such as violations of network security, computer related fraud and child pornography.

Thus far, 49 states have ratified the Cybercrime Convention.[23] Parties to the treaty are not only Member States of the Council of Europe. The treaty is also ratified by Australia, Canada, The Dominican Republic, Israel, Japan, Mauritius, Panama, Sri Lanka and the United States. Relevant for the purpose of this study is that the Cybercrime Convention criminalises offences related to child pornography (Article 9).

International Harmonisation in the Area of Abuse of Minors
Table 1.1 specifies which of the countries under investigation in this study have signed and ratified the international law instruments described above.

Criminalisation of Webcam Sex with Minors
This section reviews the criminal law offences potentially applicable to sexual abuse of minors via webcam. It does so by summarising the provisions focusing on the key types of behaviour regularly occurring in webcam abuse. It then turns to evaluate whether and how said provisions apply to virtual minors such as Sweetie.

---

[22] Convention on Cybercrime, Preamble, para 8.

[23] www.coe.int/en/web/conventions/full-list/-/conventions/treaty/201/signatures [18 May 2016].

Table 1.1 International harmonisation in the area of abuse of minors [*Source* The authors]

	UN Convention on the Rights of the Child (CRC)	Optional Protocol to the CRC on the sale of children, child prostitution and child pornography (OPSC)	CoE Convention on the Protection of Children against Sexual Exploitation and Sexual Abuse (Lanzarote Convention)	CoE Cybercrime Convention
Argentina	X	X	–	–
Australia	X	X	–	X
Belgium	X	X	X	X
Brazil	X	X	–	–
Canada	X	X	–	X
Croatia	X[a]	X	X	X
England and Wales	X	X	S (Signed, not ratified)	X
Estonia	X	X	S	X
Germany	X	X	X	X
Israel	X	X	–	X
Netherlands	X	X	X	X
Nigeria	X	X	–	–
Philippines	X	X	–	–
Poland	X	X	X	X
Scotland	X (UK)	X (UK)	S (UK)	X (UK)
South Korea	X	X	–	–
Spain	X	X	X	X
Turkey	X	X	X	X
United States	S	X	–	X

[a]Signed by Yugoslavia, ratified by Croatia

International Harmonisation

What we can deduct from Table 1.1 is that while all countries under investigation in this study have signed and ratified the CRC and the OPSC, not all countries have signed the Lanzarote Convention, which provides the most complete inventory of criminal offences related to the abuse of minors. This may pose a problem when it comes to the international harmonisation of the criminalisation of webcam sex, given that the OPSC does not specifically define or criminalise certain types of child exploitation such as webcam sex and grooming.

Criminal behaviour aimed at minors that does not qualify as prostitution or as child pornography is, for instance, not covered by the OPSC. This means that crimes such as webcam sex, grooming and the corruption of children are not fully harmonised throughout the jurisdictions under examination.

Furthermore, it may also mean that webcam sex, grooming and the corruption of children are not criminalised at the national level. Though when it comes to the topic of this research (webcam sex with minors) we have not found any proof of this. All jurisdictions that did not sign the Lanzarote Convention have still criminalised webcam sex with minors in national law, in one way or another, as the following discussion demonstrates.

Relevant Crime Descriptions

The international law instruments described above and the national criminal laws of the different countries contain a number of broadly formulated crime descriptions that may or may not cover webcam sex with minors.

Depending on the exact form and circumstances of the act, the offences listed below may come into view when a person is interacting with a minor though a webcam for the purpose of sexual gratification. For the sake of good order, we use the articles from the Lanzarote Convention as a framework for discussion. The relevant articles are:

- Article 18. Sexual abuse
- Article 19. Offences concerning child prostitution
- Article 20. Offences concerning child pornography
- Article 21. Offences concerning the participation of a child in pornographic performances
- Article 22. Corruption of children
- Article 23. Online solicitation of children for sexual purposes

Sexual Abuse

Sexual abuse may cover a range of sexual activities that take place between the perpetrator and the minor, such as rape, assault and the commission of lewd/lascivious acts. Article 18 of the Lanzarote Convention defines sexual abuse as:

a. engaging in sexual activities with a child who, according to the relevant provisions of national law, has not reached the legal age for sexual activities;
b. engaging in sexual activities with a child where: – use is made of coercion, force or threats; or – abuse is made of a recognised position of trust, authority or influence over the child, including within the family; or – abuse is made of a particularly vulnerable situation of the child, notably because of a mental or physical disability or a situation of dependence.

Paragraph 1a criminalises engaging in sexual activities with a person who has not reached the age at which it is allowed to engage in sexual activities with him or her. This age is established in domestic law:

2. For the Party shall decide the age below which it is prohibited to engage in sexual activities with a child.

Paragraph 1b criminalises engaging in sexual activities with a child where use is made of coercion, force or threats, or when this person abuses a recognised position of trust, authority or influence.

All of the countries under investigation have provisions in their domestic law that criminalise sexual abuse of minors. In the Lanzarote Convention the term 'sexual activities' has not been further defined. The negotiators preferred to leave it to the States to further define the meaning and scope of the term.[24] In the domestic law of the countries under investigation, 'sexual activities' generally cover acts whereby there is direct physical contact (including by force, under threat or through other forms of coercion) between the perpetrator and the victim, such as rape and assault. A position of trust, authority or influence over the minor is an aggravating circumstance, which generally carries higher penalties.

Offences Concerning Child Prostitution

Child prostitution covers a number of criminal acts whereby a minor is used for sexual activities in exchange for some form of remuneration. Article 2 para b of the OPSC defines child prostitution as:

> (…) the use of a child in sexual activities for remuneration or any other form of consideration.

Article 19 para 2 of the Lanzarote Convention defines child prostitution as:

> (…) the fact of using a child for sexual activities where money or any other form of remuneration or consideration is given or promised as payment, regardless if this payment, promise or consideration is made to the child or to a third person.
>
> Offences concerning child prostitution
>
> 1. Each Party shall take the necessary legislative or other measures to ensure that the following intentional conduct is criminalised:
> a. recruiting a child into prostitution or causing a child to participate in prostitution;
> b. coercing a child into prostitution or profiting from or otherwise exploiting a child for such purposes;
> c. having recourse to child prostitution.

Criminal liability is extended to both the person(s) prostituting the minor and the customer(s).

In relation to the topic of this research, this article is of particular importance. In most cases of webcam sex with minors, a minor from a developing country is forced or coerced by a third party (parents, criminals) to participate in a webcam session with a perpetrator (the 'webcam sex tourist'). This webcam sex tourist generally wants to watch (and indirectly participate in) a pornographic performance involving a minor in exchange for money.

Given the broad definition of 'sexual activities' there are no a priori limitations to applying this article in the context of webcam sex tourism.

[24] *Explanatory report to the Council of Europe Convention on the Protection of Children against Sexual Exploitation and Sexual Abuse. CETS 201*, para 127.

Offences Concerning Child Pornography

Offences concerning child pornography are also highly relevant in the context of webcam sex with minors and webcam sex tourism, given that the images streamed and captured via the webcam will generally qualify as child pornography.

Child pornography is criminalised in the OPSC, the Lanzarote Convention and the Cybercrime Convention.

The OPSC uses the following definition of child pornography:

> Child pornography means any representation, by whatever means, of a child engaged in real or simulated explicit sexual activities or any representation of the sexual parts of a child for primarily sexual purposes.

In the Lanzarote Convention child pornography is defined as:

> (…) any material that visually depicts a child engaged in real or simulated sexually explicit conduct or any depiction of a child's sexual organs for primarily sexual purposes.

Finally, according to the Cybercrime Convention:

> the term 'child pornography' shall include pornographic material that visually depicts:
>
> a) a minor engaged in sexually explicit conduct;
> b) a person appearing to be a minor engaged in sexually explicit conduct;
> c) realistic images representing a minor engaged in sexually explicit conduct.

In the international law instruments at hand both the production and consumption of child pornography is criminalised. The Lanzarote Convention lists the following acts as criminal:

> a. producing child pornography;
> b. offering or making available child pornography;
> c. distributing or transmitting child pornography;
> d. procuring child pornography for oneself or for another person;
> e. possessing child pornography;
> f. knowingly obtaining access, through information and communication technologies, to child pornography.

When it comes to illegal webcam sex, sub c and f are of particular interest. Before the advent of webcam sex, most of the child pornography was either distributed by means of physical carriers (photos, magazines) or obtained digitally and subsequently stored on local media such as hard drives and DVDs. In these cases procurement and possession of child pornography could be proven more easily. With webcam sex, however, by default there is no local storage of the streamed data, unless the offender records the stream or takes screenshots. The ephemeral nature of this type of child pornography consumption means that procurement and possession is difficult, if not impossible to prove. Therefore, Article f criminalises the access to child pornography in itself. As such, webcam sex with minors may also fall under the heading of (attempting) to access child pornography.[25] On the 'production side' making a webcam sex stream available may fall under the heading of sub c (transmitting child pornography).[26]

Virtual Child Pornography

Apart from actual child pornography, many countries also criminalise the production, sale, possession of and access to 'virtual' child pornography. Virtual child pornography refers to images whereby it realistically *appears* that a minor is engaged in sexually explicit conduct. This is particularly relevant for Sweetie, which is a virtual minor that may appear to be engaged in sexually explicit behaviour.

Both the OPSC and Lanzarote Convention use broad definitions of child pornography that may also include images of virtual minors engaged in sexually explicit conduct or the computer-generated depiction of a child's sexual organs for primarily sexual purposes. After all, they see to 'any representation' or 'any material that visually depicts', which may include computer-generated representations. However, while the Lanzarote Convention clearly intends to include it within the definition's scope (see Article 20 para 3), for the OPSC this is not evident. One could also argue, therefore, that virtual child pornography is not explicitly referred to, it is not included in the OSPC definition.

The Cybercrime Convention, in contrast, does use an explicit reference to virtual child pornography, as long as the images are realistic:

> … the term 'child pornography' shall include pornographic material that visually depicts:
>
> (…)
>
> c) realistic images representing a minor engaged in sexually explicit conduct.

In the Explanatory Memorandum to the Cybercrime Convention virtual child pornography is further described as:

> images, which, although 'realistic', do not in fact involve a real child engaged in sexually explicit conduct. This latter scenario includes pictures which are altered, such as morphed images of natural persons, or even generated entirely by the computer.[27]

Under both the Lanzarote Convention and the Cybercrime Convention it is optional to criminalise virtual child pornography (see Article 20 para 3 and Article 9 para 4 respectively) (Table 1.2).

---

[25] Please note that para 4 of Article 2 OSPC gives countries the possibility to not apply para 1f in national law.

[26] In Canada for instance, subs. 163.1(4.2) of the Canadian Criminal Code criminalises the viewing of webcam performances by attaching liability to those who knowingly access child pornography by knowingly causing child pornography to be viewed by or transmitted to himself or herself. Following the interpretation of the Supreme Court, Article 383*bis* § 2 of the Belgian Criminal Code also penalises knowingly and without a right accessing child pornographic images through information and communication technologies.

[27] *Explanatory Report to the Council of Europe Convention on Cybercrime*, para 101.

Table 1.2 Criminalisation of (virtual) child pornography [*Source* The authors]

	Child pornography	Virtual child pornography
Argentina	X	
Australia	X	X[a]
Belgium	X	X
Brazil	X	–[b]
Canada	X	X[c]
Croatia	X	X
England and Wales	X	+/–[d]
Estonia	X	X
Germany	X	X[e]
Israel	X	–
Netherlands	X	X
Nigeria	X	X
Philippines	X	X
Poland	X	+/–[f]
Scotland	X	+/–[g]
South Korea	X	–
Spain	X	X[h]
Turkey	X	X
United States	X	X

[a]*McEwen v. Simmons & Anor* [2008] NSWSC 1292 at paras 38–39, confirming convictions for possession of virtual child pornography under both Commonwealth and State legislation. For more on the matter see Urbas 2016, p. 19

[b]Article 241-E of the ECA refers to 'real or simulated' sexual activities. However the official interpretation of the provision is understood as implying the participation of *real* minors in the simulated activities and not of virtual victims

[c]It appears that virtual child pornography, although not explicitly regulated, is criminalized as a matter of statutory interpretation of the definition given to the term, 'child pornography' at s. 163.1 of the Canadian Criminal Code

[d]May fall under the heading of 'pseudo-photograph' of a child but if it does not, virtual images are also classed as child pornography under a different piece of legislation. They would be known as prohibited images of children (see country report on England and Wales)

[e]Virtual child pornography is covered by §184b (1) no. 2 and (3) StGB and covers realistic depictions of children that an average informed person could not tell apart from the depictions of real children

[f]Child pornography is not further defined in Polish law. However, a broad interpretation of the term should be taken; see Skorvanek 2016 (Polish report)

[g]May fall under the heading of 'pseudo-photograph', but not entirely clear (see Scotland country report)

[h]Article 189.1(d) speaks of images that 'appear to be of a child' and of 'realistic' images. In general, based on the letter of the law, virtual child pornography is penalised too

Offences Concerning the Participation of a Child in Pornographic Performances

Article 21 of the Lanzarote Convention criminalises pornographic performances with minors:

> (…) a. recruiting a child into participating in pornographic performances or causing a child to participate in such performances;
>
> b. coercing a child into participating in pornographic performances or profiting from or otherwise exploiting a child for such purposes;
>
> c. knowingly attending pornographic performances involving the participation of children.

Article 21 para 1 sub a and b focus on those organising the performance, whereas paragraph c focuses on the attendance of such performances.

The OPSC does not specifically criminalise pornographic performances with a minor, but these might be covered by the broader notion of child prostitution:

> (…) (b) Child prostitution means the use of a child in sexual activities or remuneration or any other form of consideration;

The Cybercrime Convention does not cover pornographic performances, although the recording of broadcasting of such a performance will be considered producing and distributing, offering, and/or transmitting child pornography.

A webcam stream in which a minor performs sexual activities can be considered a pornographic performance. The question though is if attendance of such a performance at a distance is covered in the crime description. Whether this is the case is dependent on the domestic law of the countries under investigation. The Lanzarote Convention does specifically address this issue in the explanatory report, but leaves it to the contracting states to determine whether or not they wish to include webcam sex:

> Depending on States, this provision may also cover the situation of persons who are spectators of pornographic performances involving the participation of children through such means of communication as webcams.[28]

Corruption of Children

The corruption of children is a specific offence criminalised in the Lanzarote Convention. The OPSC or the Cybercrime Convention do not cover it. Article 22 of the Lanzarote Convention criminalises:

> (…) the intentional causing, for sexual purposes, of a child who has not reached the age set in application of Article 18, paragraph 2, to witness sexual abuse or sexual activities, even without having to participate.

---

[28] *Explanatory report to the Council of Europe Convention on the Protection of Children against Sexual Exploitation and Sexual Abuse.*

In the context of webcam sex this offence is relevant as it may cover those situations whereby:

- the perpetrator performs sexual activities in front of the webcam (e.g., masturbation).[29]
- the perpetrator tries to corrupt the minor, for instance by sending pictures of sexual activities to the minor or by promoting sexual activities using online chat functions.

Given that it is possible to (video)chat with Sweetie, it is very well possible that these offences may also be committed by suspects alongside other offences. With Sweetie 1.0, suspects did indeed perform sexual activities in front of the webcam in ways that amounted to the corruption of children.

Furthermore, in some criminal systems a sexually-charged chat with a (virtual) minor may be sufficient to meet the threshold of child corruption. This is the case, for instance, in Australia, where the suspect does not necessarily have to transmit imagery to the minor to bring about the corruption.[30] The law speaks of 'indecent communication', whereas just the chat can meet this threshold provided that the communication language goes against the moral standards of ordinary people. In a similar vein, it appears that the Polish legislator has opted for a rather broad understanding of the term 'pornography', which would allow to see the indecent chat as pornographic material as well.

Online Solicitation of Children for Sexual Purposes (Grooming)

Online solicitation of children for sexual purposes, more commonly referred to as 'grooming' is only criminalised in the Lanzarote Convention:

> Each Party shall take the necessary legislative or other measures to criminalise the intentional proposal, through information and communication technologies, of an adult to meet a child who has not reached the age set in application of Article 18, paragraph 2, for the purpose of committing any of the offences established in accordance with Article 18, paragraph 1.a, or Article 20, paragraph 1.a, against him or her, where this proposal has been followed by material acts leading to such a meeting.

Grooming commonly takes place via either online chat or webcam, and as such it is related to the subject matter of our research. However, webcam child sex tourism focuses on having sexual contact through the webcam, not in physical proximity. As grooming is defined as proposing to meet (in real life), it is a form of real-life child sex tourism. Although perpetrators chatting with Sweetie may also propose a real-life meeting, this is not the primary focus of the types of cases we study. Having said that, Sweetie 2.0 may of course also be used in a national setting to lure groomers.

---

[29] Exposing oneself in front of a minor and/or masturbating may also be criminalised in domestic law under the header of 'indecent exposure'. This is for instance the case in Argentina (Article 129 Argentinian Penal Code).

[30] See Section 474.27A of the *Criminal Code Act 1995* (Cth).

1.2.2 Issues Pertaining to the Criminalisation of Webcam Sex with Minors

Having reviewed the different types of crime descriptions above we may conclude that at the international level there are different options for states to criminalise webcam sex (tourism) with minors in national law. In Table 1.3 we summarise how each of the jurisdictions under examination have criminalised webcam sex tourism.

While webcam sex has been criminalised in different ways throughout the countries presented in this report, some possibly complicating issues exist regarding the criminalisation *per se* or the harmonisation of criminalisation at the international level. Below we will briefly discuss these.

Definition of a Child/Minor

When it comes to the criminalisation of webcam sex with minors (or any other form of sexual activity with minors for that matter), there is no full harmonisation on the age below which engaging in sexual activities with a person is deemed illegal. While on the supranational level there is general consensus as to the definition of a child (or minor), there is no full harmonisation on age in crime descriptions at the national level.[31]

In the CRC, a child (minor) is defined in Article 1 as:

> (…) a child means every human being below the age of eighteen years unless under the law applicable to the child, majority is attained earlier.

This definition applies to the Convention itself and to the OPSC.[32]

Article 3(a) of the Lanzarote Convention defines a child (minor) as:

> (…) any person under the age of 18 years.

The negotiators of the Lanzarote Convention considered the possibility of harmonising criminal law in the area of child exploitation by establishing the age of sexual consent in the Convention, but it was ultimately decided to let the member States decide for themselves at what age sexual activities with a person are deemed legal. The main reason for this being that this age varies greatly in Member States of the Council of Europe because of cultural differences.[33]

Nevertheless, for most jurisdictions under investigation in this report, a minor means a person below the age of 18 years and in most cases sexual activities involving such a minor are prohibited, unless explicitly stated otherwise. Some divergences can exist regarding the age of valid consent to sexual activities.

---

[31] See also: Interagency Working Group on the Sexual Exploitation of Children 2016, *Terminology guidelines for the protection of children from sexual exploitation and sexual abuse*, Luxembourg.

[32] The OPSC refers in the preamble to Article 1 of the CRC.

[33] *Explanatory report to the Council of Europe Convention on the Protection of Children against Sexual Exploitation and Sexual Abuse. CETS 201.*

Table 1.3 Legislation applicable to webcam child sex tourism [*Source* The authors]

	18. Sexual abuse	19. child prostitution	20. child pornography	21. pornographic performances	22. corruption of children	23. Online solicitation (grooming)
Argentina			128 CC	128 CC	128 CC	131 CC
Australia	272.11 Criminal Code Act 1995 (CCA)	–[a]	474.19 CCA 474.20 CCA	272.14 CCA	272.9 CCA 474.27 A	272.15 CCA
Belgium	372 Belgian Criminal Code (BCC) 373, 375 BCC	379, 380 BCC 372, 373 and 375 BCC	383 bis BCC 371/1 BCC	379 BCC 380 BCC	385 BCC	377quater BCC 433bis/1 BCC
Brazil	217-A Brazilian Criminal Code (CPB) 241-D Children and Adolescent Statute (ECA) 213 ECA	218-B CPB	218-B CPB	217-A CPB 241-D ECA 241-E ECA	218 CPB	218B CPB
Canada	152 Canadian Criminal Code (CCC)	286.1(2) CCC. 286.3(2) CCC	163.1 CCC	163.1 CCC	173(2) CCC	170 CCC 172.2 CCC 172.1 CCC
Croatia	158 Croatian Criminal Code (CrCC) (159 CrCC)	162 CrCC	163 CrCC	164 CrCC 165 CrCC	160(1) CrCC	161 CrCC
England and Wales	8, 10, 17, SOA 2003	47, 48, 50 SOA	7(7), Protection of Children Act 1978 (?)	12 SOA 2003 or OPA 1959	12 SOA	14, 15 SOA
Estonia		175(1) EPC[b]	175^{1} EPC 178 EPC	175^{1} EPC 178 EPC	179 EPC	178^{1} EPC

(continued)

Table 1.3 (continued)

	18. Sexual abuse	19. child prostitution	20. child pornography	21. pornographic performances	22. corruption of children	23. Online solicitation (grooming)
Germany	176(4) no. 1 StGB; 176(4) no. 2 StGB	176(4) no. 2 StGB	176a (3) in conjunction with §184b StGB	176(4) no. 2 StGB	176(4) no. 4 StGB	176(4) no. 3 StGB
Israel	345–351, 203B Israeli Penal Code	203B IPC	214B (3) IPC	214B (3) IPC	208 IPC	203B IPC
Netherlands	247 Dutch Criminal Code (DCC) 246 DCC 248a DCC	246b DCC 248f DCC 250 DCC	240b DCC	248c DCC, 248f DCC	240a DCC, 248d DCC	248e DCC
Nigeria	32(1) Child Rights Act (CRA)	222A Nigerian Penal Code (NPC) 223(2) NPC 2234(4) NPC	281 NPC, 30(2) (e) CRA 23(1), (3)(c) Nigerian Cybercrimes Act (NCA)	30(2) CRA	23(3) a NCA	23(3) a NCA
Philippines	2(h) Rules and Regulations on the Reporting and Investigation of Child Abuse Cases	3(h) Anti-Trafficking in Persons Act of 2003 (ATPA)	3(a), (b). Anti-Child Pornography Act of 2009 (ACPA) 3(h), (j) ATPA 4(c) 2 Cybercrime Prevention Act of 2012 (CPA)	9 Special Protection of Children Against Abuse, Exploitation and Discrimination Act 16–18 Revised Penal Code (RPC)	340 RPC	3(h), (i) ACPA
Poland	204(4) Polish Criminal Code (PCC)	199(3), 200(1), 203, 204 PCC	202 §3, 4, 4a, 4b, 4c PCC	Article 200a PCC	Article 200 §3, 4, 5 PCC	Article 200a PCC
Scotland	–[c]	9 Protection of Children and Prevention of Sexual Offences Act 2005 (PCPSO)	–	–	23, 24, 25, 33, 34, 35 Section 23 Sexual Offences Act 2009	1 PCPSO

(continued)

Table 1.3 (continued)

	18. Sexual abuse	19. child prostitution	20. child pornography	21. pornographic performances	22. corruption of children	23. Online solicitation (grooming)
South Korea		Articles 12(2), 13 (2) of the Act on the Protection of Juveniles against Sexual Abuse	Article 44-7(1)1 of the Act on Information Promotion and Protection, and Communications Network Utilization	Articles 287 and 294 of CA	Article 13(2) of the Act on Special Cases concerning the Punishment, ETC. of Sexual Crimes Articles 287 and 294 of CA	Articles 12(2), 13 (2) of the Act on the Protection of Juveniles against Sexual Abuse
Spain	183 bis, 171, 172,183 Spanish Criminal Code (SCC)	Article 189 SCC	Article 183 bis, 189.4. 189.5. 189.6, 189.7 SCC	Article 183 bis, 189.4. 189.5. 189.6, 189.7 SCC	Articles 183 bis, 185, 186 SCC	Article 183 ter, Section 1 Section 2 SCC
Turkey	105 Turkish Penal Code (TPC)—(only harassment)	227 TPC and possibly 80 TPC	226 TPC	226 TPC	226, 105 TPC	
United States			18 U.S.C. § 2252 18 U.S.C. § 2252A 18 U.S.C. § 1466A	18 U.S.C. § 2251	18 U.S.C. § 1470	18 U.S.C. § 2422 (b)

[a]The term 'child prostitution' and the corresponding offence is more explicitly regulated under State/Territory laws. However, there is no case law applying this kind of offence to the online context, and the jurisdictional reach of such State/Territorial laws is more limited

[b]Article 175(1) EPC refers to the '*influencing* of a person of less than eighteen years of age in order to cause him or her to commence or continue the commission of a criminal offence, begging, engagement in prostitution [...]'. Usually, the provision would be an 'offline' offence. Yet, since it is possible that *influencing* is achieved through online means only, the norm should be applicable in the context of webcam child sex tourism as well. However, at the moment of writing of this report there is no case-law to confirm this interpretation

[c]Incomplete information based on country report

When it comes to sexual activities involving Sweetie, it is also relevant to consider that Sweetie depicts a person that is under the age of twelve. At this age, there will be no discussion in the jurisdictions under investigation that Sweetie is a minor and that any form of sexual activity involving a 12-year old is prohibited. Moreover, for many countries engaging in sexual activities with a minor the age of Sweetie is an aggravating circumstance (e.g. Argentina, the USA, Spain and Canada).

Relevance of Physical Presence

Given that webcam sex is a relatively new phenomenon, substantive criminal laws in many countries have not been amended to specifically include webcam sex as an offence in itself. In most cases, existing crime descriptions cover webcam sex. Most of these descriptions however stem from a time when engaging in sexual activities with a minor required physical contact with the victim. This raises the question whether the physical element in the crime descriptions is of material importance, or whether these crime descriptions also cover sexual activities taking place via a webcam, without out any physical contact or physical presence.

In several of the countries under investigation webcam sex with minors is considered sexual abuse. In the Netherlands, for instance, the Dutch Supreme Court decided that for sexual assault (Article 247 of the Dutch Criminal Code) physical contact between the assailant and the victim is not necessary.[34] This means that when the perpetrator is participating in a webcam session with a person under the age of 16 and the victim performs sexual acts (such as performing sexual acts with themselves or a third person), the perpetrator can be held accountable for sexual assault. In Canada, webcam sex may be considered sexual abuse if the perpetrator invites the victim to touch him or herself (Section 152 Canadian Criminal Code). Furthermore, physical presence is also not a requirement for the offence of indecent exposure. The court held that the element 'in any place' could also refer to the Internet.[35] In Belgium, the adult who induces or forces a minor to display breasts or genitals or to perform sexual activities in front of the webcam is committing the offence of indecent assault (Article 372 or 373 of the Belgian Criminal Code). In Turkey, however, physical contact is still considered an integral part of the offence of sexual abuse, which means that the sexual abuse of minors via the Internet could never fall within the scope of Article 103 of the Turkish Criminal Code.[36]

Apart from offences that fall into the category of sexual abuse, the Lanzarote Convention also opens up the possibility to criminalise pornographic performances viewed via a webcam:

> Article 21 incriminates certain conducts relating to the participation of children in pornographic performances. Paragraph 1 a and b are elements relating to the organisation of pornographic performances involving children while c relates to the spectator. As with child prostitution and child pornography, the provision establishes links between the supply

[34] See ECLI:NL:HR:2004:AQ0950.

[35] *R v. Alicandro* [2009] ONCA 133 (CanLII).

[36] See Önok and Bayamlıoğlu 2016, Section 2.2.3, p. 16.

> and the demand by attaching criminal liability to the organiser of such pornographic performances as well as the customer. Depending on States, this provision may also cover the situation of persons who are spectators of pornographic performances involving the participation of children through such means of communication as webcams.[37]

From the above we may conclude that while the element of physical presence does not necessarily preclude the application of existing crime descriptions such as those covering sexual abuse. However, this must be decided on a country-by-country basis.

Substantive Criminal Law Legality Principle

A third issue related to the criminalisation of webcam sex tourism is that of substantive criminal law legality (*nulla poena sine lege*). As described above, substantive criminal laws in many countries have not been amended to specifically include webcam sex as an offence in itself. Rather, webcam sex is 'read' into existing crime descriptions. As such, one could argue that the crime descriptions lack the requisite legal clarity and certainty (*lex certa*) to be applied to online contexts. Based on the examined case law for the different countries, we have no indication though that this is indeed the case.

Differences in Approach to the Criminalisation of Webcam Sex

Based on the criminal law systems of the countries under investigation, we may conclude that webcam sex with minors is generally considered criminal in one way or another. In most countries webcam sex with minors falls under the heading of offences related to child pornography. Depending on the country and the circumstances of the case, webcam sex with minors may also fall under different crime descriptions such as sexual abuse or child prostitution, offences that generally carry higher penalties. This is, however, very much dependent on the circumstances of the case and the specific jurisdiction.

While there are differences in the approach to the criminalisation of webcam sex, they do not seem very problematic in the global fight against webcam child sex tourism. All of the jurisdictions examined have a more or less complete inventory of possible offences that apply in the context of webcam sex with minors.

An issue that needs to be taken into account though is that of double criminality.[38] Given the different approaches to the criminalisation of webcam sex, combating webcam sex tourism in an international context is more difficult. For instance, in the Netherlands webcam sex with a minor may qualify as 'sexual abuse', whereas in Turkey it only qualifies as 'sexual harassment' given that physical contact is necessary to proof sexual abuse. As such, law enforcement will need to assess whether the crime descriptions are sufficiently similar as not to cause issues of (a lack of) double criminality.

---

[37] *Explanatory report to the Council of Europe Convention on the Protection of Children against Sexual Exploitation and Sexual Abuse. CETS 201.*

[38] The principle of double criminality was introduced by extradition treaties and requires the act to which a request relates to be a crime under both the criminal law of the requested state and the requesting state. For a comprehensive discussion on this see Williams 1991, p. 582.

Table 1.3 shows which criminal provisions potentially apply to webcam sex in the countries under investigation. The table fields marked in red indicate that the existing country legislation is not applicable in the context of webcam child sex tourism.[39]

1.2.3 Complicating Factors in Substantive Law Related to Sweetie

In this section, the two problematic issues will be addressed when it comes to applying existing substantive criminal laws to cases involving Sweetie. First, Sweetie is an avatar, a virtual character, programmed to appear and talk as a child but clearly no real child is ever involved in the process. Second, the avatar does not undress and therefore no sexually explicit behaviour on the part of the 'victim' takes place. This section investigates to what extent interaction with Sweetie is still deemed criminal despite the lack of involvement of both a minor and explicit sexual behaviour.

Virtual 'Victim'

Sweetie's most important asset from an ethical perspective—the fact that it does not involve real children and thus does not put actual children at risk—at the same time may be problematic with regard to some (or most) of the criminal systems at hand and the application of their provisions as discussed previously. The reason for this is that the crime descriptions in most criminal law systems deal with real victims as opposed to 'virtual' victims.

Criminal law protects society (in this case minors) against harm through threat of sanctions (general prevention). Most of the crime descriptions in criminal law therefore criminalise specific behaviour, which in most cases constitutes a direct threat to something or someone. For example, in the case of murder or homicide, human life is threatened. In the case of theft, personal property is at risk. Thus, most crime descriptions feature the main elements of the situations they aim to address, often in relation to the persons they aim to protect. Generally, if the completed act does not match a crime description, then that particular crime cannot be proven. Most of the crime descriptions examined in this research—virtual child pornography and grooming being notable exceptions—feature real persons (minors), given that they are the objects of protection. This means that without a real victim, there is no crime. Since Sweetie is an AI system and hence does not qualify as a natural person, interaction with Sweetie will most likely not fulfil most of the crime descriptions discussed previously. An interesting exception, however, is the Philippines, where 'any lascivious exhibition of sexual organs or sexual activity [...] with the aid of a computer system' is a punishable offence. This does not

---

[39] These provisions focus mostly on the actual physical contact between victim and offender, and are thus classical 'offline' offences.

require an actual victim at all, and the law can thus be applied to interactions with Sweetie where the perpetrator displays his sexual organs before the webcam.

In the laws governing the sexual exploitation of minors, two types of criminal acts exist that do not focus on actual victims, but rather on the behaviour of the suspect and the intent that the particular behaviour signals. These crimes are: (1) virtual child pornography and (2) grooming.

Virtual Child Pornography

As described in detail in Sect. 1.2.1 not only real but also virtual child pornography is considered criminal, including computer-generated images not involving real victims. There are several reasons why criminal liability is also extended to virtual child pornography. These include avoiding evidentiary problems and the fact that the materials can be used to corrupt children, as well as the idea that virtual child pornography may act as a 'stepping stone' for consumers of child pornography, prompting them to move to real child pornography or possibly even sexual abuse.

Grooming

Grooming is the act of having contact with a minor for the purpose of arranging a meeting with this minor in order to engage in sexual activities or other lascivious conduct or to create child pornography. As it provides such a clear and present danger to the well-being of minors, in most of the jurisdictions under discussion grooming is a crime in itself, regardless of whether the meeting actually takes place or leads to sexual abuse. The process of grooming typically starts with rather innocent chats and develops through time into interactions which separately may constitute, among others, indecent communication, corruption of children, sextortion et cetera.

In some jurisdictions, it is not even necessary that a real minor is groomed. In these jurisdictions, law enforcement's imperative to employ investigative powers (e.g. luring suspects) has prompted changes to substantive criminal law. The investigative method of using a lure that is ostensibly a child would not be possible without the element in the crime description allowing for adults or virtual minors to be the object of the offence. In England and Wales for instance, the fact that the offender's communications are transmitted to a virtual child is unproblematic in regard to Section 14, SOA 2003. The section's purposely wide language does not require the involvement of a child, but focuses rather on intent or belief on the part of the suspect that a child sex offence will take place.[40] The same goes for Australia where for instance Section 474.28(9) of the Criminal Code Act 1995 covers virtual children with the term 'fictitious recipients'. Noteworthy is also the Belgian offence of cyber-luring which codifies a similar rationale—according to the provision someone completes the criminal act upon communication with an apparent or probable minor.[41] The Philippine law on cybersex, while not specifically focused

[40] See Gillespie 2016 (report on England and Wales), p. 29.

[41] Cyber luring is codified in Article 344bis/1 of the Belgian Criminal Code. See Royer et al. 2016 (Belgian report), p. 21.

on grooming, declares 'any lascivious exhibition of sexual organs or sexual activity [...] with the aid of a computer system'[42] a punishable offence *per se* and does not require an actual victim at all. Finally, recent changes to Dutch criminal law have amended the grooming crime description so as to include grooming of a virtual creation of a minor (an amendment also made in the offence of inducing minors to perform or tolerate lewd acts).[43]

From these two crimes we may deduce that in the area of protection against child exploitation substantive criminal law is not only reactive in nature and focussed on actual victims, but also preventative in the sense that it criminalises behaviour (regardless whether said behaviour is aimed at an actual minor) that may provide future danger to minors, or an indication thereof.

Apart from these two offences though, directly applying substantive criminal law provisions to virtual characters is still an untested area in most of the jurisdictions investigated for the present project. This is true for instance for both the Nigerian[44] and the Scottish criminal law.[45] In Canada the issue has been litigated in relation to some of the offences with the result of conviction of an offender who thought to be communicating with a child that never existed[46]; however, this approach is complemented by the law of attempts.[47] Criminal liability through attempt also seems to be the only avenue for criminalising a sexually-charged interaction with Sweetie in Argentina, Croatia, Estonia, Israel, Poland and Turkey. Construing an attempted offence is further possible regarding all other relevant provisions of the countries discussed above and will be addressed in the section on criminalisation of attempt under Section 3.5.

No Sexually Explicit Behaviour or Nudity on the Part of Sweetie

A second complication is that Sweetie is not programmed to undress or display sexually explicit behaviour. This means that (the animations of) Sweetie cannot be qualified as child pornography given that Sweetie does not engage or seemingly engage in sexual activities or show genitalia primarily for sexual purposes.

As a result, someone interacting with Sweetie cannot complete the offence of accessing (and possibly storing) child pornography. From a law enforcement perspective this is an issue, given that in most countries accessing (virtual) child pornography would be the go-to offence in the case of Sweetie.[48]

While an attempt at accessing (virtual) child pornography might still be construed, it usually carries a lower maximum penalty, and in some jurisdictions might not be an offence at all.

---

42 Cybercrime Prevention Act of 2012, sec 4(c)(1).

43 See Chap. 11 in this Volume.

44 See Orji 2016 (Nigerian report), Section 2.3.

45 See Richardson et al. 2016 (Scottish report), p. 16.

46 See Hodge 2016 (Canadian report), p. 28.

47 See s24 of the Canadian Criminal Code, RSC 1985, c C-46.

48 Many of the other offences concerning webcam sex with minors would require additional acts on the part of the perpetrators, such as exposing themselves.

1.2.4 Criminalisation of Attempt

The two complicating factors of Sweetie (it being a virtual victim and the absence of sexually explicit behaviour or nudity) have the effect that in most jurisdictions many of the criminal offences discussed in Sect. 3.2.2 cannot be committed. But while a completed offence may be unable to be proven, an attempt might be.

For a completed criminal act, the offender must have fulfilled all components of the respective crime description, which—put in a nutshell—refer to the objective elements (a certain act or omission directed against a legally protected right, which has produced a certain result), and the subjective elements (concerning the offender's internal characteristics, in particular the intent to commit the act in question). Attempted crimes, in contrast, are prosecuted when the suspect has started to execute the crime he/she intended to commit but failed to complete the criminal act due to external circumstances. In general, the rationale behind the criminalisation of an attempt lies with the dangerousness of the offender, who has failed to complete the intended crime at the given moment, but who has nevertheless clearly manifested activities that can easily result in a completed crime on the next occasion.

It is clear that only a handful of offences could be directly applied to virtual victims like Sweetie. Most of the provisions' descriptions require the offender's interaction with a *real* person, whose rights and interests must be at stake. Evidently, Sweetie is a computer-generated character, which has no legal interests and therefore cannot fulfil said requirement. A completed offence against the avatar will therefore rarely be committed under the current legal framework. It is however worth exploring whether the missing piece of the puzzle—the crime's subject—can be found through the laws of attempt. Accordingly, the following subsections elaborate upon the difference between a completed criminal offence and an attempted one by outlining the requirements of attempt, and how these apply in Sweetie's case.

Qualification of an Attempt

The laws governing attempt vary in the jurisdictions investigated in this study. Their differences manifest themselves mainly in the origin of the doctrine, the way in which it has been codified or in its legal consequences.[49] In general, all jurisdictions criminalise attempt and employ similar considerations in outlining its main requirements.

---

[49] In the USA, for instance, there is no general crime of attempt in federal law. Instead, statutes include separate provisions regarding the criminalisation of the attempted crime, see Unikowski 2016 (US report), p. 12. In contrast, other jurisdiction like Argentina (Article 42 of the Criminal Code), Australia (an attempt to commit a crime is a distinct offence in all Australian jurisdictions and the general doctrine is regulated by their statutes), Brazil (Article 14(2) CPB), England and Wales (s.1, *Criminal Attempts Act 1981*), Germany (§§ 22, 23 StGB), The Netherlands (Article 45 DCC) and Poland (Article 13 of the Criminal Code) have a general regime of attempt liability, whose provisions are then applied in conjunction with the ones of the respective offence. Yet other systems like Croatia, Belgium (Articles 51–53 BCC), Nigeria (s.4 of the Criminal Code; s.27(1)(a) of the Cybercrimes Act) opt for both.

A criminal attempt is deemed to exist when the suspect executes an act towards the commission of the offence he/she intends to effectuate, said act is more than merely preparatory[50] and carried out according to the offender's intent.[51] Establishing the point at which mere preparation becomes a criminal attempt is a crucial part of the attempt liability, as it pinpoints both when a person becomes criminally liable and when authorities may intervene.[52] In practice, however, it may be challenging to demarcate the exact start of an attempt, but this concern is usually left to the courts to deal with and depends on the particular facts of the case and the available evidence.

Further, the laws of attempt differentiate between the reasons why the person failed to commit the crime. First, an attempt to commit a crime may fail either due to external circumstances, which are outside of the sphere of influence of the perpetrator.[53] Second, despite completing the intended actions the requirements set forth in the description of the crime might not be met. The latter is usually referred to as inadequate or inept attempt, and although recognized by all criminal systems in this study, it is applied in widely varying ways. These are of particular importance for prosecuting individuals interacting with Sweetie and will be addressed in the following subsections.

Inadequacy of an Attempt

Because Sweetie is an avatar and, therefore, from a criminal law perspective an inadequate object in relation to sexual crimes, criminal liability for interacting with Sweetie needs to be examined through the lens of the doctrine of inadequate attempt. The doctrine of impossible or inadequate attempt takes account of situations, in which the completion of the intended offence is factually or legally impossible due to the object's unsuitability, the chosen time, or the means used by the perpetrator. We can distinguish between *relative* inadequacy of an attempt and *absolute* inadequacy of an attempt.

Relative Inadequacy (Factual Impossibility)

Relative inadequacy occurs when an act does not bring about the results described in the crime description, because at the time of the attempt committing the crime

[50] Argentinian law requires a start of execution, see Article 42 of the Criminal Code. The general trend in Australia also requires 'more than merely preparatory' acts, see *Britten v. Alpogut* [1987] VR 929, at 938. The Polish criminal system requires speaks of a 'final and decisive action' towards the offence in question, see Polish report, p. 14.

[51] The Polish system refers to offender's 'decision', see p. 14 of the Polish report.

[52] Robinson 2010, p. 579. The German attempt doctrine, for instance, has recognised several moments of commencing. The attempt can begin when one of the elements of the offence description is fulfilled, but also just before fulfilling any crime element as long as the offender undertakes steps which according to his plan are prior to the crime realisation and will immediately result in the completed offence. See Hakobyan 2016 (German report), p. 65.

[53] In an online context, this would be the case if the offender's chat conversation with the victim is suddenly interrupted by a poor Internet connection or temporary unavailability of the server.

was in fact impossible. In other words, circumstances outside the knowledge of the perpetrator make it impossible for the attempt to succeed. More specifically, relative inadequacy relates to the object of the crime and/or the means with which the crime is attempted.

Examples of an attempt that is relatively inadequate are trying to shoot a person with an unloaded gun (relative inadequacy of the means) or stealing money from an empty cash register (relative inadequacy of the object).

With a relatively inadequate attempt, the concrete facts of the case make it so that the attempt fails. If however the facts were as the perpetrator believed them to be, the attempt would have succeeded, as a result of which the perpetrator can be charged with a criminal attempt. The reasoning behind criminalising this form of attempt is nicely summarised in the US case of *State v. Moretti*:

> When the consequences sought by a defendant are forbidden by the law as criminal, it is no defense that the defendant could not succeed in reaching his goal because of circumstances unknown to him.[54]

Absolute Inadequacy (Legal Impossibility)

Absolute inadequacy refers to those situations whereby an attempt can never lead to a completed offence, because what was attempted is not a criminal act (or in any case does not match with the relevant crime description). In the case of absolute inadequacy, the perpetrator intends to commit a crime but by virtue of the object or the means, the intended behaviour cannot result in a crime in reality. An example is trying to murder a corpse (inadequate object) or trying to poison somebody with very strong camomile tea (inadequate means). While criminal intent is present, the actual behaviour does not align with the crime description. Under the doctrine of absolute inadequacy, the suspect can never be held criminally liable, despite a clear display of criminal intent. The reasoning for this is that someone cannot be held criminally liable for something that is not criminal, even though that person might have thought that he or she actually was committing a crime.

Relative inadequacy is different from absolute inadequacy in that different facts could or would have made the attempt successful, while in the latter case the desired criminal outcome can never be achieved by the suspect.

Applying the Law of Attempts to Sweetie

The question is whether interacting with Sweetie would count as an attempt that is absolutely inadequate, or relatively inadequate. An answer to this question needs to take into account the circumstances of an actual case and the specifics of the doctrine of attempt in the national system. But having said that, we can make some general observations about applying the law of attempts to Sweetie.

[54] *State v. Moretti*, 244 A.2d 499.

It is not straightforward to determine whether Sweetie will lead to an absolutely inadequate attempt, or a relatively inadequate attempt. An argument in favour of an absolutely inadequate attempt is that Sweetie, being a virtual person, and programmed not to undress, is an absolutely inadequate object. The argument would then be that it is not criminal to engage in sexual activities with a virtual person, even if this virtual person is posing as a minor and the perpetrator thought that the avatar was a real person. But even if sexual activities with a virtual minor is criminalised in specific jurisdiction, then Sweetie might likely still be considered as absolutely inadequate given that she can never show sexual organs or perform sexual activities.

However, we can also take a somewhat broader perspective and qualify Sweetie as a relatively inadequate object for the crime of webcam sex with minors. The argument is than that the suspect wants to commit the crime of webcam sex with a minor and enters a chatroom with for instance twenty minors for this purpose. Unbeknownst to the suspect one of the twenty minors is Sweetie. The suspect is 'unlucky' and picks Sweetie. This case is comparable to the example of the cash register: under normal circumstance the crime would have been committed, but due to 'bad luck' on the part of the suspect, there is now a factual impossibility. If we are to follow this reasoning though, it must be clear from the behaviour of the suspect that had the suspect chosen a different child, the outcome of the behaviour would have been the same.

The attempt liability in the case of Sweetie would thus largely depend on whether the respective criminal system emphasizes the objective or subjective elements of the offence. Jurisdictions that focus on the objective elements underline the importance of the concrete crime object, while those systems that focus more on the subjective elements see the alleged offender and his mental state as a threat for the protected ideal good: minors and their overall safety and well-being. In these systems the intent of the suspect is crucial. The German law of attempts, as it will be explained below, is the only example where both objective and subjective considerations are decisive for the attempt liability of the offender.

In the following we illustrate which jurisdictions follow the first approach and which the latter.

The Law of Attempts in Different Jurisdictions

The country reports indicate that the doctrine's considerations would apply in cases where the victim of the intended crime is virtual, and thus a part of the crime description can never be fulfilled. Yet, while in some jurisdictions the suspect remains punishable or receives a lower sentence, in others the legal impossibility completely excludes criminal liability. Thus, depending on the particularities of the respective criminal system and on the circumstances of the case, a person may be found guilty of an *attempt* to commit webcam sex tourism notwithstanding the fact that he or she did not interact with a real minor, but with Sweetie.

Under Canadian, English, Israeli, Scottish and US[55] law, for instance, the reason why the offender's attempt failed is in most cases irrelevant,[56] as long as the perpetrator undertakes more than preparatory actions in pursuit of the prohibited behaviour and manifests malicious intent. In these jurisdictions the suspect's criminal intent (which has become apparent through his/her actions) weighs more heavily than the objective fulfilment of the crime description.

England's criminal attempt law applies this premise to all offences related to webcam sex with minors save for s.127 of the Communications Act 2003,[57] while its Scottish counterpart does not foresee any limitations in application.[58] Canadian case law shows that Section 24 of the Canadian Criminal Code, which regulates attempt, is not always applied. However, this does not have to do with normative restrictions on the application of the impossible attempts doctrine, but results from the fact that some offences carry included offences that are attempts.[59] Thereby, the law addresses the activities surrounding the core offence and circumvents possible loopholes. Accordingly, a conviction can be pursued for any of the webcam offences in relation to Sweetie and it will depend on the particular circumstances of the case whether the prosecution relies on s.24 or on the included offence. In Israel the law of impossible attempt can be applied to all criminal charges regardless of the reason of impossibility.[60] However, since the avatar does not undress, attempted aggravated indecency would be the most serious crime within the Israeli criminal framework.[61] In addition, both Israeli[62] and Canadian[63] law place the burden of proof upon the defendant when it comes to demonstrating that he was not aware of actually communicating with a minor. Under Australian law, in principle, the common law doctrine of attempts applies as well, and a suitable offence (under

---

[55] In US law, the impossibility to victimise Sweetie is relevant as a defence that excludes punishability. Yet, such arguments of impossibility have not been successful in the prosecution of individuals attempting to commit a crime in cases where they believed that they were interacting with a child, but were in fact interacting with an undercover law enforcement officer. For more on the matter see the country report on the US, p. 12.

[56] See on the Australian doctrine *Haughton v. Smith* [1975] AC 476, which was followed on the Scottish case law *Docherty v. Brown* [1996] JC 48 at 50; see on the law of attempt in England and Wales *s.1, Criminal Attempt Act 1981*; see also s.24 (1) of the Canadian Criminal Code.

[57] The offence relates to the *sending* of an indecent or obscene message. It is irrelevant whether or by whom it is received, therefore the law of attempts is not required.

[58] See country report on Scotland, Section 2.3, p. 17. However, covert operations where police Internet investigators have posed as children have not been challenged or judicially considered yet.

[59] For instance, the offence of obtaining sexual services from a minor includes also the offence of communicating with a person under 18 in order to obtain sexual services for consideration. See on that Canadian report, p. 29.

[60] The Israeli law of attempt is embedded in the Introductory Part of the Israeli Penal Code, Chapter 5, Article 1.

[61] See Harduf 2016 (Israeli report), p. 11.

[62] See the Israeli Supreme Court decision in the case of *Ktiei v. Israel,* LCrimA 1201/12 [9 January 2014].

[63] See for instance s.171.1(3) and s.172.2(3) of the Criminal Code of Canada, RSC 1985, c C-46.

State/Territory laws) might be an attempt to procure a child for a sexual act. However, it should be noted that many of the Commonwealth offences illustrated in the tables above are explicitly excluded from attempt liability.[64] The rationale behind the legislator's choice is the fact that many of the offences are already preparatory in nature, so that adding an attempt would over-extend criminal liability. Further, as indicated above, the preferred approach in Australia has been to explicitly allow for fictitious recipients of illegal communications, so that an attempt, although possible to construe in some if not in all cases, is not necessary for prosecution. Attempt liability in Germany[65] is regulated by § 22 StGB, while inadequate attempts are subject to the provision of § 23(3) StGB.[66] According to the norm, when the suspect targets an inappropriate object, his liability depends on whether he evidently acted in 'gross ignorance'.[67] In case the gross ignorance was evident to a reasonably informed person the court may mitigate the sentence or entirely leave out the punishment. Otherwise a punishable inadequate attempt is given.

Applying the doctrine outlined above to the case at hand means the offender would be liable for attempted sexual abuse pursuant to § 176 StGB if Sweetie looks like a real child to every informed person. Thus, even though the avatar is an inadequate crime object because the chat interaction with the offender cannot be perceived by a real child, if the offender's sexual performance is transferred through a webcam to the chatbot, an attempted sexual child abuse in accordance with § 176 (4) no. 1 in combination with §§ 22, 23 StGB would likely be given. The same applies to the provision of § 176(4) no. 2 StGB when the offender tries to induce Sweetie to perform a sexual act. In such a scenario, the suspect's inducement would be completed and would amount to an attempt, once the offender is convinced that he has done everything necessary to influence a child and the latter is just about to start performing.[68]

In Poland, the law of impossible attempts in Article 13(2) of the Criminal Code would allow construing a punishable attempt for the offences outlined in Article 202a(1 and 2) and Article 200(4) of the Criminal Code, i.e. the solicitation of minors for sexual purposes and corruption of minors respectively.[69]

The Estonian legislation (Section 26 of the Penal Code) and case law suggest that an impossible attempt regarding a webcam interaction with Sweetie is also

---

[64] See Australian report, p. 19.

[65] See the German report.

[66] 'Section 23 Liability for attempt. (…) 3) If the offender due to gross ignorance fails to realise that the attempt could under no circumstances have led to the completion of the offence due to the nature of its object or the means by which it was to be committed, the court may order a discharge, or mitigate the sentence as it sees fit (Section 49(2)).'

[67] Gross ignorance (*grober Unverstand*) is given when the suspect does not realise that under no circumstances could his acts complete the offence.

[68] In the context of webcam child sex tourism, an indication for that could be that Sweetie agrees to perform and provides the means for the transfer of rewards (e.g. bank account number).

[69] See country report on Poland, p. 14.

likely to be punishable, especially in relation to the offences of sexual enticement of a child and agreement to meet a child for sexual purposes.[70]

In Nigeria, a person who is approaching Sweetie for webcam sex would be punishable for an attempt to indecently interact with a boy or a girl (Sections 216 and 222 of the Nigerian Criminal Code respectively), and under Section 23(3)(a) of the Nigerian Cybercrimes Act. In relation to the latter, Section 27(1)(a) of the Cybercrimes Act clarifies that an attempted offence is punishable as a principal offence under the Act. The impossibility to commit webcam child abuse of a virtual minor is irrelevant pursuant to Section 4 of the Nigerian Criminal Code.

The laws of attempt in Argentina, Belgium and Croatia are not as straightforward as the ones already discussed and therefore somewhat more challenging.

In general, the laws on attempt in Argentina are applicable to all types of crimes,[71] including all offences against sexual integrity. Article 44 of the Argentine Criminal Code regulates the defence of impossibility, which reduces the sentence or even acquits the defendant when granted. It is in the light of this provision that courts consider (attempts to commit) impossible crimes. However, it appears that there are no clear criteria on how courts engage in an assessment of the action's efficacy in bringing about the intended crime.[72] Therefore, while the provision's rationale would apply to the case of Sweetie, it is unclear how and with what outcome.

In Belgium, the criminal liability for attempting a particular offence will depend on the offence itself.[73] The attempt to commit a crime is always punishable, while the attempt to commit a misdemeanour is punishable if the law explicitly provides it. In the case of Sweetie one of the potentially relevant offences is attempting indecent assault. However, a prosecution's outcome will very much depend on the court's approach towards impossible attempts. If the protected object (here minors, who are not at risk through the act since Sweetie is not a minor) is deemed more relevant than the subjective state of mind of the offender, acts involving Sweetie will not be punishable.[74] Furthermore, an attempt to access (virtual) child pornography is also not punishable in Belgium.

The Croatian law of attempts in its turn does not substantively differ from the already discussed doctrines.[75] Inadequate attempts are punishable as ordinary ones, but here it is the interpretation of the attempts provision that turns out to be problematic. A perpetrator attempting to commit an offence by inappropriate means or against an inappropriate subject may be exempted from liability due to 'rough

---

[70] Kala 2016 (Estonian report), pp. 22–23.

[71] Ferrante 2010, p. 29.

[72] Ferrante 2010, p. 31.

[73] Attempts to possess or access virtual child pornography are not punishable, see Article 383bis BCC. The same applies to the attempt to commit the offence of publicly outraging morality by actions who offend modesty, Article 385 BCC. See for more on the matter Belgian report, p. 22.

[74] See Belgian report, p. 22.

[75] See Bojić 2016 (Croatian report), p. 20.

irrationality'.[76] However, Croatian law does not provide a definition of 'rough irrationality' and courts have not dealt with the term either. Whether an offender is exempted from punishment in each case of rough irrationality, or only when his/her criminal intent results from the rough irrationality, remains unclear. To what extent impossible attempts have an impact on cases involving Sweetie is still to be determined.

A final remark concerns impossible attempts under Dutch, Spanish and Turkish criminal law. Dutch (case) law distinguishes between relatively inadequate and absolutely inadequate attempts. An attempt to have webcam sex with Sweetie would most likely constitute an absolutely inadequate attempt, which is not punishable regardless of the criminal intent of the alleged perpetrator.[77] A possibility to use Sweetie under current Dutch criminal law might be found in the article on child pornography (Article 240b Sr). As indicated previously,[78] the crime of child pornography does not require that a person below the age of 18 is pictured. The inclusion of 'seemingly involved' in the provision means that virtual child pornography is punishable as well. The addition of the text 'seemingly involved' negates the issue that Sweetie is considered an absolute inadequate object given the fact she is not a real girl. As such, trying to get access via webcam to child pornography, or trying to produce and possess child pornography by recording and storing the webcam stream, even if involving an avatar, could be construed as a criminal attempt at accessing (or producing or possessing) child pornography. While this negates the issue of the virtual nature of Sweetie, it does not necessarily mean Sweetie is then an adequate object. Given that Sweetie will never portray sexual behaviour or any nudity for that matter, as it is not part of her programmed features, she might still qualify as absolutely inadequate on those grounds. If Sweetie were programmed to show sexual behaviour or sexual organs this would make proving the suspects' intent to have criminal webcam sex more straightforward, but this is not the case at present.

The Spanish criminal system in general does not punish impossible attempts.[79] However, recent case law has opened up the possibility for punishing impossible attempts by stipulating that impossible attempts should be punished when an *ex ante* evaluation of the suspect's conduct leads a reasonable person to believe that the consummation of the offense was possible even though an *ex post* study of the facts reveals that it was impossible for the actor to consummate the offense.[80] Consequently it could be argued that if a reasonable person sees Sweetie in the

---

[76] The respective provision reads: 'The perpetrator who attempts to commit a criminal offense by inappropriate means or against an inappropriate object, *due to the rough irrationality*, may be exempted from punishment.' See Croatian report, p. 20.

[77] See Dutch report, pp. 5 and 16.

[78] See Section 3.2.1 Virtual 'victim' of this report.

[79] See Agustina and Valverde 2016 (Spanish report), p. 14. For more on the matter see also Gómez-Jara and Chiesa 2010, p. 503.

[80] Gómez-Jara and Chiesa 2010, p. 503. SSCS February 16, 2007, www.westlaw.es, Ref: RJ 2007 \2381.

webcam stream and believes it to be a real ten-year old, an attempt would still be construed even if in the aftermath it is revealed to the person that the child he/she saw was a computer-generated fiction. It remains to be seen how this approach will be implemented in prosecuting suspects of online sex crimes involving an inadequate object, such as Sweetie.

In Turkey, impossible attempts are recognised as a legal construct, the consequences of which are unclear. Some argue that impossible attempts should be punishable, but others do not agree. So far, the courts have not settled the dispute,[81] as a result of which it is unclear whether an interaction with Sweetie might amount to an attempted sexual harassment pursuant to Article 105 of the Turkish Criminal Code.

Table 1.4 gives an overview of the possibilities for criminalising (attempted) webcam sex with Sweetie. The fields marked in green refer to provisions which can be applied without further ado, while those in orange relate to provisions whose use in the context at hand has not been confirmed in case law yet. Red indicates that no provision applies.

1.2.5 Summary and Conclusion

Most of the criminal systems discussed in the previous sections seem well-equipped to combat webcam sexual crimes against *real* minors. Some have introduced new legislation to specifically tackle these types of online crimes, while others do not distinguish between online and real-world offences, and thus do not encounter systematic difficulties in applying the criminal norms to the cyber environment.

However, we do see a divergence in how webcam sex is criminalised. As will be shown in the following chapters, this finding impacts the possibilities for cross-border law enforcement, and often impacts the outcome of transnational investigation procedures.

In some countries, webcam child sex tourism is considered sexual abuse or sexual harassment, while in others it is a form of child pornography but does not fall under the heading of sexual abuse. Generally speaking though, most countries under examination have a full inventory of crimes that can apply to webcam sex tourism. What is relevant to consider in this context is that different types of interactions between victim and perpetrator may trigger different articles under (domestic) criminal law. For instance, merely talking to a minor without sexual content or hinting at sexual activities,[82] does not yet trigger any criminal law provisions, but sending indecent pictures or promoting sexual activities falls under

[81] See country report on Turkey, Section 2.2.3, p. 18.

[82] Please note that a sexualised conversation could be for instance an offence in England and Wales under the Communications Act 2003. Further, a typed conversation could amount to an offence under the Obscene Publications Act 1959. Also, when s.15A, Sexual Offences Act 2003 comes into force it would be illegal under that provision.

Table 1.4 Criminalisation of (attempted) webcam sex with Sweetie [*Source* The authors]

	18. Sexual abuse	19. child prostitution	20. child pornography	21. pornographic performances	22. corruption of children	23. Online solicitation (grooming)
Argentina	Possibly attempt	Possibly attempt	Possibly attempt	Possibly attempt	Possibly attempt	131 ACC
Australia	272.11 Criminal Code Act 1995 (CCA) (attempt)		474.19 CCA 474.20 CCA		474.27A	474.28(9) CCA
Belgium						433 bis 1 BCC Article 143 § 3bis Act Regarding Electronic Communication
Brazil						
Canada	152 CCC (attempt)	286.12 CCC (attempt) 286.32 CCC (attempt)	163.1 CCC (attempt)			
Croatia	Possibly attempt	Possibly attempt	163 CrCC	Possibly attempt	Possibly attempt	Possibly attempt
England and Wales	8, 10, 17, SOA 2003 (attempt)	47, 48, 50 SOA (attempt)	7(7), Protection of Children Act 1978 (possibly attempt)		12 SOA (attempt)	14, 15 SOA (attempt)
Estonia	Possibly attempt	Possibly attempt	178 EPC	Possibly attempt	179 EPC	178[1]EPC
Germany	Possibly attempt	Possibly attempt	Possibly an attempt of 184b (3) StGB		Possibly attempt	Possibly attempt

(continued)

Table 1.4 (continued)

	18. Sexual abuse	19. child prostitution	20. child pornography	21. pornographic performances	22. corruption of children	23. Online solicitation (grooming)
Israel	Possibly attempt	Possibly attempt	Possibly attempt	Possibly attempt	Possibly attempt	Possibly attempt
Netherlands	248a DCC		Possibly attempt of 240b DCC			248e DCC
Nigeria	216, 222 NCC	222NCC		23(3) (c) NCA jo 27(1) NCA (attempt)	23(3)(c) jo 27(1) NCA (attempt)	23(3)a jo 27(1) NCA (attempt)
Philippines					Possible attempt at cybersex (showing genitalia)	3(h), 3(i) Possibly attempt
Poland	204(4) Polish Criminal Code (PCC), possibly attempt	199(3), 200 (1), 203, 204 PCC, possibly attempt	202 § 3, 4, 4a, 4b, 4c PCC. Possibly attempt	200a PCC. Possibly attempt	200 §3, 4, 5 PCC. Possibly attempt	200a PCC. Possibly attempt
Scotland		9 PCPSO (attempt)			23, 24, 25, 33, 34, 35 Section 23 Sexual Offences Act 2009 (attempt)	1, 9 PCPSO (attempt)
South Korea		12(2), 13(2) APJSA (possibly attempt)	44-7(1)1 of the Act on Information Promotion and Protection, and Communications Network Utilization (possibly attempt)	287 and 294 of CA (possibly attempt)	13(2) of the Act on Special Cases concerning the Punishment, ETC. of Sexual Crimes (attempt); 287 and 294 of CA (possibly attempt)	12(2), 13(2) APJSA (possibly attempt)

(continued)

Table 1.4 (continued)

	18. Sexual abuse	19. child prostitution	20. child pornography	21 pornographic performances	22. corruption of children	23. Online solicitation (grooming)
Spain			Article 183 bis, 189.4, 189.5, 189.6, 189.7 SCC (attempt)			
Turkey	105 Turkish Penal Code (TPC), possibly attempt	227, 80 TPC, possibly attempt	226 TPC, possibly attempt	226 TPC, possibly attempt	226 TPC, possibly attempt	
United States			18 U.S.C. § 2252, 2252A, § 1466A (possibly attempts)	18 U.S.C. § 2251 (possibly attempt)	18 U.S.C. § 1470 (possibly attempt)	18 U.S.C. § 2422(b) (attempt)

the heading of corruption of a minor and possibly even grooming. When the victim is engaged in sexual activities and/or genitalia are exposed via webcam, this may trigger offences such as child pornography and pornographic performances, and possibly sexual assault. If payment is involved, provisions regarding child prostitution come into play. Finally, if the perpetrator exposes him/herself and masturbates or forces or coerces the victim to do or undergo sexual activities, this may constitute corruption of minors or even sexual assault.

While all countries under investigation have criminalised webcam sex in one way or another, the situation is less clear-cut when Sweetie, a chatbot rather than a person, is involved in the interactions. In a few countries webcam sex with a virtual person is criminalised (see Table 1.5). In these countries, the criminal law provisions focus on the behaviour and the intent thereof, rather than on the behaviour in relation to the result it brought about.

In those countries where the crime descriptions only mention real minors as the object of protection, interaction with Sweetie may still qualify as an attempt at

Table 1.5 Criminalisation of webcam sex [*Source* The authors]

	Webcam sex criminalised	Webcam sex with virtual person criminalised	Attempt at webcam sex with virtual person criminalised
Argentina	X	–	X (only likely for grooming)
Australia	X	X	X
Belgium	X	X (for the offence of cyber luring)	X (depending on the interpretation)
Brazil	X	–	–
Canada	X	X	X
Croatia	X	X	X (no case law yet)
England and Wales	X	–	X
Estonia	X	X (no case law yet)	X (no case law yet)
Germany	X	–	X
Israel	X	–	X
Netherlands	X	X (for some forms of sexual abuse)	–
Nigeria	X		X
Philippines	X	–	X
Poland	X	–	X
Scotland	X	X (not tested yet)	X
South Korea	X	–	X (no case law yet)
Spain	X	–	X (child pornography only)
Turkey	X (only regarding sexual harassment)	–	X (possible, not tested yet)
USA	X	–	X

illegal webcam sex. In this area we see a further divergence. In some countries an attempt at webcam sex with Sweetie can be construed,[83] in other countries an attempt at webcam sex with Sweetie is not deemed criminal because it qualifies as an impossible attempt. What is clear from our investigation that for most countries it still very much unclear whether prosecuting an attempt would be successful. It also relevant to note that an attempt generally carries a lower maximum penalty than a completed offence.

Finally, in those jurisdictions where neither a completed offence nor an attempt can be construed because of the virtual nature of Sweetie, interaction with Sweetie may still be qualified as an attempt to access to child pornography. However, in these countries we should also take into account the fact that an attempt may still be impossible given the fact that Sweetie will never show sexual organs or perform sexual acts. Furthermore, in some jurisdictions an attempt to access child pornography is not criminal in general.

1.3 Criminal Procedural Law Aspects

Since Sweetie is intended as an investigative tool for law enforcement in online investigations, its implementation in law enforcement operations is governed by the laws of criminal procedure.[84] In this section we investigate what (online) investigatory powers are available in the jurisdictions involved in this study, and how these powers apply to Sweetie.

1.3.1 Human Rights Protection in (Online) Investigations

Human rights serve first and foremost as a control tool for modern state bureaucracy against structural injustices.[85] In this role, they perform a 'check and balance' function for states using coercive investigative powers. While international human rights law imposes on states the negative obligation to refrain from interfering with the exercise of human rights, this duty is balanced by the exigencies of everyday life and the positive state obligation to proactively protect the individual's rights from violations.[86]

---

[83] It must be noted though that in most of these countries the possibility of webcam sex with virtual minors has not been tested in court.

[84] Note that Sweetie may also be used as a deterrent. However, if Sweetie is used as a deterrent, it is likely only effective if there is a risk for the perpetrator that his behaviour may ultimately be exposed. In other words, if Sweetie is used as an investigative method.

[85] Bielefeldt 2012, pp. 14–15.

[86] Bielefeldt, 2012, pp. 14–15.

In procedural criminal law, the interplay of these obligations is reflected in the enactment of special investigatory powers on the one hand (which temporarily limit the constitutionally guaranteed rights of the citizen in the fight against crime), and in the set-up of procedural rules that regulate, oversee and limit said special powers on the other. In either case, specific requirements must be met, some of which depend on the specific nature of the right itself, while others, as will be shown below, are of a more general nature.

It should be noted that in the European context, the ECtHR is particularly present in the human rights context and has assumed an active role in the interpretation of the ECHR. The Court's finding that both negative and positive human rights obligations stem from the ECHR in the context of law enforcement has been also widely followed and adopted by other human rights mechanisms, such as for instance the Human Rights Committee (HRC). The consideration of both positive and negative state obligations is thus a largely recognised tenet when it comes to the effective enforcement of individual rights.

In this research, we will focus specifically on the negative human rights obligations, and the extent to which states are permitted to infringe human rights for the purpose of combating webcam sex tourism.

Criminal Procedure Law Requirements

Human rights have been decisive in shaping constitutional guarantees and hence have implications for the reach of state powers and the use of special investigative techniques. These powers have to abide by certain requirements to ensure the integrity of the criminal procedure and the reliability of the obtained evidence. In addition, special investigative powers often have to remain the exception to the rule.[87] Consequently, state agencies may interfere with the individuals' rights only in exceptional circumstances and pursuant to the requirements of proportionality and subsidiarity. Furthermore, the use of intrusive state methods is only allowed if they are in accordance with the legality principle, in other words, if they have a specific legal basis in the law of criminal procedure.

The European Convention on Human Rights (ECHR), more specifically 8 ECHR, gives a good overview of the elements needed when doing the balancing test between law enforcement requirements and human rights. We will therefore use it as our point of departure for the discussion on criminal procedure law aspects. The elements the ECHR stipulates are the following:

Necessity

Necessity refers to the test of 'necessary in a democratic society'. The latter implies in its core a finding of proportionality,[88] while at the same time considering the particular circumstances of the case, including 'the nature, scope and duration of the

[87] The country studies drafted for the purpose of this reports reveal that this is true for the majority of the civil law countries.

[88] *Klass and others v. Germany*, Application no. 5029/7149, § 50; *Weber and Saravia v. Germany*, Application no. 54934/00, § 116–118.

measures, the grounds required for ordering them, the authorities competent to authorise, carry out and supervise them and the kind of remedy provided by the national law'.[89] The proportionality test ensures that investigative tools and methods impair the legitimate interests of the alleged offender in a way that is reasonably proportionate to the harm committed or threatened.[90]

Subsidiarity
The principle of subsidiarity requires law enforcement to employ a more intrusive means of investigation only where less intrusive solutions would be substantially less or not at all effective and would thus jeopardise the aspired operation. We found that this tenet has been widely adopted by the investigated jurisdictions.

In Accordance with the Law (Principle of Legality)
Last but not least, criminal procedures can only take place, if and in a manner provided by the law. The rationale behind this principle of legality is strictly connected with legal certainty and aims to ensure that norms are available and accessible prior to the procedure that is being set in motion. Further, criminal procedures need to be foreseeable and predictable. This means that national laws need to be sufficiently clear as to under what circumstances law enforcement can make use of special investigatory methods. Individuals should also be able to determine which authority or mechanism implements and oversees the investigations.

1.3.2 Use of Investigative Powers in an Online Context

In order to protect human rights and to safeguard the integrity of the criminal investigation, the principle of legality as described above stipulates that special investigative methods require a basis in the law. This has led to an inventory of special investigative powers codified in the laws of the jurisdictions under examination. Below (see Table 1.6) we provide an overview which investigatory powers are available in the selected countries.

The specific investigative powers used in an online context are codified in the jurisdictions under examination more or less along the lines of those described in the Cybercrime Convention. Therefore, as it provides the most harmonised international framework governing the use of investigative powers, we use the Council of Europe Convention on Cybercrime as the framework for discussing investigative powers in an online context.

Given the close resemblance of Sweetie to the work of undercover agents, we will also discuss this investigative power, even though the Cybercrime Convention does not cover it.

---

89 *Weber and Saravia v. Germany*, § 106.

90 Ashworth and Horder 2013, p. 56.

Table 1.6 Codification of investigative powers in the selected jurisdictions [*Source* The authors]

	Preservation of stored computer data	Preservation of traffic data	Production orders	Search and seizure of stored computer data	Real-time collection of traffic data	Interception of content data	Undercover operations
Argentina	–	–	X	X	–	X	X
Australia	X	X	X	X	X	X	X
Belgium	X	X	X	X	X	X	X
Brazil	X	X	X	X	X	X	X[a]
Canada	X	X	X	X	X	X	X
Croatia	X	X	X	X	X	X	X
England and Wales	X	X	X	X	X	X	X
Estonia	X	X	X	X	X	X	X
Israel	–	–	X	X	X	X	–
Netherlands	X	X	X	X	X	X	X
Nigeria	X	X	X	X	X	X	–
Philippines	X	X	X	X	–	–	–
Poland	X	–	X	X	–	X	X
Scotland	/[b]	/	/	/	X	X	/
South Korea	–	–	–	X	–	–	–
Spain	X	X	X	X	X	X	X
Turkey	X	X	X	X	X	X	X
USA	X	X	X	X	X	X	X

[a]Only regarding infiltration operations, see Law n.° 12.850/2013, www.planalto.gov.br/ccivil_03/_ato2011-2014/2013/lei/l12850.htm [9 August 2016]
[b]Insufficient information available from country report

Council of Europe Convention on Cybercrime
Article 16. Expedited preservation of stored computer data
Article 17. Expedited preservation and partial disclosure of traffic data
Article 18. Production order
Article 19. Search and seizure of stored computer data
Article 20. Real-time collection of traffic data
Article 21. Interception of content data
Other: *Undercover operations conducted on the internet*

The countries examined in this report have codified the use of investigative powers in different ways.

The table above shows that the studied jurisdictions—with the exception of South Korea—have largely accommodated procedural provisions regarding undercover policing and online investigative techniques within their respective laws. In the following, we elaborate upon the necessary safeguards and admissibility conditions in relation to investigative powers in order to protect the rights and interests of individuals.

1.3.3 Sweetie as an Investigative Method

Before applying the procedural framework to Sweetie, we need to establish what kind of an investigative tool the avatar actually presents. We therefore first elaborate upon the nature of the chatbot/avatar as an investigation tool and then establish whether and how its features fit the legal framework of criminal procedure described above.

As described at the beginning of this study, Sweetie is an AI tool developed to facilitate the work of law enforcement agencies in online operations. As an AI agent, the chatbot would operate in open (public) online systems without direct human intervention, thereby enjoying a certain autonomy in conduct. Further, the chatbot/avatar will be used as a lure for the alleged offender but will also be capable of interacting with the suspect and recording and storing their interactions as well as available information on the offender, such as for instance his IP address.

The use of Sweetie as an investigative tool is not specifically covered in any of the jurisdictions in this study. In fact, no explicit rules exist on the use of AI agents for the purpose of criminal investigations. However, undercover investigations usually make use of a number of technical tools and coercive powers that resemble Sweetie's features. For instance, when placing a lure, law enforcement may either use physical objects, such as cars or bikes (or other goods depending on the crime), or stage an officer with a fake identity. Undercover agents that interact with suspects within a criminal organisation also make use of fake identities and usually do so in the framework of an infiltration operation. Further, in order to get access to the suspect's communications, state agencies use wiretapping and interception devices, and additional technological means capable of processing and storing the obtained data. Accordingly, given the lack of similar examples and against the background of its technical features, for the purposes of the present study the chatbot will be characterised and dealt with as a 'hybrid' investigation tool that combines the capacities of the different investigative tools mentioned above.

1.3.4 Authorised Use of Sweetie by Law Enforcement

The fact that Sweetie is a new investigative technique and existing investigative powers do not explicitly refer to any software or technology comparable to it, does not *per se* exclude the use of AI for investigative purposes. The use of the chatbot is possible as long as its application stays within the boundaries established by the law.

As Sweetie is designed to identify and engage suspects in a manner comparable to undercover investigators, the rules regulating the latter will be decisive for its application. Further, the chatbot will collect certain information on the alleged offender and its devices, and store the content of the communications between that person and Sweetie for investigation purposes. Consequently, the rules authorising these different investigative powers would conjointly be applicable in the case of the chatbot.

Whether the use of Sweetie is allowed depends on an answer to the following questions:

(1) Does Sweetie lead to a substantial risk of infringement of human rights in the context of a criminal investigation?[91]

If so,

(2) Is the use of Sweetie 'necessary in a democratic society'?
(3) Is the use of Sweetie 'in accordance with the law', that is to say, should there either be a specific legal basis regulating its use, or should its use be otherwise governed by procedural requirements?

Below we will answer these questions.

1.3.5 Possible Human Rights Infringements through Sweetie

Before we discuss the investigative powers that might come into play when regulating Sweetie, we need to establish whether a specific investigative power is

[91] If the answer to this question is negative, than there is no need to further question and codify the use of Sweetie as an investigative method from a human rights and criminal law perspective.

actually necessary. This means determining whether and how human rights infringements may take place when using Sweetie. When we observe the use of Sweetie, two fundamental rights of suspects may be particularly at risk:

- privacy (given that Sweetie may engage in conversations and record any communications), and
- the right to a fair trial (given that Sweetie may entrap suspects).

If these rights are infringed upon by using Sweetie, a specific investigative power that satisfies the procedural guarantees of legality, proportionality and subsidiarity would have to be in place.

Privacy

The right to privacy is recognised throughout the world. Even though with differing degrees of relevance, both in common and civil law traditions distinctions are made between the public sphere and the private sphere, and case law has evolved to facilitate the differentiation. It is generally held that in the public sphere, a suspect has less of a reasonable expectation of privacy. The level of protection provided though differs from jurisdiction to jurisdiction.

Australia, for instance, has a comparatively undeveloped right to privacy.[92] The cogency and relevance of the obtained evidence is usually prioritised over the privacy interest of the suspect even if said evidence has not been obtained in a public communication. Under the ECHR, on the other hand, persons can have a privacy interest in the public sphere, in particular when their behaviour is recorded.[93] Under the US Constitution, however, a reasonable expectation of privacy in public spaces and with regard to information divulged to third parties is more limited.[94]

Yet, while the distinction between public and private is relatively straightforward in the physical world, it is much less so on the Internet. This raises questions with regard to the use of Sweetie by law enforcement.

Whether or not Sweetie's use would lead to a substantial infringement of the suspect's privacy rights depends on the particular circumstances of the case. When it comes to Sweetie we can distinguish two situations from a privacy perspective: (1) Sweetie being present in a public chatroom, and (2) Sweetie directly interacting with a suspect one-on-one in a private (video)chat.

---

[92] Australia does not have a constitutional Bill of Rights, which impacts the scope of the *due process* protection of the suspect. The common law protection against excessive privacy intrusions is more property-oriented, and usually the threatened public interests trump the individual ones. For instance, in the case of *O'Neill v. R* ([1995] 81 A Crim R 458) the court considered the use of listening devices a desirable methodology against the risk of untrue confessions by untrustworthy informers, instead of raising privacy concerns and how these affect the suspect's interests.

[93] See e.g. *Von Hannover v. Germany*, Application no. 59320/00, and *Peck v. United Kingdom*, Application no. 44647/98.

[94] See e.g., *Katz v. United States*, 389 U.S. 347 (1967), *United States v. Miller*, 425 U.S. 435, 443 (1976), *Smith v. Maryland*, 442 US 735, 744 (1979); *Hoffa v. United States*, 385 U.S. 293, 302 (1966); *Lopez v. United States*, 373 U.S. 427 (1963).

Sweetie Being Present in a Public Chatroom

With regard to Sweetie merely being present in a public chatroom, the privacy infringement seems limited. But of course, Sweetie just being present does not yet serve a clear law enforcement purpose. This may change though if Sweetie records (logs) the conversations in the chat. In these cases, there might be a substantial infringement of privacy. However, as it stands this is still the subject of debate.[95]

In general, information that is made publicly available on the Internet may be gathered by law enforcement as evidence in a criminal investigation without the need for a specific legal basis.[96] However, the question becomes more difficult if the information is stored, or if publicly available information is monitored for an extended period. In these cases, the privacy infringements may be of a different nature because the scope, scale and duration are different when compared to offline cases. In the jurisdictions examined, there is as of yet limited case law or specific legislation governing this issue. In the Netherlands, for instance, there is an ongoing discussion to what extent open source data (e.g. blogposts, public Facebook profiles, Twitter feeds) may be monitored on a more structural basis.[97] A similar debate can also be witnessed in the UK, accentuating that reading something that is behind password protection, even if its end result is an open source post, would normally engage privacy expectations covered by Article 8 ECHR.[98]

In any case the goal of Sweetie 2.0 is not to monitor public chatrooms and discussions. Rather, Sweetie is deployed as a lure in the public chatroom in order to engage with potential child sex offenders. As such, the possible privacy infringements that take place in the context of one-one-one conversations and interactions are likely to be more relevant.

One-on-One Conversations and Interaction

Once Sweetie has been approached for a chat there is the possibility to log the conversation (text, audio and video). These chats can subsequently be used as evidence.

Private chatrooms and one-on-one conversations are generally regarded as more privacy-sensitive.[99] Participating in or eavesdropping on conversations as law enforcement official is generally considered an interference with the suspect's private life, requiring the use of special investigative powers subject to authorisation by a public prosecutor or (investigative) judge.[100]

---

[95] See for instance Koops 2013, pp. 654–665.

[96] See for instance US country report, Section 3.4.1, p. 22.

[97] Koops 2013, pp. 654–665.

[98] England and Wales report, Section 3.3, p. 46.

[99] See Koops 2013, pp. 654–665.

[100] A notable exception is the United States. US citizens have no reasonable expectation with regard to information they voluntarily disclose to another person, even if this person turns out to be an undercover law enforcement agent (see e.g. *Hoffa v. United States*, 385 U.S. 293, 302 (1966)).

Personal communications and information stored on or transmitted through personal electronic devices are generally protected by the right to privacy[101] as described in amongst others, Article 17 ICCPR, Article 8 ECHR and the 4th Amendment to the United States Constitution. These instruments protect the individual's 'freedom from unwarranted and unreasonable intrusion into activities [...] belonging to the realm of individual autonomy'.[102]

An interference with the right to privacy may be justified under Article 8(2) ECHR as long as the public authority's actions are 'in accordance with the law' and 'necessary in a democratic society' and pursue 'legitimate aims'. The 'legitimate aims' include, among others, also crime and disorder prevention and protection of the rights of others. Article 17 ICCPR does not include an explicit constraint clause, providing instead that 'no one should be subjected to arbitrary or unlawful interference'. In the years of practice, the two bodies guarding the conventions, the ECtHR and the HRC respectively, have aligned their approaches and have established very similar assessment criteria of how a permissible limitation of the right looks like.[103] They examine in the first place whether the interference with the individual's privacy is lawful/in accordance with the law, which means that the investigative power in question should have a proper legal basis. They further consider criteria of proportionality and subsidiarity that must be satisfied as well.

The tests of legality, proportionality and subsidiarity have been widely translated into the criminal systems of the studied countries.[104] Following its considerations, most of them have enacted legislation that allows law enforcement to (temporarily) infringe upon privacy rights of citizens for investigative purposes.[105] In most cases, an independent body is to authorise the measures of a more intrusive character provided that the latter have already been enacted and specified by the law.[106] The

---

[101] Article 8 ECHR uses the term 'private life', while Article 17 ICCPR speaks of 'privacy'. Here both terms are used synonymously, as despite the linguistic differences in the two English texts it is widely recognized that 'privacy' and 'private life' mean the same thing.

[102] Wilhorn 1997, p. 833.

[103] Georgieva 2015, p. 104.

[104] In Brazil, the right to privacy is constitutionally guaranteed and said guarantee can be temporarily suspended in matters of public interest, especially in criminal investigations with judicial authorization. The investigative authority in Canada is constrained by the Canadian Charter of Rights, and in particular by its s.8 that covers the freedom from unreasonable search and seizure, from which the common law has deduced a right to privacy and a doctrine of reasonable expectations of privacy. See also Section 26(2) of the Estonian Constitution and Articles 10–13 of the Dutch Constitution. In Israel, the constitutional limitations of investigative powers are comparatively weak, see country report on Israel, p. 14. In Spain, respect for the right to privacy and the confidentiality of communications is reflected in the guiding principles of Article 588bis a). The Nigerian Constitution guarantees the right to privacy (Section 37), but it is not absolute and can be restricted (Section 45(1) of the 1999 Constitution).

[105] Under Croatian law it is Article 332 of the Criminal Procedure Act; s.111 of the Estonian Electronic Communications Act (ECA).

[106] In a recent landmark case, *R v. Spencer* [2014] 2 SCR 212, 2014 SCC 43 (CanLII), the Supreme Court of Canada held that subscriber information can no longer be obtained by the authorities from Internet service providers without a corresponding court authorization. Article 332

criminal procedural laws thus clearly distinguish between intrusive and less intrusive investigation powers, the latter of which do not always require a specific legal basis.

In summary, we can say that Sweetie may indeed infringe the privacy of suspects, and will do so in particular in one-on-one conversations, and even more so if such conversations are logged/recorded. Therefore, its use must both be necessary in a democratic society and in accordance with the law. This will be further analysed below (see s1.3.6 and s1.3.7).

Fair Trial (Entrapment)

In the context of criminal procedure, the right to a fair trial protects individuals against arbitrary application of state power and guarantees the effective realisation of other fundamental rights and liberties through fair judicial proceedings.[107] Fair trial rights may also extend to the pre-trial phase. Consequently, it also covers situations in which law enforcement officials use a fake identity, simulate a sale or purchase, or offer simulated business services to trap a suspect, as in such cases the suspect's fate is 'surrendered' to the power advantage of the state. The right to a fair trial and its considerations balance the power relation between the individual and the state.

From a procedural law perspective, the use of Sweetie for engaging suspects raises two issues regarding the fair trial rights of the suspect. They are: (1) operating Sweetie in public chatrooms upon a general suspicion may constitute non-targeted entrapment that is considered random virtue-testing, and (2) its direct interaction with suspects may amount to unlawful incitement of crimes.

---

of the Criminal Procedure Act of Croatia stipulates that restrictions of privacy rights can be legally implemented only upon a court order, leaving the state attorney a discretion to issue a warrant in urgent cases for the duration of 24 hours. See Croatian report, pp. 25–27. In Belgium, according to Article 88*bis* and Article 88*ter* CCP the intervention of the investigatory judge is necessary whenever traffic or localization data is required, or when a network search has to be performed. In England and Wales, S.65(1) RIPA has (controversially) created the Investigatory Powers Tribunal, which supervises the techniques awarded by RIPA to police and security services. In Argentina, a judicial authorization is needed whenever investigatory powers imply a violation of privacy rights, see Salt and Dupuy 2016 (Country report on Argentina), Section 3.2.2. In Australia, the statutes mirror the protection thresholds of the Cybercrime Convention on personal information by, among others, requiring a warrant to allow access to existing or prospective computer data, see country report on Australia, pp. 21–23. In Brazil, it is also the judicial authority that is considering the proportionality of the measure and the factual reasoning supporting the request for an interception warrant. In Nigeria, law enforcement authorities must obtain judicial authorization before carrying out an interception, see s.39 of the Cybercrimes Act. The 1987 Constitution of the Philippines foresees in its Article III sec 3(1) a judicial warrant or a court order as well. The Polish Code of Criminal Procedure balances investigative powers with the requirement of a court order as well, see Article 237 CCP and Article 218 CCP. In the US, under the Fourth Amendment, law enforcement must generally obtain a warrant in order to conduct a search in a computer owned by the suspect, see US country report, p. 15.

[107] Ballin 2012, p. 55.

Operating Sweetie in Public Chatrooms
Operating Sweetie upon a general suspicion means that the avatar would not target a particular suspect, but an area or space (in the present case particular cyber-areas such as chatrooms). This raises concerns in terms of the fair trial rights of the suspects, as undercover powers are usually the exception to the rule, and generally aimed at suspects against whom there is a prior suspicion.

Direct Interaction
A further concern is Sweetie's interaction with the suspects in chat conversations. A direct interaction bears the risk of influencing the suspect and thus leading him/her into committing an offence he/she would have otherwise not committed. Consequently, Sweetie may lead to the facilitation of the crimes it actually intends to prevent.

In both scenarios (use in public chatrooms and direct interaction) there is a risk that the right to a fair trial is violated.

1.3.6 *Necessity in a Democratic Society*

From the above we can surmise that Sweetie brings with it the risk of privacy infringement. As such, it is important to determine whether or not the use of Sweetie is necessary. In other words, does the end (protecting children) justify the means (using an AI to engage suspects and potentially infringe upon their rights)?

The substantial test of 'necessary in a democratic society' is in the centre of the discussion when assessing whether states are allowed to interfere with individual interests in order to address relevant societal matters. This means that when state agencies use infringing investigative powers, just *a* reason for using the power is not sufficient, as the interference must be 'necessary'.[108] The ECtHR in its case law has clarified what 'necessity means':

> … the notion of necessity implies that an interference corresponds to a pressing social need and, in particular, that it is proportionate to the legitimate aim pursued.[109]

Proportionality is considered an essential test in relation to criminal procedure, especially when courts are assessing the appropriateness of governmental measures, such as secret surveillance operations, interception and wiretapping that interfere with the individual's right to privacy and private life protected by Article 8 ECHR and Article 17 ICCPR.

The ECtHR considers the proportionality of the measure in the light of the specific circumstances as a whole, and in particular whether the authorities had 'relevant and sufficient reasons' for taking the coercive measure in question.[110]

---

[108] *Handyside v. the UK*, Application no. 5493/72.

[109] *Olsson v. Sweden*, Application no. 10465/83.

[110] *Olsson v. Sweden*, Application no. 10465/83.

The process of deciding includes a number of factors, among others, the suspect's interest to be protected from the interference with his/her rights, the severity of the infringement and the pressing social need the authorities seek to fulfil. The more far-reaching the interference is, the stronger the reasons required to justify it must be.[111] The Court, however, affords states a certain margin of appreciation[112] in choosing the means to best address the urgent needs of their society, and said margin of appreciation has a different scope depending on the actual circumstances and the subject matter. It has been held to be particularly wide in areas such as child protection.[113]

In addition, in cases dealing with children and vulnerable groups, the ECtHR has found that positive state obligations can trump negative ones when it comes to securing the physical and moral welfare of children.[114] In said scenarios, states are required to have in place effective criminal law provisions that would not only protect minors, but also effectively deter against grave acts committed towards them.[115]

Considering the above, in order to deem the chatbot's use necessary in a democratic society, the interests it seeks to protect should outweigh the interests of the potential suspects it would investigate. This means that the rights and interests of children who are or could potentially fall victim to webcam sex abuse should outbalance the interests of people who engage in a private chat conversation with the avatar. Furthermore, less infringing cannot yield the same results.

The threat webcam sex tourism poses to the well-being of children is clear. Webcam sex sessions directly hurt the victim, and the fact that webcam streaming sessions can easily produce pictures or videos of the victims, causes additional harm. Furthermore, recorded abuse images and videos can easily lead to the subsequent distribution of child pornography, causing additional harm to the child.

The threshold to engage in webcam sex tourism is low and the chances of getting caught are as of yet minimal. Perpetrators can further reduce the chances of being caught by using fake identities, but also various anonymisation services and hidden servers (to name just a few) to prevent detection. As such, the chances of identifying a suspect after the webcam sex stream has been concluded is likely low. The best chance of finding a suspect, is thus to catch them in the act. This entails luring the suspect, and subsequently interacting with them. While this can be done using actual law enforcement officers, the scale of the webcam sex tourism problem means that this 'traditional' way of investigation is not effective. In order to effectively combat webcam sex tourism, the use of scalable investigation methods such as Sweetie may therefore be necessary.

---

[111] Kilkelly 2003, p. 34.

[112] *Klass and others v. Germany*, § 48; *Leander v. Sweden*, Application no. 9248/81, § 59.

[113] Kilkelly 2003, p. 34.

[114] See *KU v. Finland*, Application no. 2872/02, § 46.

[115] Ibid. at § 43.

It is important to consider that the chatbot does not indiscriminately collect all available information in public chatrooms, but only records communications with, and gathers data of, suspects who engage the avatar in a sexually charged conversations. By facilitating the offender's identification, the chatbot would contribute to the effective investigation of serious crimes against minors, which in its turn corresponds to the ECtHR's standards on proactively defending vulnerable groups by effective and deterrent criminal procedure means.

In line with the above, the question of prioritising the investigation of such offences by means of the chatbot seems necessary to successfully police public chatrooms, and other online venues for child sexual abuse, and fight the online abuse of children. The lasting harm experienced by minors in cases of sexual abuse combined with the ever-growing danger of webcam sex in public chatrooms tips the balance in favour of Sweetie. Minors need to be safe to freely express themselves on the Internet without being monitored by offenders.[116] As such, we argue that depending on the circumstances of the case, there are strong arguments that the use of Sweetie is necessary in a democratic society.

1.3.7 Legitimacy of the Use of Sweetie

Having established that Sweetie can potentially interfere with the right to privacy means that such an interference has to be covered by an investigative power, which requires a clear legal basis. None of the jurisdictions we examined have specific legal provisions that authorise and govern the use of Sweetie. Rather, the use of Sweetie must be 'read' into the existing investigative powers. When translating the existing investigative powers into the context of Sweetie, their interpretation has to match the rationale of the original provisions. A too far stretched interpretation would contravene the legislator's will, but also rob the provisions of their foreseeability.

It is dependent on the criminal procedure law provisions of the individual country whether or not the application of Sweetie is in accordance with the law. However, drawing inspiration from amongst others the Cybercrime Convention and the European Convention on Human Rights we can give an indication of (1) what special investigative powers could apply in relation to privacy, and (2) what procedural requirements must be followed in the case of entrapment.

Privacy Considerations

As briefly discussed above, the situation in which Sweetie is merely present in a chatroom is not yet a substantial infringement of privacy. Incidental observation and/or recording of public data will likely also not yet amount to a substantial infringement of privacy. Provisions such as those of Article 32(a) of the Cybercrime Convention and domestic regulations regarding the general tasks of law enforcement could cover this.

---

116 Vendius 2015, p. 18.

However, if Sweetie starts systematically observing profiles and recording public chats (regardless of the chatroom it is present in), and particularly if she moves onwards to one-on-one conversations, then this could amount to a substantial infringement of privacy that requires a specific basis in the law in many jurisdictions.

Articles 16 through 19 of the Cybercrime Convention are not really applicable to the case of Sweetie.[117] Article 20 (the real time collection of traffic data) and Article 21 (the real time collection of content data) could be relevant in the context of Sweetie, given that Sweetie may record both traffic data (e.g. IP addresses) and content data (e.g. chats and files sent).

One complicating factor might be the definition of communications: not all jurisdictions may consider a person interacting with a chatbot to constitute 'communication' if this is interpreted as exchanging of messages between persons. Whether use of Sweetie is or can be interpreted as recording of communications therefore may depend on the specifics of national law. However, since many countries also consider interacting with a machine to be communications, where the machine (e.g., an ATM) can be seen to serve as a proxy for a person (e.g., a bank),[118] we will consider interactions with Sweetie to constitute communications, where the chatbot serves as a proxy for the investigation officer putting it into operation.[119]

The explanatory report to the Cybercrime Convention gives a broad definition of 'interception and 'technical means' that could also cover the use of Sweetie:

> Interception by 'technical means' relates to listening to, monitoring or surveillance of the content of communications, to the procuring of the content of data either directly, through access and use of the computer system, or indirectly, through the use of electronic eavesdropping or tapping devices. Interception may also involve recording. Technical means includes technical devices fixed to transmission lines as well as devices to collect and record wireless communications. They may include the use of software, passwords and codes.[120]

As such, the signatories to the Cybercrime Convention should already have in place legal provisions that could be applied in the context of Sweetie, at least for offences related to child pornography. Table 1.6 confirms this assumption.

In the few countries not party to the Convention, however, we see a divergence in how the real-time interception of communications is dealt with. Argentina and

---

[117] The articles regulate the expedited preservation of stored computer data and traffic data, production orders and the search/seizure of stored computer data respectively.

[118] Cf. *Explanatory report to the Convention on Cybercrime*, § 227, which identifies visiting a website to be communications in the context of the power of collecting traffic data, and by extension (see § 230) also of interception.

[119] Since interception of communications is often regarded to be one of the most intrusive investigation powers (see, e.g., the Cybercrime Convention), if a jurisdiction conceptualises interactions with Sweetie as non-communicative, then presumably some other, less intrusive, investigation power may be applied to the situation.

[120] *Explanatory report to the Convention on Cybercrime*, § 54.

South Korea, for instance, have not developed a formal distinction between traffic and content data yet, but refer to both types of communication information as 'data'. Further, these jurisdictions have not enacted particular legislation on real-time data collection and apply therefore the existing provisions on telephone wiretapping accordingly. The latter applies for Brazil and the Philippines which, although having introduced comprehensive legislation on cyber-investigation powers not so long ago, also lack provisions on real-time collection of both traffic and content data.[121] Among the non-signatory countries, Nigeria is the only example with explicit legislation on the matter.[122]

However, as telephone interception and wiretapping practices are largely comparable to the interception of online communications, these countries may still meet the legality requirement if the application of the respective national norms to investigative means comparable to Sweetie has been confirmed by the courts. Yet, in situations in which suspects engage in conversation with the avatar, this will likely not be (fully) covered by the norms regulating the interception/management of data but will also largely depend on the rules surrounding the use of undercover agents. Therefore, and for the sake of comprehensiveness, the applicability of the wiretapping norms mentioned above will be discussed together with the laws on undercover agents under Sect. 4.8.3.

Fair Trial Considerations (Entrapment)

In the following, we provide an overview of the criteria used by the investigated jurisdictions to draw a line between permissible luring and not-permissible entrapment. Entrapment is ordinarily an issue relevant in court proceedings, and as such not part of an investigative power. Depending on the particular system and the facts of the case, it can be pertinent in various ways—it can either lead to the exclusion of the obtained evidence,[123] be raised as a defence and thus exclude guilt or liability,[124] or even entirely stay the proceedings due to an abuse of process.[125]

As indicated above, Sweetie brings about issues of both targeted and non-targeted entrapment. This means that in both cases certain procedural requirements must be met that have been established in case law.

---

[121] See Mendes Saldanha 2016, pp. 20–21; See also Dizon 2016 (Country report on the Philippines), p. 31. Brazil applies the provisions of the *Wiretapping Act* accordingly, while the Philippines rest to the rules on search and seizure.

[122] S.39(a) of the Cybercrimes Act allows law enforcement to intercept content data and/or traffic data provided that the operation has been authorised by a judge.

[123] As for instance in Australia—see section 138 of the *Evidence Act 1995* (Cth), www.austlii.edu.au/au/legis/cth/consol_act/ea199580/ [23 May 2016].

[124] This is the case in the US system. See country report on the US, Section 3.2.3, p. 16; in Spain entrapment is a judicially crafted defense, which excludes liability. See on the matter SSCS April 18, 1972 and SSCS June 22, 1950.

[125] *R v. Loosely Attorney General's Reference (No 3 of 2000)* is considered the landmark decision in the UK establishing entrapment as a procedural bar; Canadian courts also make use of procedural bars, see *R v. Mack* [1988] 2 SCR 903, as well as their counterparts in Scotland—*Jones v. HM Advocate* [2009] HCJAC 86, para 88.

Targeted Entrapment

The compared criminal systems employ as a central consideration of whether unlawful entrapment has actually taken place, the reasons underlying the operation and the conduct of the authorities executing it. It is generally recognised that law enforcement officials are entitled to merely provide the suspect with an opportunity to commit the crime, but the latter should have been able to independently form or abandon the intention to commit it.[126] The suspect should have been thus 'allowed' to commit the crime he/she intended to from the very beginning.

The ECtHR refers to this requirement as the '*essentially* passive'[127] standard and examines whether the suspect has in any way experienced pressure by the authorities' deceit to commit the crime, be it by pro-active solicitation,[128] prompting or reiteration of the offer despite an initial refusal,[129] or by making the offer very hard to refuse.[130] The Court also considers whether there were objective indications that the suspect has been involved in criminal activity or was predisposed to commit the crime,[131] and how familiar the latter is with the criminal environment.[132]

This reasoning, however, is not unique to the ECtHR or European courts in general. The Canadian,[133] Australian[134] and South Korean[135] courts have adopted

---

[126] In Dutch criminal procedure this approach is known as the Tallon criterion, named after a landmark case from 1979 (HR 4 December 1979, NJ 1980, 356 m.nt ThWvV). The case concerned a drugs purchase from two undercover agents. The Supreme Court found that the suspect had formed his intent independently. The Belgian general prohibition of entrapment is embedded in Article 30 of the Preliminary Title of the Code of Criminal Procedure and declares criminal procedures based on entrapment inadmissible. In the UK *R v. Loosely Attorney General's Reference (No 3 of 2000)* the crucial question was whether the police have done more than present the defendant with an unexceptional opportunity to commit a crime. A criminal court would consider admissibility under Section 78 Police and Criminal Evidence Act 1984 (PACE).The Estonian Supreme Court has established in its judgement of 2 December 2004, case number 3-1-1-110-04, para 11.3 that the authorities' actions cannot be directed against persons who did not have the slightest intent of committing a crime. The Supreme Court of Croatia considers the persistent, long-term prompting of the suspect incitement when it turns out to be the decisive factor in forming the perpetrator's will to commit the crime, VSRH, I Kž-1255/04 of 16 February 2006.

[127] *Ramanauskas v. Lithuania* [GC], Application no. 74420/01, § 55. Emphasis added.

[128] *Burak Hun v. Turkey*, Application no. 17570/04, § 44.

[129] *Ramanauskas v. Lithuania* [GC], § 67.

[130] *Malininas v. Lithuania*, Application no. 10071/04, § 37.

[131] *Bannikova v. Russia*, Application no. 18757/06, § 38; *Case of Constantin and Stoian v Romania*, Applications nos. 23782/06 and 46629/06, § 55; *Texeira do Castro v. Portugal* (44/1997/828/1034), §§ 37–38.

[132] *Shannon v. the United Kingdom*, Application no. 67537/01.

[133] Supreme Court of Canada, *R v. Mack* [1988] 2 SCR 903. In this particular case the Court stayed the proceedings.

[134] See *R v. Priest* [2011] ACTSC 18 at [86] and *Ridgeway v. The Queen* [1995] CLR 19, in which the court rejects a defense of entrapment.

[135] For the South Korean doctrine see for instance Supreme Court Decision, 2008DO7362, Oct. 23, 2008.

similar approaches in outlining entrapment, stipulating that law enforcement should not go beyond offering an opportunity by crossing the line and creating the offence for the purpose of prosecuting it. The Supreme Courts of Argentina and the Philippines,[136] although phrasing their entrapment tests in slightly different terms, also see the difference between lawful trapping and unlawful entrapment in the origin of the criminal intent. The legal doctrine in Nigeria appears to be developing in a similar direction.[137]

The common law doctrine of the United States considers the predisposition element of the suspect together with the trickery, persuasion or fraud techniques performed by the law enforcement officers (or private persons acting on behalf of law enforcement).[138] According to the so-called subjective approach, entrapment is committed when law enforcement has induced the crime and the defendant had no predisposition to engage in a criminal conduct.[139]

Non-Targeted Entrapment

The case law and regulations discussed above relate to targeted entrapment, but scenarios in which law enforcement officials do not engage a particular suspect and operate upon a general suspicion need to be considered as well. The issue with said conduct is that it may be considered 'random virtue-testing'[140] that puts the integrity of the criminal investigations at risk. Criteria that govern non-targeted operations must therefore be in place.

The studied jurisdictions show, however, somewhat more perceptible differences in their approaches towards non-targeted entrapment. While in Australia, the Philippines, Poland, the UK and the US the fact that law enforcement targets a location or an area, but not a person, triggers the application of the same criteria as described above, other criminal procedure codes, as the one in Croatia for instance, limit the use of special investigatory tools to cases where a particular person has already been identified as a suspect.[141] Yet, the deployment of undercover agents would be legally permissible under the Croatian Law on Police Duties and Powers, which allows location-related undercover operations provided that there are valid grounds for suspicion of criminal behaviour. Similarly, Canada's doctrine also provides additional safeguards in this regard. Where undercover operations concern a particular place or a location, but not a given suspect, the authorities must reasonably suspect that criminal activities are occurring there. The police activity must either take place on the basis of a reasonable suspicion or over the course of a *bona fide* inquiry. In this way the doctrine ensures that no random virtue-testing is being

136 *Araneta v. Court of Appeals*, G.R. No. L-46638 [July 9, 1986]; *People v. Gatong-o*, G.R. No. 78698 [December 29, 1988].

137 See country report on Nigeria, Section 3.2.3, p. 46.

138 See country report on the US, p. 16. In the US, the defense of entrapment has no statutory basis in Federal law, but has been developed by the courts.

139 *Mathews v. United States*, 485 U.S. 58 (1988).

140 Bronitt 2004, p. 37.

141 See Article 332(1) and Article 334 of the Criminal Procedure Act of Croatia.

practiced.[142] Scottish law requires state agents to obtain an authorisation before executing an operation when the suspects are not identified.[143]

Yet other approaches focus on the ordinariness of the law enforcement behaviour. Following the *Tallon criterion*[144] in the Dutch procedural context, non-targeted entrapment would be allowed if the lure does not substantially change the original situation or location in which it is employed, so that it cannot exert any significant impact or influence on the decision-making process of the suspect.[145] The Belgian Court of Cassation handles non-targeted entrapment in a similar manner. The decisive criterion in their view is that the luring procedure imitates or portrays a daily life scene without exaggeration.[146]

Fair Trial Requirements Applied to Sweetie

Considering the substantive entrapment tests, Sweetie's conduct must remain essentially passive to avoid investigative impropriety. Further, once the suspect has been engaged in a conversation, the chat script should also leave room for the suspect to retreat. If the avatar does not proactively solicit potential suspects, but waits to be approached, and avoids pressure or any inducement to steer the conversation in a particular, sexually charged, direction, it would act beyond legal reproach. It appears that an exception to this test can only be made within the US framework, where courts are likely to close an eye on a more provocative behaviour if the suspect has a demonstrable predisposition towards child abuse offences.[147]

Further, with regard to the location of the undercover operation, that is to say when it targets an area (presently a chatroom) and not a particular suspect, it is relevant that the use of the chatbot does not significantly alter the existing circumstances. This, of course, would be largely dependent on the targeted chatroom.

If Sweetie is used in a regular chatroom for users under the age of 18, and her profile resembles that of the majority of minors there, her online presence would abide by this rule, if it does not make itself more visible than the mere entering a chatroom with a common-looking chat name. In addition, for some of the jurisdictions described above,[148] law enforcement agencies would need to substantiate their initial suspicion or get a particular authorisation from an independent body to proceed with a non-targeted operation in a chatroom. Disregarding these criteria would likely compromise the admissibility of the gathered evidence or of the investigation as a whole.

142 See Supreme Court of Canada, *R v. Mack* [1988] 2 SCR 903, paras 113, 119.

143 See country report on Scotland, p. 21.

144 Ibid. at note 140.

145 These contemplations originate from two important cases of the Dutch Supreme Court, HR 28 October 2008, ECLI:NL:HR:2008:BE9817 (*Lokfiets-arrest*) and HR 6 October 2009, ECLI:NL:HR:2009:BI7084 (*Lokauto-arrest*). In both cases bait was set by law enforcement without suspicion of a specific person.

146 Cass. 17 maart 2010, AR P100010F; Brussel 14 maart 2007, *RABG* 2008, afl. 1, 63, note L. Delbrouck.

147 See US country report, p. 18.

148 See the references made under Section 4.8.2.2 of this report.

If, however, Sweetie enters an 'above 18' chatroom, already her logging-in is likely to significantly alter the situation. In such cases, the particular assessment of the avatar's placement will depend among others on the number of real children present there. While there might be indeed some room to employ the chatbot in online areas intended as 'adults only'-fora, but demonstrably being used for the webcam prostitution of real children, Sweetie's appearance in a regular adult chatroom would likely constitute non-targeted entrapment.

An additional question arises in relation to recurring offenders (persons returning to the same chatroom and repeatedly engaging the avatar in a sexually charged conversation) and whether a different entrapment test (or a suspicion threshold for that matter) should apply to them, since the chatbot would be 're-encountering' them. However, as of the writing of this report, there is no information that such situations are being handled differently. It appears that the rules on non-targeted entrapment are applicable in such cases as well, especially since the avatar's presence in the chatroom would not aim to engage offenders with whom the chatbot has already communicated, but potentially everyone who has an interest in an illegal webcam interaction with a child.

It is, however, doubtful whether we can adequately assess Sweetie's entrapment implications at all, since the entrapment tests discussed above have their origin in police operations, which are substantially different from online investigations of sexual exploitation. This in its turn is closely connected to the question of Sweetie's authorisation as an investigation tool, which will be discussed shortly.

Although some entrapment tests refer to the situation of placing an opportunity for potential suspects in 'hotspot' areas, said techniques merely stage a (conventional) subject in an everyday environment (a car, a bike or even a person in a park that however does not interact), which does not require further police intervention until the suspect takes the bait. Sweetie, however, seems to exceed this mandate, as it will undeniably interact with the suspect. As such, Sweetie resembles more an undercover agent, and its interactive capacity would fall under the authority of more comprehensive undercover operations, which typically involve pseudo-purchases of illegal goods (the so-called *buy and bust* approach), the infiltration of a criminal environment or both at the same time. In these scenarios undercover agents are allowed to act as offenders or to be in direct contact with offenders, and thereby to actively take part in the crime (if national law allows such operations, e.g., for drugs investigations).

Yet, these controlled operations start from preconditions that differ from Sweetie. Sweetie does not act as an offender, but as an 'interacting' child victim. Accordingly, the rules on undercover operations in which police act under a false identity while posing as a victim would apply. Here, considering the chatbot's mandate to tackle webcam child sex tourism on a global level, it is pertinent to establish whether there is any international legislation that can guide undercover agents in such operations.

In the European context especially, the EU has long proactively sought to find a solution to streamlining particular undercover techniques, including the use of undercover agents and their operational capacities.[149] The European Investigation Order (EIO)[150] reflects said objective, and Article 29 stipulates:

> An EIO may be issued for the purpose of requesting the executing State to assist the issuing State in the conduct of investigations into crime by officers acting under covert or false identity ('covert investigations').

However, the article also clearly establishes that covert investigations should take place 'in accordance with the national law and procedures of the Member State on the territory of which the covert investigation takes place'.[151] In other words, the use of undercover agents remains a strictly national matter.

In terms of other international instruments on undercover agents, at the time of writing this report, no binding guidelines exist. Further authorisation is thus to be sought in the country specific norms on investigative powers.

Overview of Country-Specific Rules in Relation to Privacy and Entrapment
In Argentina, the undercover agent figure is codified under Law 24424 (the so-called Drugs Law) but is explicitly recognised only for operations investigating the trafficking or smuggling of narcotics.[152] In addition, there is no general regulation on undercover operations online, or surveillance for that matter,[153] so it appears that investigations concerning online activities resort to the rules on obtaining physical evidence and apply them in analogy. Having said that and considering that under current Argentinian law an inappropriate interaction with Sweetie can only be dealt with under the figure of impossible attempt (which as explained above appears to have no clear assessment criteria),[154] finding a legal basis for the avatar's use seems challenging.

The findings of the Australian report indicate that there are no obvious impediments to Sweetie's investigatory use. The country's regulations on covert policing[155] have been mainly developed to oversee covert operations, in which officers would perform otherwise unlawful acts, such as the delivery of narcotics. Covert online investigations of child grooming (in contrast to infiltration of online child pornography rings) usually do not require the police to resort to these

---

[149] See Vendius 2015, p. 20.

[150] Parliament and Council Directive 2014/41/EU of 3 April 2014 regarding the European Investigation Order in criminal matters.

[151] See Article 29 Directive 2014/41/EU.

[152] See Argentinian report, p. 23.

[153] Ibid., pp. 23–27.

[154] See Section 3.4.2 of this report.

[155] See the *Crimes Act 1912 (Cth)*, www.austlii.edu.au/au/legis/cth/consol_act/ca191482/ [9 August 2016].

mechanisms,[156] and no court has ruled on whether it is improper to conduct such investigations without the use of a controlled operation or assumed identity authority in the absence of otherwise unlawful conduct. Following this premise, Sweetie can be used both as a lure and an undercover agent that gathers information. The evidence gathering features of the chatbot would be qualified as a surveillance device.

In Belgium, there are possible avenues for using the chatbot following an evolutionary interpretation of the provisions on infiltration and wiretap operations.[157] These, however, are not applicable to investigations concerning the offences of cyber-luring and cyber-grooming, which are excluded from the list of offences triggering said investigative powers. Accordingly, law enforcement will have to rely on offences like cyberstalking (Article 145, § 3*bis* Act Regarding Electronic Communications) to obtain the necessary authorisation.

Brazilian criminal procedures on the interception of communications, the collection of traffic data and other similar investigation techniques require a court authorisation as well.[158] While it appears that said means have been used in the course of operation 'Darknet', known for successfully bringing child pornography offenders to justice, the details of the investigation are still sealed.[159] It is therefore difficult to find examples of their application in an online environment. In addition, while the law on infiltration operations[160] can also be taken into consideration as authorisation for Sweetie's recording of webcam streams and chat logs, employing Sweetie as bait is considered illegal. Further, as a crime against an avatar cannot be prosecuted in Brazil, there is no legal basis for its investigation either.

S.184(2)(a) of the Canadian Criminal Code, that applies to both wiretapping and undercover operations, regulates the interception of communications where one party consents to recording. Depending on the interpretation of the provision, consent could be reasonably established by Sweetie's operators/the Canadian police. Should that not be the case, a judicial authorisation would be necessary for every instance of Sweetie's interaction with a potential suspect pursuant to s.184(2)(b) CCC. As for the overall regulation of undercover operations in Canada, these are subject to the provisions on statutory police powers, whose foundation appear to be common law principles and the existing extensive case law on undercover policing.[161] Following their rationale, the legality of Sweetie's use as an undercover agent appears to be quite plausible.

---

[156] *R v. Priest* [2011] ACTSC 18 (11 February 2011) at [90]. See also country report on Australia, p. 24.

[157] Belgian report, Section 3.4, p. 37.

[158] See Brazilian country report, p. 25.

[159] Ibid., Section 3.4, p. 26.

[160] Law n.º 12.850/2013, Article n.º 3º, IV, V and VII, and Articles 10–17. Available at: www.planalto.gov.br/ccivil_03/_ato2011-2014/2013/lei/l12850.htm [9 September 2016].

[161] Said operations are limited by constitutional guarantees and controlled (where necessary) by judicial oversight.

The Croatian Criminal Procedure Act contains provisions on the use of undercover investigators and informants, simulated sale and purchase of items, and simulated bribe-giving and business services,[162] but it makes no explicit reference to the use of software or technologies for investigative purposes.[163] Yet, given that the law on special investigatory procedures allows operations in which undercover investigators set fake profiles on social networks and other Internet fora in order to communicate with potential perpetrators, the use of Sweetie for the same purpose appears possible.[164]

Under English law, the rules on undercover police operation, surveillance and on the interception of content and communications data have been tested in cases where officers would pose as children online, and it appears that said rules provide a sufficient legal basis for Sweetie. Interestingly, under those provisions, since one part of the communication is known (presently Sweetie), exchanging messages between the avatar and the alleged offender would not amount to interception.[165]

The laws on undercover surveillance activities carried out by police agents cover luring and interacting with a suspect in Estonia.[166] Yet, the provisions explicitly refer to a 'police agent', implying the involvement of a human being. This is problematic given that the rules at hand are quite recent, meaning that the laws have been consciously enacted without consideration of virtual undercover agents or other AI technological means, and there is no case law (yet) that would advocate a different interpretation.

Since under Israeli procedural law no provision explicitly authorises law enforcement to employ undercover agents, but police operations make use of such means nevertheless, according to the country report it does not matter whether said unauthorised power is executed in the context of offline or online investigations; nor does it matter whether the police use a human or a computer agent. Therefore, investigation powers in Israel can be applied to Sweetie.[167]

Given that under Dutch law many of the investigatory powers are formulated in a technology-neutral manner, they are applicable to the online context as well.[168] Sweetie's use could fall under Article 126g DCCP and 126j DCCP, which regulate systematic observations and the method of systematically gathering intelligence online respectively.[169] Further, wiretapping as stipulated in Article 126m DCCP

---

[162] Article 332(1) point 5, 6, 7 of the Criminal Procedure Act, Official Gazette no. 152/08, 76/09, 80/11, 121/11, 91/12, 143/12, 56/13, 145/13, 152/14.

[163] However, the law does not exclude them either.

[164] See the findings of the Croatian country report, Section 3.4, p. 32.

[165] See report on England and Wales, p. 47.

[166] See Section 3.3.2.2.2 of the Estonian report.

[167] See Israeli report, p. 16.

[168] Country report on the Netherlands, Section 3.3, p. 24.

[169] These provisions cannot be applied upon a general suspicion. In Sweetie's case, their application would depend on the particular circumstances of the case and on how law enforcement substantiates their conduct.

may also be used in the context of Sweetie, for instance, to record the webcam stream, or the associated chats.

The Nigerian rules that regulate the interception of communications upon a judicial authorization (online observation and electronic infiltration embedded in Sections 39 and 39(b) of the Cybercrimes Act respectively) seem to refer to technological means such as the installation of software or a device, but not to artificial intelligence agents. The conduct of undercover agents is also not explicitly regulated.[170] Therefore, there appears to be no explicit legal basis for using Sweetie in Nigeria.[171]

Under Philippine law, while there seems to be some legal basis for the use of Sweetie as a lure,[172] other investigation practices, such as the search, seizure, preservation and production of computer systems and data, are principally only applicable to cases involving real children. Exceptions can be considered with regard to the few crimes directly applicable to Sweetie, which as indicated above are grooming, luring, cybersex, and attempted cybersex. Short of this, there will be no legal basis for law enforcement to search, intercept or collect data from or about suspected sex offenders, since the latter would not be involved in the commission of any crime.[173]

The avatar's use in Poland can be submitted for one part of the process under Articles 19a and 19b of the Police Act, which regulate controlled operations or the so-called 'police provocation'.[174] These would govern the avatar's interaction with and luring of the suspect, and cover, for instance, the receipt of material containing child pornography. The collection of information could then fall under Articles 19 and 20 of the Police Act. They regulate, on the one hand, the collection of the communication's content and its preservation, including webcam footage and other computer data, and the obtaining and processing of personal information on the suspect, on the other.[175] Yet, Articles 19a and 19b usually authorise operations where a suspect has already been identified, and not to the other way around, which markedly reduces the situations in which Sweetie can be legally used.

Provided that law enforcement obtains the necessary authorisation for undercover operations, Sweetie appears to be usable under Scots law.[176]

---

[170] See Nigerian report, p. 44.

[171] See Nigerian report, Section 3.4, p. 48.

[172] The findings in the country report on the Philippines indicate that there are no specific laws on undercover policing and operations. Such activities are subject to general laws (i.e. the rights and protections granted under the Philippine Constitution) and the relevant case law on entrapment. See on the matter Section 3.2.3 of the country report.

[173] See country report on the Philippines, p. 36.

[174] See country report on Poland, Section 3.4, p. 35.

[175] Both provisions implement a three-step safeguarding system, which requires the approval of the high-rank police officers, the prosecutor and the district court. See for more information the Polish report, p. 35.

[176] See country report on Scotland, p. 21.

In Spain, Sweetie's use will only be possible under Article 282bis of the Criminal Procedure Act, which codifies the figure of the so-called cyber-undercover agent. This refers to a police officer who acts under a false identity in an online setting. The provision allows the agent's interaction with a suspect and the recording of the communication, provided that the necessary authorisations have been obtained.[177] However, since under Spanish substantive criminal law a crime against Sweetie cannot be committed and an attempt is not punishable, the chatbot could only be employed as a form of intercepting device against already identified suspects.

In Turkey, Article 139 of the Criminal Procedure Code regulates the appointment of undercover agents, while the interception, recording and evaluation of electronic signals transmitted through telecommunication channels are embedded as a method of investigation in Article 135(1) of the same statute.[178] However, undercover agents may only be used to investigate the crimes listed exhaustively in Article 139(7), which does not include the crimes against the sexual integrity embodied in Articles 102–105 or Articles 226–227 of the Turkish Penal Code.[179] In a similar manner, data interception and recording powers are only applicable to Article 103 and Article 102(1) of the Penal Code (sexual exploitation of children and sexual assault respectively), the former of which requires physical contact between the offender and the victim, while the latter refers to an adult victim. Considering the above, Sweetie would have no legal basis under the current criminal law framework in Turkey.

The US procedural system allows the recording of private one-on-one chat communications, given that Sweetie could be considered a party to the conversation.[180] As to its authorisation as a lure or an undercover agent, it is likely that courts will simply consider Sweetie a way to automate undercover operations, and the rules that apply to the chatbot are those that regulate undercover operations.[181]

Reasonable Suspicion

The previous subsection showed that in about half of the investigated jurisdictions, as of the writing of this report, the chatbot cannot be employed because there is an insufficient legal basis. Furthermore, as discussed in the chapter on substantive criminal law, the use of Sweetie may be blocked because having a sexually charged

---

[177] See Spanish report, p. 34.

[178] See country report on Turkey, Section 3.4, p. 30.

[179] See country report on Turkey, Section 3.2.3, p. 27.

[180] 18 U.S.C. § 2511(2)(d), see also 18 U.S.C. § 2511(2)(c). However, the law speaks of a 'person' and not of an AI agent. It appears that the provisions could be applied to Sweetie if the avatar is seen as a proxy of the law enforcement officer supervising its communication with the suspect. See the country report on the US, Section 3.4.1, p. 21.

[181] The rules covering undercover operations have no specific statutory basis, and as a result have been developed by the courts.

interaction with a virtual minor is not considered criminal. For the latter category, a rationale for using Sweetie nonetheless could be that interacting with Sweetie may give law enforcement reasonable suspicion/probable cause that someone is guilty of a(nother) crime, enabling law enforcement to employ other investigative techniques, such as searching a suspects' home.

In these instances, the gathered evidence cannot be used to prosecute the suspect for crimes against a digital character, but the information could yield a reasonable suspicion or probable cause that the person in question may have committed another relevant crime, be it webcam-facilitated abuse of real children, or the possession, creation, or accessing of child pornography.

If the chat, for instance, clearly shows that the person has experience in discussing webcam sex with children, then that might be sufficient for assuming he has likely committed a webcam sex crime before. In a similar vein, if the person sends to the avatar child pornographic material in order to corrupt or groom it, this would be a strong indication that the suspect is indeed in possession of such material. These indications would then trigger investigation procedures coupled with the prevention/investigation of sexual crimes against real minors, which would be guided by the already existing and regulated investigation means.

For the legality of the above approach, the use of Sweetie must be allowed in some shape or form in domestic criminal procedure law. Furthermore, the possibility of taking this road will inevitably depend on the particularities of the criminal system, and on the interpretation of the standards of reasonable suspicion[182] and probable cause,[183] which is a 'value judgement'.[184] In Turkey, for instance, an official investigation can already be triggered by 'simple suspicion',[185] while coercive powers such as search or seizure of computer devices can only be set in motion upon 'strong grounds of suspicion'.[186] The Nigerian Penal Code in its turn speaks of 'reasonable grounds' when it comes to authorising intrusive investigation procedures in relation to serious crimes.[187] We found similar formulations in most of the investigated jurisdictions.

However, it is beyond the scope of this report to discuss the exact scope of these standards and how they would apply to information obtained through the use of Sweetie. At this point it suffices to have drawn the reader's attention to the matter.

---

[182] In the US, for instance, reasonable suspicion is considered less than probable cause both in quality and quantity, but courts have failed to provide further guidelines. See *Alabama v. White*, 496 U.S. 325, 330–331 (1990).

[183] Probable cause is considered by some an 'articulable belief that a search will more likely than not *produce* significant evidence of wrongdoing', see Slobogin 2012, pp. 12–22.

[184] Taslitz 2013, p. 887.

[185] See country report on Turkey, p. 19.

[186] See Article 134 of the Turkish Criminal Procedure Law.

[187] See country report on Nigeria.

1.3.8 Summary and Conclusions

Based on the legislation and case law discussed above, we conclude that there are still serious legal impediments to (widely) employ the methodology of Sweetie in about half of the studied countries (see Table 1.7). This is mainly due to the absence of a clear legal basis.

The necessity for a specific legal basis stems directly from Sweetie's intrusive nature as an investigation tool that interferes with fundamental rights. However, we found that none of the jurisdictions at hand has enacted legislation that would explicitly consider artificial intelligence software systems as a means of investigation. Some jurisdictions compensate this lack of provisions with the analogous or direct application of other investigatory powers, while in other jurisdictions the existing investigatory powers appear insufficient to allow such an approach.

The inability to use existing investigative powers has in large parts to do with Sweetie's hybrid nature, combining a variety of investigatory methods that interfere with privacy (by intercepting communications and recording them) and fair trial

Table 1.7 Applicability of existing investigative powers to Sweetie [*Source* The authors]

	Explicit laws on AI-investigatory tools	Laws on special investigatory powers	Application of the laws to AIs such as Sweetie
Argentina	–	X	–
Australia	–	X	X
Belgium	–	X	X (for some crimes)
Brazil	–	X	–
Canada	–	X	X
Croatia	–	X	X (no case law yet)
England and Wales	–	X	X
Estonia	–	X	–
Israel	–	X[a]	X
Netherlands	–	X	X (no case law yet)
Nigeria	–	X	–
Philippines	–	X	X (for a few crimes only)
Poland	–	X	X (if suspect identified)[b]
Scotland	–	X	X (possibly)
South Korea	–	X	X (possible if Sweetie is not considered a coercive measure)
Spain	–	X	–
Turkey	–	X	–
USA	–	X	X

[a]As indicated above Israel's practice is not based on case-law or statutory laws, but on the exercise of unwritten powers

[b]This interpretation of the Polish Police Act still has to be confirmed by the courts

rights (considering its capacity to actively participate in said communications by interacting with suspects and potentially entrapping them). In several countries, the investigated jurisdictions have legislation in place to authorise the first conduct, but not the second, and are therefore incapable of comprehensively implementing Sweetie. This is the case in Argentina, Brazil, Israel, Spain and Turkey. These countries would have to enact particular legislation to fulfil the criteria laid down in international and national legislation on fundamental rights. On the one hand, these countries do not criminalise the (attempted) abuse of virtual characters, and therefore the conduct does not warrant use of investigation powers. On the other, undercover operations are either not covered by the law at all, or only with regard to a short list of offences, excluding crimes against the sexual integrity or other offences that could be potentially relevant in the context of webcam child sex tourism. It remains to be seen whether these insufficiencies can be circumvented by interpreting the information delivered by Sweetie as a reasonable suspicion/probable cause to authorise investigative measures.

As for the rest of the studied countries, Sweetie's application in investigations is likely to fit the legality standards only in regard to the coercive powers authorising it in the first place. While said application scope seems rather limited in Belgium, the Philippines and Poland, which focus on particular online offences, the chatbot and its features appear to satisfy the constitutional requirements in Australia, Canada, Croatia and England and Wales. These countries have developed substantive and procedural means to tackle the online abuse of children. Especially, the enactment of laws against the online grooming of children has prompted a perceivable shift in the investigation techniques as well, allowing for more proactive policing of online communications and for applying existing tenets on entrapment techniques to undercover investigations when targeting an area and not a particular suspect.

1.4 Digital Forensics

Undercover operations that would use Sweetie to apprehend webcam child sex offenders require not only a legal framework that authorises such an investigation tool. As the chatbot is primarily intended to facilitate the identification of perpetrators, it brings about questions on the evidentiary rules applicable to the collection of evidence from the chat communications and webcam streams. Therefore, in the following subsections we elaborate upon the general requirements applicable to (digital) evidence and how the countries in this study implement these.

1.4.1 Generally Accepted Standards

Criminal investigations aim to follow the trail that offenders leave while committing the crime and to link suspects to the crime.[188] The information gathered with this purpose, i.e. the evidence of the crime, has to be preserved and examined in a particular manner to maintain the objectivity called for by the investigative process,[189] and to be introduced in court accordingly. While this is a general tenet from traditional forensic disciplines, it applies to online investigations and digital evidence as well and means that the cyber-trail is to be rigorously followed. In this regard, progress has been made by law enforcement in the digital evidence gathering, be it by furthering the expertise of police officers in handling technology or by involving IT experts who oversee or perform the information gathering themselves. By now, courts are also used to dealing with digital evidence and deem it admissible, provided that its authenticity and integrity are ensured.

Authentication and Chain of Custody
The authentication of evidence ensures that the obtained information is the same as the originally seized. One of its most important objectives is maintaining and recording the chain of custody, which requires that each person who handles the evidence including the handling itself must be documented and may be summoned to testify on the originality of the evidence in court.[190] An improper chain of custody may result in the contamination or loss of evidence. The chain of custody requirements apply likewise in the context of digital evidence.

Evidence Integrity and Digital Fingerprints
Integrity checks further support the authentication process by ensuring that the evidence has not been altered since the time of its collection. In digital forensics, this is usually done by a comparison of the digital fingerprint of the evidence at the time of collection with the digital fingerprint of the evidence in its current state.[191]

1.4.2 Implementation of Digital Forensics in the Compared Jurisdictions

When dealing with the standards of digital forensics, we observed that all of the jurisdictions at hand admit electronic evidence in court. While some, following specific incentives introduced by the Cybercrime Convention, have already enacted particular legislation, others still lack provisions on the technical requirements and

---

[188] Casey 2011, p. 16.

[189] Idem, p. 19.

[190] Idem, p. 21.

[191] Idem, p. 22.

expertise that need to be met when gathering digital evidence.[192] In the case of the latter, the law deals with electronic evidence in the same way as with any other type of evidence. Although this means that the general rules on the authentication and integrity of evidence are observed, the fact that no formal technological standards exist may compromise the gathered information.

Many have adopted diverging approaches. In Australia, evidence from electronic sources is routinely adduced through the questioning of qualified expert witnesses, such as the particular analyst who conducted the forensic examination.[193] Also in the Polish law enforcement practice, the expert witness report would likely be the piece of key evidence.[194] In Estonia, the role of the expert witness is prescribed by the Forensic Examination Act, which states that 'information technology examinations' regarding materials in relation to the sexual abuse of minors should be performed by the Estonian Forensic Science Institute'.[195]

Yet other countries foresee a more active role of law enforcement officers in the handling of digital evidence and supplement their functions with additional technological tools. In Spain, while expert opinions will still be indispensable to identify the true origin of a communication[196] in a troublesome case, Article 588ter f) of the Criminal Procedure Act ensures the integrity of the digital information by introducing electronic signatures that affirm the origin and destination of the communication.[197] Similarly, shortly after the introduction of electronic signatures[198] by Provisional Measure (MP) 2.200/2001,[199] the Brazilian legal system has adopted notarial minutes[200] as a supportive evidential standard. While said minutes are regulated under the Code of Civil Procedure, the criminal system recognizes them as appropriate for the purpose of criminal proceedings.[201] Put together with

[192] This is true for Belgium (see country report on Belgium, p. 44), Canada (country report, p. 53), Croatia (country report, p. 34), Israel (country report, p. 16), Nigeria (see country report, p. 50), the Philippines (country report, p. 37), and Scotland (country report, p. 21).

[193] See country report on Australia, p. 26.

[194] See country report on Poland, p. 36.

[195] Regulation on the list of examinations conducted in EFSI, subsection 5(3)(1). Available from: https://www.riigiteataja.ee/akt/13365049 [28 September 2016].

[196] Judgement of the Supreme Court of Justice 300/2015 delivered on 19 May.

[197] Spanish report, p. 40.

[198] This is a digital certificate issued by the Brazilian Public Infrastructure Key (ICP-Brazil), which awards electronic files a presumption of veracity. For more on the matter see country report on Brazil, p. 24.

[199] In the Brazilian legal system provisional measures have a temporary validity and are issued by the President of the Republic in urgent situations. See https://www.oas.org/juridico/mla/en/bra/en_bra-int-des-ordrjur.html. Provisional Measure (MP) 2.200/2001 is available at www.planalto.gov.br/ccivil_03/mpv/Antigas_2001/2200-2.htm [14 September 2016].

[200] Regulated in Article 384 of the CPC. Notarial minutes enhance the credibility of images, audios, videos and writings taken from web pages that could have been easily amended or deleted.

[201] See Brazilian report, p. 27.

electronic files or digital documents, the notarial minutes support the credibility of any content taken from the web. However, notarial minutes apply to publicly available content only.

In the US, when it comes to online chat conversations officers are generally encouraged to use a screenshot or a browser's saving function to limit their interference with the content.[202]

1.4.3 Storing Data from Online Chats and Chatroom Activity

Given that webcam streams, unlike downloads, are not by default stored on the personal computer of the victim or the suspect, the criminal systems that lack provisions concerning content interception of online communications may experience practical issues to prove illegal webcam sex, as witness statements or seized electronic devices may not be available.

However, it is possible to gather other forms of evidence related to the webcam streams, such as for instance chat logs that have been saved. When Sweetie is used by law enforcement, then streams can also be recorded. Investigation agents can further take screenshots, when deemed necessary. Such records would be crucial to make the evidence available in court.

In addition, the chat scripts of the avatar should also be added to the other relevant information for the consideration of the judges before sentencing. This way the court would be able to gain an insight into how the avatar has approached the communication and whether any impropriety could be witnessed on the part of the AI agent, for instance, whether it complied with the 'essentially passive' standard required to avoid entrapment.

1.4.4 Summary and Conclusion

While the investigated jurisdictions do follow the generally accepted forensic standards such as observing the chain of custody and taking precautions to maintain the integrity of the evidence, only some of them have taken particular steps towards the implementation of digital forensics agendas. Therefore, the available lessons from the practice vary and law enforcement agencies (would likely) approach the collection of evidence from chat communications and webcam streams differently. The lack of uniformity bears risks for the comprehensiveness of future investigations of webcam crimes (against minors) and may present challenges for the handling of the gathered information and for its sharing in trans-border operations.

---

[202] See *United States v. Jackson*, 488 F. Supp. 2d 866 (D. Neb. 2007).

1.5 Jurisdictional Concerns with the Application of Sweetie

As indicated in Table 1.4, depending on the exact conduct of the suspect and how he would approach Sweetie during the chat conversation, there are some avenues for prosecuting him for soliciting (or attempting to have) webcam sex. However, as webcam child sex tourism crosses national borders and implies scenarios in which, as a rule, victim and offender reside in different countries, jurisdictional conflicts inevitably present themselves.

The latter can be best illustrated by means of the following example. Let us assume for the sake of the study that UK law enforcement authorities use Sweetie to conduct a luring operation in a public chatroom for users under 18 years. Through Sweetie's solicitation by suspects and its communication with the latter, they manage to collect enough information such as IP addresses and Skype handles to identify a number of individuals, some of which reside in Australia and Spain. The computer devices of those alleged offenders and (parts of) the potential evidence of their interactions with Sweetie would be stored beyond the physical reach of the UK officials, which means the latter will have to resort to remote transnational information searches. In so doing they would primarily face two legal difficulties: what legal grounds would entitle them to start an investigation, and under what circumstances can the investigation be conducted (also abroad)?

The first question is closely connected to the states' jurisdictional capacity to prescribe rules, but also with our finding that the different actions of the offender in front of the webcam or as a part of a chat conversation with Sweetie trigger different norms under domestic law. The second issue on how to enforce investigative procedures in relation to offences against Sweetie has to do both with the available coercive powers discussed in Sect. 1.4, and with the general jurisdictional tenets on enforcing state laws. Therefore, to answer these questions and explain what the likely outcome of our example would be in practice, in the following sections we apply the international law rules on state jurisdiction to trans-border cyber-investigations, and subsequently to the case of Sweetie.

1.5.1 Grounds for the Exercise of Jurisdiction in Cybercrime Investigations

Across frontiers, powers are differently distributed and restricted, and it is international law that governs the legal framework of interaction. Thus, it comes as no surprise that the state's operation of coercive powers abroad is subject to strict international limitations. Said operational limitations are inherent in the rules of state sovereignty, which in their turn have affected a differentiation between a state's power to regulate or otherwise impact people, property and circumstances[203] (usually referred to as prescriptive jurisdiction), and its ability to enforce it (enforcement jurisdiction).

Both these competences, although originating from the same notion, have an impact of their own in the criminal context at hand, and are conditioned upon a certain link to the state willing to assert them.

Prescriptive Jurisdiction

States' jurisdictional claim to prescribe law can be premised on a number of factors.[204] The state's direct proximity to the crime scene has enhanced the traditional idea that a crime is best punished locally.[205] The principle of territoriality is, therefore, one of the main grounds for the exercise of criminal jurisdiction, and gives states the right to exercise it regardless of the offender's nationality.[206] However, the territoriality principle is more comprehensive than one assumes it to be at first glance, as it applies also to crimes committed only partially on the territory of a state.[207] This is the case where, for instance, the criminal conduct has been initiated in one state, but completed in another.[208] Under such circumstances, both states would have criminal jurisdiction to try the perpetrator—the first under the principle of subjective territoriality, the second one under the objective territoriality principle.

Further, the principle of nationality, the state's special link to its population, allows a state to domestically regulate the conduct of its nationals wherever they are. Therefore, states are also entitled to claim jurisdiction over offences committed by their nationals abroad,[209] and can do so even if the suspect has a dual nationality.[210] Often, countries opt for additional statutes that regulate precisely which offences trigger jurisdiction over nationals abroad.[211]

In addition, a state can seek to regulate conduct under the passive personality principle and the protective principle.[212] The former, a disputed practice, allows

---

[203] Shaw 2014, p. 469.

[204] Idem, p. 474; Klabbers 2013, p. 91.

[205] August 2002, p. 534.

[206] Gillespie 2012, p. 153.

[207] Shaw 2014, p. 475.

[208] The classical example explaining such a situation is where a person shoots at someone across a border and kills them in the neighbouring state.

[209] Shaw 2014, pp. 479 ff.

[210] Harris 2010, p. 230.

[211] This is often the case in common law countries. See on this issue Shaw 2014, p. 481.

[212] A further possible base for the assertion of state jurisdiction—the universality principle—will not be discussed here. Although webcam sex crimes against children are a highly serious matter, they do not form part of the group of war crimes and crimes against peace or humanity that trigger universal jurisdiction due to the particular danger their represent for the international community as a whole. See on this Shaw 2014, pp. 485ff.

states to exercise jurisdiction over anyone who harms their nationals. The latter justifies regulating conduct that produces harmful effects within a state's territory and thus endangers national (security or economic) interests. The 'effects' theory is widely accepted in the international community and often used in treaties.[213]

Since technology-specific rules on state jurisdiction do not exist, the exercise of jurisdiction over cybercrimes remains largely based on the approaches outlined above, among which the principle of territoriality plays the most pivotal role. Although this may appear odd when dealing with criminal behaviour portrayed in the 'de-territorialised'[214] space of the Internet, international law so far has imposed very few limitations on states when they claim criminal jurisdiction over cyber-crimes and relies on their willingness to resolve positive jurisdictional clashes by rather informal means.

Jurisdiction to Enforce

Enforcement jurisdiction is another matter. Although a state may have jurisdiction to prescribe rules that even portray a certain extraterritorial reach, it generally cannot enforce said laws (whether by judicial or executive organs) outside its territory without the affected state's consent. As states are independent from each other and possess territorial sovereignty, the capacity of a state to operate within the borders of another state is essentially restricted by the sovereign powers of the latter. Translating this into the criminal context means that although a state may claim criminal jurisdiction based on rules enacted according to its prescriptive jurisdiction, law enforcement usually sees itself constrained to finding evidence and apprehending the alleged offender within their own territory or acquiring these from other states' territory through mutual legal assistance.

Extra-Territorial Application of Investigative Power

The observations made above mean that trans-border investigations are not allowed unless expressly consented by the host state[215] or established by an international agreement. The same applies to investigative techniques conducted in cyberspace such as remote information searches that aim, for instance, at accessing and securing information from a server located abroad. Opposing opinions that advocate a different understanding when investigators do not physically enter the other state's territory[216] are not tenable under the current state of international law. While admittedly this is not so much a question of substantively threatening the territorial integrity of the state hosting the data, such practices interfere with the state's sovereign control over its citizens and their rights as it subjects them to foreign

---

[213] See for instance the 1979 Hostages Convention, the Aircraft Hijacking Conventions and the 1994 Safety of UN and Associated Personnel Convention.

[214] Ryngaert 2015, Section 3.5.

[215] Shaw 2014, p. 473.

[216] Johnson and Post 1996, pp. 1367–1402; See also *United States v Gorshkov*, No CR00-500C (W D Wash May 23, 2001), where the US claimed that they did not violate Russian sovereignty because the FBI agents never left US soil.

legislation and law enforcement. The state sovereignty breach may therefore easily result in a breach of the laws on criminal procedure and the corresponding procedural guarantees.

Therefore, in the event of unauthorised remote investigations, some of the countries studied for the purposes of this report render criminal procedures against a national inadmissible if the procedure has been based on evidence obtained by entrapping the suspect, or the use of other coercive techniques by foreign police agents.[217] However, in some of the criminal systems such information may also be considered a reasonable suspicion that triggers investigative procedures on the national level.[218]

Yet some other criminal systems at hand are willing to take into account evidence that has been brought about by the investigative powers of law enforcement agencies that may have trespassed their enforcement powers and thus violated the territorial sovereignty of other states. In the Netherlands for instance, the *Schutznorm* principle allows the use of such evidential material against a suspect, since the violated norm does not serve to protect his interest (but the interest of foreign states).[219] A further interpretation of the same principle allows for the use of the evidential information if the foreign agency ends up obtaining the information on the individual without actually targeting him, but its own nationals.[220]

In any event, the exercise of investigative powers outside of national borders and its consequences would depend on the jurisdiction(s) involved and the case in question. For the purposes of this study it suffices to have highlighted that it interferes with the targeted state's sovereign competence to govern its population, and with individuals' privacy and *due process* rights.

Mutual Legal Assistance

States typically address the gap between their capacity to regulate and to enforce by relying on mechanisms for legal assistance, which are usually referred to as treaties on mutual legal assistance or MLATs. There are countless multilateral and bilateral agreements between states that establish procedures for obtaining and providing assistance in transnational criminal matters.[221] Normally, a request for assistance, for instance, to locate or to arrest a person, to produce documents or records, or to perform a search, can only be denied on the grounds specified in the respective treaty.

---

[217] That is the case to a certain extent in Belgium. See Belgian report, p. 35.

[218] This is for instance the case in Argentina, see country report on Argentina, p. 27.

[219] Rb. Alkmaar 19 February 2004, LJN A05509, case no. 14.060137-02, to be found at www.rechtspraak.nl; Rb. Groningen 16 October 2003, LJN AM1882, case no. 18/076010-01, published in 2003 *Vakstudienieuws*, 56.4 and at www.rechtspraak.nl. See also Hock and Luchtman 2005, p. 4. The Israeli system also suggests that evidence against nationals obtained by foreign agents in an unauthorized trans-border operation would be admissible.

[220] Ibid., at note 209.

[221] Bellia 2001, p. 50. See for instance Commonwealth of Independent States Agreement, Article 5; Council of Europe Cybercrime Convention, Article 23; Shanghai Cooperation Organization Agreement, Articles 3–5. Article 28(2) of the AU Convention on Cyber Security and Personal Data Protection, EX.CL/846(XXV), Malabo, 27th June 2014.

Further, it is important to note that MLATs are rather context-specific or offence-specific and may therefore never cover the entire range of circumstances leading to a particular investigation. That is to say, if the requested procedure or coercive power is not explicitly covered by an agreement, a request for assistance will be (formally) devoid of any prospects from the very beginning. The same applies if the alleged criminal conduct is not listed in the agreement. Any further steps of the investigation would then lie with the good will of the agencies across the border.

Where MLATs do not exist, the investigation authorities employ informal means of cooperation. These tend to be even more difficult to use, as they may involve multiple states with varying legal systems, and therefore different understandings as of what constitutes a criminal offence. If one of the states involved has not criminalised the alleged conduct under its national law, the requirement of double criminality[222] cannot be fulfilled. Said state will likely have no interest in contributing to the investigation.

1.5.2 Translating the Jurisdictional Rules to the Context of Sweetie

Let us now apply the jurisdictional tenets described in the preceding sections to the case of Sweetie. Since prescriptive jurisdiction can be claimed on a number of grounds, it means that also in our example several states will be able to do so over crimes against Sweetie, as the Internet amplifies the existing options. The initial act's location (presently Australia or Spain, from where the alleged child offenders have accessed the chatroom) or where it has its effect (Sweetie's location), as well as the location of the chatroom servers or other hardware (which may be in any country around the globe) can establish a sufficient link to a country to claim jurisdiction; there are states that even use the location of anything remotely connected to the crime to claim jurisdiction.[223] Thus, although it is the UK that has initiated the investigation in question, it may be difficult to establish who has the stronger jurisdictional claim.

The enforcement side of the question brings about additional challenges. As remote information searches are 'not distinguishable in legally relevant ways'[224] from physical searches, directly pursuing the digital trail of the webcam offenders would mean to engage with the territorial sovereignty of Australia or Spain, or any

222 The principle of double criminality was introduced by extradition treaties and requires the act to which a request relates to be a crime under both the criminal law of the requested state and the requesting state. For a comprehensive discussion on this see Williams 1991, p. 582.

223 See, for instance, the cybercrime jurisdiction provision of Malaysia, Article 9 Malaysia Computer Crimes Act 1997. More on that in Brenner and Koops 2004, pp. 21–23.

224 Bellia 2001, p. 62.

other state that could potentially harbour (parts of) the searched data. Consequently, in order to obtain computer data physically located on Australian or Spanish territory, UK's law enforcement must either obtain the consent of the state authorities to continue the search or resort to traditional procedures of mutual legal assistance.

Mutual Legal Assistance in the Case of Sweetie

UK law enforcement may be able to resort to an existing MLAT on cybercrime. Surveys, however, show that cybercrime MLATs tend to focus on matters of extradition rather than on evidentiary procedures.[225] Thus, there is a chance the requested investigative act would not be explicitly covered by the respective treaty, which as explained above would lead law enforcement to resort to informal means of cooperation.

In the context of webcam child sex tourism, the lack of a concrete MLAT is not unproblematic, since, as described in Sect. 1.2[226] states tend to criminalise webcam sexual abuse through different legal constructions. While the offender's conduct in the UK may fall under the offence of attempted child prostitution, the same conduct against Sweetie would fall in Australia under the crime description of attempted sexual abuse or grooming, whereas in Spain an inappropriate interaction with Sweetie would not be criminal at all. This would hinder the establishment of double criminality, which is essential for triggering investigative procedures abroad.

In addition, electronic evidence may simply be lost after a short period of time. This is critical considering the fact that legal assistance mechanisms require time to be set in motion. The time needed to issue a request and to eventually execute it on the other side of the border usually costs the authorities their opportunity to secure volatile electronic evidence.[227] In the case at hand this consideration is of particular importance given that webcam streams produce mainly volatile data that can be lost completely once the suspect has powered down his computer.

The Cybercrime Convention

Against the background of the challenges of standard MLATs and their potential interplay with Sweetie's case outlined above, a legal instrument that deserves our attention is the Cybercrime Convention. This tackles some of the shortcomings that result from out-dated and lengthy MLA procedures.

The Convention establishes mandatory contact points (24/7 networks) between national agencies, which are meant to ensure the immediate assistance in investigation matters.[228] The treaty further addresses consent and procedural difficulties

---

[225] United Nations Office on Drugs and Crime 2013, *Comprehensive Study on Cybercrime*, p. 200.

[226] Compare Table 1.4.

[227] Current reports indicate that responding to a request can actually be a matter of months. For more on this: United Nations Office on Drugs and Crime 2013, pp. 197ff.

[228] See Article 35 CCC.

between national agencies by streamlining investigation procedures and defining four methods for securing computer data, namely expedited preservation of stored computer data, expedited disclosure of preserved traffic data, real time collection of traffic data, and interception of content data.[229] In this context, signatory parties are obliged to confer said competences to their national authorities so the latter can both obtain and request the disclosure of data. However, the Convention, which can also be acceded by non-Council of Europe states and has therefore gained in importance globally, leaves jurisdictional clashes unresolved, as it does not provide guidance, or sets up mechanisms, for prioritising competing jurisdiction claims.[230]

Moreover, although this approach and the procedures it introduces clearly go far beyond ordinary mutual assistance mechanisms,[231] they are still confined within the borders of the state where the data is physically located. The Convention thus does not bring about new approaches for dealing with *trans-border* investigations, but largely relies on traditionally known and well-recognised jurisdictional concepts. The power to search, seize, or intercept digital evidence remains in the hands of the host state, and transnational investigations are not welcomed under the Convention's provisions.[232] The only exception is Article 32, which allows cross-border access to publicly available data (which is not very relevant in the case of Sweetie, except perhaps for securing data from publicly accessible chat-rooms),[233] as well as cross-border access to data with voluntary consent from someone who has the lawful authority to consent.[234] That will usually be the foreign state, although it may also include service providers (e.g., chatroom providers) if data protection and contract law allow them to consent to law enforcement accessing the data at issue (which is unlikely to be the case in our example). Efforts to draft an additional protocol to the Convention on trans-border access to data have not been successful and seem on hold at the moment; it will likely take a long time before countries will be ready to agree on some international agreement on trans-border access to data.[235]

Thus, although all three countries from our example have signed and ratified the Convention, the UK law enforcement officials have to reach out to their counterparts in Australia and Spain to formally request the securing of the data needed for an investigation of an offence against Sweetie.

---

[229] Council of Europe Cybercrime Convention, Articles 29–31, 34.

[230] See Article 22(5) CCC. The provision merely stipulates that states should be the one to determine 'the most appropriate jurisdiction for prosecution', but a further elaboration upon this 'appropriateness test' is missing.

[231] Bellia 2001, p. 59.

[232] The Convention's provisions stipulate clearly that the respective procedural rules are applicable only within a state's territory. See Articles 18, 19, 20, 21 CCC, which explicitly refer to national territory.

[233] See Article 32(a) CCC.

[234] See Article 32(b) CCC.

[235] See, extensively, Koops and Goodwin 2014.

1.5.3 Conclusion

By highlighting the existing tension between, on the one hand, a global communications network where webcam child sex tourism can take place across borders, and, on the other, law enforcement procedures that remain tightly restricted to national territory, this chapter has called attention to the problematic nature of the enforcement of criminal law. Law enforcement, but more relevantly regulators and policy-makers, should be aware of the jurisdictional challenges in the context of Sweetie. We conclude that as states and their law enforcement agencies continue to move within a consent-based legal framework, a more effective way of cross-border investigation is necessary. In the case of Sweetie, the lack of international harmonisation makes itself especially felt, because of the divergence of criminal provisions and instruments,[236] which affect the scope of and possibilities for international cooperation.

With regard to webcam child sex offenders, law enforcement agents will have to rely on the willingness and expedited proceedings on the part of their foreign colleagues when further investigating a suspect's digital trail. Be it following the Cybercrime Convention's standards or MLAT procedures, when it comes to securing digital evidence,[237] the agent's law enforcement powers end at their respective national borders. As with other forms of cybercrime, this fact may significantly undermine the effectiveness of investigations against suspects using Sweetie.

1.6 Effective and Legitimate Use of Sweetie: The Way Forward

In the following we discuss the main substantive, procedural and jurisdictional issues in Sweetie's implementation identified in the present report, and possible ways to address these.

1.6.1 Substantive Law Restrictions

Based on our research we have identified several issues that need to be remedied in order to effectively and legitimately combat webcam sex using Sweetie. This would in most cases entail changes to substantive criminal law. Whether or not countries want to actually adapt their substantive criminal law in order to facilitate the use of Sweetie is a question of a political nature.

[236] United Nations Office on Drugs and Crime 2013, p. 208.

[237] Koops and Goodwin 2014, p. 14.

Clarifying Substantive Law

In most jurisdictions under examination we see that webcam sex with minors is criminalised in one form or another. However, given that in most jurisdictions this relatively new form of crime is 'read' into existing crime descriptions there are questions regarding the extent to which this behaviour is criminalised.

In order to avoid stretched legal interpretations that might be at odds with the principles of legality and legal certainty, it is recommended that legislators include in their crime catalogues (more) explicit definitions of ambiguous terms such as 'pornographic' and 'sexual activity/abuse' and more guidance on what kinds of behaviours associated with webcam sex fall within which crime descriptions.

Changing Substantive Criminal Law in Order to Facilitate the Use of Sweetie

Sweetie is first and foremost an innovative investigation tool. Its innovativeness entails that in order for it to be used legitimately, changes to substantial criminal law will most likely need to be made. As it stands, several jurisdictions may not deem interacting with Sweetie in a sexually charged way a criminal offence at all. In these jurisdictions it will be hard to justify the application of Sweetie by law enforcement, because the behaviour Sweetie elicits and exposes is actually not criminal at all.

If these jurisdictions wish to allow the use of Sweetie by law enforcement, it stands to reason that they change their substantive criminal law systems in such a way that the intention of the suspect is the determining factor in establishing criminal liability. This will mean a shift from an 'act based' criminal law system towards a more 'intention based' criminal law system. Whether combating child sex abuse using Sweetie necessitates such a shift in the approach to criminal law is a matter of ethics and politics.

Jurisdictions that already criminalise virtual child pornography and/or the grooming of virtual characters, or those considering criminalising these acts, should also consider including subjective elements in provisions that relate to child (webcam) sex abuse. An inconsistency in the approach to criminalisation of child (webcam) sex abuse may create normative gaps, so from a legal-systematic viewpoint it makes sense to extend criminal liability to related offences.

International Harmonisation

Last, but not least, it is recommended to discuss a global approach to dealing with child webcam sex tourism using tools such as Sweetie, in order to avoid crime and penalty havens and to create more legal certainty. If consensus is reached, this must be reflected in international legal instruments such as the Lanzarote Convention and the OPSC. International investigations and mutual legal assistance procedures would benefit from domestic systems that criminalise webcam child sex exploitation in similar terms and through similar crime descriptions.

1.6.2 Procedural Law Restrictions

In terms of criminal procedure, we have found that the jurisdictions at hand have all introduced coercive investigative powers to address serious and organised crime. While a fair number of these investigative powers may also be applied in an online context, most of them are still 'traditional'. That is to say, they were written in large parts for the 'offline world' and do not readily accommodate the use of innovative investigative tools such as Sweetie.

A particular issue when it comes to the application of Sweetie is its 'hybrid' nature as a lure, an apparatus for recording conversations and video and an intelligent undercover agent. If law enforcement wants to use Sweetie, it is important to determine whether existing investigative powers used either alone or in conjunction, cover the application of Sweetie. Given the possible infringement of privacy, both in cases where the use of Sweetie is covered by existing investigative powers and in cases where new legislation is introduced, the application of Sweetie must be in accordance with the law. That is to say the laws governing the use of Sweetie must be accessible and of sufficient quality.

With regards to the issue of entrapment, it is relevant that the application of Sweetie follows existing guidelines on targeted and non-targeted entrapment. Law enforcement should carefully consider in which chatrooms Sweetie is placed and how she will interact with suspects via her chat script. Particular attention needs to be devoted to the hybrid character of Sweetie as a lure and as an undercover agent. In more traditional settings these investigative functions are not combined. In this sense the existing use of human lures (e.g. law enforcement officers posing as minors in a chatroom) is instructive.

If the use of Sweetie necessitates changes to criminal procedure law, it is also relevant to include explicit standards in the handling of digital evidence.

1.6.3 Addressing Jurisdictional Constraints

A way to possibly avoid more complex jurisdictional questions and competing jurisdictional claims (as outlined in the preceding section) would be to primarily use Sweetie to investigate nationals or residents of the respective country. This can be done by focusing on local chatrooms (intended for and frequented by national users), which would lessen the difficulties of obtaining the necessary authorization for using coercive powers, and of physically securing and investigating the devices used by offenders if needed. This task could be further facilitated if following the examples set by Australia, Canada and the UK on the matter, states legislate the criminality of a country's citizens' committing crimes against children extra-territorially.

A possible way to facilitate cooperation and to alleviate jurisdictional conflicts in relation to trans-border investigations of webcam child sex tourism and other child exploitation offences would be to adopt an Optional Protocol (OP) to the Lanzarote Convention on the matter.

OPs have the advantage of introducing additional provisions, procedures and mechanisms to the original treaty by maintaining the latter's scope and integrity. Human rights treaties for instance, oftentimes provide in their OPs for complaint procedures that address alleged human rights abuses or regulate substantive law areas not considered previously. States have no obligation to ratify those protocols but can do so if they think that said instruments enhance their national interests or broader policy and international cooperation agendas.

An OP to the Lanzarote Convention that regulates cross-border investigations would have following advantages:

- It could provide guidance on how to deal with positive jurisdictional conflicts in relation to trans-border investigations of (Internet) child sex/abuse offences only, thereby avoiding broader commitments which are not likely to be accepted by sovereign states;
- The OP would bindingly stipulate the forensic standards required for handling data searches and the resulting evidence, prompting states that have not yet introduced digital agendas to do so, and preventing the loss of digital evidence due to improper handling.

1.7 Summary and Conclusion

Webcam sex tourism, the act of engaging children in webcam prostitution, is a growing international problem. Not only does webcam sex tourism provide easy access to child abuse and child abuse images for child abusers, it also a crime that has a comparatively low risk for the offenders. Live webcam performances leave few traces and little evidence that law enforcement can use. Further difficulties arise from the fact that webcam sex tourism often has a cross-border character, which causes jurisdictional conflicts and makes it more difficult to obtain evidence or even launch an investigation.

The Dutch children's rights organization Terre des Hommes (TdH) was the first NGO to actively tackle webcam child sex tourism by using a virtual character called 'Sweetie' to identify offenders in chatrooms and online forums. An agent of the organisation operated the Sweetie avatar, posing as a ten-year old Filipino girl, in order to gather information on individuals who contacted Sweetie and solicited webcam sex. The gathered information was subsequently handed over to the authorities, who thereupon were able to launch investigations in various countries.[238]

One of the major drawbacks of Sweetie 1.0 (and law enforcement in general) is that the avatar could not be deployed at scale. Human operators can only engage in a limited number of conversations with suspects, while the (potential) solicitations

[238] Further information on the project known as 'Sweetie 1.0' can be found at www.terredeshommes.nl/en/sweetie-face-webcam-child-sex-tourism [28 September 2016].

addressed at Sweetie 1.0 far exceed that number. Sweetie 2.0 aims to solve this problem. Sweetie 2.0, being an artificial intelligence, is far more scalable because multiple instances of Sweetie can be deployed simultaneously.

But using an artificial intelligence like Sweetie raises serious legal questions. Sweetie as an investigative tool is so innovative, that it is unclear whether its use is actually covered by the existing rules of criminal procedure. However, the question of criminal procedural legality of Sweetie is preceded by a prior substantive criminal law question: is interacting with Sweetie in a sexually charged way a criminal offence in the first place, given that Sweetie is not a person, but a virtual avatar? An answer to this question is important, because if webcam sex tourism with a virtual avatar is not considered criminal, it will be much harder to make the case that Sweetie is an acceptable investigative method.

We will discuss the substantive criminal law issues and the criminal procedure law issues separately below.

1.7.1 Substantive Criminal Law Issues

In our research we have identified the following issues that impact the application of substantive criminal law:

- Sweetie is not an actual person and as such sexually charged interactions with Sweetie may not be considered criminal in jurisdictions criminalising only certain interactions with real persons.
- Sweetie is deliberately programmed not to perform sexual acts or to show sexual organs.

In most of the jurisdictions we examined webcam sex with real minors has been criminalised in one form or another (see Table 1.3), however the same cannot be said for webcam sex with a virtual person such as Sweetie (see Table 1.4).

In some jurisdictions someone can never be convicted for committing or attempting to commit an offence against Sweetie (e.g. Brazil). In other jurisdictions, criminal liability is only limited to the specific offence of grooming (e.g. Argentina and Belgium), which is in many cases not applicable to Sweetie, or some particular form of sexual abuse (e.g. the Netherlands). For most jurisdictions however, it is quite uncertain given an absence of case law (and even literature) on the matter (e.g. Germany, Israel, Poland).

In most jurisdictions the crime descriptions applicable to webcam sex tourism contain a specific mention of ‘a person under the age of X years’. Given the fact that Sweetie is not a real person, this element of the crime can never be proven. Furthermore, because Sweetie is not programmed to undress, perform sexual acts or show sexual organs, many crime descriptions such as those related to sexual performances or child pornography, cannot be fulfilled.

Although the crime descriptions related to webcam sex tourism in most jurisdictions can never be fulfilled, there might be room to qualify the behaviour of the suspect as an attempt. In this regard, the doctrine of the inadequate attempt (legal or factual impossibility) is relevant, as it will determine whether an attempt is punishable or not.

When we speak of a legal impossibility, the behaviour, even if completed never leads to a criminal offence, regardless of the criminal intentions of the suspect. The reason for this is that in these cases the behaviour on display itself is not criminal. This can either be the case because the means are absolutely inadequate (e.g. trying to shoot someone by pointing a banana at them), or because the object at which the act is aimed is absolutely inadequate (e.g. trying to murder a corpse).

With a factual impossibility a suspect's intended behaviour would constitute a crime, but the suspect fails to complete the crime, because of a circumstance unknown or beyond his or her control. In these cases, the inadequacy of the means or the object are relative. An example of a relatively inadequate means would be an unloaded pistol used to try and shoot a person. An example of a relatively inadequate object is an empty cash register in which someone puts his hand to grab money. Under normal circumstances this would have resulted in the theft of money, but in this concrete case, the attempt fails. In contrast to a legal impossibility, a factual impossibility is punishable.

In the case of Sweetie, the question is whether we should regard Sweetie as an absolutely inadequate object or a relatively inadequate object. Looking solely at the behaviour in relation to the crime description, we may argue that Sweetie is an absolutely inadequate object: the crime of webcam sex with a minor can never be completed, because Sweetie is not, nor ever will be a real minor. The fact that Sweetie cannot show sexual organs or display sexual activities, further adds to the argument for an absolutely inadequate object.

However, we can also take a somewhat broader perspective and qualify Sweetie as a relatively inadequate object for the crime of webcam sex with minors. The argument would be that the suspect wants to commit the crime of webcam sex with a minor but is 'unlucky' and picks Sweetie rather than a real minor. This case is comparable to the example of the cash register: under normal circumstances the crime would have been committed, but due to 'bad luck' on the part of the suspect, there is now a factual impossibility.

There is merit to both arguments and when we look at the jurisdictions we have examined, we see that they take different approaches. For instance, there are jurisdictions that take an objective approach and look at the actual act and thus lean towards legal impossibility, and systems that take a more subjective approach, attaching more weight to the intention of the suspect, leaning more towards factual impossibility.

In particular, countries that come from a common law tradition seem to take a more subjective approach, either in statutory law itself, or in case law (e.g. Australia, Canada, the UK and the US). In these jurisdictions the subjective element of the crime (i.e. the intention of the suspect) plays a more important role than the objective act. If the suspect is under the (false) impression that he/she is communicating with a minor, this is the determining factor for criminal liability.

So as it stands, approaches to criminalising webcam sex tourism vary throughout the world and in many jurisdictions it is still uncertain whether an attempt at webcam sex tourism can be construed at all. Only in those countries that take the intent of the suspect as the determining factor in criminal liability can Sweetie 2.0 be clearly employed as an investigative tool.

For those countries where it is impossible or substantially uncertain to find a crime description that can be used to criminalise webcam sex tourism with a virtual minor, legislative changes are needed in order to enable the use of Sweetie. The choice to move further away from an 'act-based' criminal law system towards a more 'intention-based' criminal legal system in order to combat webcam sex tourism is of a fundamental nature and would require careful ethical and political deliberation.

Finally, from the perspective of law enforcement it may be worthwhile to explore if Sweetie can be used to investigate the crime of webcam sex with a real person, or related crimes. While the interaction with Sweetie may not be considered criminal in itself, it could provide a reasonable suspicion that someone is or has been involved in webcam sex tourism with real minors, which would then provide the legal basis for further investigating the suspect using other, more traditional investigative methods. The legitimacy of such an approach is very much dependent on the circumstances of the case and the criminal procedure law of the individual jurisdiction and would also require careful deliberation.

1.7.2 Criminal Procedure Law Issues

If the use of Sweetie is possible in light of substantive criminal law, its application will also raise criminal procedure law questions. We have identified these two main questions:

- Is the use of Sweetie in accordance with the law?
- Does Sweetie respect fair trial principles in the pre-trial phase, more specifically the rules on entrapment?

Use of Sweetie in Accordance with the Law

Sweetie is an innovative investigation tool that actually combines three distinct investigative functions into one package, namely: (1) a lure (comparable to for instance a bait car), (2) an (undercover) agent that can engage in conversation with a suspect, (3) a device that can record information such as conversations, pictures and videos. The hybrid nature of Sweetie raises questions whether its application is in accordance with the law. We have established that Sweetie can infringe on the privacy of the suspect. As such, in most if not all of the jurisdictions under examination, criminal procedure law that provide procedural safeguards must govern the use of Sweetie.

In order for Sweetie to be applied legitimately, there must a legal basis that is sufficiently accessible and foreseeable. This means that either a specific legal basis for the use of Sweetie must be established in the law of criminal procedure (which is not the case in the jurisdictions examined), or its use must be covered by existing investigative powers and practices such as those on systematic observation, undercover work and the recording of confidential information and communication. As can be judged from Table 1.7, in about half of the jurisdictions we examined, the use of Sweetie is not covered by existing legislation or it is not sufficiently clear that it is. The reasons for this are that (1) the investigative techniques employed by Sweetie may not be used for crimes related to webcam sex tourism, (2) the use of Sweetie clearly does not fit the existing powers, or (3) the existing powers might be usable, but there is no legal precedent.

Clearer rules on the application of Sweetie for investigative purposes will serve both the interest of legal certainty and those of effective law enforcement. By providing more clarity on the legal status of Sweetie, either through legislation, or by testing its legality in court, the proper balance can be found between protecting children and the rights of potential suspects.

Entrapment

Sweetie can be used for the non-targeted and targeted luring of suspects. Basically, the use of Sweetie starts out in a non-targeted form (i.e. Sweetie is a passive 'lure' in a chatroom) and moves to a targeted form (once Sweetie is solicited, she interacts directly with the suspect). Whether these forms of engaging with suspects are legitimate is dependent on the circumstances of the case. Using Article 6 ECHR (fair trial) as a point of departure, we have examined the legality of Sweetie from this perspective.

When it comes to non-targeted entrapment it is important that Sweetie does not alter the existing circumstances (i.e. the chatroom and public chat) in such a way that it provides an opportunity to potential perpetrators that would not have otherwise presented itself. Furthermore, depending on the jurisdiction law enforcement must substantiate that area is a crime hotspot and/or that they have a reasonable suspicion that the crime under investigation is taking place in that area.

When it comes to targeted entrapment it is important that Sweetie does not incite or entice the suspect to commit acts that were not already his/her intention. More specifically the chat script of Sweetie must—amongst others—adhere to the following rules: (1) Sweetie may not propose webcam sex herself, or steer the suspect in that direction, (2) Sweetie may not appeal to the suspect's conscience (e.g. telling the suspect she is a poor kid and needs the money), (3) if a suspect backs down, she may not re-engage the suspect.

1.7.3 *Jurisdiction*

Since webcam sex tourism is a global phenomenon, cross-border investigations are part and parcel of combating webcam sex tourism. This inevitably leads to jurisdictional issues. When it comes to prescriptive jurisdiction, we mainly see a difference between the examined jurisdictions in terms of criminalisation. The global fight against webcam sex tourism would benefit from more harmonisation of substantive criminal law. On the whole though, we do not expect significant issues with prescriptive jurisdiction in terms of crime and penalty havens nor substantial issues surrounding double criminality and mutual legal assistance.

When it comes to enforcement jurisdictions the issues are potentially bigger. We have found that there are significant differences in terms of the regulation of investigative powers throughout the different jurisdictions. In particular, the rules on the use of undercover agents differ from jurisdiction to jurisdiction. This might lead to issues when Sweetie is used extra-territorially, for instance, using Sweetie from the United States in order to catch Dutch webcam sex offenders. Addressing the issue of enforcement jurisdiction and the (unilateral) extra-territorial application of enforcement powers is no small matter. It is therefore more practical to use Sweetie mainly in a domestic context. In other words, using Sweetie only to catch national subjects, not foreigners. Another option is to use the existing mutual legal assistance procedures and to hand over investigations to local law enforcement of suspects' countries.

Acknowledgements The authors of the present study are indebted to Marcos Salt and Daniela Dupuy, Gregor Urbas, Sofie Royer, Gaëlle Marlier and Charlotte Conings, Paloma Mendes Saldanha, Rowan Hodge, Ines Bojić, Alisdair A. Gillespie, Kaspar Kala, Haykush Hakobyan, Asaf Harduf, Uchenna Jerome Orji, Michael Anthony C. Dizon, Ivan Škorvánek, Andrew Richardson, Matthew Kerr and Eamonn Keane, Jose Agustina and Roberto Valverde, Yong Chul Park, Jonathan Unikowski, Murat Önok and Emre Bayamlıoğlu for sharing their valuable expertise and providing the basis for this comparative legal study—the country reports. A heartfelt thank you to all of them for their contribution and cooperation. We would also like to thank the Danish police for giving us insight into the Danish situation (which is not featured as a country study in this report).

References

Ashworth A, Horder J (2013) Principles of criminal law. Oxford University Press, Oxford
August R (2002) International cyber-jurisdiction: a comparative analysis. American Business Law Journal, 39(4): pp. 531–574
Ballin M F H (2012) Anticipative Criminal Investigation. T.M.C. Press, The Hague
Bellia P L (2001) Chasing bits across borders. University of Chicago Legal Forum, pp. 35–101
Bielefeldt H (2012) Philosophical and historical foundations of human rights. In: Krause C, Scheinin M (eds) International protection of human rights: a textbook. Åbo Akademi University Institute for Human Rights, Åbo, pp. 3–18
Brenner S W, Koops B J (2004) Approaches to Cybercrime Jurisdiction. Journal of High Technology Law, 4(1): pp. 189–202
Bronitt S (2004) The law in undercover policing: A comparative study of entrapment and covert interviewing in Australia, Canada and Europe. Common Law World Review, 33(1): pp. 35–80.
Casey E (2011) Foundations of Digital Forensics. In: Casey E (ed) Digital Evidence and Computer Crime, pp. 3–34, Elsevier, London, pp 3–34
Committee of Ministers of the Council of Europe (2001) Explanatory report to the Convention on Cybercrime. CETS Nr. 185, Budapest

Committee of Ministers of the Council of Europe (2007) Explanatory report to the Council of Europe Convention on the Protection of Children against Sexual Exploitation and Sexual Abuse. CETS 201, Lanzarote

Committee on the Rights of the Child (2011) General comment No. 13: The right of the child to freedom from all forms of violence, CRC/C/GC/13

Ferrante M (2010) Argentina. In: Heller K, Dubber M (eds) The handbook of comparative criminal law, pp. 12–48, Stanford University Press, Stanford, pp 12–48

Georgieva I (2015) The right to privacy under fire foreign surveillance under the NSA and the GCHQ and its compatibility with Art. 17 ICCPR and Art. 8 ECHR. Utrecht Journal of International & European Law, 31: pp. 104–130

Gillespie A A (2012) Jurisdictional issues concerning online child pornography. International Journal of Law and Information Technology, 20(3): pp. 151–177

Goldstein R D (1999) Child abuse and neglect: Cases and materials: Cases and Materials (American Casebook Series). West Group

Gómez-Jara C, Chiesa L E (2010) Spain. In: Heller K, Dubber M (eds) The handbook of comparative criminal law, pp. 488–530, Stanford University Press, Stanford, pp 488–530

Harris D (2010) Cases and materials on international law. Sweet & Maxwell, London

Interagency Working Group on the Sexual Exploitation of Children (2016) Terminology guidelines for the protection of children from sexual exploitation and sexual abuse, Luxembourg

Johnson D R, Post D (1996) Law and borders: The rise of law in cyberspace. Stanford Law Review, 48(5): pp. 1367–1402

Kilkelly U (2003) A guide to the implementation of Article 8 of the European Convention on Human Rights. Human rights handbooks, No. 1, Strasbourg

Klabbers J (2013) International Law. Cambridge University Press, Cambridge

Koops B J (2013) Police investigations in Internet open sources: Procedural-law issues. Computer Law & Security Review, 29(6): pp. 654–665

Koops BJ and Goodwin M (2014) Cyberspace, the cloud, and cross-border investigation: The limits of international law. WODC/TILT, The Hague/Tilburg

Lovejoy T P (2007) A new playground: sexual predators and pedophiles online: criminalizing cyber sex between adults and minors. St. Thomas Law Review, 20: pp. 311–358

Robinson P H (2010) United States. In: Heller K, Dubber M (eds) The handbook of comparative criminal law, pp. 563–592, Stanford University Press, Stanford, pp 563–592,

Ryngaert C (2015) Jurisdiction in International Law. Oxford University Press, Oxford

Shaw MN (2014) International Law. Cambridge University Press, Cambridge

Slobogin C (2012) Making the Most of United States v. Jones in a surveillance society; a statutory implementation of mosaic theory. Duke Journal of Constitutional Law & Public Policy, Forthcoming: pp. 12–22

Taslitz A E (2013) Cybersurveillance without Restraint: the meaning and social value of the probable cause and reasonable suspicion standards in governmental access to third-party electronic records. Journal of Criminal Law & Criminology, 103: pp. 839–905

United Nations Office on Drugs and Crime (2013) Comprehensive study on cybercrime, Vienna

Van Hock A A, Luchtman M J (2005) Transnational cooperation in criminal matters and the safeguarding of human rights. Utrecht Law Review, 1: pp. 1–39

Vendius T T (2015) Proactive undercover policing and sexual crimes against children on the internet. European Review of Organised Crime, 2: pp. 6–24

Wilborn S E (1997) Revisiting the public/private distinction: employee monitoring in the workplace. Georgia Law Review, 32: pp. 825–888

Williams S A (1991) The double criminality rule and extradition: a comparative analysis. Nova Law Review, 15: pp. 581–623

Country Reports

Argentina: Salt M, Dupuy D, Substantive and procedural legislation in Argentina to combat webcam-related child sexual abuse, May 2016
Australia: Urbas G, Substantive and procedural legislation in Australia to combat webcam-related child sexual abuse, April 2016
Belgium: Royer S, Marlier G, Conings C, Substantive and procedural legislation in Belgium to combat webcam-related child sexual abuse, updated July 2016
Brazil: Mendes Saldanha P, Substantive and procedural legislation in Brazil to combat webcam-related child sexual abuse, updated May 2016
Canada: Hodge R, Substantive and procedural legislation in Canada to combat webcam-related child sexual abuse, updated June 2016
Croatia: Bojić I, Substantive and procedural legislation in the Republic of Croatia to combat webcam-related child sexual abuse, updated May 2016
England & Wales: Gillespie AA, Substantive and procedural legislation in England & Wales to combat webcam-related child sexual abuse, March 2016
Estonia: Kala K, Substantive and procedural legislation in Estonia to combat webcam-related child sexual abuse, updated May 2016
Germany: Hakobyan H, Webcam sex with (virtual) children: Legislative gaps or criminalised conduct? A legal analysis of Sweetie 2.0 under German substantive criminal law, 2016
Israel: Harduf A, Substantive and procedural legislation in Israel to combat webcam-related child sexual abuse, May 2016
The Netherlands: Schermer B W, Koops B J, Van der Hof S, Substantive and procedural legislation in the Netherlands to combat webcam-related child sexual abuse, 2016
Nigeria: Orji UJ, Substantive and procedural legislation in Nigeria to combat webcam-related child sexual abuse, updated June 2016
The Philippines: Dizon M A, Substantive and procedural legislation in the Philippines to combat webcam-related child sexual abuse, updated May 2016
Poland: Škorvánek I, Substantive and procedural legislation in Poland to combat webcam-related child sexual abuse, May 2016
Scotland: Richardson A, Kerr M and Keane, E, Substantive and procedural legislation in Scotland to combat webcam-related child sexual abuse, May 2016
Spain: Agustina J R, Valverde R, Substantive and procedural legislation in Spain to combat webcam-related child sexual abuse, May 2016
South Korea: Park Y C, Substantive and procedural legislation in South Korea to combat webcam-related child sexual abuse, July 2016
Turkey: Önok M, Bayamlıoğlu E, Substantive and procedural legislation in Turkey to combat webcam-related child sexual abuse, June 2016
USA: Unikowski J, Substantive and procedural legislation in United States of America to combat webcam-related child sexual abuse, May 2016

Legal Instruments

Convention for the Protection of Human Rights and Fundamental Freedoms, CETS no. 194, Rome, 4.XI.1950
International Covenant on Civil and Political Rights (adopted 16 December 1966, entered into force 23 March 1976) 999 UNTS 171
Convention on the Rights of the Child (adopted 20 November 1989, entered into force 2 September 1990) 1577 UNTS 3

Optional Protocol to the Convention on the Rights of the Child on the sale of children, child prostitution and child pornography (adopted on 25 May 2000, entered into force 18 January 2002) A/RES/54/263
Council of Europe Convention on Cybercrime, CETS No.185, Budapest, 23.XI.2001
Council of Europe Convention on the Protection of Children against Sexual Exploitation and Sexual Abuse, CETS 201, Lanzarote, 25.X.2007
Parliament and Council Directive 2014/41/EU of 3 April 2014 regarding the European Investigation Order in criminal matters

Case Law: Australia

Britten v Alpogut [1987] VR 929
Ridgeway v The Queen [1995] CLR 19
O'Neill v R [1995] 81 A Crim R 458
McEwen v Simmons & Anor [2008] NSWSC 1292
R v Priest [2011] ACTSC 18

Case Law: Belgium

Brussel 14 maart 2007, *RABG* 2008
Cass. 17 maart 2010, AR P100010F

Case Law: Canada

R v Mack [1988] 2 SCR 903
R v Alicandro [2009] ONCA 133 (CanLII)
R v Spencer [2014] 2 SCR 212, 2014 SCC 43 (CanLII)

Case Law: Croatia

VSRH, I Kž-1255/04 of 16 February 2006
(Judgement of the Supreme Court)

Case Law: ECtHR

Handyside v the UK, Application no. 5493/72, judgment of 7 December 1976
Klass and others v Germany, Application no. 5029/71, judgement of 6 September 1978
Dudgeon v the UK, Application no. 7525/76, judgment of 22 Oct. 1981
Leander v Sweden, Application no. 9248/81, judgement of 26 March 1987
Olsson v Sweden, Application no. 10465/83, judgment of 24 March 1988
Texeira do Castro v Portugal, Application nos. 44/1997/828/1034, judgement of 9 June 1998
Halford v the UK, Application no. 44787/98, judgement of 25 September 2001
Peck v the UK, Application no. 44647/98, judgement of 28 January 2003
Shannon v the UK, Application no. 67537/01, admissibility decision of 6 April 2004
Von Hannover v Germany [GC], Application no. 59320/00, judgement of 24 June 2004

Weber and Saravia v Germany, Application no. 54934/00, judgement of 29 June 2006
Ramanauskas v Lithuania [GC], Application no. 74420/01, judgement of 5 February 2008
Malininas v Lithuania, Application no. 10071/04, judgement of 1 October 2008
KU v Finland, Application no. 2872/02, judgement of 2 December 2008
Constantin and Stoian v Romania, Applications no. 23782/06 and 46629/06, judgement of 29 December 2009
Burak Hun v Turkey, Application no. 17570/04, judgement of 15 March 2010
Bannikova v Russia, Application no. 18757/06, judgement of 4 February 2011

Case Law: England and Wales

Haughton v Smith [1975] AC 476

Case Law: Estonia

RKKKo 3-1-1-110-04
(Judgment of the Supreme Court of 2 December 2004, case number 3-1-1-110-04)

Case Law: Israel

Ktiei v Israel, LCrimA 1201/12 [9 January 2014]

Case Law: The Netherlands

HR 4 December 1979, NJ 1980, 356 m.nt ThWvV
HR 30 Nov 2004, ECLI:NL:HR:2004:AQ0950
HR 28 Oct 2008, ECLI:NL:HR:2008:BE9817
HR 6 Oct 2009, ECLI:NL:HR:2009:BI7084
(Supreme Court Judgements)
Rb. Groningen 16 October 2003, LJN AM1882, case no. 18/076010-01
Rb. Alkmaar 19 February 2004, LJN A05509, case no. 14.060137-02

Case Law: The Philippines

Araneta v Court of Appeals, G.R. No. L-46638 [9 July 1986]
People v Gatong-o, G.R. No. 78698 [29 December 1988]

Case Law: Scotland

Docherty v Brown [1996] JC 48
Jones v HM Advocate [2009] HCJAC 86

Case Law: South Korea

DO7362, Oct. 23, 2008
(Judgement of the Supreme Court)

Case Law: Spain

Judgement of the Supreme Court of Justice 300/2015, of 19 May (Spain)
SSCS June 22, 1950
SSCS April 18, 1972
SSCS February 16, 2007, available at www.westlaw.es, Ref: RJ 2007\2381.

Case Law: United States

Lopez v United States, 373 U.S. 427 (1963)
Hoffa v United States, 385 U.S. 293 (1966)
Katz v United States, 389 U.S. 347 (1967)
United States v Miller, 425 U.S. 435 (1976)
Smith v Maryland, 442 U.S. 735 (1979)
Mathews v United States, 485 U.S. 58 (1988)
Alabama v White, 496 U.S. 325 (1990)
State v Moretti, 244 A.2d 499 (N.J. 1968)
United States v Gorshkov, No CR00-500C (W D Wash May 23, 2001)
United States v Jackson, 488 F. Supp. 2d 866 (D. Neb. 2007)

Bart Schermer is an associate professor at eLaw, the Center for Law and Digital Technologies at Leiden University, and a fellow at the E.M. Meijers Institute for Legal Studies. He specializes in privacy, data protection and criminal law. Apart from his work at the University Bart is Chief Knowledge Officer at Considerati, member of the Cybercrime Expert Group for the Dutch judiciary and member of the Human Rights Committee of the Advisory Council on International Affairs.

Ilina Georgieva is a Ph.D. candidate of The Hague Programme for Cyber Norms. In her research, Ilina is focusing on the capacity of networks of intelligence agencies to shape the international community's perception of what is normal in cyberspace. For that purpose, she investigates the networks' normative power by looking into their practice of foreign electronic surveillance. Prior to joining the Institute of Security and Global Affairs, Ilina served as a researcher on the Sweetie Project at eLaw, the Center for Law and Digital Technologies at Leiden University. Her research encompassed a comparative legal study concerning the trans-border investigation of Internet sexual crimes against children. Before joining eLaw's team, she worked as an editor at the Utrecht Journal of International and European Law (October 2013–September 2014). Ilina was also a part of Heidelberg University's Cluster of Excellence "Asia and Europe in a Global Context" (December 2012–August 2013) and of the Austria Institute for European and Security Policy (summer of 2012) in her capacity as a research assistant. From January 2009 to June 2010 she worked at the Max-Planck Institute for Comparative Public Law and International Law in Heidelberg. She also served as a Senior Research Associate and later on as a Counsel for the Public International Law and Policy Group (PILPG) from September 2014 to October 2016.

Simone van der Hof is the director of the Center for Law and Digital Technologies (eLaw) at Leiden Law School, programme director of the Advanced Studies Programme in Law and Digital Technologies and one of the directors of the Leiden Law School research profile area Interaction between legal systems. She coordinates and teaches various courses, amongst which 'Regulating online child safety' (Master Youth Law), 'Digital Child Rights' (Advanced Master Law and Digital Technologies), 'The Rights of the Child in the Digital World' (Advanced Master International Children's Rights). She is a key lecturer at the Cyber Security Academy. Simone's particular academic interest is in the field of online privacy, digital child rights and regulation of online child safety. She was involved in the Sweetie 2.0 project on online webcam child sex abuse, commissioned by children's rights organization Terre des Hommes as well as a project on the levels of protection of personal data of European citizens. She participates in the SCALES project (big data and privacy) and leads the ethics by design work package of the Gamechangers project on the development of health games for children.

Bert-Jaap Koops is Professor of Regulation and Technology at the Tilburg Institute for Law, Technology, and Society (TILT), the Netherlands. His main research fields are cybercrime, cyber-investigation, privacy, and data protection. He is also interested in topics such as DNA forensics, identity, digital constitutional rights, 'code as law', and regulatory implications of human enhancement, genetics, robotics, and neuroscience. With a personal postdoc (1999), VIDI (2003) and VICI (2014) grant, Koops is one of the few Dutch researchers who received all three stages of NWO's (Netherlands Organisation for Scientific Research) personal research-grant scheme. Koops studied mathematics and general and comparative literature at Groningen University, and received his PhD in law at Tilburg University in 1999. From 2005 to 2010, he was a member of De Jonge Akademie, a young-researcher branch of the Royal Netherlands Academy of Arts and Sciences. In 2016/17, he was Distinguished Lorentz Fellow at the Netherlands Institute for Advanced Study (NIAS). He co-edited 13 books in English on technology regulation and published many articles and books in English and Dutch on a wide variety of topics.

Chapter 2
Sexual-Orientated Online Chat Conversations—Characteristics and Testing Pathways of Online Perpetrators

Manon Kleijn and Stefan Bogaerts

Contents

Abstract Children all over the world are constantly at risk of becoming a victim of webcam child sex tourism (hereafter WCST). Law enforcement agencies are struggling with tackling the problem. Little is known about the behaviour of online

M. Kleijn (✉) · S. Bogaerts
Tilburg University, Tilburg, The Netherlands
e-mail: m.kleijn@uvt.nl

S. Bogaerts
e-mail: s.bogaerts@uvt.nl

S. van der Hof et al. (eds.), *Sweetie 2.0*, Information Technology and Law Series 31,
https://doi.org/10.1007/978-94-6265-288-0_2

perpetrators who persuade minors to engage in webcam sex. In order to gain more information about the behaviour of this group, the ten-year-old Philippine avatar Sweetie was deployed as a tool. In this first preliminary study 200 chat conversations were analysed. Based on the first findings, almost all online perpetrators were male with an average reported age of 29 years old living in Europe, North America and Asia. Two types of perpetrators were found: the Clint-type, who had clear intentions to talk about sex related topics or about starting a webcam conversation to persuade the minor to engage in (online) sexual activities; and the Small talk type, who used small talk to persuade the minor to engage in (online) sexual activities. This study shows that online perpetrators can be classified into perpetrators who display avoiding and approaching behaviour, which is comparable to the offline offending pathways of Ward and Hudson.

Keywords Online Perpetrators · Webcam Child Sex Tourism · Child Sexual Exploitation · Offending Pathways · Internet · Chat Conversations

2.1 Introduction

In the last decades, the Internet has become a strong positive driver to strengthen social contacts. It functions as a web of networks and allows us to connect through social media platforms with people at any moment regardless of the geographical location in the world. Unfortunately, this web of networks also has a down side. Internet can be used for online sexual solicitation and exchange of illicit material. Not only for sexual purposes between people who consent voluntarily in these activities, but also for adults to approach vulnerable children for sexual activities.[1] It is a dangerous environment for children who make themselves vulnerable by posting personal identifiable information on public websites, such as MySpace and Facebook, and who engage in conversations with strangers online.[2]

Before the advent of the Internet, perpetrators found their victims in playgrounds or schools. In the digital age, they search for profiles of minors online and sexually solicit them to engage in sexual activities, such as webcam sex.[3] WCST includes men who offer money to minors to perform sexual activities on the Internet via webcam.[4] Precise numbers of minors being approached to do these unwanted

[1] Schermer et al. 2016, p. 6.

[2] Wolak et al. 2008, p. 111.

[3] Marcum 2007, p. 100.

[4] Schermer et al. 2016, pp. 6, 18.

sexual activities are unknown, but it can be assumed that it takes place on an enormous scale. Rough estimates from the Federal Bureau of Investigation in 2009 indicate that every day more than 750,000 men worldwide are online searching for sexual activities with minors in more than 40,000 chatrooms[5] and this number has increased ever since.[6] Instead of thousands of convictions, law enforcement agencies are struggling with identifying the involved perpetrators.[7] The Darknet, an encrypted part of the Internet, increases the security of these perpetrators.[8] Besides, the transnational aspect of the phenomenon and the conflicting legal framework of countries make it even more difficult to trace perpetrators.[9] Additionally, the online child pornography industry becomes more abroad with the involvement of criminal organisations that particularly operate in underdeveloped countries, such as the Philippines and Kenya, where high poverty, inadequate laws and lack of educational opportunities for children exist. Family members in these countries are often involved in the industry by forcing their children to contribute to the financial and economic survival of their family by making money in the webcam child sex industry.[10] In short, the Internet has transformed the child pornography industry radically when it comes to nature and scale, leading to various new challenges. Research on the behaviour of online perpetrators is therefore more needed than ever before.

2.1.1 Characteristics of Online Perpetrators

Although the black box still exists and descriptive studies about the behaviour and strategies of online perpetrators are limited,[11] it must be acknowledged that there is already important knowledge about these perpetrators. Online perpetrators are typically adult white men between 25 and 45 years old, educated and employed.[12] They can be divided into several groups, based on the type of offending, such as the fantasy-driven versus contact-driven offender,[13] based on the motivation behind the offending, such as the situational and preferential offender[14] and based on the social

---

[5] Further information on the project known as 'Sweetie 1.0' can be found on https://www.terredeshommes.nl/en/sweetie-face-webcam-child-sex-tourism.

[6] United States Department of Justice 2010, p. 36.

[7] Taylor and Quayle 2003, pp. 204–208.

[8] Broséus et al. 2016, p. 7.

[9] Schermer et al. 2016, p. 6.

[10] Regional Overview: Sexual exploitation of children in Southeast Asia. Report retrieved from: http://www.ecpat.org/wp-content/uploads/2018/02/Regional-Overview_Southeast-Asia.pdf.

[11] Houtepen et al. 2014, p. 467; Van Wijk et al. 2009, p. 49.

[12] Briggs et al. 2011, p. 75; Van Wijk et al. 2009, p. 49.

[13] Briggs et al. 2011, pp. 14, 15; Merdian et al. 2013, pp. 124–126.

[14] Lanning 2010, pp. 32, 33.

environment of the offending, such as the browser type and the online groomer.[15] According to the strategies to approach minors online, Malesky[16] found that perpetrators search for attractive profiles of minors online to identify potential victims. They use several communication platforms, such as chatrooms and instant messages,[17] and try to build a relationship with the minors, also known as the grooming process.[18] They often begin with discussing hobbies and other non-sexual topics. Besides, they use youthful expressions and emoticons to flatter minors and manipulate them by promising love and marriage.[19] After some chatting, perpetrators bring sexual content into the conversation. They might even send child pornography material.[20] To further examine how online perpetrators approach minors online, we should build a bridge between knowledge about the characteristics of online perpetrators on the one hand and concrete strategies (or pathways) of online perpetrators on the other hand.

2.1.2 Pathway Model for Perpetrators

The model that might clarify online child sex offending is the offending pathways model of Ward and Hudson.[21] They suggest four basic pathways to explain how perpetrators control their sexual offending behaviour according to their goals towards sex (avoidant vs. approach) and the strategies they use to achieve these goals (passive vs. active). The *avoidant-passive* pathway suggests that the perpetrator has the desire to avoid sexual offending but cannot control it from happening because of the lack of adequate self-regulation skills. He lacks coping skills, acts impulsively and makes irrelevant decisions, such as buying a newspaper in a shop full of children as a child abuser.[22] The *avoidant-active* pathway suggests that the perpetrator wants to avoid sexual offending. He tries to control his deviant sexual thoughts and fantasies, for example by masturbating on these fantasies. However, because of the lack of insights and knowledge, instead of reducing his urges, he increases the likelihood of offending.[23] The *approach-automatic* pathway suggests that the perpetrator responds to situational and emotional cues to regulate his arousal. However, with poorly planned behaviour he acts impulsively and commits

15 Briggs et al. 2011, pp. 16, 17; Krone 2004, p. 4; Wortley and Smallbone 2006, pp. 16, 17.

16 Malesky 2007, p. 28.

17 Wolak et al. 2008, p. 112.

18 Briggs et al. 2011, p. 4; Marcum 2007, p. 108.

19 Briggs et al. 2011, p. 75; Dehart et al. 2016, p. 78; Kloes et al. 2017, pp. 570–571; Marcum 2007, p. 100.

20 Malesky 2007, p. 28.

21 Ward and Hudson 1998.

22 Bickley and Beech 2002, p. 375.

23 Ward and Hudson 1998, pp. 710–713.

an offence.[24] The *approach-explicit* pathway suggests that the perpetrator wants to sexually abuse others and has perfect self-regulation skills to do it.[25] In the beginning not everyone was convinced of the model of Ward and Hudson. However, over the years sufficient support was found for their model. Bickley and Beech[26] validated the model and concluded in their study that all 87 child abusers could be reliable categorised into one of the four pathways. Webster[27] tested the model with 25 sex perpetrators and found that 21 individuals could be categorised according to the four pathways. As far as we know, this pathway model has not been empirically tested for online perpetrators. Understandably, because investigation on victim-perpetrator interactions and offence approaches can only be made if online chatroom conversations are available.

2.1.3 Modus Operandi: Opening the Black Box of Online Sexual Behaviour Strategies

It has become clear that perpetrators develop different offending pathways. However, whether Ward and Hudson's pathways of offline offending are also applicable to online perpetrators still remains unclear. Until today, it has not previously been tested whether the pathways model also applies to online perpetrators. Testing offline offending pathways in an online environment is a challenge that will be part of this study. Increasingly, chat conversations become available for researchers as a result of the open source character of many chatrooms in which children are active and come into contact with adults having undesired sexual intentions. These available conversations can be very helpful in providing information about online strategies of perpetrators. Also, there is a growing awareness and pressure of organisations and law enforcement in the fight against sexual online abuse of vulnerable minors allowing online chat conversations to be made available. A pioneer in this area is the Dutch children's rights organisation Terre des Hommes, which has taken action worldwide against online sexual abuse of vulnerable children, especially in underdeveloped countries.[28] The accessibility of chat conversations allows researchers to gain insights into online strategies of perpetrators.

---

24 Ibid, pp. 710–713.

25 Ward and Hudson 1998, pp. 710–713.

26 Bickley and Beech 2002, pp. 380–381.

27 Webster 2005, pp. 1192–1193.

28 Further information can be found at www.terredeshommes.nl.

2.1.4 Current Study

This explorative preliminary study focussed on examining the behaviour of perpetrators in terms of avoiding and approaching behaviour as previously studied by Ward and Hudson.[29] The first aim of the study was to analyse the conversation styles of online perpetrators by focussing on their characteristics and strategies. The second aim of the study was to test whether the specific offending pathways of Ward and Hudson[30] could be found. We assumed that online perpetrators would apply more approach-active strategies in chatrooms than avoiding-passive strategies to entice minors to engage in online sexual activities.

2.2 Method

2.2.1 Perpetrator Demographic Characteristics

The sample consisted of 199 male perpetrators and one female perpetrator between the reported age of 14 and 60 years old (M = 28.9, *SD* = 9.7). The majority of the perpetrators lived in Europe (32.5%), mostly in England, Germany and Turkey; North America (28.0%), mostly in Canada and the United States, and Asia (14.5%), mostly in India and Dubai. The demographics are presented in Table 2.1.

2.2.2 Material and Procedure

In 2013, Terre des Hommes and the animation company Motek Entertainment BV designed the virtual avatar called *Sweetie*. The avatar was profiled as a ten-year-old girl living in the Philippines and forced by her family to engage in conversations with men who searched for webcam sex with minors online. Four operators were located in a warehouse in the capital of the Netherlands, Amsterdam. They used the avatar to communicate with perpetrators in different chatrooms. The operators used several usernames, such as 'Julia_10_f', presuming to be a 10-year-old female girl called Julia. The operators communicated with more than 20,000 perpetrators in 19 different chatrooms during a period of 10 weeks. Of these 20,000 perpetrators, 1,000 perpetrators were identified after cross-referencing the personalia of the men with open sources. The collected information was anonymised and merged into cases. Each case consisted of chat logs between the operator and the perpetrator. For this study, 200 cases were randomly selected and examined in this study.

[29] Ward and Hudson 1998.

[30] Ibid.

Table 2.1 Perpetrator demographic characteristics

	Perpetrators (N = 200)
Age N = 183	
Mean	28.9
Minimum	14.0
Maximum	60.0
SD	9.7
Gender % (N)	
Male	99.5 (199)
Female	0.5 (1)
Continental % (N)	
Africa	3.0 (6)
Asia	14.5 (29)
Australia	
Europe	32.5 (65)
North America	28.0 (56)
Oceania	8.0 (16)
South America	2.5 (5)
Missing	11.5 (23)

[*Source* The authors]

2.2.3 *Data Coding*

The cases were examined using a coding scheme developed by the researchers based on a deductive (review of relevant literature) and inductive (analysing the cases) approach. The collected data was quantified in SPSS by using dichotomous variables to indicate absence (0) or presence (1) of particular variables. Two main categories of variables were coded: demographic characteristics and perpetrator strategies. Demographic characteristics were divided into three variables, including age, gender and location of the perpetrator. The perpetrator strategies were divided into five variables: whether the perpetrator had clear intentions to talk about sex related topics, about a webcam conversation or about continuing to Yahoo or other private chatrooms, whether the perpetrator started the conversation with small talk, whether the perpetrator acknowledged the age of the minor, whether he wanted to have a webcam sex show and whether he was willing to pay money for it. According to the main categories, we examined if pathways could be found.

2.2.4 *Analytic Approach*

All statistical analyses were conducted using SPSS 23. Data was screened for missing values, outliers and normality. All variables were normally distributed using the Skewness and Kurtosis tests. Summarising descriptive statistics and frequencies were provided to describe the demographic characteristics and strategies of the perpetrators. To test the correlation between the different variables, Pearson correlations were performed. To examine whether different offending pathways could be found, logistic regressions were performed.

2.3 Results

2.3.1 *Perpetrator Strategies*

Table 2.2 shows information about the perpetrator strategies. In 62.5% of all cases, the perpetrator had clear intentions to immediately talk about sex related topics, about starting a webcam conversation or about identifiers, such as Yahoo and Skype. The perpetrator asked the minor questions such as "Do you have Skype or Yahoo?" and "Do u want to cam?" and asked sex related questions, such as "Do u do dirty cam", "I want to lick you small girl?", "Whats ur size? Are u wear something under ur shirt" and "Will my baby get undressed?". In 33.0% of all cases, the conversation started with small talk, such as about the weather, hobbies or other non-sex related topics, before sex related topics were discussed. In 95.0% of all cases, the perpetrator acknowledged the age of the minor. One perpetrator said: "Are u 10 years? Mmm I love that. Describe yourself". In 75.5% of all cases, the perpetrator wanted to have a webcam sex show and in 31.5% of the cases, the perpetrator wanted to pay money for it. One perpetrator said: "Are u still a virgin? If I pay u, can I break you in? Have full sex with you. I will pay good money". In 53.0% of the cases, there was no discussion about money. One perpetrator did not want to pay money and said: "Why pay for something u can get for free".

2.3.2 *Perpetrator Strategies with Correlations*

Table 2.3 shows the correlations of the perpetrator strategies. Significant negative and positive correlations were found between the clear intentions of the perpetrator and the clear intentions of Sweetie ($r = -0.27$, $p < 0.01$), small talk ($r = -0.88$, $p < 0.01$), the perpetrator who asks for webcam ($r = 0.33$, $p < 0.01$) and the perpetrator who wants a webcam sex show ($r = -0.22$, $p < 0.01$). When the perpetrator had clear intentions to talk about sex related topics, webcam or identifiers, the conversation was less likely to start with small talk, it was more likely that the perpetrator asked for webcam and it was less likely that he admitted that he wanted to have a webcam sex show. Negative correlations were found between the perpetrator asking for webcam, and the perpetrator asking for identifiers ($r = -0.22$, $p < 0.01$), age, sex and location ($r = -0.18$, $p < 0.05$) and small talk ($r = -0.21$, $p < 0.01$).

Table 2.2 Perpetrator strategies

Total	N = 200
Clear intentions perpetrator % (N)	
Total	62.5 (125)
Sex related topics	29.5 (59)
Webcam	22.5 (45)
Identifiers	10.5 (21)
Small talk % (N)	33.0 (66)
Acknowledges age minor % (N)	
Yes	95.0 (190)
Unknown	5.0 (10)
Wants webcam sex show % (N)	
Yes	75.5 (151)
No	2.0 (4)
Unknown	22.5 (45)
Wants to pay money % (N)	
Yes	31.5 (63)
No	8.0 (16)
Maybe	7.5 (15)
Missing	53.0 (106)

[*Source* The authors]

When the perpetrator asked for webcam, it was less likely that he asked about identifiers, age, sex and location and used small talk. Significant positive and negative correlations were found between the perpetrator who wants a webcam sex show and asking for age, sex and location ($r = 0.21$, $p < 0.01$), small talk ($r = 0.18$, $p < 0.05$), asking for webcam (r = −0.15, $p < 0.05$) and wanting to pay money for webcam sex ($r = 0.28$, $p < 0.01$). When the perpetrator wanted to have a webcam sex show, it was more likely that he asked about the age, sex and location of the minor, it was less likely that the conversation started with small talk, it was less likely that the perpetrator asked for webcam himself and that he reported that he wanted to pay money for it. There was a positive correlation between the perpetrator asking for age, sex and location and asking for identifiers ($r = 0.24$, $p < 0.01$). Perpetrators who asked for age, sex and location were more likely to ask for identifiers as well.

2.3.3 Perpetrator Strategies with Qualitative Examples

A first global overview of the data resulted in the discovery of two types of perpetrators with different strategies to approach minors online. The first type is the Clint-type, who has clear intentions to talk about sex related topics or about starting a webcam conversation to entice the minor to engage in (online) sexual activities. The second type is the Small talk type, who starts the conversation with discussing hobbies and non-sex related topics, before introducing sex related topics and enticing the minor to engage in (online) sexual activities.

To illustrate these two types of perpetrators, examples of chat conversations are shown between the perpetrator (P) and the operator (O). The Clint-type immediately introduced sex related topics into the conversation. For example, he wanted to show his genitals via webcam or asked the minor to give a webcam sex show. The perpetrator in Example 1 wanted to masturbate in front of the webcam. The operator asked a few non-sex related questions, but the perpetrator was only focussed on one thing: to masturbate in front of the webcam.

Example 1 The Clint-type

P:	Age?
O:	10 u?
P:	24
P:	U like older?
O:	Yes is ok
P:	U wanna watch me jerk off?
(…)	
P:	Go on now
P:	We try here?
P:	I wanna see u
(…)	
P:	Its ok I just wanna show u my cock.
(…)	
P:	I wanna cum for you.

The Small talk-type started the conversation with non-sex related topics, such as the weather and hobbies. Some perpetrators wanted to build a relationship with the minor by flattering her and telling her that she was cute and showing her love and affection. For example, one perpetrator said: "how are u cutie? U sound cute". Another perpetrator said: "Can I pick u up & hug u and kiss u?". Example 2 is an example of a perpetrator who used small talk and eventually talked about sex related topics and offered to visit the minor in real life.

Example 2 The Small talk-type

P:	Im at home
P:	What u do?
(…)	
P:	What u study?
P:	Ur in vacation?

(continued)

Table 2.3 Perpetrator strategies with correlations

	Clear intentions perpetrator	Perpetrator asks identifiers	Perpetrator asks age, sex, location	Small talk	Perpetrator asks webcam	Acknowledges age minor	Wants webcam sex show	Wants to pay money
Clear intentions perpetrator	–	−0.01	−0.12	−0.88**	0.33**	−0.08	−0.22**	−0.11
Perpetrator asks identifiers		–	0.24**	0.01	−0.22**	0.05	−0.03	−0.01
Perpetrator asks age, sex, location			–	0.13	−0.18*	0.08	0.21**	0.03
Small talk				–	−0.21**	0.11	0.18*	0.09
Perpetrator asks webcam					–	−0.03	−0.15*	0.02
Acknowledges age minor						–	0.05	0.00
Wants webcam sex show							–	0.28**
Wants to pay money								–

[*Source* The authors]
Note Ns = Not significant
$^{**}p < 0.01$, $^{*}p < 0.05$

(continued)

(…)	
P:	Do u see dick before?
(…)	
P:	Mmmm do u fingering ur pussy?
(…)	
P:	Can I cum and lick?
(…)	
P:	Do u like im visit u?

Perpetrators often asked questions about sex related topics, such as questions about sexual features (e.g. breast size, nipples and pubic hair). They acknowledged the age of the minor and had no problem with it. One perpetrator said: "Do you have no boobs? Don't worry. You will have in 2 years. I love to lick no boobs". Example 3 is an example of a perpetrator who wanted to know everything about the sexual features of the minor.

Example 3 Sex related topics

P:	You have small boobs?
O:	No boobs yet
P:	ur nipples what color?
P:	You know nipples?
O:	Pink
P:	Waw
P	I like suck small nipples
P:	Can I suck ur nipples?
(…)	
P:	I wanna see you
(…)	
P:	Show me ur webcam

Some perpetrators had the intention to meet the minor in real life by visiting her in the Philippines. One perpetrator asked: "I am going to the Manila in July… How much would u like for 5 days with me when I am in the Manila?". Most of the perpetrators believed that they were talking to a minor girl. However, a few perpetrators did not believe she was a real girl. One perpetrator said: "No it is obvious that you are a man pretending". Also, some perpetrators lied about their age (Example 4). After cross-referencing the given personalia of the perpetrators with open sources, some perpetrators appeared to be older than they reported during the conversations.

Example 4 Lying about age

P:	Oh
P:	I'm 17
O:	Only 17?
O:	I like older
P:	Haha
P:	No. I'm 23 really

Some perpetrators sent child pornography material or a picture of their own genitals. One perpetrator said: "U like older guys? Want to get naked with me;)?". After that, he sent a picture of his genitals. According to the usernames of the perpetrators, some perpetrators used general names, such as 'User70' or 'Male'. Others used usernames that indicated that they had clear sexual intentions, such as 'SexyBigC**K', 'Dadpedo', 'ILoveBadLittleGirlsM' or 'DAD NEEDS DAUGHTER CAM2CAM'.

2.3.4 Perpetrator Pathways

Table 2.4 shows the results of the logistic regression analysis. All 200 cases were included in the analysis and the model correctly classified 94.5% of the clear intentions of the perpetrators. The independent variables 'small talk' and 'perpetrator asks webcam' contributed significantly to the dependent variable 'clear intentions perpetrator'. Perpetrators who used small talk in the beginning of the conversation were less likely to have clear intentions to talk about sex related topics compared to those who did not have clear intentions. Perpetrators who asked for webcam during the conversation, were more likely to have clear intentions to talk about sex related topics compared to those who did not have clear intentions.

2.4 Discussion

In this study, we examined the behaviour of online perpetrators. The first aim of the study was to analyse the conversation styles of online perpetrators. Overall, we found that most of the perpetrators were males between the age of 14 and 60 years old living in Europe, North America and Asia. In more than half of all cases, the perpetrators had clear intentions to immediately talk about sex related topics, about starting a webcam conversation or about identifiers. These perpetrators used several sexual and aggressive strategies to entice the minor to engage in sexual activities. They did not use small talk. Their only goal was to engage in online sexual activities, for example by asking the minor for a webcam conversation. Some of the

Table 2.4 Perpetrator pathways

	95% CI for the exp. b				
	B (SE)	Sig.	Lower	Exp. b	Upper
Included					
Constant	2.79 (1.91)	0.14			
Perpetrator asks identifiers	0.56 (0.80)	0.48	0.37	1.75	8.36
Perpetrator asks age, sex, location	0.96 (0.81)	0.24	0.54	2.61	12.70
Small talk	−8.75** (1.59)	0.00	0.00	0.00	0.004
Perpetrator asks webcam	3.87* (1.16)	0.00	4.87	47.68	466.87
Acknowledges age minor	−0.52 (1.47)	0.72	0.03	0.59	10.56
Wants webcam sex show	−1.01 (0.87)	0.25	0.07	0.37	2.02
Wants to pay money	−0.41 (0.52)	0.43	0.24	0.66	1.85

[*Source* The authors]
Note $R^2 = 0.65$ (Cox & Snell)
$^{**}p < 0.01$, $^{*}p < 0.05$

perpetrators almost immediately asked for identifiers, such as Yahoo or Skype, to go to this 'safer' chat platform to continue the conversation and to propose to engage in sexual activities via webcam. This is in line with Briggs et al.[31] and Kloes et al.[32] who concluded in their studies that online perpetrators use directive conversation styles to discuss sex related topics and to show their genitals and perform masturbation via webcam. In one third of all cases, the perpetrators started the conversation with small talk. In these cases, the perpetrators for example talked about the weather and hobbies with the intention to build trust and to make the minor feel comfortable. They showed love and affection towards the minor and some of them lied about their age or pretended to be a minor to win the minor's trust. Some of the perpetrators even offered to start a relationship with the minor. Eventually, almost all perpetrators started to talk about sex related topics and about a webcam sex show. This is in line with Malesky[33] and Marcum[34] who concluded in their study that perpetrators often groom their victims by discussing hobbies and non-sex related topics and by showing love and affection, to build trust and to eventually persuade their victims to engage in sexual activities. In almost all cases, the perpetrators acknowledged the age of the minor. In two third of all cases, the perpetrators wanted to have a webcam sex show and in some of these cases they were willing to pay money for it. In some cases, the perpetrators even wanted to visit the minor in the Philippines and wanted to pay money for it. Some perpetrators used usernames that indicated that they had clear intentions to have sexual activities

[31] Briggs et al. 2011, p. 11.

[32] Kloes et al. 2017, pp. 570–571.

[33] Malesky 2007, p. 28.

[34] Marcum 2007, pp. 108, 112.

with minors and some perpetrators sent child pornography material or pictures of their genitals.

The second aim of the study was to test whether specific offending pathways could be found. We found indications for two pathways used by the perpetrators. The first pathway found in this study, the Clint-pathway, can be seen as a more active goal-directed pathway. This perpetrator has clear intentions to talk about sex related topics or about starting a webcam conversation. He uses a highly sexual and aggressive approach, as he shows his genitals, sends child pornography material and/or wants to have online sexual activities, such as a webcam sex show. This pathway is consistent with the *approach-explicit* pathway of Ward and Hudson,[35] since the perpetrator wants to sexually abuse the minor and has the skills to do it. He manipulates the minor and persuades her to engage in sexual activities. The second pathway found in this study, the Small talk-pathway, can be seen as a more mixed pathway of avoidant and approach strategies. This perpetrator uses non-sex related topics and tries to build a relationship with the minor. He flatters her and makes her feel comfortable. When the minor feels comfortable, he starts discussing sex related topics and eventually has the goal to entice the minor to engage in online sexual activities. This pathway is consistent with the *approach-explicit* pathway of Ward and Hudson,[36] since the perpetrator wants to sexually abuse minors and knows how to do it by using small talk, flattery and manipulation to entice the minor to engage in sexual activities.

2.5 Limitations

The results of the study should be viewed with several limitations. First, we should recognise that it is not clear whether the findings of this study are based on facts. It is impossible to verify if all the information the perpetrators provided is based on the truth. However, the aim of this study was not to find the truth. The aim was to analyse the conversation styles of online perpetrators and examine their characteristics and strategies. All findings are based on the information provided by these perpetrators.

A second consideration is the fact that we used dichotomous variables to analyse the data. This might be criticised for missing answer categories and missing information about the perpetrators. However, the participants did not know that they were participating in this study. Besides, the only available information we had was the information the perpetrators shared during the chat conversations. Therefore, we could only make a distinction between absent and present information. Caution must be exercised when generalising the results of this study. To

[35] Ward and Hudson 1998, pp. 712–713.

[36] Ward and Hudson 1998, pp. 712–713.

get a better understanding of the strategies of online perpetrators, a qualitative data-analysis is recommended.

A third limitation relates to the offending pathways by Ward and Hudson[37] and whether these pathways would also relate to online sexual perpetrators. Until today, these pathways were only tested on offline perpetrators. Since there was no other theoretical model about pathways for online perpetrators available, we tested these offline pathways on our online environment. Since chatroom conversations become more available for researchers, we can affirmatively say that this will help us to gain more insight in the strategies of online perpetrators in future research.

2.6 Conclusion

The Internet functions as a worldwide platform for online perpetrators to commit WCST. These online small talkers and clear intention types use different strategies, from building trust and showing love to aggressive tactics to talk about sex and to persuade minors to engage in online sexual activities. In order to stop WCST, pro-active policing is needed. We need to focus more on the profiles of the perpetrators who groom minors for sexual activities. Besides, in order to minimise the risk of sexual solicitation of minors online, minors (and more importantly their parents) need to understand the dangers of online interaction in chatrooms. Minors should not share personal identifiable information on public websites and should not engage in sexual conversations with people from whom there is no possible way to verify their real identity and intentions. The impact of becoming a victim of WCST can be disastrous. Therefore, we need to become aware of the strategies of online perpetrators and support children in how to behave online. This will have benefits for police investigations, especially in underdeveloped countries, such as the Philippines and Kenya. The world is in need of awareness of the problem, conviction of the perpetrators and, most importantly, the protection of vulnerable minors.

References

Bickley JA, Beech AR (2002) An investigation of the Ward and Hudson pathways model of the sexual offense process with child abusers. Journal of Interpersonal Violence 17: 71–393

Briggs P et al (2011) An exploratory study of Internet-initiated sexual offenses and the chat room sex offender: Has the Internet enabled a new typology of sex offender? Sexual Abuse: A Journal of Research and Treatment 23: 72–91

Broséus J et al (2016) Studying illicit drug trafficking on Darknet markets: structure and organisation from a Canadian perspective. Forensic Science International 264: 7–14

[37] Ward and Hudson 1998, pp. 711–713.

Dehart D et al (2016) Internet sexual solicitation of children: a proposed typology of offenders based on their chats, e-mails, and social network posts. Journal of Sexual Aggression 23: 77–89

Houtepen JABM et al (2014) From child pornography offending to child sexual abuse: a review of child pornography offender characteristics and risks for cross-over. Aggression and Violent Behavior 19: 466–473

Kloes JA et al (2017) A qualitative analysis of offenders' modus operandi in sexually exploitative interactions with children online. Sexual Abuse 29: 563–591

Krone T (2004) A typology of online child pornography offending. Trends and Issues in Crime and Criminal Justice 297: 1–6

Lanning KV (2010) Child Molesters: A behavioral analysis, 5th edn. National Center for Missing & Exploited Children, Alexandria VA

Malesky LA (2007) Predatory online behavior: modus operandi of convicted sex offenders in identifying potential victims and contacting minors over the Internet. Journal of Child Sexual Abuse 16: 23–32

Marcum CD (2007) Interpreting the intentions of Internet predators: An examination of online predatory behavior. Journal of Child Sexual Abuse 16: 99–114

Merdian HL et al (2013) The three dimensions of online child pornography offending. Journal of Sexual Aggression 19: 121–132

Schermer BW et al (2016) Legal aspects of Sweetie 2.0 Center for Law and Digital Technologies/ Tilburg Institute for Law, Technology, and Society, Leiden/Tilburg

Taylor M, Quayle E (2003) Child Pornography: An Internet Crime. Brunner Routledge, New York

United States Department of Justice (2010) The national strategy for child exploitation prevention and interdiction: A report to Congress. USDOJ, Washington DC

Van Wijk APH et al (2009) Achter de schermen. Een verkennend onderzoek naar downloaders van kinderporno. Bureau Beke, Arnhem

Ward T, Hudson SM (1998) A model of the relapse process in sexual offenders. Journal of Interpersonal Violence 13: 700–725

Webster SD (2005) Pathways to sexual offence recidivism following treatment: an examination of the Ward and Hudson self regulation model of relapse. Journal of Interpersonal Violence 20. 1175–1196

Wolak J et al (2008) Online 'predators' and their victims: myths, realities, and implications for prevention and treatment. American Psychologist 63: 111–128

Wortley R, Smallbone S (2006) Child pornography on the Internet. Problem-oriented guides for police, problem specific guides series, 41. US Department of Justice, Washington DC

Manon Kleijn is one of the researchers of Tilburg University working on the Sweetie Project, where she examines the characteristics and strategies of online child pornography offenders. She was a teacher at the department of Criminology at the Erasmus University in Rotterdam. Currently, she is a Ph.D. student at Fivoor Science and Treatment Innovation and Tilburg University. Her Ph.D is about online and offline grooming and child pornography in sex offenders.

Stefan Bogaerts is a psychologist, criminologist and psychotherapist PGG. He is full professor of forensic psychology and is affiliated with the department of developmental psychology at Tilburg University. He is chair of the department and academic director of the Master Program Psychology and Mental Health. He is also associated to the forensic organizations Fivoor, FPC Gent and FPC Antwerpen as director Fivoor Science and Treatment Innovation. His research interests lie at the intersection of emotion-regulation, personality disorders, personal growth and development, and risk and protective factors of antisocial and aggressive behavior. His current research concerns emotion-regulation in psychopaths, violent and sex offenders, implicit theories of sexual offenders, effectiveness of virtual reality and self-control, personal growth and development of forensic psychiatric patients. For more than 15 years, he has been engaged in research into sexual offenders, both online and offline. He is co-developer of several risk assessment instruments (e.g., HKT-R & FARE) and treatment programs (Aggression Regulation Clinical; Aggression Regulation Outpatient, Sex Offender Treatment Program and Virtual Reality Aggression Prevention Training). He has written more than 200 publications of which over 100 articles are included in Web of Science.

Chapter 3
Sweetie 2.0 Technology: Technical Challenges of Making the Sweetie 2.0 Chatbot

Hans Henseler and Rens de Wolf

Contents

H. Henseler (✉)
Digital Forensics & E-Discovery, Hogeschool Leiden Leiden, Leiden, The Netherlands
e-mail: henseler.h@hsleiden.nl

H. Henseler · R. de Wolf
Tracks Inspector, The Hague, The Netherlands
e-mail: rensdewolf@gmail.com

S. van der Hof et al. (eds.), *Sweetie 2.0*, Information Technology and Law Series 31,
https://doi.org/10.1007/978-94-6265-288-0_3

Abstract In 2015, Tracks Inspector, a Dutch software company, was asked to develop the software for the Sweetie 2.0 project which was going to be an important tool for mapping, measuring and combating online child abuse. One of the most innovative ideas in the grant proposal for Sweetie 2.0 was the automation of responsive communications with online chat partners using a personified chat robot in various communication channels. For this idea, Terre des Hommes was awarded the 2015 Accenture Innovation Award. The chatbot is a virtualized minor child (aka Sweetie) who will engage in a dialogue using a chatbot engine and 3D imagery. All chat conversations are recorded by the system which assists analysts in classifying 'online predators' by analyzing the data. The chat data is processed per chat and a profile is created for each chat partner to assist with the identification of recurring chat partners. The system will be used in a study by the department of Forensic Psychology of the University of Tilburg to investigate the most effective manner of responding to predators to prevent this behavior in the future. This chapter describes the technology behind the Sweetie 2.0 system. We start by explaining the biggest challenges and the functional requirements that needed to be met. As an introduction we present an infographic that presents an overview of the final system as it is operational now. Then we shortly present the state of the art in chatbot technology in 2015 that we assessed before designing and building Sweetie 2.0. This assessment helped us to solve the challenges. In the remaining part of this chapter we highlight some of the solutions we invented to give the Sweetie chatbot a strategic character, to play 3D avatar video clips in the chatroom, how we implemented some basic rules of engagement and how we are capturing the data streams. This is followed with a short explanation how the Sweetie system is being used, how we designed the underlying architecture, and how analysts are assisted with the identification of persons of interest. We conclude by providing some statistics that have been collected during the first half of 2017. During that period the system was used by operators from Terre des Hommes and researchers from the department of Forensic Psychology of the University of Tilburg.

Keywords Chatbots · Artificial Intelligence · 3D Avatar · Chatrooms · Data Analysis · Identity Analysis

3.1 The Challenge

A Sweetie chatbot character must be built based on the experiences, work instructions and chat conversation logs from the initial Sweetie project conducted in 2013. Using results from the past, the conversation model will simulate as realistically as possible a fictitious 10/11-year-old girl who lives in the Philippines and who wants to earn money by performing a naked webcam show. The visual imagery that was developed for the first Sweetie project in 2013 has been further refined. At a later stage new virtual characters will be constructed (different gender/race/age).

An important aspect of the conversation model is that it needs to have a strategy to determine if a chat partner displays indecent/illegal intentions and if so, to obtain additional information that can be used to send him an intervention message. The chat function of the chat robot will be assessed during the project, in both private as well as public appearances. Depending on the outcomes of these tests it may be decided that the automatic chat function needs to be supplemented with a manual function. This hybrid solution will enable 'human operators' to manually takeover and reply to questions that could not be answered by the chatbot. Eventually this may not be necessary if an acceptable level is achieved by incrementally improving the various fictional characters and the associated question/answer database.

3.2 Functional Requirements

A software solution has been developed enabling the chatbot to communicate across a variety of different chatrooms. The software consists of several interconnected software components, each delivering a part of the required functionality, such as:

- automated chat functionality for the chat rooms and direct chat;
- functions to drive the generated 3D imagery;
- management functionality for the chat rooms, characters, chat structure and corresponding 'question/answer' combinations;
- storage of all chats and related details;
- processing of identifiable material from the chats for each chat partner;
- detection functionality to recognize repeating chat partners, indecent proposals and/or explicit materials;
- a dashboard for graphical presentation of all required actions, chat results, as well as statistics for operational, tactical and strategic insight;
- an intervention module to confront persons of interest with their online behavior. This module will also follow up with relevant advice, deterrent warnings and/or possible threat of identification, based on the findings of current academic research for the project.
- Logging of all chat reports and extracts of chats to a standard which facilitates the exchange of cases. This will consider generic storage and data exchange methods used by various national and international (investigation) agencies such as Interpol and Europol to simplify matching with other (online) child abuse cases.

New functionality like clickbait services (customized URLs to gather for instance IP and browser information from chat partners), capturing of received video/webcam feeds from chat partners and following usually time-restricted URLs provided by chat partners, are just a small example of unforeseen features that have been added during the project. Even more features may be put in the software framework based on operator feedback and data collected during the project.

3.3 Overview of the Sweetie 2.0 Platform

Figure 3.1 depicts an infographic that presents an overview of the Sweetie 2.0 platform. The details of platform and underlying software will be explained in more detail in the following paragraphs. The overview in the infographic explains how the system works in five steps:

1. The operator can create and maintain profiles. A profile represents a virtual Sweetie identity and set properties such as name and age. The profile also contains user credentials for communication channels and chatrooms., it points to video clips for this identity, a chatbot script that determines the rules for conversation, and means for identification of targets.
2. The operator activates a Sweetie chatbot by scheduling a profile on a certain date and time, for a number of minutes (or hours) in one of the chatrooms of a chat server. When the chatbot becomes active at the scheduled time, it also monitors the communication channels for which it is has user credentials (e.g. email and instant messaging).
3. The chatbot passively waits until it is approached by one of the chatroom visitors. It starts a conversation and tries to determine if the other person has

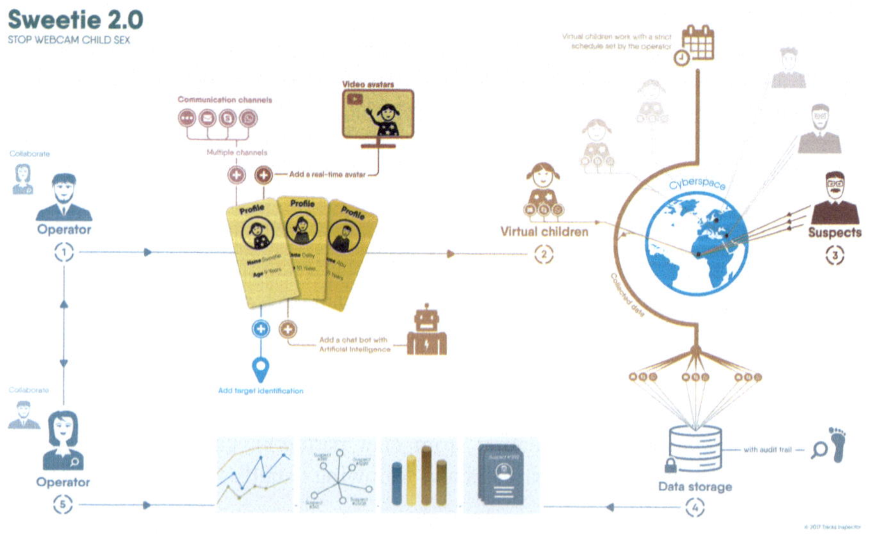

Fig. 3.1 Infographic for the Sweetie 2.0 platform [*Source* Sweetie 2.0 project]

wrong intentions. If this is the case the person is identified as a person of interest (POI). The chatbot can have multiple conversations simultaneously. The operator can override a conversation with a POI and continue to chat manually.

4. All conversations are stored in a database. The graphic window of the chatroom is recorded for the entire session. Web cam streams and messages that arrive when a chatbot is online are also stored in the database or filesystem.
5. Both operators and analysts can analyze data in the database. For example, for evaluating conversations, identifying POIs, retrieving statistics and improving the chatbot conversation. Under specific circumstances, analysts may decide to send an intervention message to a POI as a warning.

3.4 Building Sweetie 2.0

When we started in 2015 the challenge was to build a working chatbot within 12 months in which we could encode the chat script examples that were provided to us as screenshots that were produced as evidence in the 1,000 cases. In addition to these examples, Terre des Hommes and the forensic psychologists from the University of Tilburg explained to us the core elements of the strategy that needed to be encoded in the Sweetie 2.0 chatbot.

At the start of the Sweetie 2.0 project in May 2015, chatbot technology had not changed much since 2005. We shortlisted several existing solutions and identified two solutions that looked promising. One was a US based firm providing chatbot engine as a service and the other a Dutch firm. We selected the Dutch solution which already existed for more than ten years and that we could run on premise. Although the US solution was probably more advanced, we felt that the on-premise capability and the fact that we would have an experienced chatbot engineer working in our team would give the best chance of success.

The technology we used was purely rule-based artificial intelligence. For every response we needed to create one or more replies, trying to achieve the objectives. The objectives are straightforward, namely: does the chatting person know Sweetie is a minor, does this person want to see Sweetie nude behind the web camera and does this person want to pay money to Sweetie to do this? Once two or more of these objectives have been positively answered the chatbot should continue asking questions that can help identify the person. Under certain circumstances Sweetie will also try to end the conversation, e.g. when the person turns out to be a minor just looking for a friendly chat.

In 2015 and 2016 the chatbot landscape changed drastically. In June 2015 Google launched a chatbot that debated the meaning of life. On 18 March 2016 rumors suggested Facebook would unveil a messenger chatbot store at the upcoming F8 conference, on 30 March Microsoft announced ambitious bot plans at their developer conference, on 29 September Amazon launched a '2.5M Alexa Prize' for a chatbot that could converse intelligently for 20 min and meanwhile IBM was pushing Watson's cognitive chatbot. These announcements have all become true. Not only has it become much easier to build a chatbot, chatbots have been equipped with speech recognition. More importantly, chatbots are perceived as the next big step in user interface design. With more than a million apps in both Google Play and the Apple App store, it is becoming increasingly difficult for any app to become installed. Chatbots embedded in popular chat apps, e.g. WhatsApp, Telegram, WeChat will have a much better chance to assist users with functions.

These platforms are based on chatbots that recognize user intent and respond to an intent following a structured dialog. The Sweetie chatbot is focused on one single intent, namely the intent to see webcam sex, or rather to see Sweetie in front of we webcam without clothes. The strategy of the chatbot is not only to verify each of the objectives (minor, nudity and pay money) but also to establish features that can be used to identify a predator and to recognize a in future chat conversations. This differs from most chatbots and requires that Sweetie actively attempts to identify additional properties of a chat counterpart that might not considered relevant for the discussion.

We used the chat evidence from the 1,000 cases as examples to construct conversation rules. Typical chatroom language was insert so that the chatbot would blend in with other chatroom users. One example of typical chatroom language is ASL. Chatroom users will ask for ASL which means they want to know age, sex and location of the person they are chatting with. Such questions are easy to handle and easy to ask.

Figure 3.2 illustrates part of the rule-base that is used by the Sweetie chatbot. The example shows a simplified part of the ASL topic. This is the primary topic of the chatbot that accepts any input. It looks for questions regarding its location, age and name and then tells its location, age and name respectively. If no trigger is found, the trigger is forwarded to the generic random topic. If the ASL topic was addressed before, a verification is performed whether we know the counterpart's age and if too young, we end the conversation. Otherwise we redirect to a topic where we stress we are a minor child.

The current rule base for the chatbot consists of 20 topics that are described in nearly 2,000 lines of code with 50 concepts. A concept is a list of words that all represent the same meaning. Concepts are used to simplify chat bot rules, to avoid repetition and improve consistency. For example, the concept *like* is defined as:

concept: ~sw_like (like enjoy enjoyed enjoi enjoid love "i like" "i like it" "i love" "i love it" "like it" "love it" "gusto ko")

The script is not perfect, but this is not a problem. Sweetie imitates a 10 or 11-year-old girl from the Philippines and chatroom users are not suspicious if Sweetie is not chatting in perfect English. In fact, if she would speak perfect English that would probably raise suspicion. Also, we learned that the audience is not too concerned about logical conversation since they only want one thing which is to see Sweetie without clothes behind the webcam.

The model presented in Fig. 3.2 is simplified and the actual topic contains many more stages. The script of the ASL topic covers 184 lines of code.

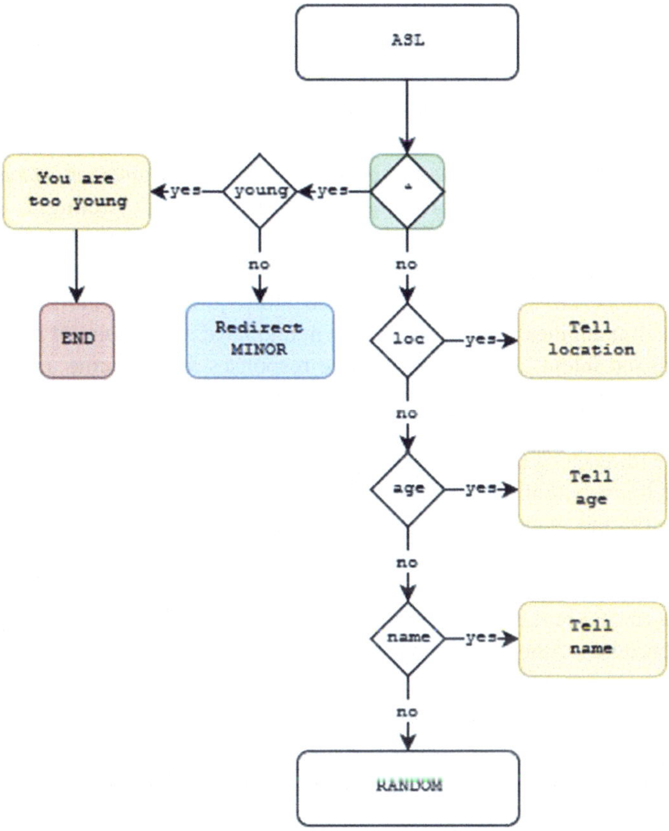

Fig. 3.2 Part of the Sweetie chatbot dialog model [*Source* Sweetie 2.0 system]

3.5 A Man-in-the-Middle

We would soon learn that the selected commercial chatbot engine could not accommodate all the features that our Sweetie 2.0 chatbot requires. For example, to obtain clues that assist with the identification of a chat counterpart, Sweetie will suggest continuing a more private channel such as instant messenger or to send a picture via email. Consequently, our chatbot needs to understand such conversations and be able to handle such requests. The chatbot engine was not equipped to handle this type of requests. There turned out to be many other cases that needed supervision. Another example is that, the chatbot should respond in a human-like way which means it cannot simply reply full sentences in a matter of seconds. We needed the chatbot to simulate human behavior so that counterparts get the illusion of chatting with a real person and not with a computer.

These unexpected technical requirements led to the implementation of a man-in-the-middle solution. This augmented the chatbot engine with new behavior not part of the standard chatbot engine functionality. For instance, we can handle communication to the chatbot via multiple channels (e.g., chatroom, instant messenger and email). Furthermore, we implemented variables so a single chatbot script can be used while imitating different characters. These variables are parameters in the chatbot script to express for instance $name, $age, $siblings. The man-in-the-middle component will automatically substitute values for the chatbot character defined by the operator in the active schedule. The same approach enabled us to handle non-verbal input such as a webcam requests and emoticons.

Most chatbot engines recognize user intent and have a repertoire of responses for each recognized intent. They have multiple responses for the same user input and will randomly present a response to prevent repetition. However, in the case of Sweetie the chatbot has its own intentions and it has a strategy to reach the goals associated with each of the intentions. It turned out difficult to program the chatbot script in such a way that it does not repeat itself. Repeating (part of) a conversation feels unnatural and, hence, it is important for Sweetie to avoid repetition. Using a man-in-the-middle approach to avoid repetition is difficult as it is mostly stateless. Eventually, we decided to switch to an open source chatbot engine that still relies on the man-in-the-middle component but that also could be modified to avoid repetition because of the implemented strategy.

For example, the chatbot asks the user for an email address or ASL and the user ignores the question or answers in an unexpected format. As the chatbot is programmed to complete this information, it does not give up and keeps trying. But if the user keeps ignoring these questions, the chatbot should stop asking at some point. However, it is usual in a script that a question can be asked once or infinitely. To address this type of problem we migrated to a new chatbot engine in 2017 supporting custom functions that can, e.g., count the number of times a question has been asked.

3.6 Sweetie 3D Avatar

The 3D avatar video is probably the most well-known feature of the first Sweetie project in 2013, see Fig. 3.3. It has become an icon for the Save-Sweetie-Now campaign. The initial reason for using avatars is that chatroom visitors are suspicious. They want to get visual confirmation that they are chatting with a real girl. The reason for using a *simulated* avatar is that it is not acceptable to use child actors to play the role of Sweetie. To make this avatar as realistic as possible, motion capture was used with adult actors who played short scenes of 10–20 s imitating a person behind a computer with a screen, webcam and keyboard. Motion capture records the movements of the head and shoulders in space and of the muscles in the face. This data is then fed into a computer graphics algorithm that animates a 3D

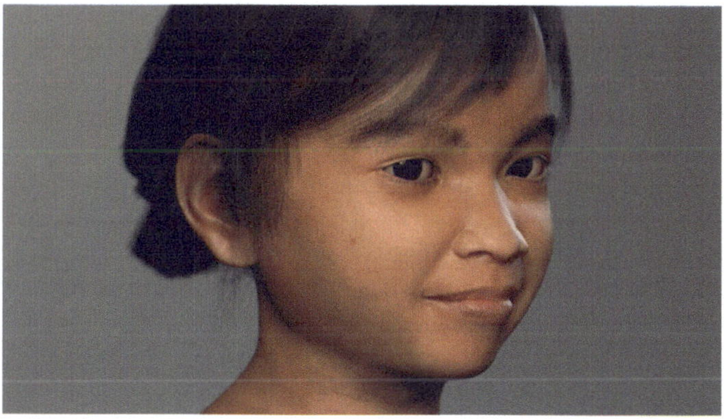

Fig. 3.3 The 3D avatar of Sweetie from 2013 [*Source* Terre des Hommes Sweetie avatar]

artificial head and face and then renders a photorealistic video with a carefully prepared (artificial) background.

Once Sweetie video clips are ready, they need to be integrated in the system. In the beginning of a conversation Sweetie will be reluctant to accept a webcam request. She will only accept a request after some progress has been made. Even when a webcam request is accepted, the system may still decide to broadcast a black screen claiming that the internet connection is poor while recording any video stream that is broadcasted by the other person. If the system decides to play one of the manufactured video clips, the stream is blurred so that quality is degraded. The blurring and the fact that the video is shown only for a few seconds reduces the chance of discovery. The 3D video of Sweetie is very realistic but when a user can study the video in full resolution for a longer period they might discover it is not real.

3.7 Rules of Engagement

The Sweetie chatbot has a few very simple rules that apply when engaging in a chat. First, the chatbot never starts a conversation. A bot that has been scheduled to enter a chatroom, simply enters the public chatroom and waits until it is approached by another user for a private chat. This sounds like a passive strategy, but this strategy was tested in the first Sweetie project and turns out to be very effective. Anyone that has ever entered a teenage chatroom with a guest account and that registers with a girl name and a typical girl avatar (e.g. a unicorn, horse or bunny) will quickly learn that other chatroom visitors are very interested in a chat. Experiments and demonstrations over the last 3 years have showed repeatedly that within 5–10 min you are chatting with at least 5 persons. And some of these persons make no secret of their intentions and are very quick to submit a webcam request to Sweetie.

This simple rule of engagement (e.g. not starting a conversation) has various advantages. First, from a legal perspective, it makes it harder to position Sweetie as entrapment since the chatbot has not started the conversation. Second, this rule avoids a Sweetie chatbot to start a conversation with another Sweetie chatbot. This may not be an issue now, with only one Sweetie system in operation, but it may become an issue if multiple agencies decide to start using the Sweetie 2.0 system. This is not an unlikely scenario since the legal research (see Chaps. 1 and 4 ff.) has concluded, amongst other things, that law enforcement should focus on using a system like Sweetie in their own jurisdiction only since it will be hard for them to prosecute offenders operating in other jurisdictions.

3.8 Capturing Data Streams

The evidence on the 1,000 cases in the first Sweetie project in 2013 was gathered manually. This evidence was mostly produced in the form of text documents containing screenshots of the chat sessions that had been conducted with the predators. There are several disadvantages to this. First, screenshots of chat conversions cannot be searched for text strings. This makes it also difficult to process the examples when we wanted to compile the examples into a rule base for the Sweetie 2.0 chatbot. Second, and more importantly, this must have been a very cumbersome and error prone process. Not only did the operators have to manage multiple chats at the same time, they also needed to record sessions either by manually taking screenshots or by recording all sessions and editing them afterwards.

The Sweetie 2.0 system has been designed in such a way that all chat sessions and video streams are recorded automatically. This data is stored in a database for data analysis and evidence purposes. There are two types of video capture. The first type is a video that has 2 frames per second. Each frame is a screenshot of the complete chat application. These frames are also used to enable operators to watch the chat application in real-time and not only the text streams. The frames are also stored in a streaming video format that can be replayed later to get an overall impression of what actually happened in this chatroom. The disadvantage of this recording is that the chatbot runs many active chats simultaneously meaning that windows are overlapping and, consequently, this video may not be useful to document individual chats. The other type of video is derived by capturing the video stream that is communicated during an active webcam session. This video is very specific to the chatting person and is stored along with the text of the chat conversation.

3.9 Using the Sweetie 2.0 Platform

The Sweetie 2.0 platform has been implemented as a distributed server-based application with a collaborative web-based front end. The architecture of the distributed server architecture is explained in the next section. Here we will explain how the different user roles can use the system. Figure 3.4 is a screenshot of the Sweetie 2.0 user interface that is displayed after the user has logged in. The left side of the screen lists the main functions that are available to the users. The right side of the screen displays simple statistics that reflect the number of conversations that have been captured in the last month.

The Sweetie 2.0 system has been designed to automate the process of chatting with and exposing predators. For the operation of the system we can distinguish three different user roles:

- the operator role for creating, maintaining and scheduling chatbots and virtual identities,
- the analyst role for analyzing the data, identifying persons of interest and producing reports, and
- the administrator role for monitoring system health, maintaining user accounts, etcetera.

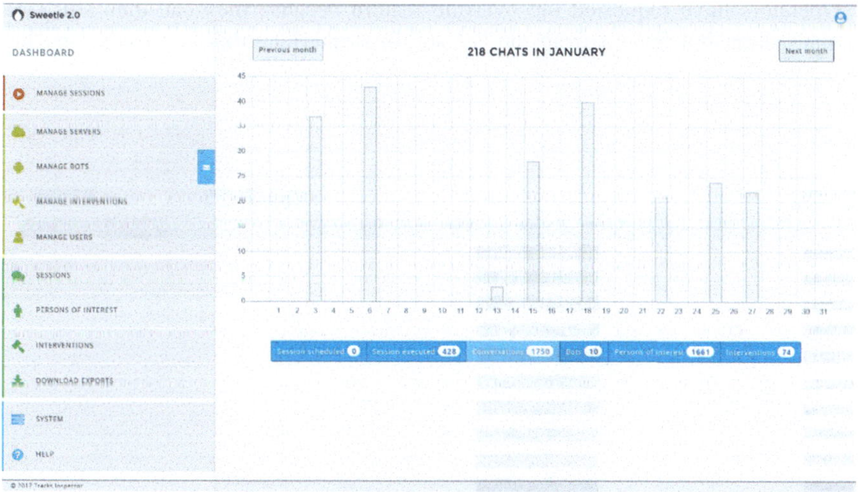

Fig. 3.4 Screenshot of the Sweetie 2.0 interface [*Source* Sweetie 2.0 system]

3.9.1 Operator

Operators schedule chatbots in online chatrooms so that chat conversation data is collected. Scheduling means the operator selects a chatroom from a predefined list, selects a virtual identity that has also been preconfigured and schedules a start and end time plus date indicating when the virtual identity should be active in the selected chatroom. When it is active it is also collecting messages from the other channels that it has access to (e.g. email, instant messenger). Operators can maintain the chat scripts, but this is a task that is typically outsourced to a team of people that have a good understanding of the chatroom audience, the behavior of Sweetie and the goals that must be achieved.

The operators from Terre des Hommes typically monitor that chatbot when it is in action. They can do this by opening the dashboard that displays active chat conversations. Conversations that have already ended are listed in the History tab. This is illustrated in Fig. 3.5.

The system has a feature that enables an operator to override a chatbot and manually take over the conversation. The typical workflow has become that the operator uses the chatbot for the initial communication. While the system is conducting tens of simultaneous conversations, the operator can quickly flip between conversations and see what's going on. If a conversation becomes interesting or the bot appears to get stuck, the operator can take over and manually try to persuade the other person to give identifying information. Chats that have been overridden are also stored in the database.

Operators have requested the override feature because they can then still take manual action but benefit from the "clean" user interface that is provided by the

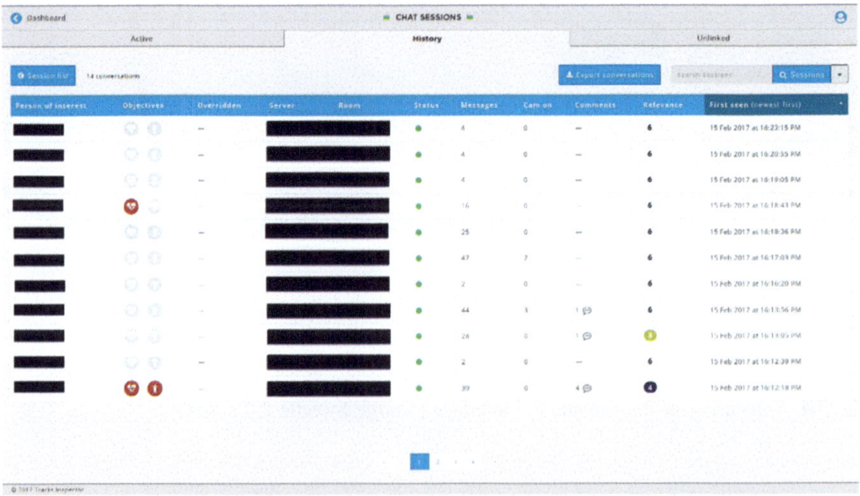

Fig. 3.5 List of conversations in a chat session (redacted) [*Source* Sweetie 2.0 system]

system. By this they mean that they are not distracted or intimidated by unexpected webcam feeds and advertisements. This may seem trivial but operators that spend hours chatting feel this can be a very demanding job which eventually wears them out psychologically.

3.9.2 Analyst

The conversations that have been collected in the database are analyzed by the analysts. The analysts have the same overview as depicted in Fig. 3.5 that is also available to the operators. Analysts examine chat conversations that have been finished. To get an idea of what conversations have been active, they can open a spider diagram that places the chatbot in the center and lists all persons that have chatted with the chatbot. The spider diagram is illustrated in Fig. 3.6.

The analysts from Terre des Hommes typically read conversations and add annotations. The system supports the following type of annotations:

- Relevance: chat sessions are flagged in a scale from 1 to 6 where 1 indicates a predator and 6 indicates a person who showed no bad intentions.
- Comments and custom attachments.
- Age, sex, location (ASL).
- Person wants to pay money.
- Person wants Sweetie to show herself naked on the web cam.
- Person knows Sweetie is a minor.
- Identifier (e.g. name, email address, instant messenger id).
- Investigated flag, to indicate the POI has been researched using OSINT techniques.
- Favorite flag, to indicate this POI is a good example for demonstration purposes.
- Flags to indicate there is room for chatbot improvement, e.g. bot is suspected, the bot's conversation is not logical, or the bot's conversation contains repetition.

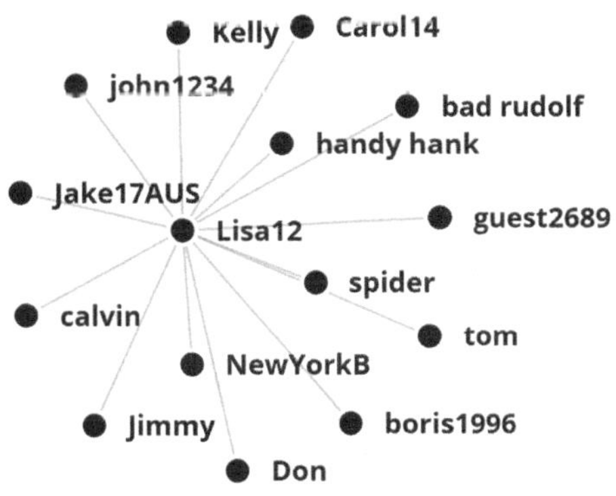

Fig. 3.6 Example Spider diagram displaying all identities (based on nickname) that have chatted with the chatbot in a session [*Source* Sweetie 2.0 system]

Some of these annotations will be set automatically by the system. The analyst should identify false positives as well as false negatives that have wrongly been classified by the computer. A false negative could be an email address that is given but where the system failed to recognize the format because it has been intentionally malformed. For example, a person can answer 'my email is JohnDoe@*mail.com' and say in the next line 'replace * by g'. This is common practice because some chatrooms do not allow the exchange of communication addresses to prevent users from continuing their conversation outside the chatroom.

In some cases, analysts will search in open source (e.g. website, social media) for additional information. This way they may find for instance an email address on social media for a person that did not want to give his address when the chatbot asked. Such additional information can be entered in a comment field or inserted as a new identifier that can be used by the system to find relations with other persons in the system (Fig. 3.7).

Analysts from the Forensic Psychology department of the University of Tilburg use the system to send out interventions i.e. warning messages. The system supports a double-blind testing feature. This means the operator may decide if an intervention should be send to a specific person. In a double-blind test the system will randomly decide in 50% of the cases not to send the intervention message. The idea is that later the scientists can objectively measure if the intervention message has led to significant lower reoccurrence rate. The hypothesis is that persons of interest that have received an intervention message will not be encountered any more or less than persons of interest that should have received such a message but did not because of the random decision.

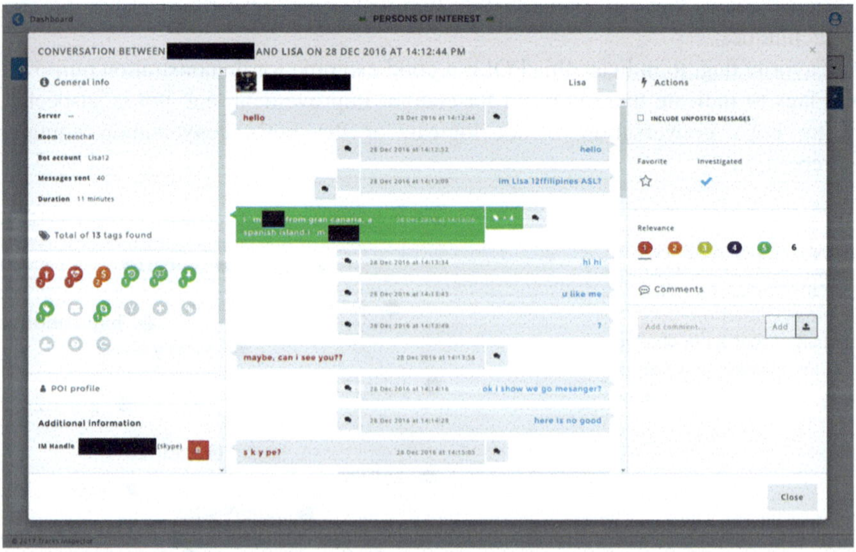

Fig. 3.7 Analyzing chat conversations (redacted) [*Source* Sweetie 2.0 system]

3.9.3 Administrator

The system is maintained by administrators. The user interface provides functions for creating new users specifying username and role. New users will receive an email that enables them to login and set their password. The administrators can also monitor the system health, i.e. are processes still running, memory usages, processor load etc.

The system uses various VPN connections to hide its logical (and physical) location. The Sweetie character pretends to be online from the Philippines by using a VPN connection to the Philippines for all online activity in chatrooms, email and instant messaging services. The VPN infrastructure is custom-built and maintained for the Sweetie 2.0 project.

In addition, the system makes daily backups of the system state and data collected so far. These backups are encrypted and stored on a remote server.

3.10 Architecture Design

The Sweetie 2.0 platform has over a dozen services and it uses its own browsers, instant messaging (IM) clients and creates virtual webcam devices for online streaming of avatar video files. The components can be grouped in six different categories:

- Connectors: Connections to chat, IM and email servers.
- Frontend: Operator, analyst and administrator access.
- Interventions: Sending intervention messages.
- Databases: Storage of data, user settings, audit logs, templates.
- Controller: System and session controllers.
- Chatbot: The chatbot engine.

Figure 3.8 shows the various components based on the categories mentioned and the interaction with the web browsers (clients) and third-party services.

3.10.1 Connectors

The connectors provide connections to a chatroom, instant messaging or email service(s). They can process input from the chatbot interface or receive manual input from the operator. Connectors can also display 3D avatar visuals if a webcam is supported in the chatroom. The connections are created through (remote) proxy or VPN servers to hide the location of the Sweetie 2.0 system.

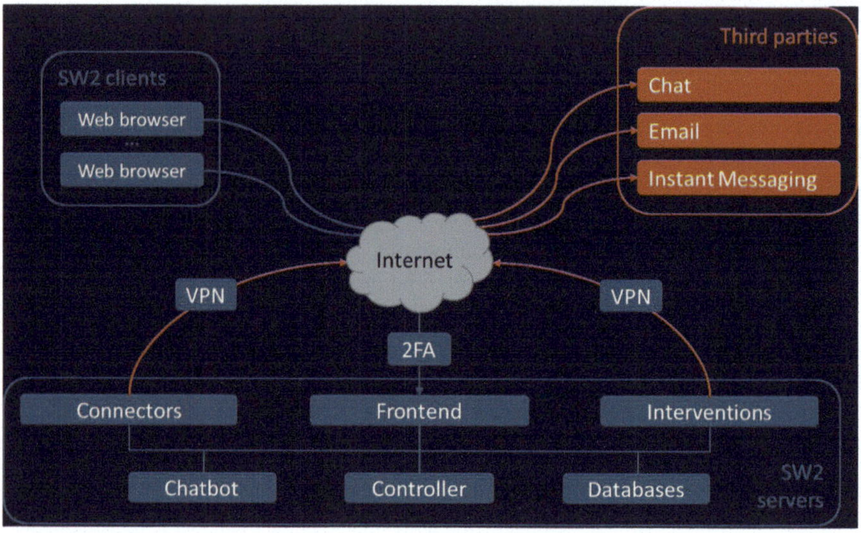

Fig. 3.8 High-level overview of components and interactions [*Source* Sweetie 2.0 project]

3.10.2 Front-End

The front-end components are responsible for the web-based user interface for operators and administrators. Users login to the web server running the frontend service to configure, review and monitor the system. Two-factor authentication (2FA) and encryption of the communications provides security to the web interface of the Sweetie 2.0 environment.

3.10.3 Intervention

The intervention component can repeatedly send messages to persons of interest based on a pre-defined model and schedule using email. The content of the messages can be customized using the web interface of the system.

3.10.4 Databases

The databases comprise a data store and system store. The data store component stores chat sessions, metadata and person of interest profiles and relations. The data store is also used to store automatically generated information such as audit logs, system status and pre-coded information (templates). The separate system store keeps configurations that have been entered by administrators and operators (users, avatars, chat schedules, templates).

3.10.5 Controller

The controller is a central control component linking all components together. It polls each component for its status. The controller is closely linked with a user session manager that is responsible for user authentication, stored user session state and audit logging.

3.10.6 Chatbot

The chatbot component provides an abstract interface to third party chatbot AI engines. The interface supports multiple chatbots and can distinguish between multiple concurrent sessions per bot. The chatbot interface to the external chatbot engine is also responsible for the man-in-the-middle filtering that was explained earlier.

3.11 Identifying Persons of Interest

The analysts analyze chats by annotating and adding additional information to recorded chat sessions (see Fig. 3.7). The system will automatically group chat sessions that probably belong to the same person because they share one or more identities that were collected during the conversation (e.g. nickname, email address, instant messenger id). An analyst may also decide that chat sessions that do not have any information in common, do belong to the same person because of (other) chat content (e.g. use of words, remarkable expressions, pictures that were shared or webcam stream). This overview is called the Persons of Interest (POI) dashboard. An example of the POI dashboard is presented in Fig. 3.9.

How interesting a POI really is, depends on the number and nature of the chat conversations that belong to that POI. To assist analysts in their investigation, they can rank POIs based on one of the following criteria:

- Relevance: This is the level that was set manually by the analysts.
- Date: Rank persons of interest by date of most recent or old activity (e.g. first seen or last seen).
- Objectives: The number of objectives that a POI matches.
- Recurrence: Rank persons of interest based on the number of conversations that have been captured that (very likely) belong to this person of interest.
- Name: Simple sorting based on the POI's primary chat handle.

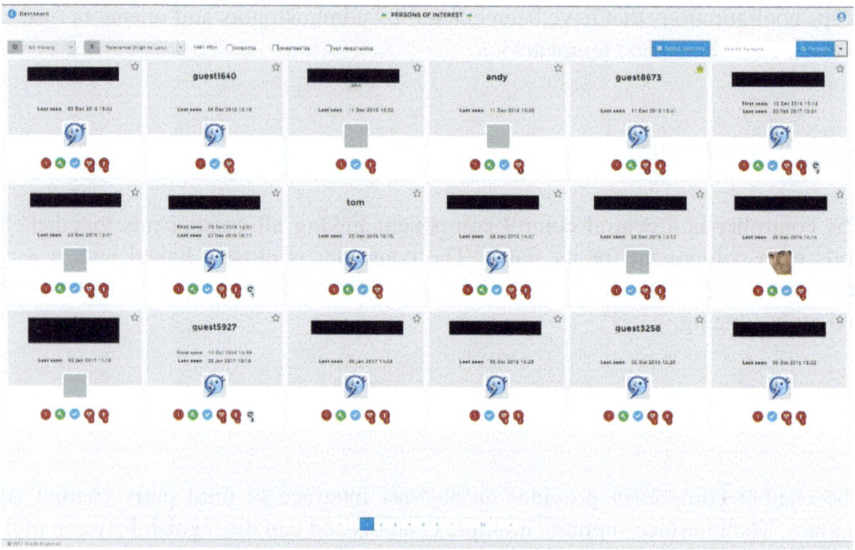

Fig. 3.9 Analyzing persons of interest (redacted) [*Source* Sweetie 2.0 system]

3.12 Results

At the end of 2016 the first version of the Sweetie 2.0 system was ready. By then Terre des Hommes had already been performing various acceptance tests by operating Sweetie in a restricted number of chat rooms under close supervision of one of their operators. These tests were the first-time contact between the target group and the automated chatbot functionality. During the first tests (in the summer of 2016) results showed that the chatbot was never discovered. At worst, a person who chatted with the chatbot would conclude that our chatbot was stupid. But never would they say that they believed they were chatting with a chatbot. Nevertheless, the operators and developers felt there was still room for improvement.

From these tests we learned how the chat script needed to be improved. First, the story needed to be expanded in some areas. Second, the chatbot needed to become more focused on achieving the objectives, i.e., ask for identifiable information. Third, the chatbot did not end the conversation in chats that we not going anywhere or that were technically completed. Fourth, we originally suggested to the operators that different variations of the same chatbot script could be made by copying the master script and editing basic details in the script. With more than 1,000 rules and with the script constantly being improved, this was not practical. Hence, we decided to introduce parameters in the script so that we could use a single script and store typical character information such as name, age, place of birth, siblings and hobbies in a separate table so that we could define different versions of Sweetie without having to maintain different scripts.

Table 3.1 Statistics for the Sweetie 2.0 system for January–May 2017 (number of chat sessions, number of conversations, total number of messages, number of persons of interest, number of chatbots, number of chatrooms)

	January	February	March	April	May	Total
Sessions	19	14	8	3	16	60
Conversations	218	158	125	62	108	671
Messages	4,512	3,709	3,294	1,393	2,110	15,018
POIs	166	25	0	17	93	301
Bots	5	5	3	3	3	19
Rooms	6	4	2	1	6	19

Since the beginning of 2017 the Sweetie 2.0 system is live and it has been operated by Terre des Hommes. Table 3.1 shows the main statistics for the first 5 months of 2017. This initial period was mainly used to study how the system was performing and establish a base line.

In August–September 2017 a 4-week pilot was conducted with the National Bureau of Investigation in the Philippines. This pilot was led by the operator from Terre des Hommes. Before starting the pilot, all existing data had been backed up so that the pilot could start with a clean setup. Interestingly, during the first days of the pilot we discovered that the public chat servers had been changed their software to another chat technology. This meant we had to develop a new version of our chat server plugin script. Fortunately, our framework turned out flexible enough to create a new version of the plugin without having to modify the framework. Still, the reverse engineering of the new chat server software and modification of the script took about 2 days to get a first working version.

During the pilot we also discovered that the commercial VPN services we were using, were not always working as expected. We performed various tests and decided to host our own VPN servers in locations that correspond with the background story of our chatbots. The pilot was concluded successfully and a follow up is planned for December 2017.

3.13 Conclusions and Recommendations

The Sweetie 2.0 software has been developed to meet the functional requirements. The system has been deployed and is in operation since the beginning of 2017. Terre des Hommes is now capable of interactively monitoring online chatrooms for webcam-sex predators with less effort and in a more defensible manner than in 2013. Researchers of the University of Tilburg will be able to use the system to study the effectiveness of different intervention strategies as a preventive measure and the system has given a new impulse to the Save-Sweetie-Now media campaign. Last, but not least, the National Police from the Philippines and police forces from several other countries have expressed interest in conducting a pilot with the system to see if it can also be used for law enforcement purposes.

There are several recommendations to further improve the system using new technologies that have emerged since 2015. Especially the advancement of machine learning algorithms (e.g. deep learning) and hardware have potential to further improve the Sweetie 2.0 system. We have several recommendations for the application of deep learning in the system.

First, deep learning has been used successfully to detect the intent of a user.[1] Currently, Sweetie 2.0 is based on rules that use string matching to determine the topic that is relevant for the conversation. Using the same method, we can train the system based on examples so that it is more flexible in selecting the most appropriate context.

Second, with the current (rule-based) approach we found that automatic detection of objectives (i.e. under age, wants to pay money, wants to see a webcam) results in a high number of false positives. Hence, we have implemented a flagging mechanism to enable analysts to manually classify each chat on a scale of 1–6. We believe that by supervised training a deep neural network with the current chat conversions that have been manually classified, will give an automatic classifier that is far more accurate than the rule-based classifier we initially designed.

Third, after having some experience with analyzing chats human analysts can recognize if two chats from persons without obvious common identifier do belong to the same person. This can be based on the choice of words, webcam content or picture content. We believe deep learning technology is very well suited to identify relevant features that can be used to distinguish different authors, and, consequently, to measure the similarity between two authors.

One more alternative application of deep learning may be the automatic modelling of a conversation model that learns automatically from new conversations. This would indeed be an important feature to speed up the development of new chatbot characters. In the summer of 2015 Google published an article about the development of a chatbot that learned conversation from existing conversations.[2] This research showed that recurrent neural network could be trained on movie subtitles so that the bot responds to input with phrases from movies.

We have doubts if this technology would be able to capture the strategy the is needed in the Sweetie 2.0 chatbot. Also, research from Microsoft in 2016 show that a chatbot that automatically learned new conversation from Twitter ended up learning unwanted conversation.[3] For this reason, Microsoft decided to take the chatbot offline. Its current successor, named Zo, is running on the Kik messenger platform and is trained to avoid topics like politics.[4] With Zo, Microsoft is focused on advancing conversational capabilities within their AI platform.

[1] GK 2017.

[2] Vinvals and Quoc 2015.

[3] Lee 2016.

[4] Johnson 2016.

Finally, there are many other improvements that can made to the system. For example, the maintenance of virtual identities is now left to the operators. We think it is possible to extend the system with special functions to assist operators with the creation and maintenance of virtual identities. Also, the current system only supports a limited number of channels. Terre des Hommes has indicated other channels such as Facebook Messenger are becoming increasingly important.

Acknowledgement Tracks Inspector would not have been able to perform the technical development of the first version of the Sweetie 2.0 system without the help of third parties for creating the 3D virtual avatars by Motek Entertainment and the development of the first Sweetie chatbot rule-base in the KMS of eCreation. The feedback from and whiteboard sessions with the operators and analysts from Terre des Hommes and researchers from the Forensic Psychology department of the University of Tilburg have been essential for the further optimization of the chatbot and development of additional system features that were not in-scope but that have greatly improved the usability of the system. We also want to acknowledge the driving force of Hans Guyt, project leader of Sweetie 2.0 at Terre des Hommes, who, in addition to all his other tasks in the project, relentlessly made sure everyone that was involved in the technical development and operation kept focus on the end goal of the Sweetie 2.0 system.

References

GK (2017) Contextual chatbots with tensorflow. Chatbots Magazine, May 6, 2017. https://chatbotsmagazine.com/contextual-chat-bots-with-tensorflow-4391749d0077. Accessed 6 June 2018

Johnson K (2016) Microsoft's Zo chatbot refuses to talk politics, unlike its scandal-prone cousin Tay. https://venturebeat.com/2016/12/05/microsofts-zo-chatbot-refuses-to-talk-politics-unlike-its-scandal-prone-cousin-tay. Accessed 6 June 2018

Lee P (2016) Learning from Tay's introduction. Official Microsoft Blog, 25 May 2016. https://blogs.microsoft.com/blog/2016/03/25/learning-tays-introduction. Accessed 6 June 2018

Terre des Hommes (2015) Accenture Innovation Award voor Sweetie. https://www.terredeshommes.nl/nieuws/accenture-innovation-award-voor-sweetie-20. Accessed 6 June 2018

Vinyals O, Quoc V Le (2015) A neural conversation model. https://arxiv.org/abs/1506.05869. Accessed 6 June 2018

Hans Henseler is managing director at Tracks Inspector and responsible for marketing and sales. Hans is also professor of Digital Forensics and e-Discovery at University of Applied Sciences Leiden, Chair of the Board of Directors of DFRWS and member of the board of the Netherlands Organization of Court Experts. In 1992 he founded the Forensic Computer Investigation Department at the NFI. He was business unit manager Forensics and E-Discovery at Fox-IT, director of forensic technology at PricewaterhouseCoopers and CTO at ZyLAB. Hans received his Ph.D. in Artificial Neural Networks at Maastricht University.

Rens de Wolf is operations director at Tracks Inspector and responsible for daily operations, finance and projects. Rens has worked with Fox-IT for 12 years in various positions starting as a security expert and digital forensics expert. He then became responsible for the audits department and was the information security officer for 4 years until he moved to the UK in 2007. There he founded Fox-IT UK and later became VP International Sales with focus on lawful interception and high-security solutions. After his return to the Netherlands, he took on Tracks Inspector sales and the partner network. Rens received his M.Sc. in Computer Science, specializing in Information Security, and currently holds CISSP, CISA and CISM certifications.

Chapter 4
Substantive and Procedural Legislation in Australia to Combat Webcam-Related Child Sexual Abuse

Gregor Urbas

Contents

Abstract Australia has a range of substantive and procedural laws that can be used to combat webcam-related child sexual abuse. This includes legislation at the national (Commonwealth) level as well as in the various States and Territories. In fact, Australia was the first jurisdiction to prosecute an offender identified through the initial Sweetie operation conducted by Terre des Hommes. Numerous others have been convicted and sentenced as a result of their online, including webcam-based, sexual exploitation of children in other countries such as the Philippines. Crucial to this success is the ability of law enforcement investigators to infiltrate online groups, or to interact with individual child predators, using covert techniques. These include posing as a child online, successfully used in child grooming investigations. The legislation used in various Australian jurisdictions is

G. Urbas (✉)
ANU Cybercrime Observatory, Australian National University, Canberra, Australia
e-mail: Gregor.Urbas@anu.edu.au

S. van der Hof et al. (eds.), *Sweetie 2.0*, Information Technology and Law Series 31,
https://doi.org/10.1007/978-94-6265-288-0_4

intentionally written so as to allow the use of fictitious identities, and numerous courts have accepted as lawfully obtained the evidence gathered in such investigations, allowing it to be admitted. These principles can be readily extended to the hypothetical use of a Sweetie-style chatbot. However, further regulations governing the use of controlled operations and assumed identity authorities for this kind of covert activity by law enforcement investigators would clarify the legitimacy and limitations of the technique.

Keywords Webcam · Child Exploitation · Substantive Laws · Procedural Laws · Australia · Child Grooming · Covert Investigation · Controlled Operations · Sweetie

4.1 Introduction

4.1.1 General Description of the Legal Framework

Australia is a common law country with a federal structure that governs legal powers and responsibilities. The constituent parts of the federation are the Commonwealth, the States and Territories. The pre-existing colonies became States in the new federation which was formed in 1901 with the adoption of the *Commonwealth of Australia Constitution Act 1900* (the Commonwealth Constitution) as the foundational legal document. The States have their own written Constitutions which operate alongside the Commonwealth Constitution. The territories are also self-governing to varying degrees.

Commonwealth

The highest level of legal responsibility at the national level lies with the Commonwealth. The Parliament of Australia, also referred to as the Commonwealth or Federal Parliament, has powers to make laws for Australia as a whole, including those with extra-territorial operation. However, unlike the States and Territories, which have full powers to make laws for their respective jurisdictions without limitation as to subject matter, the Commonwealth has legislative powers only with respect to specified matters or 'heads of power'. The main heads of power specified in the Commonwealth Constitution that provide a basis for cybercrime legislation are in relation to:

(a) *postal, telegraphic, telephonic, and other like services*—modern telecommunications services such as the Internet are considered to fall within the scope of "like services";[1] and

[1] As discussed in the case of *R v. Brislan* [1935] HCA 78; (1935) 54 CLR 262.

(b) *external affairs*—this enables laws to be made in order to give effect to international agreements such as the *Convention on Cybercrime*, which Australia acceded to in 2012.[2]

While the States and Territories have some cybercrime laws also, these are often in similar terms to Commonwealth laws, or may play a more limited role in prosecutions. Where there is any inconsistency between Commonwealth laws and those of the States or Territories, the former prevails over the latter.[3]

States

The six Australian States are New South Wales, Queensland, South Australia, Tasmania, Victoria and Western Australia. State Parliaments may make laws on any subject matter, and thus have traditionally been responsible for criminal laws relating to offences against persons, property and the peace. Some older State offences, such as theft and fraud, or more recent ones such as stalking or voyeurism, have application to the online environment. However, the recent trend has been for the Commonwealth to take a leading role in legislating against cybercrime, particularly using its legislative powers with respect to telecommunications.

Territories

The two mainland territories are the Australian Capital Territory, in which the national capital (Canberra) is located, and the Northern Territory. While the Commonwealth has constitutional power over both territories,[4] these have been given increasing independence by way of self-government, and make laws in much the same way that States do, including on cybercrime.

Table 4.1 indicates the nine jurisdictions within the Australian federal system as well as the main criminal legislation of each.

The most important legal principles that operate in Australia are based on the common law tradition. The doctrine of precedent ensures that lower courts follow the rulings of higher courts in similar cases, but the doctrine of parliamentary supremacy means that the law as enacted in legislation prevails over judge-made law. Thus, it is not uncommon for Parliament to enact legislation specifically in response to, indeed to override, the decision of a superior court on some legal point. Generally, the legal system is based on principles of fairness, equality before the law and due process.

The criminal justice system is adversarial rather than inquisitorial. Public prosecutors have responsibility for proving a criminal charge beyond reasonable doubt. More serious charges are prosecuted on indictment (known as 'indictable offences'), which may involve trial before a jury, while less serious charges are dealt with in lower courts

[2] The Convention came into force for Australia on 1 March 2013.

[3] By virtue of s109 of the Commonwealth Constitution.

[4] By virtue of s122 of the Commonwealth Constitution.

Table 4.1 Main Australian criminal legislation

Jurisdiction	Abbreviation	Main criminal statute(s)
Commonwealth	Cth	Criminal Code Act 1995 (Cth) Crimes Act 1914 (Cth)
Australian Capital Territory	ACT	Crimes Act 1900 (ACT) Criminal Code 2002 (ACT)
Northern Territory	NT	Criminal Code (NT)
New South Wales	NSW	Crimes Act 1900 (NSW)
Queensland	Qld	Criminal Code (Qld)
South Australia	SA	Criminal Law Consolidation Act 1935 (SA)
Tasmania	Tas	Criminal Code (Tas)
Victoria	Vic	Crimes Act 1958 (Vic)
Western Australia	WA	Criminal Code (WA)

[*Source* Australasian Legal Information Institute (AustLII)]

('summary offences') by Magistrates.[5] Cybercrime cases may involve either indictable or summary offences, or a mixture of the two. Further, given the fact that more than one jurisdiction may have coverage, it is possible for both Commonwealth and State/ Territory offences to be dealt with in the same proceeding.

4.1.2 Relevant Treaties and Cybercrime Laws

Australia has acceded to several international treaties:

Council of Europe Convention on Cybercrime
Australia was not among the early signatories of this Convention, but its coverage and drafting were influential in legislative reforms made at the Commonwealth level, particularly the *Cybercrime Act 2001* (Cth). This statute consolidated existing computer offences protecting Commonwealth computers and data, which at that time were in the *Crimes Act 1914* (Cth), and moved them to the *Criminal Code Act 1995* (Cth) with an expanded application based on the Commonwealth's power to regulate telecommunications. It also added new computer-related search and seizure provisions to the *Crimes Act 1914* (Cth). New offences concerning child online pornography and abuse material, along with other telecommunications misuse provisions, were added to the *Criminal Code Act 1995* (Cth) in 2004. Thus, although Australia did not formally acceded to the Convention on Cybercrime until a decade later, its substantive and procedural laws were already largely compliant in the decade before this accession. Some further amendments were made under the *Cybercrime Legislation Amendment Act 2012* (Cth), the long title of which was "An

[5] Under s80 of the Commonwealth Constitution, trials on indictment for Commonwealth offences must be by jury.

Act to implement the Council of Europe Convention on Cybercrime, and for other purposes", and Australia then proceeded to sign the Convention with ratification on 30 November 2012 and entry into force for Australia on 1 March 2013. Australia is not, however, a signatory to the Additional Protocol on racist and xenophobic acts committed through computer systems (Urbas 2013).

Lanzarote Convention

Australia is not a signatory to the Council of Europe *Convention on Protection of Children against Sexual Exploitation and Sexual Abuse* (the Lanzarote Convention), but does criminalise relevant exploitative conduct through Commonwealth, State and Territory laws, including child grooming and procuring and use in the production of pornography (see Tables 4.2, 4.3, 4.4 and 4.5).

UN Convention on the Rights of the Child 1989

Australia is a signatory to the major United Nations agreements on human rights, including the *Convention on the Rights of the Child* (in force for Australia from 16 January 1991) and the *Optional Protocol on the Sale of Children, Child Prostitution and Child Pornography* (ratified and in force from 8 January 2007).

In addition, Australian law enforcement agencies are actively involved in international co-operation on cybercrime responses, including against child exploitation. For example, Australia is a member of the Virtual Global Taskforce (VGT) through High Tech Crime Operations within the Australian Federal Police (AFP).[6] The recently established Australian Cybercrime Online Reporting Network (ACORN)[7] and the Office of the eSafety Commissioner[8] also have extensive international co-operation networks.

4.2 Analysis of Substantive Criminal Law

There are numerous Australian laws criminalising the sexual exploitation of children, both through contact offending and online. Such laws can be found in Commonwealth, State and Territory legislation, making a brief summary difficult. However, it should be noted that physical contact offences are mostly dealt with at the State and Territory level,[9] unless the contact occurs overseas e.g. the

[6] The Australian Federal Police (AFP) hosted the international VGT conference in 2010: http://virtualglobaltaskforce.com/.

[7] The Australian Cybercrime Online Reporting Network (ACORN) is a national policing initiative that allows members of the public to report instances of cybercrime securely, as well as providing advice to help Australians recognise and avoid common types of cybercrime: https://www.acorn.gov.au/.

[8] The Office of the eSafety Commissioner was established by the *Enhancing Online Safety for Children Act 2015* (Cth) and has been recently re-named to ensure that its role extends beyond the online safety of children to that of all Australians: https://www.esafety.gov.au/.

[9] Urbas and Grabosky 2006.

Table 4.2 Australian legislation relating to sexual offences against minors

Lanzarote treaty	Australia
Article 18. Sexual abuse	Commonwealth: Division 272 of the Criminal Code Act 1995 (Cth) relates to sexual offences against children outside Australia[a] and includes these offences 272.8 Sexual intercourse with child outside Australia 272.9 Sexual activity (other than sexual intercourse) with child outside Australia 272.10 Aggravated offence—child with mental impairment or under care, supervision or authority of defendant 272.11 Persistent sexual abuse of child outside Australia 272.12 Sexual intercourse with young person outside Australia—defendant in position of trust or authority 272.13 Sexual activity (other than sexual intercourse) with young person outside Australia—defendant in position of trust or authority 272.14 Procuring child to engage in sexual activity outside Australia 272.15 "Grooming" child to engage in sexual activity outside Australia A 'child' is a person under 16 years old, and a 'young person' is 16–18 years old States and Territories: Each of the States and Territories has criminal offences relating to sexual intercourse or other sexual activity with a person under 16, with more severe penalties where the minor is younger such as under 10 years e.g. s55 of the Crimes Act 1900 (ACT)
Article 19. Offences concerning child prostitution	Commonwealth: Division 270 of the Criminal Code Act 1995 (Cth) concerns slavery and slavery-like conditions, while Division 271 concerns trafficking and debt bondage. Both have numerous offences that may apply to child prostitution and similar exploitation States and Territories: Each of the States and Territories has criminal offences relating to child prostitution e.g. s91 of the Crimes Act 1900 (NSW) criminalises promoting or engaging in acts of child prostitution, where a 'child' is a person under 18 years old
Article 20. Offences concerning child pornography	Commonwealth: Division 273 of the Criminal Code Act 1995(Cth) contains offences involving child pornography material outside Australia,[b] while Division 474 contains offences relating to the use of carriage (i.e. telecommunications) services for child pornography, including: 474.19 Using a carriage service for child pornography material 474.20 Possessing, controlling, producing, supplying or obtaining child pornography material for use through a carriage service[c]

(continued)

Table 4.2 (continued)

Lanzarote treaty	Australia
	The definition of 'child pornography material' relates to persons under 18 years old and can include fictional representation, including cartoons[d] States and Territories: Each of the States and Territories has criminal offences relating to child pornography, where the relevant age varies from under 16 to under 18 years of age (see Tables 4.3 and 4.4 for more detail)
Article 21. Offences concerning the participation of a child in pornographic performances	Commonwealth: Division 273 of the Criminal Code Act 1995 (Cth) contains offences involving child pornography material outside Australia, while Division 474 contains offences relating to the use of carriage (i.e. telecommunications) services for child pornography. Both may apply to the use of technology e.g. to film and transmit child pornography material online using a webcam States and Territories: Each of the States and Territories has criminal offences relating to child pornography e.g. s91G of the Crimes Act 1900 (NSW) criminalises using a child for the production of child abuse material, where a 'child' is a person under 16 years old
Article 22. Corruption of children	Commonwealth: Division 474 of the Criminal Code Act 1995 (Cth) contains offences relating to the use of carriage (i.e. telecommunications) services[e] involving sexual activity with a person under 16 years old, including 474.25A Using a carriage service for sexual activity with person under 16 years of age 474.25B Aggravated offence—child with mental impairment or under care, supervision or authority of defendant[f] 474.25C Using a carriage service to prepare or plan to cause harm to, engage in sexual activity with, or procure for sexual activity, persons under 16 years of age[g] 474.27A Using a carriage service to transmit indecent communication to person under 16 years of age

(continued)

Table 4.2 (continued)

Lanzarote treaty	Australia
	States and Territories: Each of the States and Territories has criminal offences relating to acts of indecency, which may include sexual acts in the presence of or directed at children
Article 23. Solicitation of children for sexual purposes	Commonwealth: Division 474 of the Criminal Code Act 1995(Cth) contains offences relating to the use of carriage (i.e. telecommunications) services involving sexual solicitation of a person under 16 years old, including 474.26 Using a carriage service to procure persons under 16 years of age 474.27 Using a carriage service to "groom" persons under 16 years of age States and Territories: Each of the States and Territories has criminal offences relating to child pornography e.g. s66EB of the Crimes Act 1900 (NSW) criminalises procuring or grooming a child under 16 for unlawful sexual activity (see Table 4.5 for more detail)

[a]Division 272 was added by the *Crimes Legislation Amendment (Sexual Offences Against Children) Act 2010* (Cth)

[b]Division 273 was added by the *Crimes Legislation Amendment (Sexual Offences Against Children) Act 2010* (Cth)

[c]Section 4.474.21 contains statutory defences, while Sections 4.474.22 to 474.24 contain analagous offences and defences in relation to 'child abuse material'. Sections 4.474.24A to 474.24C were added by the *Crimes Legislation Amendment (Sexual Offences Against Children) Act 2010* (Cth) to provide for aggravated offences where offending involves more than one person or occasion, and to require the Attorney-General's consent for the prosecution of a person under 18 years of age for child pornography material or child abuse material offences. This requirement provides some safeguard against the prosecution of minors for "sexting": see Urbas and Fouracre (2013)

[d]See e.g. the case concerning "The Simpsons": *McEwen v. Simmons [2008] NSWSC 1292*, discussed further below

[e]Division 474 provisions including telecommunications offences were added by the *Crimes Legislation Amendment (Telecommunications Offences and Other Measures) Act (No. 2) 2004* (Cth)

[f]These offences were added by the *Crimes Legislation Amendment (Sexual Offences Against Children) Act 2010* (Cth)

[g]This offence was added by the *Criminal Code Amendment (Protecting Minors Online) Act 2017* (Cth) and is known as "Carly's Law" to honour a teenaged girl who was murdered in South Australia by a child sex predator who had groomed her online using a fake identity. A joint media release on 9 August 2017 by the Australian Federal Police and South Australia Police announced that the law had been used for the first time in charging a convicted child offender with a range of offences: https://www.afp.gov.au/news-media/media-releases/police-use-carly%E2%80%99s-law-first-time

Table 4.3 Australian child pornography definitions [*Source* Urbas 2015a, b]

	Provision	Main definitional elements	Additional elements
CTH	Criminal Code Act 1995, s473.1 ("child pornography material"); s473.4 ("offensive")	Material that depicts or describes a person under 18 in a sexual pose or engaged in sexual activity, or that depicts for a sexual purpose the private parts of the person	Material must be offensive by the standards of reasonable adults, taking into account its literary etc. nature
ACT	Crimes Act 1900, s64(5) ("child pornography")	Anything that represents the sexual parts of a child or a child engaged in or present at an activity of a sexual nature, where "child" is a person under 18	Material must be substantially for the sexual arousal or sexual gratification of someone other than the child
NSW	Crimes Act 1900, ss91FA ("child), 91FB(1) ("child abuse material")	Material that depicts or describes a person under 16 in a sexual pose or engaged in sexual activity, or that depicts for a sexual purpose the private parts of the person	Material must be offensive by the standards of reasonable adults, taking into account its literary etc. nature
NT	Criminal Code, s125A ("child abuse material")	Material that depicts or describes a person under 18 engaging in sexual activity or in a sexual, offensive or demeaning context	Material must be likely to cause offence to a reasonable adult
QLD	Criminal Code 1899, s207A ("child exploitation material")	Material that depicts or describes a person under 16 engaging in sexual activity or in a sexual, offensive or demeaning context	Material must be likely to cause offence to a reasonable adult
SA	Criminal Law Consolidation Act 1935, s62 ("child pornography")	Material that depicts or describes a child under 17 engaging in sexual activity or the bodily parts of such a child	Material must be intended to excite or gratify sexual interest or sadistic or other perverted interest in violence or cruelty
TAS	Criminal Code Act 1924, s1A ("child exploitation material")	Material that depicts or describes a person under 18 engaging in sexual activity or in a sexual, offensive or demeaning context	Material must be likely to cause offence to a reasonable person
VIC	Crimes Act 1958, s67A ("child pornography")	Material that describes or depicts a minor engaging in sexual activity or depicted in an indecent sexual manner or context, where "minor" is a person under 18	
WA	Criminal Code, s217A ("child exploitation material" and "child pornography")	Material that describes or depicts a child engaging in sexual activity or in a sexual context, or that is likely to offend a reasonable person, where "child" is a person under 16	

Table 4.4 Australian child pornography offences [*Source* Urbas 2015a, b]

	Provision	Physical elements	Fault elements or defences	Maximum penalty
CTH	Criminal Code Act 1995, s474.19 (Using a carriage service for child pornography material)	Accessing, causing transmission to self, transmitting, making available, publishing, distributing, promoting or soliciting child pornography material, using a carriage service	Intentional accessing etc. and recklessness as to the nature of the material; absolute liability applies to using a carriage service (but note s474.21 defences)	Imprisonment for 15 years
	Criminal Code Act 1995, s474.20 (Possessing etc. child pornography material for use through a carriage service)	Possessing, controlling, producing, supplying or obtaining child pornography material	Intention that the material be used in committing an offence against s474.19 (but note s474.21 defences)	Imprisonment for 15 years
ACT	Crimes Act 1900, s64 (Using child for production of child pornography etc.)	Using, offering or procuring a child for the production of child pornography or for a pornographic performance	Absolute liability where the child is under 12; strict liability otherwise in relation to the element of the child's age	Imprisonment for 15 years (child under 12) or 10 years (child 12 years old or more); fines also apply
	Crimes Act 1900, s64A (Trading in child pornography)	Producing, publishing, offering or selling child pornography		Imprisonment for 12 years; fines also apply
	Crimes Act 1900, s65 (Possessing child pornography)	Possessing child pornography	Intentionally possessing; absolute liability applies to the nature of the material but with a statutory defence (s65(3))	Imprisonment for 7 years; fines also apply
NSW	Crimes Act 1900, s91G (Children not to be used for production of child abuse material)	Using a child for the production of child abuse material, or casing, procuring or allowing to such use	A defence of innocent production etc. applies (s91HA)	Imprisonment for 14 years (child under 14) or 10 years (child 14 years old or more)
	Crimes Act 1900, s91H (Production, dissemination or possession of child abuse material)	Producing, disseminating or possessing child abuse material		Imprisonment for 10 years

(continued)

Table 4.4 (continued)

	Provision	Physical elements	Fault elements or defences	Maximum penalty
NT	Criminal Code, s125B (Possession of child abuse material)	Possessing, distributing, producing, selling, offering or advertising child abuse material	Defences apply for law enforcement, health and medical purposes	Imprisonment for 10 years; fines also apply for a corporation
	Criminal Code, s125E (Using child for production of child abuse material etc.)	Using, offering or procuring a child for the production of child abuse material or for a pornographic or abusive performance		Imprisonment for 14 years; fines also apply for a corporation
QLD	Criminal Code 1899, s228A (Involving child in making child exploitation material)	Involving a child in making child exploitation material	Statutory defences apply for genuine artistic, legal etc. purposes (s228E)	Imprisonment for 14 years
	Criminal Code 1899, s228B (Making child exploitation material)	Making child exploitation material	Statutory defences apply for genuine artistic, legal etc. purposes (s228E)	Imprisonment for 14 years
	Criminal Code 1899, s228C (Distributing child exploitation material)	Distributing child exploitation material	Statutory defences apply for genuine artistic, legal etc. purposes (s228E)	Imprisonment for 14 years
	Criminal Code 1899, s228D (Possessing child exploitation material)	Knowingly possessing child exploitation material	Statutory defences apply for genuine artistic, legal etc. purposes (s228E)	Imprisonment for 14 years
SA	Criminal Law Consolidation Act 1935, s63 (Production or dissemination of child pornography)	Producing or disseminating, or taking any step in producing or disseminating, child pornography	Knowing its pornographic nature	Imprisonment for 10 years; 12 years if aggravated
	Criminal Law Consolidation Act 1935, s63A (Possession of child pornography)	Possessing or intending to obtain access to, or taking any step in obtaining access to, child pornography	Knowing its pornographic nature; but with a statutory defence (s63A(2)) of unknowingly getting possession and taking immediate steps to get rid of the material	Imprisonment for 5 years; 7 years if aggravated (first offence); or 7 years; 10 years if aggravated (subsequent offence)

(continued)

Table 4.4 (continued)

	Provision	Physical elements	Fault elements or defences	Maximum penalty
TAS	Criminal Code Act 1924, s130 (Involving person under 18 years in production of child exploitation material)	Involving or facilitating the involvement of a person under 18 years in the production of child exploitation material	Knowledge of the nature of the material; statutory defences apply for genuine artistic, legal etc. purposes (s130E)	A general maximum penalty of 21 years applies to Criminal Code offences
	Criminal Code Act 1924, s130A (Distribution of child exploitation material)	Distributing, or facilitating the distribution of, child exploitation material	Knowledge of the nature of the material; statutory defences apply (s130E)	A general maximum penalty of 21 years applies to Criminal Code offences
	Criminal Code Act 1924, s130B (Production of child exploitation material)	Producing, or facilitating the production of, child exploitation material	Knowledge of the nature of the material; statutory defences apply (s130E)	A general maximum penalty of 21 years applies to Criminal Code offences
	Criminal Code Act 1924, s130C (Possession of child exploitation material)	Possessing, or facilitating the production of, child exploitation material	Intending to access such material; statutory defences apply (s130E)	A general maximum penalty of 21 years applies to Criminal Code offences
	Criminal Code Act 1924, s130D (Accessing child exploitation material)	Accessing child exploitation material	Knowledge of the nature of the material; statutory defences apply (s130E)	A general maximum penalty of 21 years applies to Criminal Code offences
VIC	Crimes Act 1958, s68 (Production of child pornography)	Printing, making or otherwise producing child pornography	Statutory defences apply (s68(1A)-(4)) as well as exceptions for minors (s70AAA)	Imprisonment for 10 years
	Crimes Act 1958, s69 (Procurement etc. of minor for child pornography)	Inviting, procuring or causing a minor to be concerned in the making or production of child pornography	Statutory defences apply (s69(2)) as well as exceptions for minors (s70AAA)	Imprisonment for 10 years
	Crimes Act 1958, s69 (Procurement etc. of minor for child pornography)	Knowingly possessing child pornography	Statutory defences apply (s70(2)-(6)) as well as exceptions for minors (s70AAA)	Imprisonment for 5 years

(continued)

Table 4.4 (continued)

	Provision	Physical elements	Fault elements or defences	Maximum penalty
WA	Criminal Code, s217 (Involving child in child exploitation)	Inviting, procuring, offering or causing a child to be involved in the production of child exploitation material	Statutory defences and exclusions apply (s221A)	Imprisonment for 10 years
	Criminal Code, s218 (Producing child exploitation material)	Producing child exploitation material	Statutory defences and exclusions apply (s221A)	Imprisonment for 10 years
	Criminal Code, s219 (Distributing child exploitation material)	Distributing child exploitation material (including by electronic means)	Statutory defences and exclusions apply (s221A)	Imprisonment for 10 years
	Criminal Code, s220 (Possession of child exploitation)	Possessing child exploitation material	Statutory defences and exclusions apply (s221A)	Imprisonment for 7 years

Commonwealth has extensive child sex tourism laws. In relation to online offending, there are laws at all levels, but the most important offences are increasingly based on Commonwealth laws regulating the use of telecommunications services.

4.2.1 *Possibly Relevant Criminal Offences*

Succinct Overview of Sexual Offences Involving Minors

Table 4.2 lists the possibly relevant provisions of criminal law in Australia, grouped together by the provisions of the Lanzarote Convention, which gives the most comprehensive catalogue of sexual child-abuse offences available.

Additional information regarding child pornography offences under Australian law is provided by the Tables 4.3 and 4.4.

Additional information regarding child procuring and grooming offences under Australian law is provided by Table 4.5.

Overview of Sexual Offences Related to Webcam Child-Sexual Abuse

Webcam-based child sexual abuse may be prosecuted at all levels of the Australian criminal justice system, but for serious offending transcending national borders it is most likely that Commonwealth offences under the *Criminal Code Act 1995* (Cth) would be used. However, because Commonwealth offence proceedings are almost always brought in State or Territory courts, usually based on the place of residence of the defendant, it is entirely possible that some State or Territory

Table 4.5 Commonwealth, State and Territory child grooming offences [*Source* Urbas 2015a, b]

	Provision	Main elements	Maximum penalty
CTH	Criminal Code Act 1995, s474.26 ("procure")	Using a carriage service to transmit a communication to another person who is, or is believed to be, under 16 years old, with the intention of procuring the recipient to engage in, or submit to, sexual activity	Imprisonment for 15 years
	Criminal Code Act 1995, s474.27 ("groom")	Using a carriage service to transmit a communication that includes indecent material to another person who is, or is believed to be, under 16 years old, with the intention of making it easier to procure the recipient to engage in, or submit to, sexual activity	Imprisonment for 12 years; or 15 years imprisonment if s474.27(3) applies (grooming a child for another person)
ACT	Crimes Act 1900, s66(1)	Using electronic means, suggest to a young person (under 16 years) that the young person commit or take part in, or watch someone else committing or taking part in, an act of a sexual nature	Imprisonment for 10 years; or 5 years for a first offence
	Crimes Act 1900, s66(2)	Using electronic means, send or make available pornographic material to a young person (under 16 years)	Imprisonment for 5 years or 100 penalty units or both
NSW	Crimes Act 1900, s66EB(3)	An adult who engages in any conduct exposing a child (under 16 years) to indecent material, with the intention of making it easier to procure the child for unlawful sexual activity with that or any other person	Imprisonment for 12 years if the child is under 14 years of age; 10 years otherwise
NT	Criminal Code, s131	No specific provisions; see s131 (Attempting to procure child under 16 years); s132 (Indecent dealing with child under 16 years)	Imprisonment for 3 years (5 years if offender is an adult); 10 years (14 years if child under 10 years)
QLD	Criminal Code 1899, s218A	Using electronic communication (defined as "email, Internet chat rooms, SMS messages, real time audio/video or other similar communication") with intent to procure a person who is, or is believed to be, under 16, to engage in a sexual act	Imprisonment for 5 years; 10 years if the person intended to be procured is, or is believed to be, under 12
SA	Criminal Law Consolidation Act 1935	No specific provisions; see s63B (Procuring child to commit indecent act etc)	Imprisonment for 12 years for aggravated offence; 10 years for basic offence

(continued)

Table 4.5 (continued)

	Provision	Main elements	Maximum penalty
TAS	Criminal Code Act 1924, s125D	Making a communication by any means with the intention of procuring a person under the age of 17 years, or a person the accused person believes is under the age of 17 years, to engage in an unlawful sexual act	A general maximum penalty of 21 years applies to Criminal Code offences
VIC	Crimes Act 1958, s49B(2)	A person of or over 18 years of age communicating with a child under 16 years of age or a carer etc. of such a child with intent to facilitate the child's engagement or involvement in a sexual offence with that person or another adult	Imprisonment for 10 years
WA	Criminal Code, s204B	Using electronic communication (defined to include "data, text or images") with intent to procure a person who is, or is believed to be, under 16 to engage in sexual activity; or exposing a person under 16 to indecent matter	Imprisonment for 5 years (10 years if person is, or is believed to be, under 13)

offences may be added to the charges. The following are the most relevant Commonwealth offences:

Division 272—Child sex offences outside Australia

(c) 272.9 Sexual activity (other than sexual intercourse) with child outside Australia
(d) 272.11 Persistent sexual abuse of child outside Australia
(e) 272.14 Procuring child to engage in sexual activity outside Australia
(f) 272.15 "Grooming" child to engage in sexual activity outside Australia

Division 474—Telecommunications offences

(g) 474.19 Using a carriage service for child pornography material
(h) 474.20 Possessing, controlling, producing, supplying or obtaining child pornography material for use through a carriage service
(i) 474.24A Aggravated offence–offence involving conduct on 3 or more occasions and 2 or more people
(j) 474.25A Using a carriage service for sexual activity with person under 16

(k) 474.25B Aggravated offence–child with mental impairment or under care, supervision or authority of defendant
(l) 474.25C Using a carriage service to prepare or plan to cause harm to, engage in sexual activity with, or procure for sexual activity, persons under 16
(m) 474.26 Using a carriage service to procure persons under 16
(n) 474.27 Using a carriage service to "groom" persons under 16
(o) 474.27A Using a carriage service to transmit indecent communication to person under 16.

Case example 1: In the Victorian case of *R v. Goggins* [2014] VCC 1086 (7 July 2014), the defendant was sentenced on his guilty plea to numerous offences relating to his webcam exploitation of children in the Philippines. These activities were summarised by the judge as follows (Her Honour Judge Davis, Victorian County Court at [32]):

> I note that in terms of the webcam chats in schedule 1, there are many victims, particularly overseas victims. The sums paid to the victims indicates that you were willing to exploit the vulnerability of these children for your own sexual gratification, showing no regard for their welfare. You actively encouraged sexual abuse of these children and gave instructions about what you wanted to view. It was not isolated offending; it took place over a protracted period and you cultivated relationships with some of the victims, resulting in persistent sexual abuse. The age of the victims was as young as three, but mostly between the ages of six and 16. You recorded the shows and on one occasion shared it with another like-minded individual and the offending generally demonstrates your interest in under-age female children. You used an alias so that you could be anonymous on the internet. I acknowledge that your offending occurred online and that you are to be sentenced on that basis and not on the basis of physical contact, as none has occurred.

The charges under the *Criminal Code Act 1995* (Cth) included two charges of using a carriage service to access child pornography material (under s474.19); eight charges of engaging in sexual activity with a child (under s272.9); seven charges of persistent sexual abuse of a child outside Australia (under s272.11); one charge of cause child pornography material to be transmitted to himself using a carriage service (under s474.19); one charge of using a carriage service to solicit child pornography (under s474.19); and one charge of producing child pornography material for use through a carriage service (under s474.20). There were also two charges under the *Crimes Act 1958* (Vic): one charge of production of child pornography and one charge of knowingly possess child pornography. (This is a summary of [1]-[2] of the judgment).

The total effective sentence inposed was 11 years and 6 months imprisonment, with a minimum of 8 years to be served before eligibility for parole.

Case example 2: *DPP v. Le Gassick* [2014] VCC 1288 (12 August 2014) was another case in the Victorian County Court. The defendant pleaded guilty to 23 Commonwealth charges, arising from webcam child exploitation activities (at [4]–[8]):

> In short compass, the offending took place between April 2009 and January 2004. Charge 1 is a rolled up charge whereby between 5 April 2009 and 20 November 2013 you procured 54 individual children to engage in sexual activity. Essentially this related to you engaging on online chats to under aged girls in the Philippines, where in explicitly sexual terms you asked them to engage in sexual activity.
>
> On one occasion, detailed as an example, you masturbated to the point of ejaculation while engaging in sexualised chat with an 11 year old girl and showing her your penis over the webcam during the exchange.
>
> You also forwarded monies to accounts in the Philippines in 56 international transfers via Western Union accounts controlled by you between May 2011 and January 2014 for a total of $3,676.90 which were payments for online live sex shows to be performed by the children with whom you had negotiated.
>
> Chat log conversations retrieved from your computer revealed that before the money was transferred, you would negotiate the price and the activity you wanted the children to perform regularly expressing a preference for young girls and asking them whether they would meet you in person and whether they would have sex with you, if you visited the Philippines.
>
> Charge 2 relates to screen shot images you took of the children during the live sex shows, which you saved to your computer. These images described or showed children engaging in sexual activity or indecent sexual acts.

The total effective sentence imposed was 11 years imprisonment, with a minimum of 8 years to be served before eligibility for parole.

Case example 3: *Adamson v. The Queen* [2015] VSCA 194 (28 July 2015) was an appeal on severity of sentence imposed after guilty pleas in the Victorian County Court, with the main ground of appeal relating to: whether harm to child victims is to be presumed with respect to sexual offences committed via the internet (described during the appeal as 'cybersex' offences) (at [2]). The offender used disguised online identities (including an avatar and the identities of some of his victims) to persuade or blackmail mostly young victims into performing sexual acts over webcam. The list of charges and victim locations reveals widespread international victimisation (at [9]):

Count no	Count	Age of victim	Location of victim
1	Use carriage service to procure child under 16 for sexual act	15	England
2	Use carriage service to procure child under 16 for sexual act	13	England
3	Use carriage service to procure child under 16 for sexual act	13	USA
4	Use carriage service to procure child under 16 for sexual act	15	Poland
5	Procure a minor for child pornography	13 or 14	United Kingdom
6	Use carriage service to procure child under 16 for sexual act	13	Canada
7	Produce child pornography	13 or 14	Albury, NSW
8	Procure a minor for child pornography	16	Austria
9	Procure a minor for child pornography	16	Canada
10	Procure a minor for child pornography	17	France
11	Procure a minor for child pornography	16	USA
12	Use carriage service to groom child under 16 for sexual act	14	Serbia
13	Use carriage service to groom child under 16 for sexual act	15	England
14	Procure a minor for child pornography	16	USA
15	Produce child pornography	15–16	USA
16	Use carriage service to transmit indecent communications to a child under 16	13	England
17	Use carriage service to procure child under 16 for sexual act	14	Brisbane, QLD
18	Use carriage service to groom child under 16 for sexual act	15	USA
19	Use carriage service to groom child under 16 for sexual act	15	Netherlands
20	Use carriage service to groom child under 16 for sexual act	15	Netherlands
21	Use carriage service to transmit indecent communications to a child under 16	14	Queensland
22	Use carriage service to groom child under 16 for sexual act	14	USA
23	Use carriage service to procure child under 16 for sexual act	15	Queensland
24	Produce child pornography	16	England
25	Use carriage service to procure child under 16 for sexual act	15	USA

(continued)

(continued)

Count no	Count	Age of victim	Location of victim
26	Use carriage service to procure child under 16 for sexual act	14	Canada
27	Use carriage service to procure child under 16 for sexual act	15	Ballarat, VIC
28	Possess child pornography	No direct victim	

The sentence imposed by the trial judge on guilty pleas exceeded 6½ years with nearly 4 years non-parole, which was not altered on appeal.

4.2.2 *Interim Conclusion*

The following is a summary of *Criminal Code Act 1995* (Cth) provisions applicable to webcam-based child exploitation, particularly involving victims in other countries:

- If the perpetrator induces or forces the minor to display breasts or genitals or to perform sexual activities (e.g., masturbate) in front of a webcam, this may constitute:

272.9 Sexual activity (other than sexual intercourse) with child outside Australia
272.11 Persistent sexual abuse of child outside Australia—if done repeatedly
272.14 Procuring child to engage in sexual activity outside Australia
272.15 "Grooming" child to engage in sexual activity outside Australia

- If the perpetrator shows his genitals or masturbates in front of the webcam, this may constitute:

272.9 Sexual activity (other than sexual intercourse) with child outside Australia
272.11 Persistent sexual abuse of child outside Australia—if done repeatedly
272.14 Procuring child to engage in sexual activity outside Australia
272.15 "Grooming" child to engage in sexual activity outside Australia

Note: For the above offences, 'sexual activity' is defined to include either sexual intercourse (defined in s272.4) or 'any other activity of a sexual or indecent nature (including an indecent assault) that involves the human body, or bodily actions or functions (whether or not that activity involves physical contact between people)'.

- If any of this conduct is recorded and/or transmitted, so as to capture sexualised images of a child, the following telecommunications offences may also apply:

474.19 Using a carriage service for child pornography material
474.20 Possessing, controlling, producing, supplying or obtaining child pornography material for use through a carriage service
474.25A Using a carriage service for sexual activity with person under 16 years of age
474.26 Using a carriage service to procure persons under 16 years of age
474.27 Using a carriage service to "groom" persons under 16 years of age
474.27A Using a carriage service to transmit indecent communication to person under 16 years of age.

- If a child is involved in the production of child pornography material, then additional State/Territory offences may apply.
- All of the offences listed above are punishable by substantial imprisonment, the most serious by 25 years (s272.11). These maximum penalties are sufficient to allow a wide range of investigative powers (including covert techniques) and to result in convicted offenders being placed on sex offender registers.

4.2.3 *Possible Obstacles in Substantive Law Concerning Sweetie*

The main issue arising from the Sweetie scenario is the fact that no child is actually involved in either the communications from the offender or the images or communications being transmitted to the offender. The 'child' is a fictional representation, and the communications are with adults pretending to be the child.

However, there is both Australian legislation and case law that strongly suggests that this is no impediment to prosecution and conviction for at least some of the Commonwealth, State and territory offences considered above.

Child Pornography

Relevant legislative definitions typically refer to 'material that depicts a person, or a representation of a person, who is, or appears to be, under 18 years of age [or a different specified age] ... engaging in sexual activity or a sexual pose'. This has been held to include (as a 'representation of a person') entirely fictional images such as those based on "The Simpsons" cartoon characters:[10]

> Once it is accepted that the "person" may be fictional or imaginary and may be depicted by a drawing, it follows that a cartoon character might well constitute the depiction of such a "person" ... the drawing must be that of a human being and recognisable as such but no

---

[10] Adams J in *McEwen v. Simmons & Anor* [2008] NSWSC 1292 (8 December 2008) at [38]–[39], upholding convictions for possession of child pornography under both Commonwealth and NSW legislation.

particular human being needs to be depicted and even a substantial departure from realism will not necessarily mean that the depiction is not that of a person in this sense.

Child Grooming

Legislation creating the offences makes explicit that no actual child need be involved, and background materials make clear that this is deliberately in order to allow covert law enforcement investigations of grooming activities with police or other investigators pretending to be a (real or fictional) child. For example, s474.28 (9) of the *Criminal Code Act 1995* (Cth) provides:

> Fictitious recipient: For the purposes of sections 474.26, 474.27 and 474.27A, it does not matter that the recipient to whom the sender believes the sender is transmitting the communication is a fictitious person represented to the sender as a real person.

Further, the specified offences apply where communications are sent to a recipient who 'is someone who is, or who the sender believes to be, under 16 years of age', thus ensuring that the offence can still be committed (without relying on inchoate offences such as attempt) even where no child of that age was involved. Further, s474.28(3) provides:

> For the purposes of sections 474.26, 474.27 and 474.27A, evidence that the recipient was represented to the sender as being under or of a particular age is, in the absence of evidence to the contrary, proof that the sender believed the recipient to be under or of that age.

These provisions, and similar provisions of State/Territory child grooming legislation, have been interpreted by courts as clearly allowing covert investigation of child exploitation using covert or undercover means:[11]

> The evil to be confronted by this kind of investigation is of high public importance … the widespread use of the internet gives those disposed to corrupting and sexually exploiting children unprecedented access to vast numbers of potential victims. Such predators would be difficult to detect absent a complaint from an actual victim or an operation such as the present. The Gospel of St Matthew records Christ as condemning those who would corrupt the young in the following terms:
>
> 18:6 But who so shall offend one of these little ones which believe in me, it were better for him that a millstone were hanged about his neck, and that he were drowned in the depth of the sea (see *Mark 9:42; Luke 17:2*)
>
> That, I think reflects the community attitude toward such offences and such offenders. It would support the use of covert operations to detect them in a manner that does not place an actual young person at risk.

In a similar case, an ACT Supreme Court judge stated in relation to a defendant identified through a joint Federal Bureau of Investigations (FBI)—Australian Federal Police (AFP) operation, whose defence lawyer had likened to 'shooting fish in a barrel':[12]

---

[11] Higgins CJ in *R v. Stubbs* [2009] ACTSC 63 (26 May 2009) at [69], involving a New Zealand police officer posing as a 12-year-old girl online.

[12] Per Penfold J in *R v. Priest* [2011] ACTSC 18 (11 February 2011) at [65].

> The problem for Mr Priest is that the criminal law and the criminal justice system are not a game with rules designed to ensure a challenge for all participants and an enjoyable spectacle for observers. Certainly the criminal justice system involves more rules based on fairness than any game or sport I can think of, but those rules are aimed at protecting "the integrity of the administration of criminal justice" (*Ridgeway* at 33), at ensuring that police officers and other officials do not abuse their powers, and at ensuring that innocent people are not wrongly convicted.

Child Sex Tourism

Although the Division 272 offences in the *Criminal Code Act 1995* (Cth) have been applied to webcam-based child exploitation, as in the *R v. Goggins* case, there are no known cases where this has involved fictional children. However, the analogous offences involving child pornography and child grooming suggest that this may be a realistic possibility. Failing this, the inchoate offences of attempt, incitement or conspiracy may assist in prosecution of offenders engaging with a fictional childlike entity such as Sweetie.

4.3 Analysis of Criminal Procedure Law

4.3.1 General Description of Legal Framework

In Australia most criminal investigations are conducted by the police, though some public agencies also have investigative powers. Police refer almost all cases where it appears that an offence may have been committed and a person is charged to public prosecutors for the judicial process to begin. Prosecutors have no role in criminal investigations, but exercise discretion as to whether to bring or continue a prosecution based on: (i) the availability of admissible evidence with a prima facie case strong enough to support a conviction, and (ii) the public interest.

Under Australia's federal legal system, the Commonwealth, the States and Territories each have their own police forces, prosecution services and courts. However, there is some overlap in responsibilities e.g. the Australian Federal Police (AFP) investigate not only Commonwealth offences but also provide local policing to the Australian Capital Territory (ACT), and Commonwealth offences may be prosecuted alongside State/Territory offences (usually in a State or territory court as the Federal Court of Australia has limited criminal jurisdiction), with either the Commonwealth Director of Prosecutions (CDPP) or a State/Territory counterpart taking the lead according to which jurisdiction has the greater connection with the offences concerned. Thus, a typical cybercrime prosecution involving Commonwealth offences will be brought in a State or Territory court by the CDPP, who will also have carriage of any State/Territory offences involved.

The main stages of criminal procedure are: arrest, bail, committal (for more serious charges), trial, sentencing and appeals. There is a distinction between summary offences (with a lower maximum penalty, determined by a Magistrate without a jury) and indictable offences (more serious, dealt with in a higher court and possibly before a jury). Almost all child exploitation offences are indictable, and any Commonwealth offence prosecuted on indictment must be before a jury (due to a guarantee in s80 of the Commonwealth Constitution). Thus, a typical cybercrime prosecution involving Commonwealth offences will be on indictment before a jury (unless the defendant pleads guilty, in which case no jury is required and the matter proceeds to sentencing). Convicted defendants may appeal against the verdict or sentence, while prosecutors have more limited appeal rights, but can appeal against an unduly lenient sentence imposed by a trial judge or Magistrate.

Investigatory Powers

Succinct Overview of Investigatory Powers

Australian law complies with the procedural provisions of the *Convention on Cybercrime*, with recent amendments having been made under the *Cybercrime Legislation Amendment Act 2012* (Cth) in relation to data preservation so as to ensure Australia's accession with effect from 1 March 2013 (Table 4.6).

Human Rights
The common law principles on which the Australian criminal justice system is built include the right to a fair trial, which normally includes access to legal representation for serious contested charges, and certain constitutional guarantees. An explicit constitutional provision is the requirement that Commonwealth trials on indictment be before a jury. Implicit rights include equal treatment, and a requirement of proportionality in any laws curtailing freedom of political communication.

Two jurisdictions in the federation have a human rights act, i.e. the *Human Rights Act 2004* (ACT) and the *Charter of Human Rights and Responsibilities Act 2006*(Vic). Both essentially mirror *International Covenant on Civil and Political Rights* (ICCCPR) provisions, including in relation to rights in criminal proceedings. While there is no similar legislation at the Commonwealth level, there are statutory bodies including the Australian Human Rights Commission with powers to deal with complaints and to conduct investigations: *Australian Human Rights Commission Act 1986* (Cth).

Entrapment
There is no generally recognised defence of entrapment under Australian common law: *Ridgeway v. The Queen* (1995) CLR 19 (19 April 1995). However, courts have powers to exclude evidence that has been obtained illegally or improperly, since codified in s138 of the *Evidence Act 1995* (Cth) and mirror legislation in most of the States and Territories. Among the factors to be taken into account in balancing the public interest in having evidence admitted against the undesirability of condoning unlawful or improper investigative methods, is reference is s138(3) to the norms of the ICCCPR. In the context of covert investigations into child grooming activities,

Table 4.6 Australian procedural legislation relating to sexual offences against minors

Council of Europe Convention on Cybercrime	Australia
Article 16. Expedited preservation of stored computer data	Commonwealth: Part 3-1A of the Telecommunications (Interception and Access) Act 1979 (Cth) relates to preservation of stored communications by means of the following Div. 2—Domestic preservation notices Div. 3—Foreign preservation notices Part 3-3 allows access by law enforcement agencies with a warrant
Article 17. Expedited preservation and partial disclosure of traffic data	Part 4-1 of the Telecommunications (Interception and Access) Act 1979 (Cth) allows law enforcement access to telecommunications data by means of Div. 4—Authorisations for access to existing or prospective data
Article 18. Production order	A range of Commonwealth and other legislation provides for the use of production orders in legal processes
Article 19. Search and seizure of stored computer data	Part IAA of the Crimes Act 1914 (Cth) contains provisions on search and seizure of 'things' including computers 3K Use of equipment to examine or process things 3L Use of electronic equipment at premises 3LAA Use of electronic equipment at other place 3LA Person with knowledge of a computer or a computer system to assist access etc. 3LB Accessing data held on certain premises–notification to occupier of that premises Part 3-3 of the Telecommunications (Interception and Access) Act 1979 (Cth) allows access to stored data by law enforcement agencies with warrant
Article 20. Real-time collection of traffic data	Telecommunications interception with a warrant, access to telecommunications data with authorisation, are allowed by the Telecommunications (Interception and Access) Act 1979 (Cth). Delayed notification search warrants are allowed in some cases under the Crimes Act 1914 (Cth)
Article 21. Interception of content data	Telecommunications interception with warrant, access to telecommunications data with authorisation, are allowed by the Telecommunications (Interception and Access) Act 1979 (Cth). Delayed notification search warrants are allowed in some cases under the Crimes Act 1914 (Cth)
[Other (special) investigatory powers, not covered by the Cybercrime Convention, such as undercover operations]	Part IAB of the Crimes Act 1914 (Cth) governs the use of controlled operations for the covert investigation of serious Commonwealth offences, while Part IAC governs the use of assumed identities State/Territory laws mirror the Commonwealth laws

defence applications have been made to have evidence of online chats between defendants and undercover police excluded as having been unfairly or improperly obtained, but these have not usually been successful.

As noted in *R v. Priest* [2011] ACTSC 18 (11 February 2011) at [86], there will normally be no impropriety involved in undercover officers posing as children online and collecting evidence of predatory communications targeting the "child":

> Counsel for Mr Priest did not point to any particular part or aspect of Agent Chin's communications that he said went beyond offering an opportunity and crossed the line into encouraging or inciting the commission of the alleged offence. It is apparent that, although Agent Chin responded to Mr Priest's suggestions, he was careful not to initiate any particular proposal, and careful not even to respond too enthusiastically to Mr Priest's suggestions, preferring to emphasise his ignorance of sexual matters and his expectation that Mr Priest would prefer to engage with a person with more sexual experience. In particular I find that it was Mr Priest who initiated and pursued each development in the "relationship" between him and "Jamie". I am accordingly satisfied that Agent Chin did not go beyond providing an opportunity for Mr Priest to commit an offence he intended to commit, as permitted by the common law, and did not commit an offence by aiding, abetting, counselling or procuring a contravention of s 474.26 of the Criminal Code.

Succinct Overview of Investigatory Powers in an Online Context

After the *Ridgeway* case was decided, with the High Court ruling that police had acted improperly in that case by conducting a "controlled delivery" of narcotics without a statutory basis making such conduct lawful, a suite of controlled operations and related legislation was enacted. This can be found in Part IAB of the *Crimes Act 1914* (Cth) dealing with the use of controlled operations for the covert investigation of serious Commonwealth offences, and Part IAC governing the use of assumed identities. Both sets of laws, along with State and Territory counterparts, absolve police from criminal and civil liability for duly authorised covert investigations, thus removing the "illegality" limb for discretionary exclusion of evidence under s138. It should be noted, however, that there are some limits e.g. a controlled operation cannot be lawfully authorised or conducted so as to endanger the public: *Gedeon v. Commissioner of the NSW Crime Commission* [2008] HCA 43 (4 September 2008).

In the context of covert online investigations of child grooming (e.g. as opposed to infiltration of online child pornography rings), it has usually been considered unnecessary for police to resort to these mechanisms as no unlawful acts are usually engaged in, as observed in *R v. Priest* [2011] ACTSC 18 (11 February 2011) at [90]:

> Since I have found that Agent Chin did not commit an offence by aiding, abetting, counselling or procuring Mr Priest's alleged offence, I am also satisfied that there was no need for Agent Chin to have authority either for a controlled operation or for the use of an assumed identity in order to legitimise his actions.

No court has ruled on whether it is improper to conduct such investigations without the use of a controlled operation or assumed identity authority in the absence of otherwise unlawful conduct. The use of covert means to obtain admissions as to past criminal behaviour has also not generally been found to justify exclusion of evidence: *Tofilau v. The Queen; Marks v. The Queen; Hill v. The Queen; Clarke v. The Queen* [2007] HCA 39 (30 August 2007), a case involving the "Mr Big" technique. However, admissions evidence can be excluded if there are concerns about voluntariness, reliability or unfairness (under ss 84, 85 and 90 of the *Evidence Act 1995* (Cth) and mirror legislation or common law in the States and Territories).

An example of the application of these provisions in the context of covert policing of child sex offences is the case of *Pavitt v. Regina* [2007] NSWCCA 88 (2 April 2007), where by a 2:1 majority the NSW Court of Criminal Appeal dismissed an appeal based on the alleged unfairness of police using a victim to make a covertly recorded phone call to the (then suspected but not charged) defendant. After considering a series of Australian and international decisions on such "pretext phone calls", the majority (McColl J. A. and Latham J.) held that the evidence was admissible as "the conversation did not take place in circumstances which lead to the conclusion that its admission or a conviction obtained, at least in part in reliance upon it, was bought at a price which is unacceptable, having regard to contemporary community standards" (at [86]) and an "acceptable deception" (at [87]), while the dissenting judge (Adams J.) would have excluded the phone call evidence on the basis of unfairness.

Application of Relevant Investigatory Powers to the Sweetie Case

Based on the legislation and case law discussed above, there is no obvious impediment to the use of the "Sweetie" methodology in Australian child exploitation investigations. Police regularly use covert methods, including telecommunications interception and fictional identities, to identify child grooming suspects. This often results in the arrest of a suspected offender after a meeting between that person and the "child" has been arranged.[13] Forensic analysis of a suspect's computer and the records of communications stored either by police or an Internet Service Provider (ISP) then provide detailed evidence of the communications e.g. chat room logs, for which the intention of the defendant can be inferred. Convictions on guilty pleas in the face of such evidence are commonplace, and defence attempts to have such evidence excluded are rarely successful.

Importantly, both the substantive legislation and regulatory system of covert policing developed in Australia allow convictions for serious child sexual exploitation offences to occur even in the absence of a real child being groomed. Prosecutors do not need to rely on charging inchoate offences such as attempt or conspiracy. More widely, Australian police have been at the international forefront of disruption of organised online child exploitation networks, often using covert means. Where conduct needs to be engaged in during such investigations that might otherwise be illegal, controlled operations and assumed identities authorities can be

[13] Urbas 2010.

obtained to ensure both legality and admissibility of evidence. Such investigations may also involve the use of civilians, such as an alleged victim, in helping to obtain evidence.

What is relatively unexplored in the Australian context is the position of non-government organisations (NGOs) such as Terre des Hommes in developing and instigating covert investigations of online child offending. In part, this is explicable because the police have such a strong presence in this area, so that there is not a perceived need for victims or victim groups to take matters into their own hands (which is not the case in some other countries). If there were to be ongoing co-operation between law enforcement agencies and NGOs in covert investigations, courts would likely regard the latter as being "agents of the state" and apply provisions dealing with the admissibility of evidence obtained by investigating officials accordingly. However, even in this situation, there is considerable flexibility, as the *Evidence Act 1995* (Cth) and mirror legislation in the States and Territories defines 'investigating official' so as to not include covert operations:

> investigating official means:
>
> (a) a police officer (other than a police officer who is engaged in covert investigations under the orders of a superior); or
> (b) a person appointed by or under an Australian law (other than a person who is engaged in covert investigations under the orders of a superior) whose functions include functions in respect of the prevention or investigation of offences.

There is no reason to believe that the particular technology used in the Sweetie case would have any bearing on the application of these general considerations.

Relevant Aspects of Digital Forensic Evidence
Evidence from electronic sources is routinely admitted in Australian courts, using printouts or in-court demonstrations as required.[14] In most cases, forensic evidence is adduced through the questioning of an appropriately qualified expert witness, such as a particular analyst who conducted a forensic examination. Forensic examiners use standardised technologies (such as enCase) in making digital forensic copies of computer data, and prepare written reports for the court. Such reports may be admissible as affidavits or by means of a certificate of expert evidence, but in many cases the forensic examiner will be called to give testimony and be available for cross-examination. Defendants may also call their own expert witnesses.

Cases where forensic digital evidence is challenged are uncommon, though a recent example is *R v. Tahiraj* [2014] QCA 353 (19 December 2014). The defendant in this case was convicted of using a webcam to obtain pornographic images of young victims using threats, inducements and covert malware i.e. a remote administration tool (RAT) known as "Poison Ivy". The defence unsuccessfully argued at trial and on appeal, in part by reference to the evidence of their

---

[14] Urbas and Choo 2008.

own expert witness (Dr. Schatz), that the malware and offending images may have been installed on the defendant's computer by some unknown third party (the "Trojan defence"). Per McMurdo P (with the other two judges of the Queensland Court of Appeal agreeing) at [93]:

> The defence hypothesis that a malicious third party may have hacked into his computer without his knowledge and committed each of the offences was extremely unlikely. Whilst Dr Schatz's evidence made clear that this was theoretically possible, even on his evidence it remained extremely unlikely that the appellant's computer, located in his bedroom in a brick house, was interfered with in this way. He found no evidence to suggest hacking. And according to Dr Schatz it was extremely unlikely that such a hacker would plant a trace file of the kkkk.avi video on the appellant's computer without being detected. The complainant, A, identified the appellant's photo as being the person on webcam who forced her to participate in the video recording of kkkk.avi. I consider there is no real chance that the matters raised by the appellant, even in combination, may have resulted in him being convicted on any count. Rather, the jury considered the evidence and on each count and rejected the possibility that someone other than the appellant used his computer to commit the offences. The appellant has not established the matters raised in his grounds of appeal against conviction have resulted in a miscarriage of justice …

Miscellaneous

Australian courts have not yet considered the Sweetie technology explicitly, though it appears that offenders identified through the first exercise in 2013 may have been prosecuted in this country.[15] According to reports a year later, a guilty plea by a Queenslander to offences including sending obscene pictures to a child led to the first convictions arising from the use of Sweetie.[16]

4.4 Conclusions and Recommendations

Australia has adequate substantive and procedural laws to allow investigation and prosecution of online child sexual offending, including in cases where a fictitious child is used by police in detecting and identifying offenders. There is no serious obstacle to the admissibility of evidence thus obtained, though defence arguments can be made in individual cases based on illegality, impropriety or unfairness. The existing regulatory system for controlled operations and assumed identity authorities is available for online child exploitation investigations, but is not always used by police unless they perceive that there is a need based on breach of the law. Courts thus far have ruled covert online investigations to be lawful and generally acceptable.

---

[15] "Digital girl 'Sweetie' used by Dutch activists to track 1,000 alleged sexual predators, including Australians", *ABC News* (5 November 2013).

[16] M. Starr, "First man convicted in child predator sting with virtual girl Sweetie", *Cnet* (22 October 2014); see also Terre des Hommes, "Sweetie: First conviction in Australia" (22 October 2014): http://www.terredeshommes.org/sweetie-first-conviction/.

Recommendation: The legal basis on which covert online investigations of child sexual exploitation occurs in Australia appears to be adequate and results in successful prosecutions, but is somewhat unclear in resting on a mix of legislation generally applying to covert investigations and case law. It would be preferable to have more explicit and focussed legislation clearly regulating this area.

Acknowledgements This chapter was prepared while the author was an Associate Professor of Law at the University of Canberra, and during a Visiting Fellowship at the Tilburg Institute for Law, Technology and Society (TILT) during 2016–17, with final revisions in June 2018.

Relevant Legal Provisions (in Original)

These are the main provisions of the *Criminal Code Act 1995* (Cth) discussed above:

Division 272—Child Sex Offences Outside Australia

272.9 Sexual activity (other than sexual intercourse) with child outside Australia

Engaging in sexual activity with child

(1) A person commits an offence if:

(a) the person engages in sexual activity (other than sexual intercourse) with another person (the child); and
(b) the child is under 16; and
(c) the sexual activity is engaged in outside Australia.

Penalty: Imprisonment for 15 years.

Causing child to engage in sexual activity in presence of defendant

(2) A person commits an offence if:

(a) the person engages in conduct in relation to another person (the child); and
(b) that conduct causes the child to engage in sexual activity (other than sexual intercourse) in the presence of the person; and
(c) the child is under 16 when the sexual activity is engaged in; and
(d) the sexual activity is engaged in outside Australia.

Penalty: Imprisonment for 15 years.

(3) The fault element for para (2)(b) is intention.
(4) Absolute liability applies to paras (1)(b) and (c) and (2)(c) and (d).

Note: For absolute liability, see Section 6.2.

Defence—child present but defendant does not intend to derive gratification

(5) It is a defence to a prosecution for an offence against subsection (1) or (2) if:

(a) the conduct constituting the offence consists only of the child being in the presence of the defendant while sexual activity is engaged in; and
(b) the defendant proves that he or she did not intend to derive gratification from the presence of the child during that activity.

Note 1: A defendant bears a legal burden in relation to the matter in this subsection, see Section 13.4.

Note 2: For a defence based on belief about age, see Section 272.16.

272.11 Persistent sexual abuse of child outside Australia

(1) A person commits an offence against this section if the person commits an offence (the underlying offence) against one or more of the following provisions in relation to the same person (the child) on 3 or more separate occasions during any period:

(a) subsection 272.8(1) (engaging in sexual intercourse with child outside Australia);
(b) subsection 272.8(2) (causing child to engage in sexual intercourse in presence of defendant outside Australia);
(c) subsection 272.9(1) (engaging in sexual activity (other than sexual intercourse) with child outside Australia);
(d) subsection 272.9(2) (causing child to engage in sexual activity (other than sexual intercourse) in presence of defendant outside Australia).

Penalty: Imprisonment for 25 years.

(2) There is no fault element for any of the physical elements described in subsection (1) other than the fault elements (however described), if any, for the underlying offence.
(3) To avoid doubt, a person does not commit the underlying offence for the purposes of subsection (1) if the person has a defence to the underlying offence.

Offence or conduct need not be the same

(4) For the purposes of subsection (1), it is immaterial whether the underlying offence, or the conduct constituting the underlying offence, is the same on each occasion.

Certain matters need not be proved

(5) In proceedings for an offence against this section, it is not necessary to specify or to prove the dates or exact circumstances of the occasions on which the conduct constituting the offence against this section occurred.

Content of charge

(6) A charge of an offence against this section:

(a) must specify with reasonable particularity the period during which the offence against this section occurred; and
(b) must describe the nature of the separate offences alleged to have been committed by the person during that period.

Trier of fact to be satisfied of certain matters

(7) In order for the person to be found guilty of an offence against this section:

(a) the trier of fact must be satisfied beyond reasonable doubt that the evidence establishes at least 3 separate occasions during the period concerned on which the person engaged in conduct constituting an offence against subsection 272.8(1) or (2) or 272.9(1) or (2), of a nature described in the charge, in relation to the child; and
(b) the trier of fact must be so satisfied about the material facts of the 3 such occasions, although the trier of fact need not be so satisfied about the dates or the order of those occasions; and
(c) if the trier of fact is a jury and more than 3 such occasions are relied on as evidence of the commission of an offence against this section—all the members of the jury must be so satisfied about the same 3 incidents.

(8) In proceedings for an offence against this section, the judge must warn the jury (if any) of the requirements of subsection (7).

Double jeopardy etc.

(9) A person who has been convicted or acquitted of an offence against this section may not be convicted of another offence against Section 272.8, 272.9 or 272.10 that is alleged to have been committed in relation to the child in the period during which the person was alleged to have committed the offence against this section.
(10) However, subsection (9) does not prevent an alternative verdict under Section 272.28.
(11) A person who has been convicted or acquitted of an offence against Section 272.8, 272.9 or 272.10 in relation to a person (the child) may not be convicted of an offence against this section in relation to the child if any of the occasions relied on as evidence of the commission of the offence against this section includes the conduct that constituted the offence of which the person was convicted or acquitted.

272.14 Procuring child to engage in sexual activity outside Australia

(1) A person commits an offence if:

(a) the person engages in conduct in relation to another person (the child); and
(b) the person does so with the intention of procuring the child to engage in sexual activity (whether or not with the person) outside Australia; and
(c) the child is someone:

(i) who is under 16; or
(ii) who the person believes to be under 16; and

(d) one or more of the following apply:

(i) the conduct referred to in para (a) occurs wholly or partly outside Australia;
(ii) the child is outside Australia when the conduct referred to in para (a) occurs;
(iii) the conduct referred to in para (a) occurs wholly in Australia and the child is in Australia when that conduct occurs.

Penalty: Imprisonment for 15 years.

(2) Absolute liability applies to subparagraph (1)(c)(i) and para (1)(d).

Note 1: For absolute liability, see Section 6.2.

Note 2: For a defence based on belief about age, see Section 272.16.

(3) A person may be found guilty of an offence against subsection (1) even if it is impossible for the sexual activity referred to in that subsection to take place.
(4) For the purposes of subsection (1), it does not matter that the child is a fictitious person represented to the person as a real person.

272.15 "Grooming" child to engage in sexual activity outside Australia

(1) A person commits an offence if:

(a) the person engages in conduct in relation to another person (the child); and
(b) the person does so with the intention of making it easier to procure the child to engage in sexual activity (whether or not with the person) outside Australia; and
(c) the child is someone:

(i) who is under 16; or
(ii) who the person believes to be under 16; and

(d) one or more of the following apply:

(i) the conduct referred to in para (a) occurs wholly or partly outside Australia;
(ii) the child is outside Australia when the conduct referred to in para (a) occurs;

(iii) the conduct referred to in para (a) occurs wholly in Australia and the child is in Australia when that conduct occurs.

Penalty: Imprisonment for 12 years.

(2) Absolute liability applies to subparagraph (1)(c)(i) and para (1)(d).

Note 1: For absolute liability, see Section 6.2.

Note 2: For a defence based on belief about age, see Section 272.16.

(3) A person may be found guilty of an offence against subsection (1) even if it is impossible for the sexual activity referred to in that subsection to take place.
(4) For the purposes of subsection (1), it does not matter that the child is a fictitious person represented to the person as a real person.

Division 474—Telecommunications Offences

474.19 Using a carriage service for child pornography material

(1) A person is guilty of an offence if:

(a) the person:

(i) accesses material; or
(ii) causes material to be transmitted to himself or herself; or
(iii) transmits, makes available, publishes, distributes, advertises or promotes material; or
(iv) solicits material; and

(aa) the person does so using a carriage service; and

(b) the material is child pornography material.

Penalty: Imprisonment for 15 years.

(2) To avoid doubt, the following are the fault elements for the physical elements of an offence against subsection (1):

(a) intention is the fault element for the conduct referred to in para (1)(a);
(b) recklessness is the fault element for the circumstances referred to in para (1)(b).

Note: For the meaning of intention and recklessness see Sections 5.2 and 5.4.

(2A) Absolute liability applies to para (1)(aa).

Note: For absolute liability, see Section 6.2.

(3) As well as the general defences provided for in Part 2.3, defences are provided for under Section 474.21 in relation to this section.

474.20 Possessing, controlling, producing, supplying or obtaining child pornography material for use through a carriage service

(1) A person is guilty of an offence if:

(a) the person:

(i) has possession or control of material; or
(ii) produces, supplies or obtains material; and

(b) the material is child pornography material; and
(c) the person has that possession or control, or engages in that production, supply or obtaining, with the intention that the material be used:

(i) by that person; or
(ii) by another person;

in committing an offence against Section 474.19 (using a carriage service for child pornography material).

Penalty: Imprisonment for 15 years.

(2) A person may be found guilty of an offence against subsection (1) even if committing the offence against Section 474.19 (using a carriage service for child pornography material) is impossible.
(3) It is not an offence to attempt to commit an offence against subsection (1).

474.25A Using a carriage service for sexual activity with person under 16 years of age

Engaging in sexual activity with child using a carriage service

(1) A person commits an offence if:

(a) the person engages in sexual activity with another person (the child) using a carriage service; and
(b) the child is under 16 years of age; and
(c) the person is at least 18 years of age.

Penalty: Imprisonment for 15 years.

Causing child to engage in sexual activity with another person

(2) A person (the defendant) commits an offence if:

(a) the defendant engages in conduct in relation to another person (the child); and
(b) that conduct causes the child to engage in sexual activity with another person (the participant) using a carriage service; and
(c) the child is under 16 years of age when the sexual activity is engaged in; and
(d) the participant is at least 18 years of age when the sexual activity is engaged in.

Penalty: Imprisonment for 15 years.

(3) The fault element for para (2)(b) is intention.

Defence—child present but defendant does not intend to derive gratification

(4) It is a defence to a prosecution for an offence against subsection (1) or (2) if:

(a) the conduct constituting the offence consists only of the child being in the presence of a person while sexual activity is engaged in; and
(b) the defendant proves that he or she did not intend to derive gratification from the presence of the child during that activity.

Note 1: A defendant bears a legal burden in relation to the matter in this subsection, see Section 13.4.

Note 2: For other defences relating to this offence, see Section 474.29.

474.25B Aggravated offence—child with mental impairment or under care, supervision or authority of defendant

(1) A person commits an offence against this section if:

(a) the person commits an offence against either of the following provisions in relation to another person (the child):

(i) subsection 474.25A(1) (engaging in sexual activity with child using a carriage service);
(ii) subsection 474.25A(2) (causing child to engage in sexual activity with another person); and

(b) either or both of the following apply at the time the person commits the offence:

(i) the child has a mental impairment;
(ii) the person is in a position of trust or authority in relation to the child, or the child is otherwise under the care, supervision or authority of the person.

Penalty: Imprisonment for 25 years.

(2) To avoid doubt, a person does not commit the offence against subsection 474.25A(1) or (2) for the purposes of para (1)(a) if the person has a defence to that offence.

Alternative verdicts

(3) If, on a trial for an offence (the aggravated offence) against subsection (1), the trier of act:

(a) is not satisfied that the defendant is guilty of the aggravated offence; but
(b) is satisfied beyond reasonable doubt that he or she is guilty of an offence (the underlying offence) against subsection 474.25A(1) or (2);

it may find the defendant not guilty of the aggravated offence but guilty of the underlying offence, so long as the defendant has been accorded procedural fairness in relation to that finding of guilt.

474.25C Using a carriage service to prepare or plan to cause harm to, engage in sexual activity with, or procure for sexual activity, persons under 16

A person (the first person) commits an offence if:

(a) the first person does any act in preparation for doing, or planning to do, any of the following:

(i) causing harm to a person under 16 years of age;
(ii) engaging in sexual activity with a person under 16 years of age;
(iii) procuring a person under 16 years of age to engage in sexual activity; and

(b) the first person is at least 18 years of age; and
(c) the act is done using a carriage service.

Penalty: Imprisonment for 10 years.

Example: A person misrepresents their age online as part of a plan to cause harm to another person under 16 years of age.

474.26 Using a carriage service to procure persons under 16 years of age

(1) A person (the sender) commits an offence if:

(a) the sender uses a carriage service to transmit a communication to another person (the recipient); and
(b) the sender does this with the intention of procuring the recipient to engage in sexual activity with the sender; and
(c) the recipient is someone who is, or who the sender believes to be, under 16 years of age; and
(d) the sender is at least 18 years of age.

Penalty: Imprisonment for 15 years.

(2) A person (the sender) commits an offence if:

(a) the sender uses a carriage service to transmit a communication to another person (the recipient); and
(b) the sender does this with the intention of procuring the recipient to engage in sexual activity with another person (the participant); and
(c) the recipient is someone who is, or who the sender believes to be, under 16 years of age; and
(d) the participant is someone who is, or who the sender believes to be, at least 18 years of age.

Penalty: Imprisonment for 15 years.

(3) A person (the sender) commits an offence if:

(a) the sender uses a carriage service to transmit a communication to another person (the recipient); and
(b) the sender does this with the intention of procuring the recipient to engage in sexual activity with another person; and
(c) the recipient is someone who is, or who the sender believes to be, under 16 years of age; and
(d) the other person referred to in para (b) is someone who is, or who the sender believes to be, under 18 years of age; and
(e) the sender intends that the sexual activity referred to in para (b) will take place in the presence of:

(i) the sender; or
(ii) another person (the participant) who is, or who the sender believes to be, at least 18 years of age.

Penalty: Imprisonment for 15 years.

474.27 Using a carriage service to "groom" persons under 16 years of age

(1) A person (the sender) commits an offence if:

(a) the sender uses a carriage service to transmit a communication to another person (the recipient); and
(c) the sender does this with the intention of making it easier to procure the recipient to engage in sexual activity with the sender; and
(d) the recipient is someone who is, or who the sender believes to be, under 16 years of age; and
(e) the sender is at least 18 years of age.

Penalty: Imprisonment for 12 years.

(2) A person (the sender) commits an offence if:

(a) the sender uses a carriage service to transmit a communication to another person (the recipient); and
(c) the sender does this with the intention of making it easier to procure the recipient to engage in sexual activity with another person (the participant); and
(d) the recipient is someone who is, or who the sender believes to be, under 16 years of age; and
(e) the participant is someone who is, or who the sender believes to be, at least 18 years of age.

Penalty: Imprisonment for 12 years.

(3) A person (the sender) commits an offence if:

(a) the sender uses a carriage service to transmit a communication to another person (the recipient); and
(c) the sender does this with the intention of making it easier to procure the recipient to engage in sexual activity with another person; and
(d) the recipient is someone who is, or who the sender believes to be, under 16 years of age; and
(e) the other person referred to in para (c) is someone who is, or who the sender believes to be, under 18 years of age; and
(f) the sender intends that the sexual activity referred to in para (c) will take place in the presence of:

(i) the sender; or
(ii) another person (the participant) who is, or who the sender believes to be, at least 18 years of age.

Penalty: Imprisonment for 15 years.

474.27A Using a carriage service to transmit indecent communication to person under 16 years of age
(1) A person (the sender) commits an offence if:

(a) the sender uses a carriage service to transmit a communication to another person (the recipient); and
(b) the communication includes material that is indecent; and
(c) the recipient is someone who is, or who the sender believes to be, under 16 years of age; and
(d) the sender is at least 18 years of age.

Penalty: Imprisonment for 7 years.

(2) In a prosecution for an offence against subsection (1), whether material is indecent is a matter for the trier of fact.
(3) In this section:

Indecent means indecent according to the standards of ordinary people.

474.28 Provisions relating to offences against this Subdivision
Age-related issues–application of absolute liability

(1) For the purposes of an offence against this Subdivision, absolute liability applies to the physical element of circumstance of the offence that:

(a) in the case of an offence against Section 474.25A–the child is under 16 years of age; and

(b) in the case of an offence against Section 474.26, 474.27 or 474.27A–the recipient is someone who is under 16 years of age.

Note 1: For absolute liability, see Section 6.2.

Note 2: For a defence based on belief about age, see Section 474.29.

(2) For the purposes of an offence against subsection 474.25A(2), 474.26(2) or (3) or 474.27(2) or (3), absolute liability applies to the physical elements of circumstance of the offence that the participant is at least 18 years of age.

Note 1: For absolute liability, see Section 6.2.

Note 2: For a defence based on belief about age, see Section 474.29.

Proof of belief about age–evidence of representation

(3) For the purposes of Sections 474.26, 474.27 and 474.27A, evidence that the recipient was represented to the sender as being under or of a particular age is, in the absence of evidence to the contrary, proof that the sender believed the recipient to be under or of that age.

(4) For the purposes of Sections 474.25A, 474.26 and 474.27, evidence that the participant was represented to the sender as being:

(a) at least 18 years of age; or
(b) over or of a particular age;

is, in the absence of evidence to the contrary, proof that the sender believed the participant to be at least 18 years of age or over or of that age.

Determining age–admissible evidence

(5) In determining for the purposes of this Subdivision how old a person is or was at a particular time, a jury or court may treat any of the following as admissible evidence:

(a) the person's appearance;
(b) medical or other scientific opinion;
(c) a document that is or appears to be an official or medical record from a country outside Australia;
(d) a document that is or appears to be a copy of such a record.

(6) Subsection (5) does not make any other kind of evidence inadmissible, and does not affect a prosecutor's duty to do all he or she can to adduce the best possible evidence for determining the question.

(7) If, on a trial for an offence against a provision of this Subdivision, evidence may be treated as admissible because of subsection (5), the court must warn the jury that it must be satisfied beyond reasonable doubt in determining the question.

Issues relating to aggravated offence involving sexual activity

(7A) For the purposes of an offence against subsection 474.25B(1):

(a) there is no fault element for the physical element described in para (a) of that subsection other than the fault elements (however described), if any, for the underlying offence; and
(b) absolute liability applies to the physical element of circumstance of the offence that the child has a mental impairment; and
(c) strict liability applies to the physical element of circumstance of the offence that the defendant is in a position of trust or authority in relation to the child, or the child is otherwise under the care, supervision or authority of the defendant.

Note 1: For absolute liability, see Section 6.2.

Note 2: For strict liability, see Section 6.1.

Note 3: For a defence based on belief that the child did not have a mental impairment, see Section 474.29.

Impossibility of sexual activity taking place

(8) A person may be found guilty of an offence against Section 474.26 or 474.27 even if it is impossible for the sexual activity referred to in that section to take place.

Fictitious recipient

(9) For the purposes of Sections 474.26, 474.27 and 474.27A, it does not matter that the recipient to whom the sender believes the sender is transmitting the communication is a fictitious person represented to the sender as a real person.

Attempt not offence

(10) It is not an offence to attempt to commit an offence against Section 474.26 or 474.27.

474.29 Defences to offences against this Subdivision

Offences involving sexual activity–belief that child at least 16 years of age

(1) It is a defence to a prosecution for an offence against Section 474.25A if the defendant proves that, at the time the sexual activity was engaged in, he or she believed that the child was at least 16 years of age.

Note: A defendant bears a legal burden in relation to the matter in this subsection, see Section 13.4.

Offences involving sexual activity with other participant–belief that participant under 18 years of age

(2) It is a defence to a prosecution for an offence against subsection 474.25A(2) if the defendant proves that, at the time the sexual activity was engaged in, he or she believed that the participant was under 18 years of age.

Note: A defendant bears a legal burden in relation to the matter in this subsection, see Section 13.4.

Aggravated offence involving sexual activity–belief that child did not have mental impairment

(3) It is a defence to a prosecution for an offence against subsection 474.25B(1) (as that subsection applies because of subparagraph 474.25B(1)(b)(i)) if the defendant proves that, at the time the defendant committed the offence, he or she believed that the child did not have a mental impairment.

Note: A defendant bears a legal burden in relation to the matter in this subsection, see Section 13.4.

Offences involving procuring or "grooming" person for sexual activity with other participant–belief that participant under 18 years of age

(4) It is a defence to a prosecution for an offence against subsection 474.26(2) or (3) or 474.27(2) or (3) if the defendant proves that, at the time the communication was transmitted, he or she believed that the participant was under 18 years of age.

Note: A defendant bears a legal burden in relation to the matter in this subsection, see Section 13.4.

Offences involving transmission of communication–belief that recipient at least 16 years of age

(5) It is a defence to a prosecution for an offence against Section 474.26, 474.27 or 474.27A if the defendant proves that, at the time the communication was transmitted, he or she believed that the recipient was at least 16 years of age.

Note: A defendant bears a legal burden in relation to the matter in this subsection, see Section 13.4.

Trier of fact may take into account whether belief reasonable

(6) In determining whether the defendant had the belief mentioned in one of the preceding subsections of this section, the trier of fact may take into account whether the alleged belief was reasonable in the circumstances.

These are the main provisions of the *Crimes Act 1914* (Cth) discussed above:

3K Use of equipment to examine or process things

Equipment may be brought to warrant premises

(1) The executing officer of a warrant in relation to premises, or constable assisting, may bring to the warrant premises any equipment reasonably necessary for the examination or processing of a thing found at the premises in order to determine whether it is a thing that may be seized under the warrant.

Thing may be moved for examination or processing

(2) A thing found at warrant premises, or a thing found during a search under a warrant that is in force in relation to a person, may be moved to another place for examination or processing in order to determine whether it may be seized under a warrant if:

(a) both of the following apply:

(i) it is significantly more practicable to do so having regard to the timeliness and cost of examining or processing the thing at another place and the availability of expert assistance;

(ii) the executing officer or constable assisting suspects on reasonable grounds that the thing contains or constitutes evidential material; or

(b) for a thing found at warrant premises—the occupier of the premises consents in writing; or

(c) for a thing found during a search under a warrant that is in force in relation to a person—the person consents in writing.

Notification of examination or processing and right to be present

(3) If a thing is moved to another place for the purpose of examination or processing under subsection (2), the executing officer must, if it is practicable to do so:

(a) inform the person referred to in para (2)(b) or (c) (as the case requires) of the address of the place and the time at which the examination or processing will be carried out; and

(b) allow that person or his or her representative to be present during the examination or processing.

(3AA) The executing officer need not comply with para (3)(a) or (b) if he or she believes on reasonable grounds that to do so might:

(a) endanger the safety of a person; or

(b) prejudice an investigation or prosecution.

Time limit on moving a thing

(3A) The thing may be moved to another place for examination or processing for no longer than 14 days.

(3B) An executing officer may apply to an issuing officer for one or more extensions of that time if the executing officer believes on reasonable grounds that the thing cannot be examined or processed within 14 days or that time as previously extended.
(3C) The executing officer must give notice of the application to the person referred to in para (2)(b) or (c) (as the case requires), and that person is entitled to be heard in relation to the application.
(3D) A single extension cannot exceed 7 days.

Equipment at warrant premises may be operated

(4) The executing officer of a warrant in relation to premises, or a constable assisting, may operate equipment already at the warrant premises to carry out the examination or processing of a thing found at the premises in order to determine whether it is a thing that may be seized under the warrant if the executing officer or constable believes on reasonable grounds that:

(a) the equipment is suitable for the examination or processing; and
(b) the examination or processing can be carried out without damage to the equipment or the thing.

3L Use of electronic equipment at premises

(1) The executing officer of a warrant in relation to premises, or a constable assisting, may operate electronic equipment at the warrant premises to access data (including data not held at the premises) if he or she suspects on reasonable grounds that the data constitutes evidential material.

Note: A constable can obtain an order requiring a person with knowledge of a computer or computer system to provide assistance: see Section 3LA.

(1A) If the executing officer or constable assisting suspects on reasonable grounds that any data accessed by operating the electronic equipment constitutes evidential material, he or she may:

(a) copy any or all of the data accessed by operating the electronic equipment to a disk, tape or other associated device brought to the premises; or
(b) if the occupier of the premises agrees in writing copy any or all of the data accessed by operating the electronic equipment to a disk, tape or other associated device at the premises;

and take the device from the premises.

(1B) If:

(a) the executing officer or constable assisting takes the device from the premises; and
(b) the Commissioner is satisfied that the data is not required (or is no longer required) for a purpose mentioned in Section 3ZQU or for other judicial or administrative review proceedings;

the Commissioner must arrange for:

(c) the removal of the data from any device in the control of the Australian Federal Police; and
(d) the destruction of any other reproduction of the data in the control of the Australian Federal Police.

(2) If the executing officer or a constable assisting, after operating the equipment, finds that evidential material is accessible by doing so, he or she may:

(a) seize the equipment and any disk, tape or other associated device; or
(b) if the material can, by using facilities at the premises, be put in documentary form—operate the facilities to put the material in that form and seize the documents so produced.

(3) A constable may seize equipment under para (2)(a) only if:

(a) it is not practicable to copy the data as mentioned in subsection (1A) or to put the material in documentary form as mentioned in para (2)(b); or
(b) possession by the occupier of the equipment could constitute an offence.

(4) If the executing officer or a constable assisting suspects on reasonable grounds that:

(a) evidential material may be accessible by operating electronic equipment at the premises; and
(b) expert assistance is required to operate the equipment; and
(c) if he or she does not take action under this subsection, the material may be destroyed, altered or otherwise interfered with;

he or she may do whatever is necessary to secure the equipment, whether by locking it up, placing a guard or otherwise.

(5) The executing officer or a constable assisting must give notice to the occupier of the premises of his or her intention to secure equipment and of the fact that the equipment may be secured for up to 24 hours.

(6) The equipment may be secured:

(a) for a period not exceeding 24 hours; or
(b) until the equipment has been operated by the expert;

whichever happens first.

(7) If the executing officer or a constable assisting believes on reasonable grounds that the expert assistance will not be available within 24 hours, he or she may apply to an issuing officer for an extension of that period.

(8) The executing officer or a constable assisting must give notice to the occupier of the premises of his or her intention to apply for an extension, and the occupier is entitled to be heard in relation to the application.

(9) The provisions of this Division relating to the issue of warrants apply, with such modifications as are necessary, to the issuing of an extension.

3LAA Use of electronic equipment at other place

(1) If electronic equipment is moved to another place under subsection 3K(2), the executing officer or a constable assisting may operate the equipment to access data (including data held at another place).

(2) If the executing officer or constable assisting suspects on reasonable grounds that any data accessed by operating the electronic equipment constitutes evidential material, he or she may copy any or all of the data accessed by operating the electronic equipment to a disk, tape or other associated device.

(3) If the Commissioner is satisfied that the data is not required (or is no longer required) for a purpose mentioned in Section 3ZQU or for other judicial or administrative review proceedings, the Commissioner must arrange for:

(a) the removal of the data from any device in the control of the Australian Federal Police; and
(b) the destruction of any other reproduction of the data in the control of the Australian Federal Police.

(4) If the executing officer or a constable assisting, after operating the equipment, finds that evidential material is accessible by doing so, he or she may:

(a) seize the equipment and any disk, tape or other associated device; or
(b) if the material can be put in documentary form put the material in that form and seize the documents so produced.

(5) A constable may seize equipment under para (4)(a) only if:

(a) it is not practicable to copy the data as mentioned in subsection (2) or to put the material in documentary form as mentioned in para (4)(b); or
(b) possession of the equipment, by the person referred to in para 3K(2)(a) or (b) (as the case requires), could constitute an offence.

3LA Person with knowledge of a computer or a computer system to assist access etc.

(1) A constable may apply to a magistrate for an order requiring a specified person to provide any information or assistance that is reasonable and necessary to allow a constable to do one or more of the following:

(a) access data held in, or accessible from, a computer or data storage device that:

(i) is on warrant premises; or
(ii) has been moved under subsection 3K(2) and is at a place for examination or processing; or

(iii) has been seized under this Division;

(b) copy data held in, or accessible from, a computer, or data storage device, described in para (a) to another data storage device;
(c) convert into documentary form or another form intelligible to a constable:

(i) data held in, or accessible from, a computer, or data storage device, described in para (a); or
(ii) data held in a data storage device to which the data was copied as described in para (b); or
(iii) data held in a data storage device removed from warrant premises under subsection 3L(1A).

(2) The magistrate may grant the order if the magistrate is satisfied that:

(a) there are reasonable grounds for suspecting that evidential material is held in, or is accessible from, the computer or data storage device; and
(b) the specified person is:

(i) reasonably suspected of having committed the offence stated in the relevant warrant; or
(ii) the owner or lessee of the computer or device; or
(iii) an employee of the owner or lessee of the computer or device; or
(iv) a person engaged under a contract for services by the owner or lessee of the computer or device; or
(v) a person who uses or has used the computer or device; or
(vi) a person who is or was a system administrator for the system including the computer or device; and

(c) the specified person has relevant knowledge of:

(i) the computer or device or a computer network of which the computer or device forms or formed a part; or
(ii) measures applied to protect data held in, or accessible from, the computer or device.

(3) If:

(a) the computer or data storage device that is the subject of the order is seized under this Division; and
(b) the order was granted on the basis of an application made before the seizure;

the order does not have effect on or after the seizure.

Note: An application for another order under this section relating to the computer or data storage device may be made after the seizure.

(4) If the computer or data storage device is not on warrant premises, the order must:

(a) specify the period within which the person must provide the information or assistance; and
(b) specify the place at which the person must provide the information or assistance; and
(c) specify the conditions (if any) determined by the magistrate as the conditions to which the requirement on the person to provide the information or assistance is subject.

(5) A person commits an offence if the person fails to comply with the order.

Penalty for contravention of this subsection: Imprisonment for 2 years.

3LB Accessing data held on certain premises—notification to occupier of that premises

(1) If:

(a) data is accessed, in relation to a warrant, under subsection 3L(1) or 3LAA(1); and

(aa) either:

(i) if the warrant is in relation to premises—the data is held on premises other than the warrant premises; or
(ii) if the warrant is in relation to a person—the data is held on any premises; and

(b) it is practicable to notify the occupier of the premises on which the data is held that the data has been accessed under a warrant;

the executing officer must:

(c) do so as soon as practicable; and
(d) if the executing officer has arranged, or intends to arrange, for continued access to the data under subsection 3L(1A) or (2) or 3LAA(2) or (4)—include that information in the notification.

(2) A notification under subsection (1) must include sufficient information to allow the occupier of the premises on which the data is held to contact the executing officer.

References

Urbas G (2010) Protecting children from online predators: the use of covert investigation techniques by law enforcement. Journal of Contemporary Criminal Justice 26(4): 410–425

Urbas G (2013) Australia's accession to the Convention on Cybercrime. Internet Law Bulletin 16 (2): 2–3

Urbas G (2015a) Complicity in cyberspace: doctrines of accessorial liability and online groups. In: Crofts T, Loughnan A (eds) Criminalisation and criminal responsibility in Australia. Oxford

University Press; also In: Smith R, Cheung R, Lau L (eds) Cybercrime risks and responses: Eastern and Western perspectives. Palgrave Macmillan
Urbas G (2015b) Cybercrime: legislation, cases and materials. LexisNexis, Chatswood NSW
Urbas G, Choo KKR (2008) Resource materials on technology-enabled crime. Technical and Background Paper No. 28. Australian Institute of Criminology, Canberra
Urbas G, Fouracre K (2013) Legal responses to sexting: the importance of consent. Internet Law Bulletin 16(7): 171–173
Urbas G, Grabosky P (2006) Cybercrime and jurisdiction: an Australian perspective. In: Koops B J, Brenner S (eds) Cybercrime and jurisdiction: a global survey. IT and Law Series no. 11. T.M.C. Asser Press, The Hague

Gregor Urbas is an Associate Professor of Law at the University of Canberra, where he teaches Criminal Law and Procedure, Cybercrime and Evidence Law. He is also qualified as a lawyer in Australia. He has written extensively on cybercrime and other criminal justice issues. He previously held positions at the Australian National University (ANU), the Australian Institute of Criminology (AIC) and the Law Council of Australia, and is an Adjunct Associate Professor at the ANU College of Law and at Simon Fraser University in Vancouver, Canada. Dr. Urbas was recently a Visiting Fellow at the Tilburg Institute for Law, Technology, and Society (TILT) at Tilburg University in the Netherlands.

Chapter 5
Substantive and Procedural Legislation in Belgium to Combat Webcam-Related Sexual Child Abuse

Sofie Royer, Charlotte Conings and Gaëlle Marlier

Contents

Abstract The aim of this chapter is to examine whether Belgian substantive law and criminal procedure are adequate to clamp down on webcam-related sexual child abuse. After an introduction to the Belgian legal system, the chapter gives an

---

This chapter is up to date as regards legislation, case law and literature until 1 December 2018.

S. Royer (✉) · G. Marlier
Institute of Criminal Law, KU Leuven, Leuven, Belgium
e-mail: sofie.royer@kuleuven.be

G. Marlier
e-mail: gaelle.marlier@kuleuven.be

C. Conings
Stibbe, Brussels, Belgium
e-mail: charlotte.conings@stibbe.com

S. van der Hof et al. (eds.), *Sweetie 2.0*, Information Technology and Law Series 31,
https://doi.org/10.1007/978-94-6265-288-0_5

overview of the relevant criminal offences involving minors (sexual assault, rape, child prostitution, child pornography, voyeurism, corruption of children, grooming, cyber luring). We conclude that the Belgian substantive law offers many possibilities to prosecute and try perpetrators of these crimes. Criminal liability of the offender for cyber luring, for instance, is not affected when the avatar is operated by a police officer. As regards the criminal procedure, several legislative reforms have given extended powers to the public prosecutor and the investigating judge. Law enforcement authorities currently have a wide array of investigative measures at their disposal, such as the online infiltration, the IT sneak and peek operation and the covert access to private communication and computer data. This chapter argues that, as a consequence, under Belgian criminal procedure, evidence gathered by use of an avatar, is lawfully collected. A few question marks remain, as it is doubtful that the extended powers are fully consistent with human rights law.

Keywords Belgium · Criminalisation Webcam Sex Minors · Child Pornography · Cyber Luring · Electronic Stalking · Online Searches · Undercover Operations Entrapment

5.1 Introduction: Legislation in Belgium

5.1.1 General Description of the Legal Framework[1]

Civil Law Tradition

Similar to other continental countries such as the Netherlands and France, the Belgian legal system is embedded within a civil law tradition. Given the crucial role of the principle of legality, written acts are the most important source of law both for criminal law and criminal procedure. The post-revolutionary French Codes of Criminal Law and Criminal Procedure were an important source of inspiration for the Belgian Codes of the fledgling Belgian monarchy, which established its independence in 1830. As a result, both systems reveal considerable similarities to this day. The Belgian criminal law system is basically inquisitorial in scope, with an increasing number of accusatorial features. This reveals itself in the criminal process that consists of a pre-trial stage or investigation and a trial stage. In contrast with the trial stage, the pre-trial stage focuses on the collection of evidence. It is divided into a preliminary and a judicial inquiry. The public prosecutor and the investigating judge, respectively, are in charge of those investigations. Whereas the powers of the public prosecutor have gradually increased over the past decennia,

---

[1] This chapter is generally based on de la Serna 2015, pp. 7–51; Fermon et al. 2007, pp. 29–58.

the role of the investigating judge has evolved into a subsidiary position, which is likely to diminish even further in future.

Collection of Evidence

Originally, evidence that was illegally obtained during an investigation could not be used in court against the suspect. The Belgian Court of Cassation, however, limited the effects of this principle through several rulings, which were partly codified in Article 32 of the Preliminary Title of the Code of Criminal Procedure in 2013. This Article currently stipulates that it is only for the following cases that illegally obtained evidence is inadmissible: (i) the formality that was neglected is sanctioned with nullity by law; (ii) the irregularity affects the reliability of the evidence; or (iii) the use of the illegally obtained evidence would violate the right to a fair trial. The Court of Cassation spelled out some criteria which could be taken into account in determining whether the use of the evidence would violate the right to a fair trial, which includes the seriousness of the investigated offence when compared to the gravity of the breach by law enforcement and the intentional nature of the breach. The criteria are, however, neither binding nor conclusive. As a result, evidence is almost never excluded in court today.[2]

5.1.2 Relevant Treaties and Cybercrime Laws

Relevant Legislation

Belgium has ratified the following relevant treaties: the Cybercrime Convention; the Lanzarote Convention; the UN Convention on the Rights of the Child; and the Optional Protocol to the Convention on the Rights of the Child on the sale of children, child prostitution and child pornography. In addition to these treaties, the following national laws are of profound importance in relation to cybercrime and cyber investigation: the Computer Crime Act of 28 November 2000 and the Act of 13 June 2005 on Electronic Communications. The first Act is incorporated in the Belgian Code of Criminal Law and the Code of Criminal Procedure.

Legislative Reforms

Over the past years, the Belgian legal framework has considerably changed on several topics related to the Sweetie project. First of all, an Act concerning the data retention duties has been adopted in 2016.[3] Second, an Act concerning specific open and covert IT related searches has been adopted in the same year.[4] The

[2] Meese 2013, pp. 63–66.

[3] Act of 29 May 2016 concerning the collection and retention of data in the electronic communicationsector, *Off.Gazette* 18 July 2016.

[4] Act of 25 December 2016 concerning various amendments to the Code of Criminal Procedure and the Criminal Code with a view to improving the special investigation methods and certain research methods relating to the internet and electronic communications and telecommunications and the establishment of a database of voice prints, *Off.Gazette* 17 January 2017.

relevant changes which both acts entail, are mentioned below. Furthermore, the current Minister of Justice has introduced a fundamental reform to the Belgian criminal law system, which is planned to be rolled out in 2018. In this framework, two committees are working on the rewriting of both the Criminal Code (CC) and the Code of Criminal Procedure (CCP) at the time of this paper's writing.

5.2 Analysis of Substantive Criminal Law

5.2.1 Introduction

Criminal Code
The Belgian Criminal Code protects children against sexual abuse, penalizing different forms of sexual abuse. We can find most of these offences under Title VII of Book II of the Criminal Code: "felonies and misdemeanours against the order of the family and public morality".

Definition Minor
Article 100*ter* CC contains the definition of a minor:

> Where use is made of the term "minor" in the provisions of book II, it refers to the person who had not attained the age of eighteen.

5.2.2 Possibly Relevant Criminal Offences

Succinct Overview of Sexual Offences Involving Minors

Article 18 Lanzarote Treaty. Sexual abuse
Article 372 CC: Sexual assault without violence or threats against or with the aid of a person under the age of 16. Sexual assault without violence or threats by a relative in the ascending line or an adoptive parent committed on or with the aid of a minor, even if he or she has reached the age of 16, is considered a more severe offence. The same penalty applies if the offender is either the brother or sister of the minor victim or any person occupying a similar position within the family or any person who regularly or occasionally lives together with the minor and has authority over him.

Article 373 CC: Sexual assault with violence, coercion, threat, deception or by surprise or made possible by the physical or mental disability of the victim. Minority of the victim is an aggravating circumstance.

Article 374 CC: Sexual assault exists as soon as there is a commencement of execution.

Article 375 CC (Rape): Any act of sexual penetration, of any type and by any means, committed against a person without his or her permission. There is no permission if the act is forced by violence, coercion, threat, surprise or deception, or is made possible by the physical or mental disability of the victim. Minority is an aggravating circumstance. Any act of sexual penetration committed against a minor who has not reached the age of 14 is considered as rape with violence.

Article 19 Lanzarote Treaty. Offences concerning child prostitution
Article 379 CC: Violation of public morals by inciting, furthering or facilitating sexual immorality, corruption or the prostitution of a minor, of either sex, in order to satisfy the desires of another. The actual age of the minor can be an aggravating circumstance.

Article 380, § 4 CC: 1° The recruitment, transportation, transfer or harbouring, directly or through an intermediary, of a minor, for the purpose of sexual immorality or prostitution in order to satisfy the desires of another; 2° to keep, directly or through an intermediary, an establishment for sexual immorality or prostitution where minors commit prostitution or debauchery; 3° To sell, hire or make available rooms to a minor for the purpose of prostitution with the aim of unlawful profit; 4° to exploit, by whatever means, the sexual immorality or prostitution of a minor; 5° to obtain the sexual immorality or prostitution of a minor by submitting, offering, promising material or financial benefit.[5] The actual age of the minor can be an aggravating circumstance.

Article 380, § 6 CC: To witness directly, included through information and communication technologies, the debauchery or the prostitution of a minor.

Article 380*bis* CC: To incite to sexual immorality in a public place by words, gestures or signs. Minority of the victim is an aggravating circumstance.

Article 20 Lanzarote Treaty. Offences concerning child pornography[6]

Article 383*bis* CC: § 1 To display, offer, sell, rent, transmit, supply, distribute, disseminate, make available or deliver, when committed without right, child pornographic materials, or to produce, import or arrange it for import.

§ 2 Knowingly and without right obtain, possess or knowingly access, through information and communication technologies, child pornographic materials.

§ 4 For the purposes of this Article, "child pornographic materials" is understood to mean: 1° any material that visually depicts in some way or another – a minor engaged in real or simulated sexually explicit acts or that depicts the sexual organs of a child for primarily sexual purposes; 2° any material that visually depicts any person appearing to be a minor engaged in real or simulated sexually explicit conduct or that depicts the sexual organs of this person for primarily sexual purposes; 3° realistic images that depict non-existent

[5] The use of the sexual services of a child prostitute under the age of 16, is also punishable on grounds of sexual assault or rape.

[6] The production of child pornography is also punishable on grounds of other provisions, for instance in conjunction with depravity of the youth and prostitution. Hutsebaut 2000, p. 188.

minors engaged in sexually explicit acts or that depict the sexual organs of this minor for primarily sexual purposes.

Article 371/1 CC (Voyeurism):[7] 1° To observe a person or to make an image or sound recording of him or her, directly or by means of a technical or another tool, without permission[8] or without the knowledge of that person, while that person is nude or performs an explicit sexual act and while he can in all reasonableness expect that his or her privacy will not be violated; 2° To show, render accessible or distribute an image or sound recording of a nude person or a person performing an explicit sexual act, without that person's permission or without the knowledge of that person, even if that person consented to the making of those images. Minority of the victim is an aggravating circumstance. Voyeurism exists when there is a commencement of execution.

Article 21 Lanzarote Treaty. Offences concerning the participation of a child in pornographic performances[9]

- Article 379 CC: *Supra*
- Article 380, § 4 CC: *Supra*
- Article 380, § 6 CC: *Supra*
- Article 380*bis* CC:[10] *Supra*

Article 22 Lanzarote Treaty. Corruption of children

Article 385 CC (Public indecency): The fact that the offence is committed in the presence of a minor under the age of 16 is an aggravating circumstance.

Article 386, subsection 2 CC: The fact that the offence is committed against a minor is an aggravating circumstance.

Article 23 Lanzarote Treaty. Solicitation of children for sexual purposes[11]

Article 377*quater* CC (Grooming):[12] The proposal of an adult, through information and communication technologies, to meet a child who has not reached the age of 16, for the purpose of committing one of the offences mentioned in chapter

---

[7] Date of entry into force: February 29th 2016.

[8] Contrary to Article 372 CC the legislator did not establish an irrefutable presumption that minors under the age of 16 cannot consent. Spriet and Boeckxstaens 2016, p. 12.

[9] Articles 372, 373 and 375 CC apply to offenders who have sexual intercourse with a minor in this context.

[10] This provision can apply to the offender that lures a minor for the purpose of producing child pornography. Hutsebaut 2000, p. 191.

[11] The Belgian legislator has introduced a specific offence penalizing grooming. Before the introduction of this offence, the facts of grooming were punishable on grounds of other provisions. For example, an offender who created a fake Netlog- and Hotmail account, used it to get in touch with a minor and subsequently proposed that minor in a chat session to have paid sex with him, was convicted on grounds of IT-forgery (Article 210*bis* CC) and inciting sexual immorality (Article 379 CC). Cass. 12 February 2013, *TJK* 2013, 286.

[12] Grooming is furthermore an aggravating circumstance (Article 377*ter* CC) of several sexual offences, like rape and sexual assault.

V (concerning sexual assault and rape), chapter VI (concerning depravity of the youth and prostitution) or chapter VII (concerning public indecency), against the child, if this proposal has been followed by material acts leading to such a meeting.

Article 433*bis*/1 CC (Cyber luring): To communicate by means of information and communication technologies with an apparent or probable minor to facilitate the commission of a felony or misdemeanour against this minor while (i) concealing or lying about his identity, age or capacity; (ii) emphasizing the confidential nature of the conversations; (iii) offering or holding up the prospect of a gift or other advantage; (iv) tricking the minor in any other way.

Other offences, not covered by the Lanzarote Convention.

Advertising
Article 380*ter* CC: § 1 Irrespective of the means used, to directly or indirectly publish, distribute or disseminate advertising, offering services of a sexual nature, if this advertising is aimed specifically at minors or if it refers to services offered by minors or persons claiming to be minors, even if one conceals the nature of the services offered by means of artifices of language. An aggravating circumstance consist of the publicity having the purpose or effect, direct or indirect, to facilitate the prostitution or debauchery of a minor or exploitation for sexual purposes, it is.

§ 2 To make, to publish, to distribute or to disseminate publicity, by whatever means, directly or indirectly, even when concealing the nature through artifices of language, for sexual services when these services are provided by means of telecommunication.

§ 3 To make it known, by means of any publicity, even when concealing the nature of the offer or request through artifices of language, that one engages in prostitution, facilitates prostitution of others or wishes to enter into a relationship with a person engaged in debauchery.

To encourage, by any means of publicity, through the allusion that is made, the sexual exploitation of minors or adults, or use such advertising on the occasion of an offer of services.

Stalking
Article 442*bis* CC: To stalk a person while he knew or should have known that he would seriously affect by his behaviour the tranquillity of that person.

Article 145, § 3*bis* Act regarding Electronic Communication: To use electronic communication services to cause inconvenience or harm to his correspondent.

Overview of Sexual Offences Related to Webcam Sexual Child Abuse
A specific offence penalizing webcam sex with minors does not exist.[13] However, different existing criminal provisions can be applied to offenders who engage in webcam sex with minors. We focus on the most relevant provisions, in particular sexual assault, rape, possession of child pornography, inciting sexual immorality of

[13] A proposal introducing this was not accepted: Wetsvoorstel tot wijziging van het Strafwetboek wat de aanranding van de eerbaarheid via het internet betreft, *Parl.St.* Chamber 2010–11, nr. 53-1479/1.

the youth, public indecency and cyber luring. Other offences could apply to webcam sex with children, for instance stalking,[14] extortion[15] and IT-forgery,[16] but these offences will not be discussed in detail.

Sexual Abuse

Sexual Assault

A first possible criminal provision that applies to webcam sex with minors is the offence of sexual assault. The offence consists of an intentional attack on one's sexual integrity—as it is perceived by the collective consciousness at a given time in a given society—with a certain degree of severity, on or with the aid of a living person, without consent of the victim.[17] On the one hand, consent is lacking if the act is forced by violence, threat, surprise or deception, or is made possible by the physical or mental disability of the victim, regardless of the age of the victim.

On the other hand, consent is lacking if the act is carried out on or with the aid of a minor who is—according to the law[18]—incapable of consenting to sexual acts. Article 373 CC penalizes sexual assault if the act is forced by violence, threat, surprise or deception or is made possible by the physical or mental disability of the victim, while Article 372 CC penalizes sexual assault without violence or threats.

Article 372 CC

Subsection 1 proclaims an irrefutable presumption of moral coercion and thus of a lack of consent for minors under the age of 16. Further, subsections 2 and 3 of Article 372 CC create an irrefutable presumption of moral coercion if the offender has a specific relation with a minor of 16 or 17 years old, e.g. a family bond. Section 2 refers to a relative in the ascending line or an adoptive parent while section 3 refers to the brother or sister of the underaged victim or any person occupying a similar position within the family, or any person who regularly or occasionally lives with the minor and has authority over him or her.

No Requirement of Immediate Physical Contact

To be punishable, the offence of sexual assault does not require immediate physical contact between offender and victim. The law requires an attack on the sexual integrity carried out on a living person or with his or her help, without requiring actual physical contact.[19] "With his or her help" means that the offender uses the victim as a tool, forcing or inducing him or her to perform sexual acts on another person (either the offender or a third person) or on him or herself.[20] Consequently,

---

[14] Vandromme 2009, p. 180.

[15] Corr. Antwerp 20 June 2008, *T.Strafr.* 2009, 173.

[16] Cass. 12 February 2013, *TJK* 2013, 286.

[17] Cass. 24 May 2011, *Arr.Cass.* 2011, 1327, nr. 341; Cass. 31 March 2015, P.14.0293.N/2; Cass. 3 November 2015, *RABG* 2016, 530; Spriet and Marlier 2013, p. 83; Wattier 2007, p. 610.

[18] Article 372 CC.

[19] Spriet and Marlier 2013, p. 87; Vandromme 2009, p. 177.

[20] Spriet and Marlier 2013, p. 87; Vandromme 2009, p. 177.

the offence of sexual assault is committed if the offender induces the minor to undress or to perform sexual activities on him or herself. It is not required that the victim be completely naked. The partial exposure of intimate parts is sufficient.[21]

Webcam Sex

Therefore, we can conclude that visual contact can suffice to be liable for sexual assault. The offence can exist even if the victim is not physically present. The fact that bodily contact is impossible, has no effect on liability to punishment.[22] Consequently, the adult who induces or forces a minor to display breasts or genitals or to perform sexual activities in front of the webcam is committing the offence of sexual assault. Those acts violate the sexual integrity of a minor.[23] The specific liability to punishment differs depending on the age of the victim on the one hand and to the use of violence or threats on the other.

Webcam sex with a minor under the age of 16 without violence or threats

The offender who induces a minor under the age of 16 to have webcam sex is committing the offence of sexual assault. Those victims are—as mentioned above—unable to give permission for those acts. The fact that the minor *de facto* consented is irrelevant, even if the victim has encouraged or initiated the webcam sex. The offender remains liable to punishment even if the minor does not consider his or her sexual integrity to be violated.[24] Without violence or threats, the offender is punishable according to Article 372, first subsection CC. Ghent's Criminal Court of First Instance already convicted an offender who persuaded minors to perform sexual activities in front of the webcam on the basis of this criminal provision.[25]

Even if the minor does not respond to the request of the offender, the latter may remain punishable. Article 374 CC equates the attempt to commit sexual assault with the accomplished offence. Consequently, the offence of sexual assault is accomplished from the very first act that initiates the execution.[26] The judge can derive the commencement of execution for example from explicit sexually tinged conversations that reveal indecent proposals by the offender to minors under the age of 16.[27] Since minors are incapable of consenting to those acts, violence or threats

[21] Vandromme 2009, p. 177.

[22] Spriet and Marlier 2013, p. 88.

[23] Spriet and Marlier 2013, p. 88; Vandromme 2009, p. 177.

[24] Cass. 10 June 2015, *JT* 2015, 594; Cass. 14 December 1971, *Arr.Cass.* 1972, 372; Bastyns 2003, p. 6; Delbrouck 2015, 27; De Nauw 2010, p. 140; Hameeuw et al. 2001, p. 56; Spriet and Marlier 2013, p. 92; Vandromme 2009, p. 177.

[25] Corr. Ghent 10 October 2007, *T.Strafr.* 2008, 328. *Contra*: Corr. Antwerp 20 June 2008, *T. Strafr.* 2009, 173.

[26] Spriet and Marlier 2013, p. 100.

[27] Ghent's Criminal Court of Appeal, however, did not qualify indecent proposals (with threats) leading to an actual meeting as the commencement of execution of sexual assault with threats or violence, but as an attempt to commit the crime described in Article 379 CC (*infra*), inciting sexual immorality. Ghent 10 December 2009, www.juridat.be; Delbrouck 2015, p. 19.

are not required to cause the commencement of execution.[28] Different judgements can be found confirming this view.[29] The repeated offer of the defendant to visit him at his home, expressed to minors during a chat session, however, is—according to Antwerp's Criminal Court of Appeal—not enough to constitute the commencement of execution.[30]

Webcam Sex with Violence or Threats Committed against a Minor under the Age of 16

If the offender forces the minor by means of violence, coercion, threat, surprise or deception or abuses the physical or mental disability of the victim, he is punishable on grounds of Article 373 CC combined with the aggravating circumstance concerning the young age of the victim. In the case of webcam sex the offender will likely use threats (moral pressure). Given the physical absence of the victim the use of violence is unlikely.[31] In this regard, Antwerp's Criminal Court of First Instance convicted an offender on grounds of this provision because he forced a minor of 15 years of age to show her breasts and to masturbate in front of the webcam after threatening to hack her computer—if she failed to comply.[32]

Webcam Sex with Violence or Threats Committed against Other Minors or Adults

Sexual acts committed against minors of 16 or 17 years old or against adults are only punishable if those acts are accompanied by violence, coercion, threats, surprise or deception or abuse of the physical or mental disability of the victim (Article 373 CC). The mere invitation to have webcam sex is not punishable on grounds of sexual assault since these persons are capable to consent. If the offender threatens the victim, there is a commencement of execution if the victim refuses to act.[33]

Rape

In addition to the offence of sexual assault, the offender can commit the offence of rape. This is the case if the offender induces[34] or forces a minor to penetrate himself or herself with objects in front of the webcam. Like the offence of sexual assault, the offence of rape does not require any physical contact with the victim.[35] Contrary to sexual assault, the law does not equate the attempt with the accomplished

---

[28] Claus 2015, p. 20; Spriet and Marlier 2013, p. 100; Vandromme 2009, p. 178.

[29] Corr. Antwerp 25 June 2008, *T.Strafr.* 2009, 174; Corr. Hasselt 25 October 2012, unpublished. Spriet and Marlier 2013, p. 101. Corr. Tongeren 3 December 2005, unpublished. Delbrouck 2015, p. 16. *Contra*: Ghent 22 January 2007, *T.Strafr.* 2007, 204.

[30] Antwerp 19 June 2013, *Limb.Rechtsleven* 2013, 297.

[31] Vandromme 2009, p. 177.

[32] Corr. Antwerp 27 June 2008, *T.Strafr.* 2009, 175. Corr. Antwerp 2 April 2015, nr. 1722. Conings 2015a.

[33] Spriet and Marlier 2013, p. 102; Vandromme 2009, p. 178.

[34] The law creates an irrefutable presumption that minors under the age of 14 cannot consent with a sexual penetration. Article 375, subsection 6 CC.

[35] Corr. Hasselt 6 April 2012, unpublished. Spriet and Marlier 2013, p. 88; Stevens 2002, p. 451.

offence.[36] If the minor does not respond to the request of the offender, the offender can be punished for the attempt to commit a rape, as soon as the judge finds there is a commencement of execution.

Offences Concerning Child Pornography

Article 383*bis* CC – Article 383*bis*, § 1 CC penalizes the offender who displays, offers, sells, rents, transmits, supplies, distributes, disseminates, makes available or delivers, without right, child pornographic materials, or produces, imports or arranges it for import. Article 383*bis*, § 4 CC defines "child pornographic materials" as: (i) any material that visually depicts—in some way or another—a minor engaged in real or simulated sexually explicit acts or that depicts the sexual organs of a child for primarily sexual purposes; (ii) any material that visually depicts any person appearing to be a minor engaged in real or simulated sexually explicit conduct or that depicts the sexual organs of this person for primarily sexual purposes; (iii) realistic images that depict non-existent minors engaged in sexually explicit acts or that depict the sexual organs of these minors for primarily sexual purposes.[37]

Article 383*bis* CC targets any image carrier that depicts child pornography. It is limited to visual images of child pornography. Texts or sound recordings of a child-pornographic nature are, in contrast, not covered by Article 383*bis* CC. Article 383 CC (public indecency) penalizes those acts.[38] What is significant for the Sweetie case is that the involvement of a real minor is not necessary. Pseudo-child pornography, for example computer-generated drawings or illustrations, is also considered to be child pornography.[39]

Article 383*bis* § 2 CC furthermore penalizes the possession of child pornography. The term possession refers to the simple possession of images on a hard disk or any other optical or electronic medium. The internet user who downloads images on his home computer, is in possession of child pornography.[40] The offender who induces or forces a minor to show breasts or genitals or to perform sexual acts in front of the webcam and then stores these images on the computer commits the offence of possessing child pornography.[41]

Since 2012 knowingly accessing child pornographic materials through information and communication technologies also constitutes an offence according to

[36] Spriet and Marlier 2013, p. 107.

[37] This definition is based on Directive 2011/92/EU on combating sexual abuse and sexual exploitation of children and child pornography, and replacing Council Framework Decision 2004/68/JHA.

[38] De Hert and Bodard 1996, p. 106; Hutsebaut 2000, p. 193.

[39] Before the Act of 31 may 2016, this already fell within the scope of Article 383*bis* CC. De Hert and Bodard 1996, p. 105; Hutsebaut 2000, p. 193.

[40] Berneman 2009, p. 500; De Hert and Bodard 1996, p. 105; Ketels 2007, p. 7; Stevens 2002, p. 548.

[41] De Nauw 2010, p. 176; Vandromme 2009, p. 180.

Article 383*bis* § 2 CC.[42] The aim of the legislator was to punish the offender who accesses child pornography without downloading it, for example by watching it in real time via the internet.[43] "Accessing" refers to a positive act aimed at gaining access to pornographic materials.[44] In light of the foregoing, it can be argued that an offender who induces or forces a minor to show breasts or genitals or to perform sexual acts in front of the webcam, without storing these images on the computer commits the offence of knowingly accessing child pornographic materials. In addition to the possession and taking access, the Act of 31 may 2016 added the acquisition of child pornographic materials to this paragraph.

Article 371/1 CC
On the one hand, Article 371/1 CC penalizes the offender who observes a person or makes an image or sound recording of him or her (i) directly or by means of a technical or another tool, (ii) without permission or without the knowledge of that person, (iii) while that person is nude or performs an explicit sexual act, (iv) while that person can in all reasonableness expect that his or her privacy will not be violated. On the other hand, it penalizes the offender who shows, renders accessible or distributes an image or sound recording of a nude person or a person performing an explicit sexual act, without that person's permission or without the knowledge of that person, even if that person consented to the taking of said images. The offender who distributes, via social media or via internet, webcam images of the minor—without his permission or without the knowledge of the minor—is also punishable on grounds of this offence.

Offences Concerning the Participation of a Child in Pornographic Performances

Inciting Sexual Immorality
On the basis of Article 379 CC the offender who violates public morality by inciting, furthering or facilitating sexual immorality, corruption or the prostitution of a minor of either sex, in order to satisfy the desires of another, is punishable. This Article penalizes complicity in sexual immorality, corruption or prostitution of a minor in order to satisfy the desires of another. The offence does not require any effect or consequence.[45] The punishable acts consist of incitement, furthering or facilitation. The term "incitement" is defined as the attempt of the offender to entice the minor into committing the qualified behaviour in the absence of any initiative

[42] This legal change is in line with the case law of the Supreme Court who decided that the possession of child pornography does not require that the offender downloads the image. The fact that he knowingly visits a website and looks at the images, is sufficient: Cass. 20 April 2011, *RABG* 2011, 959; Cass. 3 February 2015, P.13.2070.N/1; Corr. Tongeren 25 October 2012, *Limb. Rechtsleven* 2013, 47.

[43] Bill, *Parl.St.* Chamber 2010–11, nr. 53-1639/1, 9; Huybrechts 2011, p. 1163.

[44] Bill, *Parl.St.* Chamber 2010–11, nr. 53-1639/1, 10.

[45] Stevens 2002, p. 86; Vandromme 2014a, p. 41.

deriving from the minor. "Furthering" is to encourage an inclination already present in the minor's head to proceed with the qualified behaviour. The term "facilitating" refers to helping the minor to commit the qualified behaviour, even if the minor has taken the initiative.[46]

Further, the law requires the incitement, furthering or facilitation *of the sexual immorality, corruption or prostitution* of a minor. Inducing or forcing a minor to sexually expose himself or herself or to perform sexual acts in front of the webcam can be qualified as inciting, furthering or facilitating *sexual immorality*. Sexual immorality has a broader scope than prostitution. Every form of unacceptable sexual expression is considered sexually immoral.[47] It concerns acts of grave indecency that are considered to be excessive, taking into account the age of the other party involved.[48] Whether the "Sweetie case" can constitute incitement of *prostitution* (for example if the minor is paid for the webcam sex) is uncertain, because legal doctrine seems to require physical contact.[49]

In addition, the victim must be a minor. A minor is a person who has not reached the age of eighteen (Article 100*ter* CC). It is sufficient to commit the acts against one minor;[50] whether the acts were committed with the minor's consent is irrelevant.[51]

Finally, the moral constituent component requires the intent to satisfy the desires of another. If one acts with the sole intent to satisfy one's own needs, one does not meet the moral constituent component.[52] By contrast, if one acts in order to satisfy the needs of the minor, the moral element is met. In other words, the minor can be considered as "another" as defined in Article 379 CC.[53]

As soon as the offender commits one of the acts, namely the incitement, the furthering or the facilitation of the sexual immorality of the minor, he has committed the offence.[54] The offence does not require that the minor actually commits the immoral act.[55] We found case law convicting an offender on grounds of Article 379 CC because he created a fake Netlog- and Hotmail account, used it to get in touch with a minor and subsequently proposed the minor in a chat session to have paid sex with him.[56]

[46] Stevens 2002, p. 511.

[47] Delbrouck 2014a, p. 3; Hutsebaut 2000, p. 188.

[48] Cass. 17 January 2012, *RW* 2012–13, 944.

[49] Vandromme 2009, p. 179.

[50] Stevens 2002, p. 90.

[51] Delbrouck 2014a, p. 5; Stevens 2002, p. 90.

[52] Cass. 22 May 2001, *Arr.Cass.* 2001, nr. 300.

[53] Cass. 8 September 1992, *Arr.Cass.* 1991–92, 1082; Cass. 12 February 2013, *TJK* 2013, 286; Cass. 9 December 2014 *NC* 2015, 56.

[54] Vandromme 2009, p. 180.

[55] Claus 2015, p. 21; Stevens 2013, p. 293.

[56] Cass. 12 February 2013, *TJK* 2013, 286. Ghent's Court of Appeal, however, qualified indecent proposals (with threats) leading to an actual meeting as an attempt to commit the crime described in Article 379 CC. Ghent 10 December 2009, www.juridat.be; Delbrouck 2015, p. 19.

Webcam Sex

Also (the attempt to have) webcam sex with a minor is punishable under this criminal provision. We can consider the acts stimulated by the offender as sexually immoral since those acts are performed by a minor at the request of an unknown chat partner.[57] In addition, the conditions of incitement, furthering or facilitating are met.[58] The required intent is more problematic. The offender is solely punishable under Article 379 CC if he acted with the intent to satisfy the needs of the minor or a third person.[59] Antwerp's Criminal Court of First Instance acquitted the defendant on grounds of Article 379 CC for having forced webcam sex with a minor because there was no proof that he acted with the intent to satisfy the needs of the victim, nor a third person.[60]

Article 380, § 4, 5° CC

Alternatively, we can apply Article 380, § 4, 5° CC to webcam sex with minors. This criminal provision penalizes the offender who obtains the sexual immorality or prostitution of a minor by submitting, offering or promising material or financial benefit. Through this provision, the legislator aimed at penalizing the customer of prostitution or sexual immorality when obtained through the provision of a benefit.[61]

Article 380, § 6 CC

This provision penalizes the offender who witnesses the debauchery or the prostitution of a minor. Persons who are spectators of such pornographic performances are punishable on grounds of this provision. The Act of 31 may 2016 specified that this includes witnessing through information and communication technologies. The offender who attends such a performance at a distance, for example via a webcam, is also punishable on grounds of this provision.

Corruption of Children

Article 385 CC—The most plausible provision to apply to an offender showing his genitals or performing sexual activities in front of the webcam, is Article 385 CC. This criminal provision penalizes public indecency. The expression "public indecency" is not defined.[62] The judge will compare the act to the general sexual norms.[63] This expression—subject to evolution—is determined by "the legally protected values concerning the public morals, as perceived by the collective consciousness of the moment." The judge will hold the questionable behaviour

57 Vandromme 2009, p. 179.

58 Vandromme 2009, p. 179.

59 Vandromme 2009, p. 179. Corr. Antwerp 2 April 2015, nr. 1722. Conings 2015a.

60 Corr. Antwerp 27 June 2008, *T.Strafr.* 2009, 175.

61 Vandemeulebroeke 2005, p. 57.

62 De Zegher 1973, p. 138; Stevens 2002, p. 318.

63 De Nauw 2010, p. 176.

against the perceptions of the average person in the street.[64] This perception is different when minors are involved. The context in which the acts are committed is highly significant.[65] According to case law, acts which offend public decency are, among others, certain acts of exhibitionists, nudity visible to witnesses who are not nudists, and having sex in a car on the public road.[66] Full nudity is not required.[67] It is sufficient that the nudity offends morality because of the circumstances in which it occurred. Whereas the simple exposure of a fully naked body is no longer considered as offending public decency per se, it can amount to punishable behaviour when done by an adult supervising a group of children of very conservative upbringing. [68] The fact that a woman bares her breasts will accordingly constitute the offence under Article 375 CC if it happens at the birthday party of the woman's son.[69]

The public nature is a constitutive element of the offence.[70] On grounds of the legitimate analogy between physical public space and public space on the internet, this criminal provision also penalizes the witnessing of acts via the internet.[71] Sexual acts performed in front of a webcam which can be observed by involuntary witnesses are punishable under Article 385 CC.[72] For our research, one of the relevant questions is whether or not a person is punishable when performing sexual acts in a non-accessible and private place. Sexual activities that offend public decency performed in such a place can indeed acquire a public character. This will be the case when the act is performed in the presence of involuntary witnesses.[73] An involuntary witness is a person who is a "passive participant" and does not actively participate in the behaviour.[74] Involuntary means that observation of the act is forced upon the witness. The number of witnesses is irrelevant: one witness is sufficient.[75] From this we can deduce that the person who performs unwanted sexual acts in front of the webcam in a private webcam session is punishable on grounds of Article 385 CC, because we can consider the person on the other side of

64 Cass. 15 June 1982, *RW* 1982–83, 1986; Brussels 19 March 2003, *JT* 2003, 468; De Nauw and Kuty 2014, p. 266; De Zegher 1973, p. 414.

65 Brussels 19 March 2003, *JT* 2003, 468.

66 Vandromme 2014c, p. 60; De Zegher 1973, p. 138.

67 Cass. 14 December 1971, *Arr.Cass.* 1972, 374; Vandromme 2014c, p. 60.

68 Brussels 19 March 2003, *JT* 2003, 468.

69 Delbrouck 2014b, p. 19.

70 Colette-Basecqz and Blaise 2011, p. 281; De Nauw 2010, p. 176; Stevens 2002, p. 116.

71 De Hert and Bodard 1996, p. 102.

72 Vandromme 2014c, p. 61.

73 Colette-Basecqz and Blaise 2011, p. 284; De Nauw 2010, p. 177; Stevens 2002, 331.

74 If the act, by contrast, is not committed in front of a "passive participant", but committed in the sole presence of the victim-witness of sexual assault (thus in the sole presence of the person whose physical integrity is affected by the behaviour), Article 385 CC does not apply. The criminal provisions regarding sexual assault apply to this situation. Cass. 23 April 1951, *Arr.Cass.* 1951, 483; Colette-Basecqz and Blaise 2011, p. 285; De Zegher 1973, p. 148; Stevens 2002, p. 334.

75 De Zegher 1973, p. 153; Stevens 2002, p. 331.

the webcam as an involuntary witness.[76] However, we found case law—wrongfully, in our opinion—ruling in favour of the opposite.[77]

Another relevant question to our research is whether minors can consent to such behaviour. This issue is hotly debated in legal doctrine and jurisprudence.[78] Some are of the opinion that a minor—a person under the age of 18—can never give permission. Consequently, each time a minor is witnessing such behaviour, the condition of "public nature" is fulfilled.[79] Others set the age limit at 16 (referring to the age limit in Article 372 CC).[80] One case states that there is no absolute age limit given the silence of the legislator.[81] Furthermore, some highlight that certain minors are so young and completely unaware of the sexual nature of the acts that they cannot be considered a witness.[82] Depending on the interpretation followed, the fact that the defendant shows his genitals or masturbates in front of the webcam, can (or cannot) be punished under Article 385 CC.

Solicitation of Children for Sexual Purposes

Act of April, 10th 2014—The legislator has recently introduced two new criminal offences, grooming and cyber luring, into the Belgian Criminal Code. These criminal offences may apply to abuse via a webcam.

Grooming

Article 377*quater* CC penalizes a proposal made by an adult through information and communication technologies to meet a child under the age of 16 for the purpose of committing sexual offences, in the event that this proposal is accompanied by material acts leading to such a meeting. This criminal provision does not apply if there is no proposal to meet in real life.[83] Consequently, this provision shall not apply to a mere sexually tinged conversation, even if the child is incited during the course of the conversation to perform sexual acts in front of the webcam or to watch at sexual performances of the adult. Conversely, the new offence of cyber luring provides more possibilities for punishing sexual child abuse via webcams.[84]

---

[76] *Contra:* Delbrouck 2014b, p. 18.

[77] Antwerp 20 April 2005, unpublished; Delbrouck 2014b, p. 18.

[78] Colette-Basecqz and Blaise 2011, p. 284; De Nauw 2010, p. 177; Stevens 2002, p. 335: De Zegher 1973, p. 160.

[79] For example: Liège 18 October 1948, *Jur.Liege* 1948–49, 153; Antwerp 9 January 1976, *RW* 1977–78, 936.

[80] For example: Brussels 12 April 1941, *Pas.* 1944, II, 7; Ghent 5 June 1953, *RW* 1953–54, 898; De Zegher 1973, p. 161.

[81] Corr. Bergen 12 January 1953, *JT* 1953, 188.

[82] Cass. 30 June 1958, *Arr. Cass.* 1958, 880; Corr. Tongeren 1 June 1948, *RW* 1948–49, 90; De Nauw 2010, p. 178; De Zegher 1973, p. 161.

[83] Claus 2015, p. 19; Stevens 2015, p. 849.

[84] Bill, *Parl.St.* Senate 2012–13, nr. 5-1823/1, 11; Conings and De Schepper 2014, p. 269; Stevens 2015, p. 849.

Cyber Luring

The legislator penalizes cyber luring of children through the new Article 433*bis*/1 CC. The term cyber luring refers to the use of internet to online communicating with minors in order to manipulate them with criminal intent.[85] This new criminal provision more specifically penalizes the offender who communicates by means of information and communication technologies with an apparent or probable minor to facilitate the commission of a felony or misdemeanour against this minor and who (i) lies about or withholds his identity, age or capacity;[86] (ii) emphasises the confidential nature of their conversations; (iii) offers a gift or other advantage or deludes the minor with such an offer; or (iv) tricks the minor in any other way. The offender must act with the specific intent to facilitate the commission of a felony or misdemeanour against the minor. The content of the conversation should prove the fact that the offender had this intent.[87] Even if the minor does not respond to the manipulation and the offender consequently does not commit the contemplated offence, the offender remains punishable for committing the offence of cyber luring.[88] In this case, the offence of cyber luring can be the commencement of execution of another offence.[89] Such facts could also be punishable under Article 372 CC *jo.* Article 374 CC (sexual assault), on condition that the judge accepts the commencement of execution. Thus, the conversation must bring about enough elements that demonstrate the commencement of execution (*supra*).[90] As stated above, Article 374 CC equates the attempt to commit sexual assault with the accomplished offence. If the offender by contrast commits the intended offence, he obviously remains punishable on grounds of other completed offences, like sexual assault and the possession of child pornography.[91]

Significant to our research is the fact that, in contrast to the offence of grooming, this offence does not require a proposal to meet.[92] Preparatory works highlight the fact that many offences are committed without any meeting between the child and the adult manipulating the child, for example the misuse of private information and the forwarding of pornographic images.[93]

An important limitation is that a single conversation cannot constitute the offence of cyber luring. The preparatory works indicate that the term "to

---

85 Bill, *Parl.St.* Senate 2012–13, nr. 5-2253/1, 4.

86 If the offender creates a fake profile posing as a minor, a conviction on ground of IT-forgery is also possible. Cass. 12 February 2013, *TJK* 2013, 286; Corr. Antwerp 2 April 2015. Conings 2015a.

87 Bill, *Parl.St.* Senate 2012–13, nr. 5-2253/1, 9; Conings and De Schepper 2014, p. 270; Stevens 2015, p. 855.

88 Stevens 2015, p. 845.

89 Claus 2015, p. 22.

90 Stevens 2015, p. 846.

91 Claus 2015, p. 20.

92 Bill, *Parl.St.* Senate 2012–13, nr. 5-2253/1, 9; Conings and De Schepper 2014, p. 270; Stevens 2015, p. 853.

93 Bill, *Parl.St.* Senate 2012–13, nr. 5-2253/1, 9.

communicate" requires a set of interactions between the minor and the adult offender that are spread over time. According to the legislator it must be a well-developed manipulation strategy with several objective characteristics.[94] Moreover, to be punishable the offender must use one of the four techniques[95] mentioned in Article 433*bis*/1 CC. The law thus requires a certain manipulation of the minor. In practice, online sex offenders indeed often use one of these techniques.[96]

Webcam Sex

Manipulating a child—by means of a series of chat conversations spread over time—to make him send pornographic pictures (without the intent of meeting the child) can for example constitute the offence of cyber luring.[97] The same can be accepted if the offender manipulates a child to display his or her genitals or to perform sexual activities in front of the webcam.[98] In our opinion, the fact that the offender manipulates the minor in order to commit the offence of exhibitionism (e.g., by showing genitals via the webcam), could also fall within the scope of this criminal provision.[99]

Other Offences

Stalking

Finally, the person who tries to induce a minor to have webcam sex can also be punished on grounds of stalking. Article 442*bis* CC penalizes the offender who stalks a person while he knew or should have known that he would seriously affect the tranquillity of that person with his conduct. If the offender does not leave the minor alone and he repeatedly attempts to change the minor's mind to have webcam sex, he is punishable for sexual stalking.[100] An offender who repeatedly shows his genitals in front of the webcam to an involuntary witness is also punishable on grounds of stalking.[101] Furthermore, Article 145 § 3*bis* Act Regarding Electronic Communication may also apply in this case. This provision penalizes the offender who uses electronic communication services to cause inconvenience to his

---

[94] Report of the commission, *Parl.St.* Senate 2013–14, nr. 5-2253/3, 8; Report of the commission, *Parl.St.* Chamber 2013–14, nr. 53-3450/2, 4–5; Stevens 2015, p. 854.

[95] This list is non-cumulative. Report of the commission, *Parl.St.* Senate 2013–14, nr. 5-2253/3, 5.

[96] Claus 2015, p. 18; Stevens 2015, p. 854.

[97] Report of the commission, *Parl.St.* Senate 2013–14, nr. 5-2253/3, 5; Conings and De Schepper 2014, p. 269.

[98] Stevens 2015, p. 846.

[99] Bill, *Parl. St.* Senate 2012–13, nr. 5-2253/1, 9.

[100] Vandromme 2009, p. 180.

[101] Delbrouck 2015, p. 12.

correspondent or harm. Contrary to Article 442*bis* CC a single act of conduct is sufficient and the law does not require the disruption of the victim's tranquillity.[102] However, this provision requires the specific intent to cause inconvenience or harm to the correspondent.[103] An offender who sent sexual text messages to a minor was condemned on grounds of Article 145, § 3*bis* WEC. He ceased his attempts the moment the minor made clear she was not interested. The court derived the specific intent from the content of the messages: his goal was setting up a meeting with the minor for sexual purposes.[104] In analogy to this conviction, the offender who tries to induce a minor via electronic communication services to have webcam sex, can be punished according to Article 145, § 3*bis* WEC. However, a remark to this ruling criticized the fact that the court derived the specific intent solely from the circumstance of the offender aiming for a meeting, but not precluding sexual relations. The court did not bother establishing proof of how the specific intent caused inconvenience or harm.[105]

Conclusion

If a perpetrator induces or forces a minor to display breasts or genitals or to perform sexual activities (e.g., masturbate) in front of the webcam, this may constitute:

- Sexual assault without violence or threats (Article 372 CC)
  - Physical contact is not required;
  - In principle limited to minors under the age of 16;
  - Commencement of execution (Article 374 CC) can be derived from explicit sexual conversations.
- Sexual assault with violence or threats (Article 373 CC).
- Rape (Article 375 CC)
  - Physical contact is not required;
  - Penetration is required.
- Possession of child pornography or access to it (Article 383*bis* CC)
  - A "child" is a minor under the age of 18;
  - The involvement of a real minor is not required.

102 Kerkhofs and Van Linthout 2013, p. 139; Vrielink and Van Dyck 2015, p. 782.

103 Kerkhofs and Van Linthout 2013, p. 139; Vrielink and Van Dyck 2015, p. 782.

104 Corr. Turnhout 16 May 2012, *T. Straf.* 2012, afl. 6, 474.

105 F.S. 2012, p. 476.

- Voyeurism (Article 371/1 CC)
  - If one distributes the images without permission of knowledge.
- Inciting sexual immorality of minors (Article 379 CC)
  - Inciting, furthering or facilitating sexual immorality;
  - No result is required;
  - A "minor" is a person under the age of 18;
  - With the intent to satisfy the desires of the minor or a third person.
- To obtain sexual immorality of a minor by submitting, offering or promising material or financial benefit (Article 380, § 4, 5° CC).
- Cyber luring (Article 433*bis*/1 CC)
  - Online communication with minors in order to commit an offence;
  - No requirement of a proposal to meet;
  - Use of specific techniques to manipulate;
  - Set of interactions spread over time;
  - The involvement of a real minor is not required.
- Stalking (Article 442*bis* CC and/or Article 145, § 3*bis* WEC).

If the perpetrator shows his genitals or masturbates in front of the webcam, this may constitute:

- Public indecency (Article 385 CC)
  - Public nature: even in a private location, if involuntary witness is present;
  - Discussion on whether minors can consent.
- Cyber luring (Article 433*bis*/1 CC).
- Stalking (Article 442*bis* CC and/or Article 145, § 3*bis* WEC).

Based on the principle of discretionary powers, the public prosecutor is free to decide whether or not to prosecute. Each offence may lead to the taking into custody of the suspect. Further, the victim can file a civil complaint with the investigating judge or can directly summon the suspect (if the offence is a misdemeanour[106]—this is not possible for a felony[107]). Several intrusive investigation methods can be applied (*infra*).

All offences mentioned above are punishable with correctional or criminal sentences. However, the Criminal Court will not accumulate the sanctions of different offences, if it accepts that the facts concerned constitute several criminal

[106] "Wanbedrijf", this is an offence punishable by correctional sentences.

[107] "Misdaad", this is an offence punishable by, more severe, criminal sentences.

offences. The Court will impose one single sanction, which will relate to the most serious offence (Article 65 CC).

5.2.3 *Potential Obstacle in Substantive Law Concerning Sweetie: The Use of a Virtual Minor*

Completed Offence of Cyber Luring

The offence of cyber luring requires that an adult communicates with an *apparent or probable minor*. With this broad formulation, the legislator sought to include offenders who believe they are communicating with a minor. Thus the law does not require that the facts be committed against real minors. The legislator sought to enable investigators to use a so-called "lure teenager".[108] The fact that the apparent minor afterwards turns out to be an adult infiltrator, does not affect the criminal nature of the offence. If a police officer uses a fake profile to pose as a minor on the internet the constitutive elements of the offence are met. The offender who had a sexual conversation with an apparent or probable minor can be prosecuted on the condition that the police officer did not provoke him or her (*infra*).[109] Consequently, under Belgian law, the fact that the Sweetie avatar is operated by a police officer, does not affect the criminal liability of the correspondent for cyber luring.

Completed Offence of Child Pornography

Since Article 383*bis*, § 4 CC refers to "child pornographic materials", *inter alia*, as "realistic images that depict non-existent minors engaged in sexually explicit acts or the sexual organs of these minors for primarily sexual purposes",[110] the involvement of a real minor is not required. Pseudo-child pornography, for example computer-generated drawings or illustrations, is also considered to be child pornography.[111] As a consequence, acts with virtual avatars, such as Sweetie, can engage criminal liability, according to Article 383*bis* CC.

Completed Offence of Electronic Stalking

The question arises whether the offence under Article 145, § 3*bis* Act Regarding Electronic Communication can be committed against Sweetie. This provision penalizes the offender who uses electronic communication services to cause inconvenience or harm to his correspondent. If an offender chats with Sweetie with an intent to cause inconvenience, he meets the material and moral component of the

---

[108] This is a police officer who acts on the internet like a minor in order to track paedophiles.

[109] Bill, *Parl.St.* Senate 2012–13, nr. 5-2253/1, 9; Conings and De Schepper 2014, p. 269; Claus 2015, p. 18; Stevens 2015, p. 854.

[110] This definition is based on Directive 2011/92/EU on combating sexual abuse and sexual exploitation of children and child pornography, and replacing Council Framework Decision 2004/68/JHA.

[111] Before the Act of 31 may 2016, this already fell within the scope of Article 383*bis* CC. De Hert and Bodard 1996, p. 105; Hutsebaut 2000, p. 193.

offence. If this is the case, he is, in our view, punishable for committing the completed offence of electronic stalking.

Criminal Attempt?

As regards other offences, often a real minor should be involved. Therefore, the question raises whether someone, who has a sexual conversation with an avatar representing a 10-year-old girl who does not undress herself, can be punished on grounds of criminal attempt? The first question is whether the attempt to commit the specific offence is liable to punishment.[112] The commencement of execution of sexual assault (e.g., by making indecent proposals to a minor under the age of 16), for instance, is punishable and even placed on the same footing as the completed offence and thus punished by the same penalties. *In casu*, the offence of sexual assault can never be completed, since this offence requires violation of the sexual integrity of a living person. This is a so-called "impossible offence". Whether or not such an "impossible offence" is punishable by way of attempt depends on the point of view. The objective view focuses on a protected object. If this object is never at risk, the act will not be punishable. By contrast, the subjective view emphasises an offender's subjective state of mind. If the conduct reveals the dangerousness of the offender, it should be punishable.[113] Following the subjective view's logic, the fact that an offender makes indecent proposals to an avatar of a 10-year-old girl might be punishable on grounds of sexual assault. In any event this question is open to debate. Instead of using the construction of the (punishable) attempt, it is better to have a specific crime according to which this behaviour is punishable. Since the introduction of cyber luring, the offender who chats with Sweetie is punishable on grounds of the completed offence as a result of which an appeal to the construction of the criminal attempt is unnecessary.

5.3 Analysis of Criminal Procedure Law[114]

5.3.1 General Description of the Legal Framework[115]

Pre-Trial Investigation: Preliminary Inquiry

The Belgian investigative stage of criminal proceedings is mainly inquisitorial. Two types of pre-trial investigations can be distinguished: the preliminary inquiry and the judicial inquiry. Both types are written, secret and non-adversarial. In most cases only a preliminary inquiry, led by the public prosecutor, is launched.[116]

112 For example, the attempt to possess (virtual) child pornography or access it, is not punishable.

113 Verbruggen and Verstraeten 2014, p. 92.

114 For an extended analysis of traditional and digital investigative methods, see Conings 2017.

115 This general description is mainly based on Verstraeten 2012; De Hert and Van Leeuw 2010, pp. 867–956; Fermon et al. 2007, pp. 29–58.

116 Article 28*ter* CCP.

The public prosecutor has several competences, such as seizure and interviewing of suspects or witnesses. Following the principle of prosecutorial discretion he can decide to dismiss the case or bring it directly before the court when his investigation has come to an end.[117] Almost all competences of the public prosecutor also belong to the investigating judge.

Judicial Inquiry

Certain far-reaching investigative measures, such as the covert computer search or covert access to private communications, require a judicial inquiry led by the investigating judge.[118] The investigating judge is an impartial and independent judge who searches for both incriminating and exculpatory evidence. There are, however, important exceptions to the principle of judicial control over coercive measures. For instance, in the case of *flagrant délit*, the public prosecutor has more extensive powers. After the judicial inquiry has been closed, Belgian criminal law provides a mandatory judicial supervision. The Council Chamber presided over by a judge of the Court of First Instance, reviews possible procedural issues ensuring *inter alia* that illegally obtained evidence is excluded when necessary. Moreover, this Chamber decides on whether there are sufficient grounds to refer the suspect to trial. The Indicting Chamber deals with appeals against this decision. This pre-trial court consists of three judges of the Court of Appeal.

Mini Judicial Inquiry

Given the burdensome procedural requirements of a judicial inquiry, in 1998 the legislator introduced a simplified procedure called "mini judicial inquiry".[119] In this procedure the intervention of the investigating judge is limited to a one-off authorisation. As a result, certain investigative measures no longer require a judicial inquiry and the investigation remains in the hands of the public prosecutor. Accordingly, the case does not need to pass by the Council Chamber either. However, the investigating judge retains the option of taking over the investigation. This entails the start of a full blown judicial inquiry, despite the limited request for a mini judicial inquiry. The arrest warrant, the completely anonymous testimony, the covert access to private communications, the covert computer search, the IT-sneak and peek and the sneak and peek operation, the systematic observation within a residence or equivalent location and the domiciliary visit are excluded from the mini judicial inquiry.

Special Investigative Methods

Belgian criminal procedure contains a number of special investigative methods. Due to its area of expertise, the public prosecutor's office supervises the exercise of these methods, regardless of the possible order of an investigating judge and startfof a judicial inquiry. These special investigative methods, i.e. systematic

[117] Article 28*quater*, § 1 CCP.

[118] Articles 55–56 CCP.

[119] Article 28*septies* CCP.

observation,[120] infiltration[121] and active informer management,[122] are characterised by their secretive and treacherous nature. Police officers exercising these powers have to pay attention to the general prohibition of entrapment (Sect. 5.3.2), and are not allowed to commit offences, although there are exceptions to the latter prohibition (Sect. 5.3.3). The use of systematic observation or offline infiltration entails the compilation of a confidential record, which is usually accompanied by an additional supervision by the Indictment Chamber.

Proactive Investigation

In principle, criminal investigations take place further to serious and specific indications that an offence has been or will be committed. However, once reasonable presumptions exist that offences have been committed or will be committed, a proactive investigation is allowed. The proactive investigation is limited to offences which are or will be committed in the framework of a criminal organisation or which are included in the list of serious offences of Article 90*ter*, §§ 2–4 CCP.[123] A proactive investigation can only be initiated upon the written approval of the public prosecutor, while a reactive investigation only presupposes an immediate notification to the public prosecutor. Investigative methods which belong to the sole competence of the investigating judge cannot be applied in the framework of a proactive investigation.

Federal Prosecutor

As a member of the public prosecution service, the federal prosecutor is entrusted with the same powers as the public prosecutor. His competence concerns complex and transborder cases and specific offences such as organized crime and terrorism.[124] Moreover, he interferes in the event of lack of clarity as to which local prosecutor is competent for certain offences. The federal prosecutor is also charged with matters of international cooperation.[125]

Police

The Belgian police is structured on two levels: the local police and the federal police. In principle, both the local and federal police exercise their powers under the supervision and responsibility of the public prosecutor or investigating judge. Within the police force, a specialised department deals with IT-matters and tackling cyber criminality. These departments are called the Federal and Regional Computer Crime Units.

---

120 Article 47*sexies* CCP.

121 Article 47*octies* CCP.

122 Article 47*decies* CCP.

123 Article 28*bis*, § 2 CCP. This list contains, *inter alia*, the following relevant offences: computer related forgery (7°), rape (15°), grooming (16°), some types of moral depravity of minors and prostitution (17°), child pornography (17°), kidnapping of minors (20°), cyber luring (21°) and electronic stalking (40°).

124 Article 144*ter* Judicial Code.

125 Article 144*bis* Judicial Code.

5.3.2 *Investigatory Powers*

Succinct Overview of Investigatory Powers[126]

Article 16 Council of Europe Convention on Cybercrime. Expedited preservation of stored computer data

- Article 39*ter* CCP provides the possibility for each officer of the judicial police to order a (legal) person to preserve data which are particularly vulnerable to loss or modification and which are in that person's possession or control. The duty to preserve the data can concern a period up to a maximum of ninety days. The order can subsequently be renewed.
- In addition to the expedited preservation measure, Belgium has a general obligation for providers of electronic communication services to store electronic communication data (identification data, traffic data and localisation data) during 12 months (Article 126 Act of June 13th 2005 regarding electronic communications).

Article 17 Council of Europe Convention on Cybercrime. Expedited preservation and partial disclosure of traffic data

- Expedited preservation and general storage obligation (*supra*).
- Article 88*bis* CCP determines the general cooperation duty of operators of an electronic communication network or suppliers of an electronic communication service (*sensu lato*) regarding the disclosure of traffic data and localisation data. The concept of an electronic communication service supplier has to be understood in a broad manner. It concerns each (legal) person who provides a service or makes a service available on Belgian soil, which allows its users to gather, disclose or distribute information by using an electronic communications network.[127]
- Article 88*bis* CCP empowers the investigating judge to retrieve communication traffic data and the localisation of the origin or destination of communication, the so-called localisation data. Refusal to supply the requested data mentioned in Article 88*bis* CCP (in time) is punished with a fine.
- Expedited disclosure? In case of *flagrant délit*, the public prosecutor has the power to order disclosure of the said data to a certain extent (*infra*). In exigent circumstances, the order can be given orally.

Article 18 Council of Europe Convention on Cybercrime. Production order

- Article 46*bis*, 88*bis*, 88*quater*, § 2 and 90*quater*, § 4 CCP.
  - Article 46*bis* CCP: Production order concerning identification data. This provision determines the duty of cooperation of operators of an electronic communication network or suppliers of an electronic communication service

---

[126] De Hert and Lichtenstein 2004, pp. 153–169; Kerkhofs and Van Linthout 2013, pp. 165–377.

[127] For instance: Twitter, Yahoo, Microsoft etc.

(*sensu lato*)[128] regarding the disclosure of identification data. The Article covers (i) the identification of subscribers or frequent users of electronic communication services; and (ii) the identification of an electronic communication service to which someone is subscribed or that someone generally uses. No order of the investigating judge is needed.
The Court of Cassation recently decided that Belgian authorities are not operating outside of their jurisdiction when they directly order foreign-based service providers who are economically active in Belgium, to produce identification data.[129] A lower court extended this jurisprudence to Articles 88*bis* and 90*quater* CCP.[130]
- Refusal to supply the requested data of Articles 46*bis* and 88*bis* is punishable by a fine. The refusal to cooperate according to Articles 88*quater*, § 2 CCP and 90*quater*, § 4 CCP is punishable by imprisonment and/or a fine.
- Article 88*bis* CCP: Production order concerning traffic and localisation data (*supra*).
- Article 88*quater*, § 2 CCP contains a more general production order. It empowers the investigating judge to order every qualified person to operate the IT-system or to search, to make available, to copy, to render inaccessible or to remove the relevant data. The Article explicitly excludes suspects and certain family members from this duty of cooperation.
- Article 90*ter* CCP ff. regulates the covert access to private telecommunications and covert search of computers (*infra*). This legal framework contains, among others, a duty of cooperation for (legal) persons to make relevant data available to the extent that they are able (Article 90*quater*, § 4 CCP). The said competences mainly belong to the investigating judge.

Article 19 Council of Europe Convention on Cybercrime. Search and seizure of stored computer data

- Article 39*bis* CCP provides a graduated system of powers to search computer data in a non-covert manner. According to Article 39*bis*, § 2 CCP each officer of judicial police can decide to search for computer data, stored on a computer system that is being seized. The public prosecutor has the power to order a search of a computer system that can be, but is not seized. The competence to bypass security measures or to apply technical measures to decrypt data belongs to the public prosecutor (Article 39*bis*, § 5 CCP).
- Article 39*bis*, § 3 CCP specifies the remote search or network search. This provision stipulates that searches of an IT-system, can be extended to linked IT-systems, that are situated somewhere else within or in certain cases even

[128] Supra Article 88*bis* CCP.

[129] Cass. 1 December 2015, AR P.13.2082.N.

[130] Corr. Antwerp 27 October 2016, *NjW* 2016, afl. 353, 921, *NC* 2017, afl. 1, 89.

beyond Belgian territory.[131] The network search may not go beyond the IT-systems to which the persons entitled to use the IT-system under investigation, have access. The public prosecutor is the competent authority to order the network search. The competence to bypass security measures or to apply technical measures to decrypt data belongs to the investigating judge (Article 39*bis*, § 5 CCP). If it turns out that the data, found through the extension of the search, are not situated on Belgian territory, they can only be copied. In that event the authorities of the state concerned will be notified, if this state can be reasonably identified.
Article 39*bis*, § 4 CCP stipulates that the investigating judge is the competent authority for all other types of non-covert computer searches.

- Article 39*bis*, § 6 CCP determines the conditions for data seizure. The data seizure competence empowers law enforcement authorities to copy data whenever the seizure of the support is not desirable and to block access to or remove these data.
- Article 88*quater*, § 1 CCP defines an information order, as required by Article 19, section 4 of the Cybercrime Convention. It empowers the investigating judge to order persons who he suspects to have specific knowledge of the information system under investigation or of services to encrypt or secure data, to give information on the functioning of the system or the access to the system or data. The Article does not exclude suspects from this duty of cooperation. However, the Ghent Court of Appeal decided that obliging the suspect to provide the key necessary to access encrypted data, violates the right of non-incrimination. As a result, the evidence obtained in breach of Article 6 ECHR was excluded.[132] Refusal to cooperate is punishable by imprisonment and/or a fine.

Article 20 Council of Europe Convention on Cybercrime. Real-time collection of traffic data

- Article 88*bis* CCP (*supra*). In addition to the option of collecting traffic and localisation data of communications which have taken place in the past, this provision entails the power to collect traffic and localisation data in real-time. Article 88*bis* CCP empowers the investigating authorities to collect the data autonomously or with the cooperation of the operator of an electronic communication network or supplier of an electronic communication service. An order of the investigating judge is, in principle, required.

Article 21 Council of Europe Convention on Cybercrime. Interception of content data of specified communications

[131] Based on this provision, investigators can, for example, extend the search of a computer to a webmail account.

[132] Ghent 23 June 2015, *NjW* 2016, afl. 336, 134.

- Article 90*ter* CCP ff.: The investigating judge can order the covert access by means of technical devices to (electronic) communications which are not accessible to the public.
- Additionally, in view of this investigative measure, the investigating judge can also order the secret entry of a private site or residence, the bypassing of security measures of an information system or the decryption of data by technical means.
- The legal framework contains several duties of cooperation:
  - Article 90*quater*, § 2 CCP: Duty of operators of a communication network or suppliers of an electronic communication service (*sensu lato* (*supra*)) to provide technical assistance.
  - Article 90*quater*, § 4 CCP: (i) Duty of persons suspected to have specific knowledge of the telecommunication service or of services to encrypt or secure data, to provide information on the functioning of the service or access to the full content; (ii) Duty of persons to make relevant data available to the extent possible.

Other (Special) Investigatory Powers, Not Covered by the Cybercrime Convention, Such as Undercover Operations

- Covert access to computer systems:
  - IT sneak and peek: Article 89*ter* CCP provides for the IT sneak and peek, which refers to a covert access to and a search of computer systems with a view to orienting the criminal investigation. Seizure is thus not the aim of this investigative measure. The investigating judge has the power to order this kind of searches.
  - Covert computer search: the Act on IT-related searches considerably broadened the wiretap competence in Article 90*ter* CCP. In addition to the competence to access private communications, the Article now provides for the possibility to break into computer systems in order to search for data or to monitor the use of data in real time. The competence concerns data which are not publicly accessible. The investigating judge is the sole competent authority, except for some limited competence for the public prosecutor in case of a *flagrant délit*.
  - Online infiltration: Article 46*sexies* CCP determines the conditions for online undercover operations. The public prosecutor can authorize a police officer to get in contact with a suspect online while using a fake identity.
  - Progressive interpretation of the systematic observation (Article 47*sexies* and 56*bis* CCP): as far as the intensive monitoring of information on publicly accessible websites and information systems is concerned, the Belgian legislation does not contain an IT-specific regime. We could interpret the legal regime of systematic observation progressively. However, the criteria are not apt to the online or IT-context.

Human Rights

As a Member State of the European Convention of Human Rights (ECHR), Belgium recognises the case law of the European Court of Human Rights and as a Member of the European Union it recognises the case law of the European Court of Justice. Yet, when it comes to the right to privacy, the Belgian Constitution demands a stronger protection. Whereas the ECHR determines that interferences to the right to privacy must be in accordance with the law in substantive sense, a formal Act is required to abrogate this right in compliance with the Belgian Constitution.[133] In accordance with the principle of legality's quality requirements, the investigative measures in the CCP are bolstered by a number of safeguards, depending on the more or less intrusive character of the measure. Below, we explore the main safeguards of the competences listed above in more detail. A more detailed overview of the secret investigatory measures in an online context is provided in Sect. 5.3.

Competent Authority

Each officer of judicial police, the public prosecutor and the investigating judge[134] have the power to perform or order:

- A search of an information system which is being seized (Article 39*bis*, § 2 CCP).
- Data seizure (Article 39*bis*, § 6 CCP).
- The preservation of data (Article 39*ter* CCP).
- The public prosecutor and the investigating judge[135] both have the power to perform, authorize or order:
- A search of an IT-system which can be seized, but is not being seized (Article 39*bis*, § 2 CCP);
- A network search (Article 39*bis*, § 3 CCP);
- Bypassing security measures or applying technical measures to decrypt data, except in case of a network search (Article 39*bis*, § 5 CCP);
- Data seizure (Article 39*bis*, § 6 CCP);
- The production of identification data (Article 46*bis* CCP);
- An online infiltration (Article 46*sexies* CCP).

Certain investigative measures require the intervention of the investigating judge given their intrusive character. Unless the authorisation can be given within the framework of a mini judicial inquiry, it necessarily results in the opening of a

133 Article 22 Belgian Constitution. Arbitragehof 30 April 2003, nr. 50/2003; Arbitragehof 30 April 2003, nr. 51/2003; Arbitragehof 19 January 2005, nr. 16/2005, *Computerr.* 2005, 136, *RDTI* 2005, 129; Arbitragehof 18 October 2006, nr. 151/2006; Cass. 2 May 1990, *Arr. Cass.* 1989–90, nr. 516; Cass. 21 April 1998, AR P961470N, *Arr. Cass.* 1998, 446, *RW* 1998–99, 1452; De Hert and Saelens 2009, p. 838.

134 Article 56, § 1, third section CCP.

135 Article 56, § 1, third section CCP.

judicial inquiry. The investigating judge has the exclusive power to perform or order:

- All open/transparent searches of private IT-systems not mentioned above (Article 39*bis*, § 4 CCP);
- Bypassing security measures or applying technical measures to decrypt data in case of a network search or a search on the bases of Article 39*bis*, § 4 CCP (Article 39*bis*, § 5 CCP);
- The production of traffic and localisation data (Article 88*bis* CCP);
- A general information and production order (Article 88*quater* CCP);
- An IT sneak and peek (Article 89*ter* CCP), which is excluded from the mini judicial inquiry;
- Covert access to private communications and a covert computer search (Article 90*ter* CCP), which are excluded from the mini judicial inquiry.

Under certain special circumstances the public prosecutor has broader powers, among which the power to perform, order or authorize:

- The production of traffic or localisation data (Article 88*bis* CCP) (i) in case of *flagrant délit* of one of the offences listed in Article 90*ter*, §§ 2–4 CCP[136] or (ii) if the measure appears to be indispensable to establish one of the offences meant in Article 145, § 3 and 3*bis* Act of 13 June 2005 regarding electronic communications,[137] on the complainer's request.
- The covert access to private communications and covert computer search (Article 90*ter* CCP) in case of *flagrant délit* of the terrorism offences mentioned in Article 137 CC, the taking of hostages (Article 347*bis* CC), illegal deprivation of liberty (Article 434 CC) or extortion (Article 470 CC).

Proportionality and Necessity or Subsidiarity

Second, a number of provisions require that the investigative measures are proportionate and essential to uncovering the truth, implying that no less intrusive investigative measure is more appropriate. The required proportionality of the measure is sometimes expressed in terms of offences for which the measure can be ordered.

- Article 39*bis*, § 3 CCP determines that the network search has to be (i) necessary to reveal the truth; and (ii) other investigative measures are disproportionate or evidence could be lost without the network search. The same conditions apply to the search of information systems mentioned in article 39*bis*, § 4 CCP, which belongs to the competence of the investigating judge.

---

136 The investigating judge has to confirm the measure within 24 hours, except for the terrorism offences mentioned in Article 137 CC, the taking of hostages (Article 347*bis* CC), illegal deprivation of liberty (Article 434 CC) or extortion (Article 470 CC).

137 Electronic stalking/spamming (*supra*).

- The investigative measure of Article 46*bis* CCP can be ordered for all offences, under the condition that the motivation expresses the proportionality of the measure with regard to private life and its subsidiarity to other investigative measures. The Act concerning data retention limits the possibility to retrieval of identification data to six months in relation to investigations of smaller offences, punishable with up to one year of imprisonment.
- The scope of the online infiltration is limited to offences punishable by at least one year of imprisonment. Article 46*sexies* CCP explicitly refers to the subsidiarity principle.
- Article 88*bis* CCP stipulates that this investigative measure can only be ordered if it is necessary to uncovering the truth. Since the new data retention act, the competence is limited to offences punishable by at least one year of imprisonment and the act explicitly refers to the subsidiarity principle. Furthermore, the act makes the option of delving further in the past through data, which are stored due to the data retention duty, dependent upon the severity of the offence under investigation.
- As regards the IT sneak and peek (Article 89*ter* CCP), the principle of proportionality demands that serious indications are present that the behaviour under investigation constitutes to or would constitute to an offence mentioned in the list of Article 90*ter*, §§ 2–4 CCP or an offence committed within the framework of a criminal organisation. The principle of subsidiarity demands that less intrusive means of investigation appear insufficient to reveal the truth.
- Article 90*ter* CCP mentions the requirements of necessity and subsidiarity. It limits the scope of the covert access to private communications and private data in information systems in two ways: (i) the investigative measure can only be directed at persons who, on the basis of serious indications, are suspected of having committed the offence under investigation, at (tele)communication tools or information systems which are regularly used by this person under suspicion or at places where this person is suspected to be staying or at persons that are deemed, on the basis of precise facts, to regularly be in contact with the suspect; and (ii) the investigative measure is only justified for a limited list of serious offences (Article 90*ter* §§ 2–4 CCP, *supra* footnote 122). In addition, the investigating judge must list the reasons for the necessity of the measure in his order.[138]

Limited Duration

The CCP stipulates a maximum duration for the real time investigative measures in view of limiting their impact and ensuring renewed (judicial) control of long-term investigative measures. The duty to preserve data (Article 39*ter* CCP) can concern a period up to a maximum of ninety days. The duration of the duty can be prolonged in writing. The online infiltration can be ordered for a period of three months. Once again, renewal by a new order is possible without limitation as to the maximum

[138] Article 90*quater*, § 1, 2° CCP.

duration. The investigative measure of Article 88*bis* CCP is limited to a maximum of 2 months after the investigating judge issued the order. Renewal of the measure is possible with a new, well-founded decision of the investigating judge.[139] The provision does not determine a maximum duration. The duration of the investigative measure of Article 90*ter* CCP is limited to one month. The investigating judge can prolong the duration one month at a time, for a maximum of six months.[140]

Notification Duties

The CCP requires that for some investigative measures certain persons would be informed of the execution of these investigative measures:

- Article 39*bis*, § 7 CCP stipulates that the public prosecutor or investigating judge informs the person in charge of the information system of the IT-search and provides him with a summary of the data that are copied, made inaccessible or removed. This duty applies as far as the identity or residence can reasonably be established;
- Article 90*novies* CCP contains a delayed duty of notification to each person who has been the subject of a covert communication or computer search in accordance with Article 90*ter* CCP, as far as his identity or residence can reasonably be established.

Confidentiality Duties

In order to ensure the possible covert nature of certain investigative measures and the secret nature of the pre-trial investigation, the CCP prescribes confidentiality duties for the persons who must comply with duties of cooperation. These confidentiality duties are to be found in:

- Article 39*ter*, § 3 CCP;
- Article 46*bis*, § 2, last section CCP;
- Article 88*bis*, § 4, second section CCP;
- Article 88*quater*, § 4 CCP; and
- Article 90*quater*, § 2, second section and § 4, fourth section CCP.

Entrapment[141]

General Prohibition

The concept of entrapment was first introduced in the context of the special investigative methods.[142] After the Constitutional Court nullified this provision, a general prohibition of entrapment was introduced in Article 30 Preliminary Title of

[139] Article 88*bis*, § 1, fourth section CCP.

[140] Article 90*quater*, §§ 1, 4° and 90*quinquies*, first section CCP. This maximum can be extended by two months with a view to installing the necessary technical tools.

[141] De Nauw 2012, pp. 90/01–90/20.

[142] Former Article 47*quater* CCP.

the Code of Criminal Procedure. According to this Article, entrapment exists whenever the intervention of a police officer or a third party, who acts at the explicit request of a police officer, directly creates the intent of an offender to commit an offence, reinforces his intent or confirms it, at such time that he or she wished to terminate the offence. Thus the concept of entrapment in Belgian criminal law covers a broad range of meaning. The criminal procedure based on entrapment is inadmissible regarding the offences which arose by means of the entrapment. The identification of offences before the entrapment remains valid.[143] Although jurisprudence is divided, certain scholars take the view that entrapment caused by foreign police officers also leads to the inadmissibility of the criminal procedure initiated in Belgium.[144]

Conditions

Entrapment requires the use of a fraudulent means to provoke an offence. Consequently, it needs to precede the offence. There is no question of entrapment when the police intervention is limited to the creation of an occasion for the potential offender to commit an offence. The offender should, however, remain free to abandon his criminal plan. The Belgian Court of Cassation decided that police officers can, without exaggerating, imitate a daily life scene, such as the positioning of a portable computer lying visible in a closed car.[145] In analogy with the case law on "luring cars", scholars claim that also the technique of "luring people" could be launched, if the conditions set out in jurisprudence, are fulfilled.[146]

Entrapment and Cyber Luring

The relation between the prohibition of entrapment and the abovementioned offence of cyber luring that criminalizes adults communicating with an apparent or probable minor (Sect. 5.2.2) is significant in this regard. As we explained, someone who is of the opinion that he is communicating with a minor is punishable by Belgian Criminal Law. In the preparatory works of the offence of cyber luring, we read that Parliament sought to permit the use of "lure teenagers" in this way.[147] In the Sweetie 1.0 project, there was no question of entrapment, since the avatar was operated by an adult working for *Terre des Hommes*. The aim of the Sweetie 2.0 project, however, is to establish an avatar that is operated by law enforcement authorities. In this context, scholars are of the opinion that inadmissibility for reasons of entrapment does not jeopardize the criminal procedure, on the condition that the police officer instructing the avatar is reluctant and his reactions are limited to simple answers so that the potential offender remains free to renounce criminal steps at every moment.[148]

---

143 Constitutional Court 19 July 2007, AR 205/2007.

144 De Nauw 2012, p. 90/09.

145 Cass. 17 March 2010, AR P100010F; Brussels 14 March 2007, *RABG* 2008, afl. 1, 63.

146 Berkmoes and Delmulle 2011, pp. 544–546.

147 Bill, *Parl.St.* Senate 2012–13, nr. 5-2253/1, 9.

148 Stevens 2015, p. 854.

5.3.3 Succinct Overview of Secret Investigatory Powers in an Online Context

Online Infiltration
Definition
The Act on IT-related searches introduced the online infiltration in the Code of Criminal Procedure. An online infiltration exists when a police officer maintains online contact with one or more persons, about whom serious indications exist that they have committed or will commit an offence punishable by at least one year of imprisonment.[149] The police officer can be authorised to use a fictitious identity in the framework of the online infiltration. Since the contact must be lasting, a single, directed contact is insufficient for the provisions on infiltration to apply. A short contact, directed at a simple verification or arrest does therefore not amount to an online infiltration. The use of undercover agents who are not law enforcement officers is not permitted in Belgium. However, police officers participating in an infiltration may avail themselves of the expertise of a person not belonging to the police forces for a short period of time, when strictly necessary.[150] All communications are registered and are added to the criminal file for evidentiary purposes or deposited at the registry.

Safeguards
The public prosecutor as well as the investigating judge[151] have the power to authorize an online infiltration. The public prosecutor has the exclusive power to supervise the execution of the online infiltration, even if a judicial inquiry has been started.[152] The conditions of subsidiarity and proportionality must be fulfilled. The first condition implies that less intrusive means of investigation are insufficient to uncover the truth.[153] The second condition results from the limitation as to the offences for which the infiltration is permitted (offences punishable by at least one year of imprisonment). The authorization of the public prosecutor or the investigating judge is issued in a written, well-founded decision.[154] An oral consent, which is possible in urgent circumstances, needs to be confirmed in writing as soon as possible. The authorization for an online infiltration is valid for a period up to three months.[155] Extension of the online infiltration demands a new well-founded authorization.

---

149 Article 46*sexies*, § 1 CCP.

150 Article 46*sexies*, § 1, third section CCP.

151 Article 56 CCP.

152 Article 56*bis* CCP.

153 Article 46*sexies*, § 1 CCP.

154 Article 46*sexies*, §2 CCP.

155 Article 46*sexies*, § 2 CCP.

Committing Offences

Article 46*sexies*, § 3 CCP provides for the exceptional possibility for law enforcement agencies to commit strictly necessary offences during the course of the infiltration operation, to guarantee its success or to ensure the safety of the executing officers and other people involved. The explicit consent of the public prosecutor is required. Furthermore, the offences may not be more serious than those warranting said investigative methods and must always be proportionate to the objective. If these conditions are fulfilled, the public prosecutor having granted permission for the commission of offences during the execution of the infiltration and the executing police officers will remain unpunished.

IT Sneak and Peek Operation[156]

Definition

Article 89*ter* CCP authorizes the investigating judge to order the police force to access an information system at any time without the knowledge or consent of the owner or authorized user. The investigative measure provided in Article 89*ter* CCP is excluded from the mini judicial inquiry.

Safeguards

First, a sneak and peek operation can only be used for limited purposes. The operation can only take place in order to (i) carry out reconnaissance of the property and verify the presence of "goods" that are the object of the offence, that served or were meant to serve for the commission of the offence or that proceed from the offence; (ii) gather evidence of the presence of these mentioned goods; (iii) fit a technical device for a systematic observation.[157] The last purpose is not relevant for IT sneak and peek operations since the "systematic observation" in private information systems will amount to a covert search of information systems in the sense of Article 90*ter* CCP. The goal of the sneak and peek operation cannot be the seizure of evidence, but can only be the establishment of the presence of evidence, for example by taking pictures or samples. If certain data require further investigation, some of them could be copied by way of sampling. Second, a sneak and peek operation can only take place at "locations" for which a suspicion exists that the mentioned goods can be found there, that evidence of the presence of those goods can be gathered or that they are being used by persons under suspicion. Third, the principle of proportionality demands that serious indications are present that the behaviour constitutes to or would constitute an offence mentioned in the list of Article 90*ter*, §§ 2–4 CCP or an offence committed within the framework of a criminal organisation. Fourth, the principle of subsidiarity demands that less intrusive means of investigation appear insufficient to reveal the truth. Finally, a written substantiated decision is necessary. In case of emergency, an oral decision suffices, but needs to be confirmed in writing as soon as possible.[158]

---

[156] In relation to the traditional sneak and peek operation: Berkmoes and Delmulle 2011, pp. 817–852.

[157] Article 47*sexies*, § 1, third section CCP.

[158] Article 46*quinquies*, § 1 CCP.

Covert Access to Computer Data Which Are Not Accessible to the Public
Definition
Article 90*ter* CCP used to be the legal basis authorizing the wiretap. However, the Act on IT-related searches has substantially broadened this competence. Since the beginning 2017, the investigating judge has the power to intercept, access, search and register not publicly accessible data in an IT-system or a part of it by technical means. It provides in other words the possibility to break into computer systems in order to search for private data or to monitor the use of private data in real time. This implies the power to bypass security measures and to apply technical measures to decrypt data.

Competent Authority
The investigating judge is the sole competent authority, except for some limited competences for the public prosecutor. The latter can order this measure in case of *flagrant délit* of the terrorism offences mentioned in Article 137 CC, the taking of hostages (Article 347*bis* CC), illegal deprivation of liberty (Article 434 CC) or extortion (Article 470 CC). The public prosecutor is in charge as long as the situation of *flagrant délit* continues, and with regard to the terrorism offences, during at least 72 hours after the *flagrant délit* has been discovered. In urgent circumstances, both the investigating judge and the public prosecutor can orally order the measure, but it needs to be confirmed in writing.[159] The covert access to private communications is excluded from the mini judicial inquiry. As a consequence, the opening of a judicial inquiry is required, except for the aforementioned offences in case of *flagrant délit*.

Safeguards
Article 90*ter* CCP mentions the requirements of necessity to uncovering the truth and subsidiarity, implying that no less intrusive investigative measure is more appropriate. It limits the scope of the covert access to private communications and private data in information systems in two ways. First, the investigative measure can only be directed at persons who, on the basis of serious indications, are suspected of having committed the offence under investigation, at (tele)communication tools or information systems which are regularly used by this person under suspicion or at places where this person is suspected to be staying or at persons that are deemed, on the basis of precise facts, to regularly be in contact with the suspect. Second, the investigative measure is only justified for a limited list of serious offences.[160] In addition, the investigating judge must list the reasons for the

[159] Article 90*quater*, § 1, section 3 CCP: within 24 hours as far as the investigating judge is concerned. Article 90*ter*, § 5, section 2 CCP, however, stipulates that the public prosecutor must confirm the authorization in writing *as soon as possible*.

[160] Article 90*ter* §§ 2–4 CCP. This list contains inter alia following relevant offences: computer related forgery (7°), rape (15°), grooming (16°), some types of moral depravity of minors and prostitution (17°), child pornography (17°), kidnapping of minors (20°), cyber luring (21°) and electronic stalking (40°).

necessity of the measure in his order.[161] Furthermore, the duration of the investigative measure of Article 90*ter* CCP is limited up to one month per order. The investigating judge can prolong the duration one month at a time, for a maximum of six months. This maximum can be extended by two months with a view to installing the necessary technical tools.[162] The persons whose computer data or information systems were secretly accessed pursuant to Article 90*ter* CCP, must be notified within 15 days after the end of the judicial inquiry.[163] This duty applies as far as the identity or residence can reasonably be established.

Systematic Observation in Publicly Accessible (Online) Places
Online Observation
As far as access by law enforcement to publicly accessible data is concerned, the Belgian legislation does not provide for specific investigative measures. The Court of Cassation recently accepted that Article 26 of the Act on the Police Function of 5 August 1992 does not only allow each officer of judicial police to enter "offline" places which are accessible to the public, but it also allows them to access online publicly accessible places.[164] The question remains however whether the intensive or ongoing access to publicly accessible data would qualify as a systematic observation at a certain point. Belgian legislation does not provide an online counterpart of the classical systematic observation. The existing regulation on systematic observation, however, is clearly aimed at the physical and not the digital world.

Definition
According to Article 47*sexies* CCP a systematic observation[165] is the systematic visual surveillance by a police officer of one or more persons, their presence or their conduct or of specific goods, places or events. This provision further defines the systematic character of the observation in terms of (i) the duration of the observation, in particular should it consist of more than five consecutive days or more than five non-consecutive days within a one month period;[166] (ii) technical devices that are being used; (iii) the international nature of the observation; or (iv) the performance of the observation by the special units of the Federal Police. A technical device is defined by this provision as a configuration of components which detects signals, transports them, activates their registration and registers the signals, with the exception of technical means used for the measure mentioned in

161 Article 90*quater*, § 1, 2° CCP.

162 Article 90*quater*, § 1, 4° and 90*quinquies*, first section CCP.

163 Article 90*novies* CCP.

164 Cass. 28 March 2017, AR P.16.1245.N.

165 Berkmoes and Delmulle 2011, pp. 583–680.

166 An observation lasting less than 5 days is not considered a systematic observation, to which the specific guarantees apply. The mere formal establishment of offences is not a systematic observation. An observation, which is not systematic is part of the general investigative competence as provided by Article 8 CCP and 15 Law on the Police Function of 5 August 1992. Berkmoes 2005, p. 65.

Article 90*ter* CCP. Examples of technical devices are video cameras, GPS systems and motion detectors,[167] but actually each computer could be considered as a technical device.

Safeguards
In principle, the public prosecutor and the investigating judge[168] can authorize an observation in a publicly accessible place. The public prosecutor has the exclusive power to supervise the observation, even if a judicial inquiry has been started.[169] An observation can only be ordered when the investigation requires it and less intrusive means of investigation do not appear sufficient to reveal the truth, which expresses the subsidiarity of this investigative measure.[170] The observation lasts for a maximum of three month starting from the authorization.[171] Extension of this period demands a new and well-founded authorization. The authorization is issued in a written, well-founded decision that contains several obligatory specifications.[172] In case of emergency, an oral decision suffices, but needs to be confirmed in writing.[173] Although serious indications of a committed offence or a future offence generally are sufficient to allow an observation, the use of a technical device is only allowed with regard to offences able to warrant a prison sentence of one year or more. In the framework of a proactive investigation, reasonable suspicion should exist in relation to an offence listed in Article 90*ter*, §§ 2–4 CCP or an offence committed within the framework of a criminal organization.

5.3.4 Application of Relevant Investigatory Powers to the Sweetie Case

Introduction
We can infer from the above mentioned provisions that law enforcement authorities can utilize an avatar representing a 10-year old girl in Belgian criminal procedure. Below, we provide an overview of the provisions with which law enforcement authorities launching Sweetie 2.0, must comply.

Access to Public Places
The first question is whether police officers can obtain access to online fora or chat rooms frequented by possible offenders. The answer to this question entails a distinction between public or publicly accessible places on the one hand and private

---

[167] Bill, *Parl.St.* Chamber, 2001–02, nr. 50-1688/001, 60.

[168] Article 56, § 1 CCP.

[169] Article 47*ter*, § 2, first section CCP.

[170] Article 47*sexies*, § 2, first section CCP.

[171] Article 47*sexies*, § 3, 5° CCP.

[172] Article 47*sexies*, § 3, 1°–6° CCP.

[173] Article 47*sexies*, § 5 CCP.

places on the other. It is generally accepted that investigators may investigate the first category by means of government computers, as this is equivalent to police officers patrolling streets and public places. Scholars contend, moreover, that investigators may also access these places with anonymized IP-addresses,[174] since these privacy tools are at the disposal of every internet user.[175] As nicknames are part of online environments, the mere use of a nickname while accessing these places should not present a problem, as far as this nickname is not provocative.[176] However, this could constitute an offence according to Belgian criminal law, provided that a fictitious surname is being used in relation to third parties, aimed at making them believe the fictitious identity is real.[177] Furthermore, the deceptive use of investigative measures entails a more serious interference into the right to privacy, demanding clear legal grounds. In the framework of the online infiltration, a police officer could indeed use a reliable fictitious identity since the legal provision concerned explicitly provides for this possibility. The same is true for the covert computer search mentioned in Article 90*ter* CCP, since the article explicitly mentions that security measures could be bypassed by the use of a fake identity.

Establishing Enduring Contact
Once law enforcement agencies want the avatar to start or participate in conversations on online fora or in chatrooms and establish lasting contacts, the provision on online infiltration applies. The avatar can only get in contact with persons, about whom serious indications exist that they committed or will be committing offences punishable by at least one year of imprisonment. The concept of entrapment limits the activities of undercover police officers. For instance, the fictitious identity may not be provocative,[178] nor is it permitted for the avatar to say provocative things. All communications of the avatar have to be registered and will be added to the criminal file or be deposited at the registry after ending the infiltration operation.

Access to Private Places
Should it be necessary, at a certain point in the investigation, to obtain access to an existing private online environment, to which real access barriers exist (for example a private chat box secured by login credentials), relatively new competences apply. As mentioned above, the Act on IT-related searches introduced explicit covert "hacking" powers for police officers. Depending on the objective of the search, an IT sneak and peek (Article 89*ter* CCP) or covert computer search applies. The sneak and peek would allow merely establishing the presence of, for example, the profile which is used to commit offences like the moral depravity of minors, access to child pornography, grooming or cyber luring (means of the offence) or a fake profile (the object of computer related forgery). Some scholars allege that the

---

174 For instance, via *Tor*.

175 Conings and Van Linthout 2012, p. 216; Kerkhofs and Van Linthout 2013, p. 232.

176 Kerkhofs and Van Linthout 2013, p. 234.

177 Article 231 CC; Conings and Van Linthout 2012, p. 214.

178 Article 30 Preliminary Title CCP. Kerkhofs and Van Linthout 2013, p. 234.

authorization of a sneak and peek operation implies the authorization to overcome access barriers when needed, in an upright manner.[179] When the measure is however clearly aimed at evidence gathering, the covert search for private communications and data applies. Article 90*ter* CCP explicitly provides for the possibility to enter an information system and bypass security measures (for example by using a fake identity) or apply technical measures to decrypt data. In order to obtain access to a private online environment, police officers may have to upload illegal content themselves or become part of an illegal peer-to-peer network. As mentioned before, the Belgian legislation provides for a possibility for law enforcement to commit offences in the framework of the observation and online infiltration. Furthermore, uploading illegal content could be considered as a specific method of bypassing security measures.

Identification and Localisation

Finally, in order to identify and/or locate possible offenders who were caught by means of the avatar, the Articles 46*bis* and 88*bis* CCP must be taken into consideration respectively.

5.3.5 Relevant Aspects of Digital Forensic Evidence

Belgian legislation does not contain regulations regarding technical tools which are used to perform online legislations. As regards online infiltration, Article 46*sexies*, § 4 CCP only stipulates that the communications should be registered by "suitable means". Articles 90*sexies* and 90*septies* CCP determine that intercepted communication and data should be registered and stored for further use, for example by the defence. Only a transcription of the relevant communication segments must be added to the case file. In accordance with Article 90*septies*, § 1 CCP "suitable technical means" should be used to ensure the integrity and confidentiality of the registered communications and data and, as far as possible, the translation and transcription of the relevant segments.

Regarding the duties of cooperation mentioned in the Articles 46*bis*, 88*bis* and 90*quater* CCP, a Royal Decree of January 9th, 2003 contains certain provisions on the reliability and integrity of digital evidence.[180] Article 6, § 1 of this Decree stipulates the regulations which cooperating operators of an electronic communication network or suppliers of an electronic communication service must respect. They must, for instance, transmit the requested information in a commonly available format (3°) and in a secure manner, ensuring that the information cannot be intercepted (5°). Article 6, § 3 moreover prescribes the technical specifications that must meet the standards of the European Telecommunications Standards

[179] Conings 2015b, p. 285; Kerkhofs and Van Linthout 2013, p. 243.

[180] Royal Decree 9 January 2003, *Off.Gazette* 10 February 2003.

Institute.[181] Furthermore, Article 10*bis* requires that the requested data are transmitted according to the rules and by means of an efficient technical means, which is available on the market.

The relevant legal provisions are thus vague and fragmented. Belgian criminal law lacks a general framework protecting the reliability and integrity of digital evidence.[182] The elaboration of such a legal framework is nevertheless necessary to ensure the effective use of the gathered data as evidence in court.

5.4 Conclusions

The Belgian criminal procedure framework seems to provide a proper legal basis for the use of Sweetie as a criminal investigative measure. At the time of this paper's writing in 2016 and before the adoption of the Act on IT-related searches some problems still arose. Hereinafter, we give an overview of these legal issues and explain how the Act on IT-related searches has solved most of them.

First, the existing competences had to be interpreted in a more or less progressive way, due to the lack of competences adapted to the digital era. When law enforcement agencies wished to establish a lasting contact with one or more persons using the false Sweetie identity, they would have to fulfil the legal conditions of a classic, "offline" infiltration. The legal framework of infiltration, however, was deemed to be too strict for an exclusive online execution, which entailed far less risks than its offline counterpart. Furthermore, as soon as the online chatroom was considered to be a private environment, the classic sneak and peek operation or the competence to enter a private place in the framework of a wiretap, entailed the possibility of secretly entering the chatroom, when interpreted in a progressive way. These interpretations could be problematic in light of the legality condition of Article 8 ECHR, due to a potential lack of foreseeability. The Act on IT-related searches, introducing provisions on online infiltration (Article 46*sexies* CCP), an IT sneak and peek (Article 89*ter* CCP) and a competence to covertly access computer data (Article 90*ter* CCP), seems to solve this problem. Yet, we truly doubt whether the Act on IT-related searches has provided enough safeguards to compensate for these far-reaching competences.[183]

Second, the wiretap competence was the sole competence enabling law enforcement agencies to record online communications between Sweetie and its communication partner. In contrast to the offline infiltration, the legal framework of

[181] For instance: "1° TS 101-331: "Lawful Interception (LI); Requirements of Law Enforcement Agencies"; 2° TS 101-671: "Lawful Interception (LI); Handover interface for the lawful interception of telecommunications traffic"; 3° TS 101-909-20-1: "AT Digital. Digital Broadband Cable Access to the Public Telecommunications Network; IP Multimedia Time Critical Services; Part 20: Lawful Interception; Sub-part 1: CMS based Voice Telephony Services" […]."

[182] Conings and Van Linthout 2012, p. 207; Kerkhofs and Van Linthout 2013, pp. 183–192.

[183] Conings and Royer 2017, pp. 311–338.

the online infiltration explicitly provides the possibility and even the duty to register all communications of the undercover agent. Therefore, Article 90*ter* CCP does not apply to these registrations.

Our third remark concerns the proportionality test. The different necessary competences could only be used when investigating serious offences, which are part of an exhaustive list (Article 90*ter*, §§ 2–4 CCP). The legislator defined cyber luring as an offence against an *apparent* or *probable* minor. The broad definition was aimed at ensuring that the use of undercover agents, pretending to be minors, would not stand in the way of prosecution. However, while referring explicitly to the case of the undercover agent, the legislator initially forgot to insert the offences of cyber luring (Article 433*bis*/1 CC) and grooming (Article 377*quater* CC) into the list of serious offences. Since the Act on IT-related searches has supplemented the list with these offences, a distorted application of cyberstalking, which was included in the list, is no longer needed to reach the legislator's goal. Furthermore, the online infiltration can be used to investigate offences punishable by at least one year of imprisonment. Questions, however, raise on whether it would be more appropriate to link this possibly far-reaching and intrusive investigation method to the said list of serious offences as well.

Fourth, overall, the legislator should provide more explicit safeguards to ensure the reliability of evidence. Ensuring reliability is to a large extent currently left to practice. The lack of legislative attention to this topic entails a significant opportunity for defence lawyers to question the trustworthiness of collected evidence. This is especially true for digital evidence, which is more vulnerable to manipulation. Consequently, obedience to the law does not sufficiently guarantee the use of the collected evidence in court. The Act on IT-related searches remained silent on this topic.

Annex: Relevant Legal Provisions

Criminal Code (Strafwetboek)

Article 371/1 CC

With a prison sentence from six months to five years a person is punished who:

1° observes or causes a person to observe, or makes a video or audio recording of, a person,

- directly or by technical or other means,
- without the consent of that person or without his knowledge,
- while being naked or engaged in an sexually explicit act, and
- while being in a situation if he can reasonably expect that his privacy will not be violated;

2° makes accessible or distributes the visual or audio recording of a naked person or a person who performs an explicit sexual act without his or her consent or without his or her knowledge, even if that person has consented to the making of those images.

If these offences are committed against the person or with the assistance of the person of a minor who has reached the age of sixteen, the offender shall be punished with imprisonment of five to ten years.

If the minor has not reached the age of sixteen, the sentence is imprisonment of ten years to fifteen years.

Voyeurism exists as soon as there is a beginning of execution.

Article 372 CC

Any indecent assault, without violence or threats committed against the person or with the assistance of the person of a minor of the male or female sex under the age of sixteen shall be punishable by imprisonment from five years to ten years.

Indecent assault on integrity, without violence or threats by a relative in the ascending line or adopter, committed against the person or with the assistance of a minor, even if he or she has reached the age of sixteen, shall be punishable by imprisonment from ten to fifteen years. The same penalty shall apply if the offender is either the brother or sister of the minor victim or any other person occupying a similar position in the family, or any person who regularly or occasionally lives with the victim and has authority over it.

Article 373 CC

Imprisonment from six months to five years shall be imposed for indecent assault committed against persons or with the assistance of persons of the male or female sex, by force, coercion, threat, surprise or deception, or made possible by an imperfection or physical or mental disability of the victim.

If the attack is committed against the person or with the help of the person of a minor who has reached the age of sixteen, the guilty party will be punished with imprisonment of five years to ten years.

If the minor has not reached the age of sixteen, the sentence is imprisonment of ten years to fifteen years.

Article 374 CC

Indecent assault exists as soon as there is a beginning of execution.

Article 375 CC

Rape is any act of sexual penetration of any kind and by any means committed against a person who does not consent.

There is no consent if the act has been imposed by force, coercion, threat, surprise or deception or has been made possible by an imperfection or physical or mental disability of the victim.

Imprisonment of five to ten years shall be imposed to anyone who commits the crime of rape.

If the crime is committed against the person of a minor who has reached the age of sixteen, the offender will be punished with imprisonment of ten to fifteen years.

If the crime is committed against the person of a minor who has reached the age of fourteen, but who has not reached the age of sixteen, the offender will be punished with imprisonment of fifteen to twenty years.

Any act of sexual penetration, of whatever nature or by whatever means, committed against the person of a minor who has not reached the age of fourteen shall be considered 'rape by force'. In that case, the sentence is imprisonment of fifteen to twenty years.

The sentence is imprisonment of twenty to thirty years, if the child has not reached the age of ten.

Article 379 CC
Anyone who violates public morals by inducing, favouring or facilitating the sexual immorality, spoilage or prostitution of a minor of the male or female sex in order satisfy sexual desires of someone else is punished by imprisonment of five years to ten years and a fine of five hundred euros to twenty-five thousand euros.

He is punished by imprisonment of ten years to fifteen years and by a fine of five hundred euros to fifty thousand euros if the minor has not reached the age of sixteen.

The sentence is imprisonment of fifteen years to twenty years and a fine of one thousand euros to one hundred thousand euros if the minor has not reached the age of fourteen.

Article 380, § 4 CC
Are punishable with imprisonment of ten years to fifteen years and a fine of one thousand to one hundred thousand euros:

1° he who, in order to satisfy sexual desires of someone else, either directly or through an intermediary, recruits, takes, takes away, or holds with him, even with his consent, a minor with a view to committing sexual abuse or prostitution;
2° anyone who, directly or through an intermediary, runs a house of sexual abuse or prostitution in which minors commit prostitution or debauchery;
3° he who sells, rents out or makes available rooms or any other space to a minor with a view to sexual immorality or prostitution, with the intention of realizing an abnormal profit;
4° any person who, in any way, exploits the sexual immorality or prostitution of a minor;
5° any person who, by handing over, offering or promising a material or financial advantage, has obtained sexual immorality or prostitution from a minor.

Article 380, § 5 CC
The offences referred to in § 4 are punishable by imprisonment of fifteen years to twenty years and by a fine of one thousand euros to one hundred thousand euros if they are committed against a minor under the age of sixteen.

Article 380, § 6 CC
Those who attend the debauchery or prostitution of a minor directly, including by means of information and communication technology, are punished with imprisonment of one month to two years and a fine of one hundred euros to two thousand euros.

Article 380*bis* CC
Those who use words, gestures or signs, in a public place, to incite someone to sexual immorality are punished with imprisonment of eight days to three months and a fine of twenty-six euros to five hundred euros. The sentence is doubled if the offence is committed against a minor.

Article 380*ter* CC
§ 1. Any person who, directly or indirectly, advertises, publishes, distributes or disseminates an offer of services of a sexual nature, is punished with imprisonment of two months to two years and a fine of two hundred to two thousand euros, if that advertisement is specifically aimed at minors or if it refers to services offered by minors or by persons alleged to be minors, even if one conceals the nature of the services offered by means of artifices of language.

If the purpose or effect of the advertising referred to in the first subsection is, directly or indirectly, to facilitate the prostitution or sexual immorality of a minor or his exploitation for sexual purposes, the penalty shall be three months to three years of imprisonment and a fine of three hundred euros to three thousand euros.

§ 2. Prison sentences of one month to one year and a fine of one hundred euros to one thousand euros apply to anyone who in any way, directly or indirectly, advertises, distributes or disseminates an offer of services of a sexual nature provided by any means of telecommunication, even if they conceal their offer in covert terms.

§ 3. In all cases not defined in §§ 1 and 2, imprisonment of one month to one year and a fine of one hundred euros to one thousand euros shall be imposed on those who, by any means of advertising, even if one conceals the nature of the services offered by means of artifices of language, make it known that they are engaging in prostitution, facilitate the prostitution of others or wish to deal with someone who engages in debauchery.

The same penalties shall apply to any person who induces by any means of advertising, by reference to it, the sexual exploitation of minors or adults, or who uses such advertising in response to an offer of services.

Article 383 CC
Any person who produces, displays, sells or distributes songs, magazines or other writings, whether or not printed, pictures or prints contrary to morals, shall

be punished with imprisonment of eight days to six months and a fine of twenty-six euros to five hundred euros.

He who sings, reads, recites, performs or brings out obscenities in public meetings or places referred to in Article 444, subsection 2, shall be punished with the same punishments.

The same penalties shall apply to:

Any person who, with a view to trade or distribution, produces, holds in stock, imports or has imported, transports or has transported, hands over to a transport or distribution agent, or publishes by any means of publicity, songs, periodicals, writings, illustrations or prints which are contrary to morals;

He who displays, sells or distributes phrases or objects contrary to morals, manufactures or stocks them with a view to trade or distribution, imports them or has them imported, transports them or has them transported, hands them over to a transport or distribution agent or publishes them by any means of publicity. [...]

Article 383*bis* CC

§ 1. Without prejudice to the application of Articles 379 and 380, any person who unlawfully exhibits, offers, sells, rents, transmits, supplies, divides, distributes, makes available or hands over child pornography material, or manufactures, imports or causes import of child pornography material, shall punished with imprisonment of five to ten years and with a fine of five hundred euros to ten thousand euros.

§ 2. Any person who knowingly acquires, possesses or, with full knowledge, gains access to illegal child pornography material by means of information and communication technology shall be punished with imprisonment of one month to one year and a fine of one hundred to one thousand euros.

§ 3. The offence referred to in § 1 is punishable by imprisonment of ten years to fifteen years and by a fine of five hundred euros to fifty thousand euros if it concerns an act of participation in the main or secondary activity of an association, regardless of whether the offender has a leading position or not.

§ 4. For the purposes of this Article, 'child pornography material' is understood to mean:

1° any material that visually depicts—in some way or another—a minor engaged in real or simulated sexually explicit acts or that depicts the sexual organs of a child for primarily sexual purposes;

2° any material that visually depicts any person appearing to be a minor engaged in real or simulated sexually explicit conduct or that depicts the sexual organs of this person for primarily sexual purposes;

3° realistic images that depict non-existent minors engaged in sexually explicit acts or that depict the sexual organs of this minor for primarily sexual purposes.

[...]

Article 385 CC

Anyone who commits acts of indecency in public by acts that violate dignity is punished with imprisonment of eight days to one year and a fine of twenty-six euros to five hundred euros.

If the public indecency is committed in the presence of a minor under the age of sixteen, the penalty is imprisonment from one month to three years and a fine from one hundred euros to one thousand euros.

Article 386 CC

If the offences defined in Article 383 are committed against minors, the term of imprisonment is between six months and two years and the fine between one thousand euros and five thousand euros.

In the same case, the penalties provided for in the first subsection of that Article are doubled, without prejudice to the application of subsection 2 of Article 385.

Article 377*quater* CC

An adult who, by means of information and communication technology, proposes to meet a minor under the age of sixteen with a view to committing an offence provided for in this Chapter or in Chapters VI and VII of this Title shall, in so far as this proposal has been followed by material acts leading to such a meeting, be punished with imprisonment of one year to five years.

Article 433*bis*/1 CC

Imprisonment of three months to five years shall be imposed on an adult who, through the use of information and communication technologies, communicates with an apparent or probable minor in order to facilitate the commission of a crime or an offence against him or her:

1° if he has concealed his identity, age or capacity or has lied about it;
2° if he has emphasized the discretion to be observed with regard to their conversations;
3° if he has offered or previewed any gift or benefit;
4° if he has used any other list.

Article 442*bis* CC

Anyone who has stalked a person when he knew or should have known that his conduct would seriously disturb that person's peace and quiet shall be punished with imprisonment of fifteen days to two years, and with a fine of fifty euros to three hundred euros, or with one of those penalties.

In the event that the offences referred to in the first subsection are committed to the detriment of a person whose vulnerable condition due to age, pregnancy, illness or physical or mental disability or inadequacy was obvious, or the perpetrator was known, the minimum penalty provided for in the first subsection shall be doubled.

Preliminary Title of the Code of Criminal Procedure (Voorafgaande titel Wetboek van Strafvordering)

Article 30 PT CCP

Entrapment is forbidden.

Entrapment exists whenever the intervention of a police officer or a third party, who acts at the explicit request of a police officer, directly creates the intent of an offender to commit an offence, reinforces his intention or confirms it, at such time that he wished to terminate the offence.

In the event of entrapment, the criminal procedure is inadmissible regarding these facts.

Article 32 PT CCP

An element of proof which is obtained in an irregular manner will be declared null if:

- compliance with the formal conditions in question is sanctioned with nullity by law; or
- the irregularity committed affects the reliability of the evidence; or
- the use of evidence is contrary to the right to a fair trial.

Code of Criminal Procedure (Wetboek van Strafvordering)

Article 28*septies* CCP

The public prosecutor may request the investigating judge to carry out an investigative act for which only the investigating judge is competent, with the exception of the arrest warrant referred to in Article 16 of the Law of 20 July 1990 on pre-trial detention, the completely anonymous testimony referred to in Article 86*bis*, the supervision measure referred to in Article 90*ter*, the investigative acts referred to in Articles 56*bis*, second subsection, and 89ter and the home search without a judicial investigation being initiated. After the execution of the investigative act by the investigating judge, he decides whether to send the file back to the public prosecutor who is responsible for the continuation of the investigation, or whether to continue the entire investigation himself, in which case further action is taken in accordance with the provisions of Chapter VI of this Book. This decision cannot be subject to appeal.

In the event of a new request pursuant to the first subsection in the same file, the case shall be brought before the same investigating judge if he it is still in office.

Sneak and Peek Operation

Article 46*quinquies* CCP

§ 1. Without prejudice to Article 89*ter*, the public prosecutor may, by a written and reasoned decision, authorize police forces at any time, without the knowledge of the owner or his representative or without their consent, to enter a private place and open closed objects located there, if there are serious indications that the offences constitute or would constitute a crime within the meaning of Article 90*ter*, §§ 2–4, or are committed or would be committed within the framework of a criminal organization within the meaning of Article 324*bis* of the Criminal Code, and the other means of investigation do not appear to be sufficient to establish the truth.

A private place within the meaning of this Article is the place that appears to be:

- no home;
- no inherent part of a home within the meaning of Articles 479, 480 and 481 of the Criminal Code;
- no place for professional purposes or domicile of a lawyer or a doctor as referred to in Article 56*bis*, subsection 3.

When required because of the urgency of the situation, the decision referred to in the first paragraph may be communicated orally. In such a case, the decision must be substantiated and confirmed in writing as soon as possible.

If the decision referred to in the first subsection is taken in the context of the application of the special investigation methods referred to in Articles 47*ter* to 47*decies*, the decision and all related official reports are attached to the criminal file not later than at the end of the application of the special investigation method.

§ 2. Entering the private place as referred to in para. 1, and opening closed objects that are located at this place, can only take place for the following purposes:

1° to see that place and to ensure that there are items which are the object of the offence, which have served or are intended for the commission of the offence or which are the products of the offence, which are proceeds from the offence, which are goods and values which have replaced the proceeds or which are income from the proceeds invested;
2° to collect the evidence of the presence of the goods referred to in 1°;
3° to place, repair or take back, in the framework of an observation, a technical device as referred to in Article 47*sexies*, § 1, third subsection.
4° to place back the objects taken with them in accordance with para. 5.

§ 3. A sneak and peek operation can only be ordered by the public prosecutor in respect of places if it is suspected, on the basis of precise indications, that the goods referred to in § 2, 1° are present, that evidence of those can be gathered, or that they are used by suspects.

§ 4. The use of technical tools for the purpose intended in § 2 is assimilated to entering a private place as provided for in § 1.

§ 5. If the investigation of an object referred to in para. 1 cannot be carried out on site and if the information cannot be obtained by other means, the police service is permitted to take this object with them for a strictly limited period of time. The object concerned shall be returned as soon as possible, unless this would impede the proper conduct of the investigation.

§ 6. Within the framework of the measure referred to in para. 1, entry into an IT-system is only possible for the purposes referred to in para. 2, 3°.

§ 7. The prosecutor in charge of the execution of the measure referred to in para. 1 or in Article 89*ter* §1 shall draw up an official report on the execution of the measure. If closed objects have been opened during the implementation of a measure or para. 5 has been applied, this will be mentioned in the official report. The report shall be attached to the file not later than at the end of the application of the measure.

Article 89*ter* CCP

Within the framework of the implementation of the measure provided for in Article 46*quinquies*, and under the conditions and with a view to the purposes stated therein, only the investigating judge can authorize the police service designated by the King to:

- without the knowledge of the owner or his representative, or of the occupant, or without their consent, enter at any time into a private place other than that referred to in Article 46*quinquies*, §1, including the opening of closed objects located there;
- without the owner's, the possessor's or the user's knowledge or without their consent, gain access to a computer system and search it, without prejudice to the possibility for the public prosecutor to authorize entry into a computer system within the limits laid down in Article 46*quinquies*, §6.

If the authorization referred to in subsection 1 is granted in the context of the application of special investigation methods in accordance with Articles 47*ter* to 47*decies* or 56*bis*, the authorization and any related report shall be attached to the criminal file not later than at the end of the application of the special investigation method.

A copy of the order will be sent to the public prosecutor.

Special Investigation Methods: Observation and Infiltration

Article 46*sexies* CCP
§ 1. When investigating crimes and offences, if the investigation so requires and if the other means of investigation do not appear to be sufficient to establish the truth, the public prosecutor may authorize the police services referred to in subsection 2 to maintain contact on the internet, if appropriate under a fictitious identity, with one or more persons of whom there are serious indications that they are committing or would commit offences which may result in a one-year prison sentence or a more serious penalty.

The King determines the conditions, among others with regard to the training, and the more detailed rules for the designation of the police forces competent to implement the measure referred to in this Article.

In exceptional circumstances and with the explicit authorization of the public prosecutor, the official of the police services referred to in the second subsection may, in a specific operation and for a short period of time, call upon the expertise of a person who is not a member of the police services, if this appears to be strictly necessary for the fulfilment of his task. The authorization and the identity of this person shall be kept in the file referred to in para. 3, subsection 7.

This Article does not apply to the personal interaction of police officers on the Internet, in the course of their duties as a judicial police officer, with one or more persons for the sole purpose of carrying out targeted checks or arrests, without using a credible fictitious identity.

§ 2. The measure referred to in § 1 is ordered by the public prosecutor by means of a prior, reasoned written authorization. This authorization is valid for a period of three months, without prejudice to renewal.

When required because of the urgency of the situation, the authorization may be orally granted. The authorization must be confirmed as soon as possible in the form stipulated in the first subsection.

§ 3. Police officers who, in the course of their task and with a view to the accomplishment of their task or to ensuring their own safety or that of other persons involved in the measure, commit strictly necessary offences shall remain exempt from punishment, subject to the express approval of the public prosecutor.

Such offences shall not be more serious than those for which the measure is applied and shall necessarily be proportionate to the objective pursued.

The first and second subsections also apply to the persons who have provided necessary and direct help or assistance for the performance of this assignment and to the persons referred to in § 1, third subsection.

The magistrate who, in accordance with this Code, authorizes a police officer and the person referred to in the third subsection to commit offences in the context of the execution of the measure, shall remain free from punishment.

Police officers shall notify the public prosecutor, in writing and prior to the execution of the measure, of the offences they or the persons referred to in subsection 3 intend to commit.

If such prior notice could not be given, the police officers shall immediately notify the public prosecutor of the offences committed by them or by the persons referred to in subsection 3, and subsequently confirm this in writing.

The public prosecutor shall state in a separate written decision the offences which may be committed by the police services and the persons referred to in subsection 3 in the context of this measure ordered by him. That decision shall be kept in a separate and confidential file. He shall have sole access to the file, without prejudice to the right of access of the examining magistrate and the chamber of indictment as provided for in Article 56*bis* and Article 235*ter*, § 3 and Article 235*quater*, §3 respectively. The contents of this file shall be covered by professional secrecy.

§ 4. The officer who is in charge of the investigation shall draw up minutes of the various stages of implementation of this measure, including the relevant contacts. These minutes shall be attached to the file not later than at the end of the execution of the measure.

The contacts referred to in subsection 1 shall be recorded by the appropriate technical means and shall be attached to the file or lodged at the Registry, in digital or other form, not later than at the end of the period covered by the measure.

§ 5. The public prosecutor is responsible for the execution of the authorizations for the measure referred to in § 1, first subsection, granted by the investigating judge in the framework of a judicial investigation in accordance with Article 56*bis*.

The public prosecutor shall state in a separate written decision the offences that may be committed by the police services and the persons referred to in § 3, subsection 3, within the framework of the measure ordered by the investigating judge. This decision shall be kept in the file referred to in § 3, subsection 7.

Article 47*quinquies* CCP

§ 1. Without prejudice to the provisions of § 2, the police officer in charge of the execution of the special investigation methods is prohibited from committing criminal offences in the framework of his task.

§ 2. Police officers who, within the framework of their task and with a view to their success or to ensuring their own safety or that of other persons involved in the operation, commit strictly necessary offences shall remain immune from punishment, subject to the express approval of the public prosecutor.

Such offences must not be more serious than those for which the methods are used and must necessarily be proportionate to the objective pursued.

The first and second subsections also apply to the persons who have provided necessary and direct help or assistance for the performance of this task and to the persons referred to in Article 47*octies*, § 1, second subsection.

The magistrate who, in accordance with this Code, authorizes a police officer (and the persons referred to in the third paragraph) to commit criminal offences in the context of the execution of a special investigation method, remains free from punishment.

§ 3. The police officers shall notify the public prosecutor in writing, and prior to the execution of the special investigation methods, of the offences referred to in § 2, which they or the persons referred to in § 2, subsection 3, intend to commit.

If this prior notice could not be given, the police officers immediately notify the public prosecutor of the criminal offences committed by them or the persons referred to in § 2, subsection 3, and subsequently confirm this in writing.

§ 4. The Minister of Justice and the Minister of the Interior, acting on a joint proposal from the Federal Public Prosecutor and the Attorney General responsible for specific tasks relating to terrorism and super-banditism, shall at all times take the special measures strictly necessary to safeguard the protection of the identity and security of the police officers responsible for carrying out special investigation methods while preparing and carrying out their tasks. There can be no criminal offence when facts are committed in that context.

§ 5. Remain exempt from punishment, police officers of the board of the special units of the federal police who, within the framework of their training and in order to be able to carry out the special investigation methods of observation and undercover operations, commit strictly necessary offences referred to in the Royal Decree of 1 December 1975 laying down general rules for the police of road traffic and use of public roads.

Such offences must necessarily be proportionate to the objective of training pursued, using the caution which may be expected of specialized police forces, always giving priority to road safety and taking all reasonable care to ensure that no bodily injury or damage to property is caused to third parties or to the person concerned.

Committing these offences requires the prior written consent of the Federal Public Prosecutor. This agreement shall include the days and places on which these offences, if any, may be committed, as well as the vehicle used by the police service and its license plate.

The magistrate who authorizes a police officer as referred to in the first subsection to commit criminal offences within the framework of the training referred to in this Article shall remain exempt from punishment.

Article 47*sexies* CCP

§ 1. Observation in the sense of this Code is the systematic observation by a police officer of one or more persons, their presence or behaviour, or of certain things, places or events.

A systematic observation within the meaning of this Code is an observation of more than five consecutive days or of more than five non- consecutive days spread over a period of one month, an observation using technical means, an observation of an international nature, or an observation carried out by the specialized units of the federal police.

Technical means within the meaning of this Code refer to a configuration of components that detects signals, transports them, activates their registration and records the signals, with the exception of the technical means used to carry out a measure as referred to in Article 90*ter*.

An apparatus used for taking photographs is only considered to be technical means in the sense of this Code in the case referred to in Article 56*bis*, second subsection.

§ 2. The public prosecutor may authorize an observation in the framework of a preliminary investigation if the investigation so requires and the other means of investigation do not appear to be sufficient to establish the truth.

An observation using technical means can only be authorized if there are serious indications that the offences can lead to a one-year criminal prison sentence or a more severe one.

§ 3. The authorization to carry out the observation shall be in writing and shall include:

1° the serious indications of the criminal offence justifying the observation, or, if the observation takes place within the framework of a proactive investigation as described in Article 28*bis*, § 2, the reasonable suspicion of offences committed or to be committed but not yet detected, and the special indications regarding the elements described in this latter provision, which justify the observation.

2° the reasons why the observation is indispensable for revealing the truth;

3° the name or, if this is not known, a description as accurate as possible of the person or persons observed, as well as of the goods, places or events referred to in § 1;

4° the manner in which the observation will be carried out, including the permission to use technical means in the cases provided for in § 2, subsection 2, and Article 56*bis*, subsection 2. In the latter case, the authorization of the examining magistrate states the address or location of the dwelling to which the observation relates that is as accurate as possible;

5° the period during which the observation can be carried out, which may not exceed a three-month period, to be calculated from the date of the authorization;

6° the name and capacity of the judicial police officer, who is in charge of the execution of the observation.

§ 4. The public prosecutor then mentions in a separate written decision the criminal offences that can be committed within the framework of the observation by the police services and the persons referred to in Article 47*quinquies*, §2, subsection 3.

This decision shall be kept in the file referred to in Article 47*septies*, §1, subsection 2.

§ 5. In case of urgency, the authorization to carry out the observation may be given orally. The authorization shall be confirmed as soon as possible in the form laid down in para. 3.

§ 6. The public prosecutor may at any time amend, supplement or extend his observation authorization in a well-founded manner. He may at any time withdraw his authorization. He shall verify, whenever his authorization is amended, supplemented or extended, whether the conditions set out in §§ 1 to 3 are met, and shall act in accordance with § 3, 1° to 6°.

§ 7. The public prosecutor is responsible for the execution of the observation authorizations granted by the investigating judge in the framework of a judicial investigation in accordance with Article 56*bis*.

The public prosecutor then mentions in a separate written decision the criminal offences that can be committed by the police services and the persons referred to in Article 47*quinquies*, §2 subsection, 3 within the framework of the observation ordered by the investigating judge. This decision shall be kept in the file referred to in Article 47*septies*, §1, subsection 2.

Article 56*bis* CCP

Contrary to Article 56, § 1, third subsection, the authorisations of the investigating judge ordering a measure mentioned in Articles 46*sexies* or special investigative measures, are executed by the public prosecutor, according to Articles 46*sexies* and 47*ter* to 47*novies* respectively.

Moreover, only the investigating judge can authorize an observation, as meant in Article 47*sexies*, with the use of technical devices to obtain view in a residence, or in an adjacent place surrounded by a residence in the sense of Articles 479, 480 and 481 CC, or in a location used for professional activities of or the residence of a lawyer or medical doctor under Article 56*bis*, third section, when serious indications are present that the behaviour amounts to or would amount to an offence meant in Article 90*ter*, §§ 2–4, or is being or would be committed within the framework of a criminal organisation as meant in Article 324*bis* CC.

The investigating judge can only authorize an observation meant in the previous section, an infiltration as meant in Article 47*octies* or a sneak and peek operation meant in Article 89*ter*, which concern locations used for professional activities or the residence of a lawyer or medical doctor, when the lawyer or medical doctor is suspected of having committed one of the offences meant in Article 90*ter*, §§ 2–4, or an offence being committed within the framework of a criminal organisation as meant in Article 324*bis* CC, or if on the basis of precise indications, third persons are suspected of having committed the offence, using these locations or residences.

These measures cannot be executed without informing the president of the Bar Council or the delegate of the Medical Order. These persons are bound by professional secrecy. Each breach of the professional secrecy will be punished in conformity with Article 458 CC.

During his investigation, the investigating judge has always the right to consult the confidential record concerning the execution of the special investigative methods, without mentioning the content of the confidential record in the framework of the judicial inquiry. At any time, he can modify, complete or extend his authorization. He can also at any time withdraw the authorization. For every modification, completion or extension of the authorization, the investigating judge will examine whether the conditions of the special investigative methods are still fulfilled.

The investigating judge confirms the existence of the granted authorization to execute the special investigative measures by a written order.

The proces-verbaux (written reports) and the written order mentioned in the previous section are added to the case file after the ending of the use of the special investigative methods at the latest.

Covert Communication and Computer Search

Article 90*ter*

§ 1. Without prejudice to the application of Articles 39*bis*, 87, 88, 89*bis* and 90, the investigating judge may, while willing to act covertly, intercept, take knowledge of, search and record, by technical means, communications or data from a computer system or part of a system which are not accessible to the public, or extend the search in a computer system or part of a system.

This measure may be ordered only in exceptional cases, if the investigation so requires, if there are serious indications of an offence referred to in para. 2, and if the other means of investigation are not sufficient to establish the truth.

In order to make this measure possible, the examining magistrate may order, at any time, also without knowledge or without the consent of either the occupant, the owner or his representative, or the user:

- to penetrate a home, a private place or an IT system;
- to temporarily suspend any security of the computer systems concerned, if appropriate by using technical means, false signals, false keys or false capacities;
- to introduce technical means into the computer systems concerned for deciphering and decoding data stored, processed or transmitted by those systems.

The measure referred to in this paragraph can only be ordered in order to trace information that may serve to establish the truth. It may be ordered only vis-à-vis persons suspected of having committed the offence on the basis of precise indications, or vis-à-vis means of communication or computer systems regularly used by a person under suspicion, or vis-à-vis the places where the person, who is suspected of having committed the offence, stays. The measure may also be ordered vis-à-vis persons who, on the basis of precise facts, are suspected of having a regular connection with a person under suspicion.

§2. The offences that may justify the measure referred to in para. 1 are those referred to in:

[…]

7° Article 210*bis* CC;

[…]

15° Articles 372 to 377*bis* CC;

16° Article 377*quater* CC;

17° Article 379, 380, 383*bis*, § 1 and 3 CC;

[…]

20° Articles 428 and 429 CC;

21° Article 433*bis*/1 CC;

40° Article 145, § 3 and § 3*bis* of the Act of 13 June 2005 concerning electronic stalking;

[…]

§ 3. Attempting to commit a crime as referred to in the previous paragraph may also justify the application of the measure.

§ 4. An offence, as referred to in Articles 322 or 323 of the Penal Code, can also justify the application of the measure in as far as the association was formed for the purpose of committing an attack on the persons or property referred to in § 2 or of committing the offence referred to in Article 467 first subsection of the Penal Code.

§ 5. If a suspect is caught in the act, the prosecutor can order the measure referred to in § 1 for the offences referred to in Articles 137, 347*bis*, 434 or 470 of the Criminal Code. If a suspect is caught in the act, the public prosecutor may also order the measure referred to in para. 1 for the offence referred to in Article 137 of the Criminal Code, with the exception of the offence referred to in Article 137, § 3, 6° of the same Code, within seventy-two hours of the discovery of this offence. The authorization may be granted orally and must be confirmed in writing as soon as possible.

§ 6. A proper foreign authority can, in the framework of a criminal investigation, temporarily intercept, take knowledge of and record communications or data from a computer system which are temporarily not accessible to the public, in case the person who is subject to this measure is staying on Belgian territory and under the following terms:

1° No technical support is needed of any authority residing in Belgium.

2° The foreign government informs the Belgian judicial authority of the measure.

3° This possibility has been covered by an international judicial instrument between Belgium and the requesting state.

4° The decision of the investigating judge, as meant in para. 7, has not been notified to the foreign government.

The data collected based on this paragraph can only be used if the Belgian judicial authorities approve the measure.

§7. When the public prosecutor receives the notification in para. 6, first section, 2°, he immediately brings the matter before an investigating judge.

The investigating judge who has been requisitioned with this notification can decide that the measure is allowed corresponding to this Article.

He informs the proper foreign authority of his decision within 96 hours after receiving the notification by the Belgian judicial authorities.

If he needs more time to come to a decision, he informs the requesting government, which allows him to have eight more days to decide. The judge has to give the reasons why he will delay his decision.

If the investigating judge does not allow the measures mentioned in para. 6, he informs the foreign authority of his decision and notifies them that the data which have been collected by the measure should be destroyed and may not be used before a court or during an investigation.

References

Bastyns O (s2003) Attantat à la pudeur. In: Chome P et al (ed) Droit pénal et procedure penal, Kluwer, Mechelen, pp 1–32

Berkmoes H (2005) Zien en zien is twee: vaststellen is niet observeren (noot onder Corr. Gent 29 oktober 2004). Vigiles 65–67

Berkmoes H, Delmulle J (2011) De bijzondere opsporingsmethoden en enige andere onderzoeksmethoden. Politeia, Brussels

Berneman S (2009) Navigatie op het internet en kinderpornografie. Rechtspraak Antwerpen Brugge Gent 497–501

Claus L (2015) Cyberkinderlokkerij en grooming: daadkrachtig wetgevend optreden of een kwestie van overregulering? Nullum Crimen 15–24

Colette-Basecqz N, Blaise N (2011) Des outrages publics aux bonnes mœurs. In: Les infractions contre l'ordre des familles, la moralité publique et les mineurs, Larcier, Brussels, pp 251–297

Conings C (2015a) Sweetie leidt tot eerste Belgische veroordeling. Tijdschrift voor Computerrecht 238

Conings C (2015b) Statusupdate: Belgische opsporing voelt zich #verward bij het speuren in sociale media. In: Wauters E et al (ed) Sociale media anno 2015, Intersentia, Antwerp, pp 267–304

Conings C (2017) Klassiek en digitaal speuren naar strafrechtelijk bewijs. Intersentia, Antwerp

Conings C, Van Linthout P (2012) Sociale media: een nieuwe uitdaging voor politie en justitie. Panopticon 3: 205–230

Conings C, De Schepper K (2014) Grooming en cyberkinderlokkerij strafbaar. Tijdschrift voor Computerrecht 269–270

Conings C, Royer S (2017) Het verzamelen en vastleggen van digitaal bewijs in strafzaken. Nullum Crimen 311–338

De Hert P, Bodard K (1996) Internetmisdaad: een uitdaging? Situering van de problematiek aan de hand van (kinder)pornografie. Algemeen juridisch tijdschrift 97–124

De Hert P, Lichtenstein G (2004) De betekenis van het Europees verdrag cybercriminaliteit voor het vooronderzoek en de internationale samenwerking. Vigiles 153–169

De Hert P, Saelens R (2009) Recht op bescherming van het privé-leven. Tijdschrift voor Privaatrecht 2: 838–866

De Hert P, Van Leeuw F (2010) Cybercrime legislation in Belgium. In: Dirix E et al (ed) The Belgian reports at the Congress of Washington of the International Academy of Comparative Law, Bruylant, Brussels, pp 867–956
de la Serna I (2015) Quelques principes de droit pénal et de procédure pénale. In: de Valkeneer C et al (ed) A la découverte de la justice pénale, Larcier, Brussels, pp 7–51
Delbrouck I (2014a) Bederf van de jeugd en prostitutie. In: Postal Memorialis, Kluwer, Mechelen, pp 1–35
Delbrouck I (2014b) Openbare schennis van de goede zeden. In: Postal Memorialis, Kluwer, Mechelen, pp 1–24
Delbrouck I (2015) Aanranding van de eerbaarheid. In: Vandeplas A et al (ed) Commentaren strafrecht, Kluwer, Mechelen, pp 1–38
De Nauw A (2010) Inleiding tot het bijzonder strafrecht. Kluwer, Mechelen
De Nauw A (2012) Provocatie. In: Postal Memorialis. Lexicon strafrecht, strafvordering en bijzondere wetten, Kluwer, pp 1–20
De Nauw A, Kuty F (2014) Manuel de droit pénal spécial. Kluwer, Waterloo
De Zegher J (1973) Openbare zedenschennis. Story-Scientia, Gent
Fermon J, Verbruggen F, De Decker S (2007) The investigative stage of the criminal process in Belgium. In Cape E et al (ed) Suspects in Europe. Procedural Rights at the Investigative Stage of the Criminal Process in the European Union, Intersentia, Antwerp, pp 29–58
Hameeuw B, Heribant P, Lefere P (2001) Aanranding van de eerbaarheid en verkrachting van minderjarigen. In: Vermeulen G (ed) Strafrechtelijke bescherming van minderjarigen, Kluwer, Antwerp, pp 42–80
Hutsebaut F (2000) Kinderpornografie in het Belgisch strafrecht. Tijdschrift voor Strafrecht 185–200
Huybrechts L (2011) De wet tot verbetering van de aanpak van seksueel misbruik en pedofilie binnen een gezagsrelatie. Rechtskundig Weekblad 1150–1166
Kerkhofs J, Van Linthout P (2013) Cybercrime. Politeia, Brussels
Ketels B (2007) Tentoonstellen kinderporno ook zonder bewijs van raadpleging of download strafbaar. Juristenkrant 6–7
Meese J (2013) The use of illegally obtained evidence in Belgium: a status questionis. Digital Evidence and Electronic Signature Law Review 63–66
Spriet B, Marlier G (2013) Aanranding van de eerbaarheid en verkrachting in het begin van de 21ste eeuw (2000–2012). In: Verbruggen F et al (ed) Themis Straf- en strafprocesrecht, die Keure, Brugge, pp 81–154
Spriet B, Boeckxstaens J (2016) Het nieuwe misdrijf van voyeurisme en een aanpassing van het misdrijf van aanranding van de eerbaarheid en van verkrachting. Tijdschrift voor Strafrecht 3: 207–223
Stevens L (2002) Strafrecht en seksualiteit. De misdrijven inzake aanranding van de eerbaarheid, verkrachting, ontucht, prostitutie, seksreclame, zedenschennis en overspel. Intersentia, Antwerp
Stevens L (2013) Grooming via internet. Tijdschrift voor Jeugdrecht en Kinderrechten 288–294
Stevens L (2015) Grooming en cyberlokking strafbaar. Uitbreiding van de strafrechtelijke bescherming van de seksuele integriteit van minderjarigen in cyberspace. Rechtskundig Weekblad 22: 844–855
Vandemeulebroeke O (2005) Ontucht – prostitutie – mensenhandel. In: Strafrechtelijke kwalificaties met jurisprudentie. Die Keure
Vandromme S (2009) Aanzetten van minderjarigen tot webcamseks: aanranding van de eerbaarheid en/of aanzetten tot ontucht? Tijdschrift voor Strafrecht 176–180
Vandromme S (2014a) Commentaar onder art. 379 Sw. In: De Bussher M et al (ed) Larcier Wet en Duiding. Larcier, Brussels, pp 41–43
Vandromme S (2014c) Commentaar onder art. 385 Sw. In: De Bussher M et al (ed) Larcier Wet en Duiding. Larcier, Brussels, pp 60–61
Verbruggen F, Verstraeten R (2014) Strafrecht & strafprocesrecht. Maklu, Antwerp
Verstraeten R (2012) Handboek strafvordering. Maklu, Antwerp

Vrielink J, Van Dyck S (2015) Seksismeverbod in de Strafwet. Baadt niet, schaadt wel (deel 1). NJW 770–793

Wattier I (2007) Reference points in the legal discourse about the sexual abuse of minors. In: Alen A et al (eds) The UN Children's Rights Convention: Theory meets practice. Intersentia, Antwerp, Intersentia, pp 609–620

Sofie Royer is a research and teaching assistant at the Institute of Criminal Law at KU Leuven. Since September 2015, she has been working on her Ph.D. thesis, "Criminal Seizure: Digiproof and (Multi)functional?" (Strafrechtelijk beslag: digiproof en (multi)functioneel?), under the supervision of Prof. Dr. M. Panzavolta and Prof. Dr. F. Verbruggen. Sofie is a member of the editorial board of the annotated criminal law code Strafrecht geannoteerd and frequently writes for the news section of the Belgian/Dutch journal of computer law Tijdschrift voor Computerrecht. On a regular basis, she comments on jurisprudence for the fortnightly law journal Nieuw Juridisch Weekblad, of which she was an editor for two years. Moreover, Sofie is a member of an interdisciplinary research forum on the regulation of soft drugs. Sofie studied law at KU Leuven, the Catholic University of Leuven. She spent one year at ULg, the University of Liège, for an Erasmus exchange. Sofie received her law degree in 2014. Her dissertation, "Confiscation in a Digitizing Society" (De verbeurdverklaring in de digitale wereld), was awarded second prize in the competition for the best dissertation in criminal law by the journal Nullum Crimen. In the same year she joined the Leuven Bar Association, where she practiced law in different areas during one year.

Charlotte Conings is currently working as a criminal litigation lawyer at the Stibbe law office in Brussels, specialized in corporate criminal law, criminal procedure, cybercrime and digital investigation. She holds a Ph.D. in criminal law from the University of Leuven. Her Ph.D. concerns the search for physical and digital evidence in criminal matters and was completed under the supervision of Prof. Dr. F. Verbruggen and Prof. Dr. R. Verstraeten. Her Ph.D., entitled "Klassiek en digitaal speuren naar strafrechtelijk bewijs", was published by Intersentia at the end of 2017. She contributes to the public debate both through research articles and via public speaking on all kinds of topics related to cybercrime and e-evidence gathering. She also contributes on a regular basis to the Belgian news section of the journal of computer law (Tijdschrift voor Computerrecht). In her capacity of doctoral researcher, she was also a voluntary member to the Belgian Cybercrime Centre of Excellence for Training, Research and Education. As a member of the B-CCENTRE she performed research at the request of law enforcement agencies struggling with difficulties in day-to-day cyber investigations and she co-organised expert seminars and legal trainings concerning online criminal investigations.

Gaëlle Marlier in 2011 obtained the degree of Master in law (KU Leuven). Between 2011 and 2017 she was part of the Institute of Criminal Law of KU Leuven. On Friday 16 March 2018 she defended her doctoral thesis: "The family in criminal and criminal procedural law: is its role as a cornerstone of society eroding?" under the supervision of Prof. Dr. Frank Verbruggen (supervisor) and Prof. Dr. Charlotte Declerck (co-supervisor). Since 2016 she is a member of the editorial board of the annotated code on criminal law (die Keure). Since 2017 she works as a judicial trainee in Kortrijk.

Chapter 6
Substantive and Procedural Legislation in the Republic of Croatia to Combat Webcam-Related Child Sexual Abuse

Ines Bojić and Zvjezdana Kuprešak

Contents

I. Bojić (✉)
Zagreb, Croatia
e-mail: ines.bojic@zg.t-com.hr; ines.bojic@gmail.com

Z. Kuprešak
Zagreb, Croatia
e-mail: zvjezdana.kupresak2@gmail.com

S. van der Hof et al. (eds.), *Sweetie 2.0*, Information Technology and Law Series 31,
https://doi.org/10.1007/978-94-6265-288-0_6

Abstract This report contains analysis of substantive and procedural criminal law in Croatia regarding the protection of children from web-cam sexual abuse in the light of the possibility of the use of artificial intelligence in the form of the child avatar in the investigation and prosecution of web-cam sexual offenses concerning children. The report presents existing criminal law framework applicable to the cases of web-cam sexual abuse of children and deals with the question of its adequacy to combat web-cam sexual offences. Also, the report deals with the question of limitations in criminal procedure regarding entrapment as well as the issue of applicability of criminal law to virtual victims.

Keywords Web-Cam Child Sexual Abuse · Virtual Victim · Entrapment · Incitement · Investigatory Powers · Artificial Intelligence in Criminal Proceedings

6.1 Introduction: Legislation in Croatia

6.1.1 General Description of the Legal Framework

The Republic of Croatia is a unitary and indivisible democratic welfare state. Power in the Republic of Croatia derives from the people and rests with the people as a community of free and equal citizens. The people exercise this power through the election of representatives and through direct decision-making.[1] Freedom, equal rights, national and gender equality, peace-making, social justice, respect for human rights, inviolability of ownership, conservation of nature and the environment, the rule of law and a democratic multiparty system are the highest values of the constitutional order of the Republic of Croatia.[2] In the Republic of Croatia government is organized on the principle of separation of powers into the legislative, executive and judicial branches, but also limited by the constitutionally-guaranteed right to local and regional self-government. The principle of separation of powers includes different forms of mutual cooperation and reciprocal checks and balances of legislative, executive and judicial branches, proscribed by the Constitution and law.

Croatia is a civil law country where courts administer justice according to the Constitution and law.[3] In a civil law system, legislation is seen as the primary source of the law. As such, the courts rely on the enacted legislation in their judgments, rather than on evolving case law jurisprudence as in common law countries. Lower courts are however bound to follow the decisions of superior courts.[4]

The Croatian Parliament (Sabor) is the representative body of its citizens and is vested with legislative power. It has a minimum of 100, and a maximum of 160,

---

[1] Article 1 of the Constitution of the Republic of Croatia.

[2] Article 3 of the Constitution of the Republic of Croatia.

[3] European Cross Boarder Justice, The Case Study of the EAW, 2010.

[4] Ibid.

members. Members of the Parliament have no imperative mandate and enjoy immunity. They are elected to a four year term on the basis of direct, universal and equal suffrage by secret ballot.

The Government of the Republic of Croatia exercises executive power, in conformity with the Croatian Constitution and legislation enacted by the Croatian Parliament. It is led by the President of the Government, the prime minister, and is responsible to the Parliament.

The President is the head of state and chief representative of the Republic of Croatia in the Country and abroad. The President is elected pursuant to universal and equal suffrage by direct election for a period of five years and is the holder of the highest office within the Republic of Croatia, however is not the head of the executive branch.

Judicial power is exercised by the courts, which are autonomous and independent. According to the law, courts are obliged to protect the Constitution and laws confirmed by the legal order of the Republic of Croatia and guarantee the uniform application of the law and equal rights and privileges of all before the law. The courts decide on disputes concerning basic human and civil rights and obligations, the rights and obligations of the state and units of local self-government and impose criminal and other measures upon perpetrators of criminal offences established by law.

Substantive criminal law in Croatia is codified in the Croatian Criminal Code[5] which provisions are divided in general and special part. General part contains provisions relating to all criminal offenses and regulates the general assumptions of culpability and sanctions, while the special part is essentially a catalog of criminal offenses. However, the Criminal Code does not represent an exhaustive codification of substantive criminal law. Criminal offenses are also proscribed in a few other laws, while misdemeanors, which are a part of the Croatian criminal law system in a wider sense, are regulated by a number of special laws and other regulations.[6]

The Croatian Criminal Code proscribes fundamental principles which form the basis for the implementation of criminal law in Croatia. The principle of legality[7] proscribes that no one can be punished and no criminal sanctions can be applied for conduct which did not constitute a criminal offence at the time it was committed and for which the punishment was not proscribed by the law. This principle confirms the principle of separation of powers in the sense that legislative authority decides about which conduct should be punishable, in which way legal security is guaranteed to all citizens. Article 3 proscribes mandatory application of more lenient law which means that if after the criminal offense is committed the law changes one or more times, the law that is more lenient to the perpetrator must be

[5] Criminal Code, Official Gazette no. 125/11, 144/12, 56/15, 61/15.

[6] Misdemeanour Act, Official Gazette no. 107/07, 39/13, 157/13, 110/15.

[7] Horvatić and Derečinović 2002.

applied.[8] Finally, principle of culpability established in Article 4 proscribes that no one can be punished if not culpable of the committed offense.

Procedural criminal law is codified in the Criminal Procedure Act[9] which regulates criminal procedure in accordance with the basic postulates of Croatian Constitution. The Criminal Procedure Act establishes rules which guarantee that an innocent person cannot be convicted and that a punishment or other criminal sanction can be imposed on the person who commits a criminal offense, subject to the provisions of the criminal law and in lawful proceedings before the competent court. Restrictions of fundamental rights and freedoms are allowed only if determined by the law and if the competent Court has reached decision on it. The Criminal Procedure Act is divided into three parts—general provisions, the course of proceedings and special proceedings. The general provisions of Criminal Procedure Act proscribe basic principles of criminal procedure which are used to assist judges in interpreting and applying law and can be divided into two groups, the ones concerning the initiation and commencement of criminal proceedings (principle of legality, accusatory principle and principle of mandatory criminal prosecution) and those relating to the conduct of criminal proceedings (inquisitorial principle, principle of orality and hearing, principle of immediacy, presumption of innocence, protection of personal freedom and protection of human personality and dignity).

Following the above mentioned principles, it is evident that there are two competing aspirations in the criminal procedure, one towards efficiency, which means that every perpetrator must be caught and punished, and the other aiming at protection of human rights and prevention of criminal prosecution and conviction of an innocent person.

Relevant Aspects of Criminal Law and Criminal Procedure for the Sweetie 2.0

The first aspect that is relevant from the perspective of the Sweetie 2.0 is the definition of a child in Croatian criminal law. According to Article 87 of the Criminal Code a child is a person under the age of eighteen, which implies that a child is every person that has not reached the age of majority, which is in Croatia eighteen years of age.

However, provisions of Chapter XVII of the Criminal Code, which incriminate criminal offences of sexual abuse and exploitation of children, including here relevant provisions for the Sweetie Case, differ two groups of children, under the age of fifteen, and between the age of fifteen and eighteen. The reason for this is that the age in which is allowed to enter into sexual relations in Croatia is fifteen, while the age of majority is eighteen. Consequently, certain provisions of Chapter XVII of the Criminal Act, although intended to protect children, are applicable only to persons under the age of fifteen.

---

[8] In these two articles all the contents of modern meaning of the principle of legality are included: lege stricta, lege scripta, lege praevia, lege certa; Horvatić and Derečinović 2002.

[9] Criminal Procedure Act, Official Gazette no. 152/08, 76/09, 80/11, 121/11, 91/12, 143/12, 56/13, 145/13, 152/14.

An aspect of criminal law that is also of particular relevance to the Sweetie 2.0 case is the difference between a completed criminal act and that of an attempt.

Article 34 of the Croatian Criminal Code that regulates an attempt states the following:

> Whoever, with an intention to commit criminal offence, takes an action that spatially and temporally directly precedes to the realization of criminal offence, shall be punished for the attempt, if a proscribed punishment for that criminal offence is five years of imprisonment or if a more serious penalty is prescribed by law, while an attempt of another criminal offense is punishable only if the law expressly proscribes the punishment of an attempt.

Therefore, attempt is punishable under Croatian law under the condition that the proscribed punishment for criminal offence is 5 years of imprisonment or more, or where legislator expressly proscribed punishment for an attempt. Although an attempt does not represent the infringement of the legal good, it is considered that in the case of an attempt violation of legal good is imminent. The perpetrator must make an unconditional decision to commit the crime and act with the intent to complete his work. Just the tendency to commit a crime is not enough.

In general, an attempt is punishable the same as the offence itself would be punished if it was finished, but the law provides the possibility of more lenient punishment of the perpetrator for an attempt of criminal offence.

The law does not prescribe punishment of preparatory actions, however, some preparatory actions, because of their danger, present separate criminal offence, while in some others criminal offences the legislator extended the description of the offence to all preparatory action so that they are punishable just as the offence itself, what is rather questionable from the aspect of the legality principle. In general, the trend of criminalization of preparatory actions for certain, the most serious, offences had been noticed.

Finally, it is important to mention that the statute of limitations for criminal prosecution of crimes committed against children, begins to run when the victim reaches the age of majority, since it is considered that only at that age children will be able to report and initiate proceedings for offences by which they were damaged during the childhood.

A relevant aspect from the perspective of criminal procedure law is the question of investigation of criminal offences of sexual abuse and exploitation of children. The State Attorney, who is the authorized body for prosecution of criminal offences for which the criminal proceedings are instituted ex officio, including crimes against children, has certain discretionary powers in deciding on the initiation of criminal prosecution. The State Attorney is also authorized to file a motion for conduction of special investigatory actions to the investigative judge, in order to prove that the perpetrator committed a criminal offence.

Special investigatory actions which constitute the temporary restriction of constitutional rights and freedoms during the investigation are normatively regulated by the Criminal Procedure Act. Encroachment on citizens' fundamental rights must be based on the norms of legal rank, legitimized by the gravity of the offense that is being investigated, justified by the existence of a sufficient level of grounds for suspicion that a criminal offense has been committed and limited by the prescribed time period during which the restrictions can be applied. Judicial oversight of the conducting of special investigatory actions is of particular relevance since restrictions of constitutional rights are justified only if they are applied against persons whose actions represent an infringement of the law, and if a sufficient level of suspicion exists that they have committed a criminal offence. The execution of investigatory actions is in the jurisdiction of the police, who represent a kind of subsidiary body of the court. Thus, judicial supervision is more than necessary, also bearing in mind the obligation of the court to terminate the action when the conditions upon which investigatory action was determined no longer exist.[10]

Another relevant aspect from the perspective of conduction of criminal procedure in Croatia, relevant for the Sweetie case, is the fact that Croatia within its judicial system has established special Youth Courts[11] in whose jurisdiction are cases in which victims are children, including here all criminal offences of sexual abuse and exploitation of children. Since the perpetrators are adults, the criminal procedure in such cases is carried out accordingly to the provisions of Criminal Procedure Act with the difference that the authorities who conduct proceedings (State Attorney, Judges, Police, Investigators) in such cases are specially educated for situations concerning children. In that way special attention is dedicated to the protection of rights and interests of children.

6.1.2 *Relevant Treaties and Cybercrime Laws*

Croatia is a party to Cybercrime Convention, Lanzarote Convention and the UN Convention of the Rights of the Child, including here the Optional Protocol on the sale of children, child prostitution and child pornography, which have been implemented through the Croatian Criminal Act and the Criminal Procedure Act.

Relevant are also provisions of Directive 2011/93/EU on combat on sexual abuse and sexual exploitation of children and child pornography, which substantially coincide with the provisions of the Lanzarote Convention.

Also, it should be noted that Article 141 of the Croatian Constitution proscribes that international treaties, which have been concluded and ratified in accordance with the Constitution, published, and which have entered into force, are a part of the domestic legal order of the Republic of Croatia and have primacy over domestic law. Therefore, in cases of non-compliance of national regulations with international treaties, as well as in cases where international treaties have not been yet

[10] Gluščić 2012, pp. 555–573.

[11] Law on Juvenile Courts, Official Gazette no. 84/11, 143/12, 148/13, 56/15.

implemented into domestic system, the Courts are obliged to give international treaties primacy over domestic laws.

6.2 Analysis of Substantive Criminal Law

6.2.1 Introduction

Crimes related to the abuse of children are codified in Chapter XVII of the Croatian Criminal Code under the title "Criminal offences of exploitation and sexual abuse of children", running from Article 158 up to Article 166. This part of the Criminal Code is specifically devoted to crimes against children and presents implementation of the relevant provisions of the Lanzarote Convention.

Some other offences that may coincide with the abuse of children (such as kidnapping) are dealt with in other provisions, but as they are not relevant to the subject matter of this report, they will not be discussed.

Relevant provisions of the Cybercrime Convention are implemented in Chapter XVIII of the Croatian Criminal Code which regulates criminal offences against computer systems, programs and data.

6.2.2 Possibly Relevant Criminal Offences

Succinct Overview of Sexual Offences Involving Minors

Table 6.1 lists the relevant provisions of the Croatian Criminal Code compared with the provisions of the Lanzarote Convention, which gives the most comprehensive catalogue of sexual child-abuse offences available.

Overview of Sexual Offences Related to Webcam Child-Sexual Abuse

In this section criminal offences that are potentially applicable to different forms of webcam sexual child-abuse are presented, structured by the relevant types of offences from the Lanzarote Convention.

Sexual Abuse of a Child under the Age of Fifteen (Article 18 of Lanzarote Convention)

Article 158 of the Criminal Code reads:

Whoever performs sexual intercourse or an equivalent sexual act with a child under the age of fifteen or induces a child to perform sexual intercourse or an equivalent sexual act with a third person, or to *commit* sexual act equivalent to sexual intercourse on himself, shall be punished by imprisonment from one to ten years.

Who performs a lewd act upon a child under the age of fifteen or entices a child to perform a lewd act with a third person or to commit a lewd act on himself, shall be punished by imprisonment from six months to five years.

Table 6.1 Succinct overview of sexual offences involving minors [*Source* The authors]

Lanzarote treaty	Criminal Code, Croatia
Article 18 Sexual abuse	Article 158 Sexual abuse of a child under the age of fifteen Article 159 Sexual abuse of a child over the age of fifteen Article 166 Severe Criminal Offences of sexual abuse and exploitation of children
Article 19 Offences concerning child prostitution	Article 162 Pandering
Article 20 Offences concerning child pornography	Article 163 Abuse of Children for Pornography
Article 21 Offences concerning the participation of a child in pornographic performances	Article 164 Abuse of children for the participation in pornographic performances
Article 22 Corruption of children	Article 160 Satisfying Lust in the Presence of a Child under the age of fifteen Article 165 Introducing Pornography to Children
Article 23 Solicitation of children for sexual purposes	Article 161 Enticement of children for satisfying sexual needs Article 162 Pandering
[Other offences, not covered by the Lanzarote Convention]	None

Article 158 of the Croatian Criminal Code criminalizes sexual abuse of a child under the age of fifteen. Despite the fact that the Criminal Code in Article 77 defines a child as a person under the age of eighteen, there are two groups of children who appear as victims of sexual offences, those under the age of fifteen, and those between fifteen and eighteen years of age, because legal age in which is permitted to engage in sexual relations is fifteen, while the age of majority is eighteen. First part of the description of this criminal offence criminalizes sexual intercourse or an equivalent sexual act with a child younger than 15, or performance of lewd act with a child under the age of fifteen, what requires physical presence of perpetrator and the victim, thus is not applicable to the cases of webcam sexual abuse of children. This behavior represents criminal offence regardless of whether child gave his consent or not, because children under the age of fifteen are legally not able to give consent to sexual intercourse.

However, the second part of the description of this offence which states:

> … or induces a child to perform sexual intercourse or an equivalent sexual act with a third person, or to commit sexual act equivalent with sexual intercourse on himself…" and "…or induces a child to perform a lewd act with a third person or to commit a lewd act on himself…

does not require the physical presence of the child and the perpetrator, so we consider that this criminal offense can be applicable to the cases of webcam sexual abuse since it is quite possible that the perpetrator tries to induce a child to perform such activities via webcam.

Pursuant to the Article 18 para 3 of Lanzarote Convention, if the age difference between the child and the perpetrator is three years or less, criminal liability is excluded.[12] Paragraph 4 allows punishment for avoidable error regarding the age of the victim.

Sexual Abuse of a Child over the Age of Fifteen (Article 18 of Lanzarote Convention)

Article 159 of the Criminal Code reads:

Whoever performs sexual intercourse or an equivalent sexual act with a child over the age of fifteen, who was entrusted to his care for the educational, religious or other purposes, or induces a child who was entrusted to his care to perform sexual intercourse or an equivalent sexual act with a third person, or to perform sexual act equivalent with sexual intercourse on himself, shall be punished by imprisonment from six months to five years.

Since children between the age of fifteen and eighteen years are free to engage into sexual relations, this offence incriminates sexual relations between children over the age of fifteen and persons with whom they are in specific educational or emotional relationship (e.g. relatives, teachers), in cases when that kind of sexual relation is voluntary.

Same as in the Article 158, the part of the description of the criminal offence which states: "…or induces a child, who was entrusted to his care, to perform sexual intercourse or an equivalent sexual act with a third person, or to perform sexual act equivalent with sexual intercourse on himself…" could be possible to do via webcam, however considering the special kind of relationship which the perpetrator and the victim must have, and bearing in mind that webcam sexual abuse usually includes children that are unknown to the perpetrators, it is hard to imagine that this situation could be applicable to the sexual abuse of children via webcam, considering that the element of special relationship between the perpetrator and the victim would be missing.

Considering that this provision does not contain the element of profit, it would not be applicable to the situations in which parents or other persons close to the child induce a child to engage in the webcam sexual activities for money, considering that that kind of situation would constitute criminal offense proscribed in Article 162 or 164 of the Criminal Code.

Satisfaction of Lust in the Presence of a Child under the Age of Fifteen (Article 22 of Lanzarote Convention)

Article 160 para 1 of the Croatian Criminal Code incriminates the criminal offence of the satisfaction of lust in the presence of a child under the age of fifteen and reads:

[12] Turković et al. 2013.

Whoever in the in front of a child under the age of fifteen performs acts aimed at satisfying his own lust or the lust of a third person shall be punished by imprisonment up to one year.

Paragraph 2 contains qualified form of this criminal offence for which the sentence is up to three years imprisonment and refers to commission of any of the offenses against sexual freedom and sexual morality proscribed by Articles 152–159 of the Criminal Code in front of a child under the age of fifteen. Paragraph 3 proscribes punishment for attempt, pursuant to the provisions of Lanzarote Convention.

Since the law does not explicitly proscribes that for the commission of this crime the physical presence of the perpetrator and the child is required and uses the term "in front of the child under the age of fifteen", it can be concluded that this offence can be applicable to the cases of sexual abuse of children via webcam, considering that situations in which the perpetrator performs certain acts aimed at satisfying his own lust in front of the webcam while the child is watching him from the other side, are more than likely to happen.

It should be noted that sexual behavior on the victim's side is not important for the existence of this offence, it is enough that perpetrator performs sexual acts while the child does not have to actively participate or in any way encourage or engage in that kind of behavior.

Enticement of Children for Satisfying Sexual Needs—"Child Grooming" (Article 23 of Lanzarote Convention)

Article 161 of the Criminal Code reads:

An adult with an intention that he or another person commits criminal offence from art. 158. upon a person under the age of fifteen, via information and communication technologies, or otherwise, proposes a meeting with her or another person, and who takes measures to realize that meeting, shall be punished up to three years imprisonment.

Whoever collects, gives or transmits information about the person under the age of fifteen with an intention to commit criminal offence from par. 1 shall be punished by imprisonment up to one year.

The perpetrator shall be punished for the attempted criminal offence from par. 1.

Aim of this provision is protection of children from sexual abuse of adults who they initially meet via phone or internet.[13] It is not enough to merely contact with the child via internet with an intent to commit sexual abuse, it is necessary for the perpetrator to take concrete measures in order that the meeting actually occurs.[14] Therefore, arranging the meeting with the intention of committing criminal offence, without taking concrete actions for the realization of the meeting, presents only an attempt, but is also punishable.[15]

---

13 Turković et al. 2013.

14 Ibid.

15 Škrtić 2013, pp. 1139–1170.

Paragraph 2 incriminates certain preparatory actions, for example the collection of various different information about a child (phone number, e-mail address, home address and similar), which presents one of the rare exceptions in the Criminal Code, since preparatory actions in general are not punishable under Croatian law.[16]

This criminal offense enables the protection of children in circumstances when there is no risk of sexual abuse, or when a child has not yet been brought into real danger, much earlier than it would be possible if the only option was to punish the perpetrators of attempted criminal offense of sexual abuse of a child or exploitation of children for pornography.[17]

Considering that the description of this criminal offence itself expressly mentions the use of information and communication technology, it is certainly applicable to the cases of webcam sexual abuse of children.

Pandering (Article 19 of Lanzarote Convention)

Article 162 of the Croatian Criminal Code reads:

Whoever, for profit or other benefit, entices, recruits or induces a child to offer sexual services, or organizes or enables offering of sexual services with a child, and knew or should have known that it was a child, shall be punished by imprisonment from one to ten years.

Whoever uses sexual services of a child who has reached the age of fifteen years by giving compensation or other benefit and knew or could have known that it was a child, shall be punished by imprisonment from six months to five years.

Whoever, by force, threat, deception, fraud, abuse of power or position of dependency forces or induces another person for whom he knew or should have known and could have known that it was a child to offer sexual services, or who uses sexual services of the child with charge, and he knew or should have and could have known for the above mentioned circumstances, shall be punished by imprisonment of three to fifteen years.

Whoever advertises exploitation of sexual services of a child, shall be punished by imprisonment for six months to five years.

This article prohibits "child prostitution" in the manner provided for in Article 34 of the Convention on the Rights of the Child and Article 3 of Optional Protocol to the Convention on the Rights of the Child, Sale of Children, Child Prostitution and Child Pornography and Article 19 of Lanzarote Convention.[18] Paragraph 1 incriminates enticement, recruitment, inducement, organization and enabling of sexual services for profit if the perpetrator knew or should have known that the person involved is a child, therefore this offence is punishable in the cases of avoidable error regarding the age of the child.[19] For the existence of this criminal act it is enough that the profit or other benefit was promised, it is not necessary

[16] Ibid.

[17] Ibid.

[18] Turković et al. 2013.

[19] Ibid.

that it was given to the child or another person. This provision also punishes persons who use sexual services of children, which is a novelty in Croatian Law.

This article as well is applicable to the cases of webcam sexual abuse of children since the way in which a person solicits a child to provide sexual services is not determined, therefore, can do so also via webcam. For the existence of this criminal offence element of profit or other benefit is essential.

Article 162 is also important for the purpose of Sweetie since it enables punishment for the organization of sexual tourism.[20] It also criminalizes advertising of child prostitution which is essential for effective combat against sex tourism.[21]

Abuse of Children for Pornography (Article 20 of Lanzarote Convention)
Article 163 of the Criminal Code reads:

Whoever entices, recruits or induces a child to take part in making of child pornography or who organizes or enables making of child pornography, shall be punished by imprisonment from one to eight years.

Whoever illegally records, produces, offers, makes available, distributes, exports, imports, obtains for himself or another, sells, gives, presents or possess child pornography or via information and communication technologies knowingly access to child pornography shall be punished with the punishment from par. 1.

Who forces or induces a child, by force or threat, deception, fraud, abuse of power or of a position of vulnerability or dependency, to the recording of child pornography shall be punished by imprisonment from three to twelve years.

Special devices, equipment, computer programs or data designed, adapted or used to commit or facilitate the commission of the offense shall be confiscated, and pornographic material which originated from the offense will be destroyed.

The child will not be punished for the production and possession of pornographic material that displays himself or him and another child, if they produced that material and posses it with the consent of each of them, exclusively for their personal use.

Child pornography is a material that visually or otherwise displays a real child or a real illustrated non-existent child or a person who looks like a child, in actual or simulated sexually explicit conduct or that displays sexual organs of children in sexual purposes. Material which have artistic, medical or scientific importance are not considered pornography in terms of this Article.

The above stated article incriminates child pornography, especially, in accordance to the Article 20 of Lanzarote Convention, criminal offences of child pornography committed via computer system or computer network. Conscious access to any content of child pornography via information and communications technologies is punishable, therefore the crime is already committed when the person approaches to such content and it is not necessary that the perpetrator

---

[20] Ibid.

[21] Ibid.

downloads the content on his computer or other device.[22] The term to take part "in the making of child pornography" is not defined, which may cause a problem regarding the interpretation of this provision.

It should be emphasized that this article criminalizes realistic presentation of non-existent children and persons who look younger that eighteen years. Ratio of such stipulation is protection of children in general, and prevention of child abuse, which is often performed by those who enjoy child pornography, where it is irrelevant whether images display real or non-existent children.[23]

Paragraph 5 provides a definition of child pornography where is re-emphasized that this term also includes realistic presentation of non-existent children and persons who only look younger than eighteen, although they are not.

Therefore, this provision is applicable to the cases of webcam sexual abuse. It should be noted that this is the only provision that incriminates displaying of sexual explicit conduct of virtual children, however does not contain the definition of virtual child.

Abuse of Children for the Participation in Pornographic Performances (Article 21 of Lanzarote Convention)

Article 164 of the Criminal Code reads:

Whoever entices, recruits or induces a child to participate in pornographic performances shall be punished by imprisonment from one to eight years.

Whoever gains profit from pornographic performances which include children, or otherwise exploits children for pornographic performances, shall be punished by imprisonment from one to ten years.

Whoever, by force or threat, deception, fraud, abuse of power or of a position of vulnerability or dependency, forces or induces a child to participate in pornographic shows, shall be punished by imprisonment from three to twelve years.

Whoever watches pornographic performance, live or via communication technology shall be punished accordingly to par. 1 of this article, if he knew or should have known and could have known that a child is involved.

Special devices, equipment, computer programs or data designed, adapted or used to commit or facilitate the commission of the offense referred to in paragraphs. 1, 2 and 3 will be confiscated, and pornographic material which originated from the offense will be destroyed.

Although pornographic performances will in most cases be of a commercial character, the offense would be committed even if the performance is organized in a private home.[24] Viewers of such performance will also be punished, which is essential for effective combat against child pornography. Information and communication technology in most cases includes watching these kind

---

[22] Žs Os Kž-339/08, of 23 May 2008.g., the defendant who had repeatedly visited websites containing child pornography was not convicted because access to such content alone was not punishable, Žs Os Kž-339/08. Ibid.

[23] Turković et al. 2013.

[24] Turković et al. 2013.

performances via computers and smart phones with web cams. Paragraph 4 enables punishment of the viewers who are in avoidable error regarding the age of the victim, thus, if the viewer objectively should have known that pornographic performances include children he will be responsible for this crime.[25]

Introducing Pornography to Children (Article 21 of Lanzarote Convention)
Article 165 of the Criminal Code reads:

Whoever sells, gives, presents or by public display via computer system, network or storage media of computer data or otherwise, makes available documents, images, audiovisual content or other pornographic material or shows a pornographic performance to a child under the age of fifteen, shall be punished by imprisonment up to three years.

Objects, special devices, equipment, computer programs or data designed, adapted or used to commit or facilitate the commission of this offence shall be confiscated and pornographic material originated from the offence shall be destroyed.

For the purpose of this article pornography is material that visually or otherwise shows a person in a real or simulated sexually explicit conduct or that displays sexual organs in sexual purposes. Materials that have artistic, medical or a scientific character are not considered pornography.

This article incriminates presentation of harmful materials to children and implies a crime of corruption of children.[26] Paragraph 3 contains a definition of pornography, since the definition of pornography given by the Constitutional Court for the purpose of this article is considered to be too narrow.[27]

Conclusion
Following the above mentioned criminal offences proscribed by the Croatian Criminal Code it can be concluded that sexual abuse of children via webcams could be prosecuted on the basis of several offences, depending of exact circumstances of each individual case. For example:

(a) If the perpetrator induces or forces the minor to display breasts or genitals or to perform sexual activities (e.g., masturbate) in front of the webcam, this may constitute:

    1. Sexual abuse of a child punishable under Article 158, if the child is younger than fifteen, or Article 159, if the child is older than fifteen and there is a special kind of relationship between the child and the perpetrator
    2. Pandering punishable under Article 162, if the perpetrator forces the child to perform sexual activities for the purpose of making profit or other benefit

---

[25] Ibid.

[26] Ibid.

[27] U-III-279/1998 of 9 October 1998.

3. Abuse of children for pornography punishable under Article 163 if the purpose is making of child pornography
4. Abuse of children for the participation in pornographic performances under Article 164 if the purpose is participation of a child in pornographic performances

(b) If the perpetrator shows his genitals or masturbates in front of the webcam, this may constitute:

- Satisfaction of lust in the presence of a child under the age of fifteen punishable under Article 160, if the child is younger than fifteen.

Even though the above cited provisions of the Croatian Criminal Code offer several different options for prosecution of webcam sexual abuse of children and minors, penalties for several of those offences are quite low, for example maximum proscribed punishment for criminal offence of satisfaction of lust in the presence of a child under the age of fifteen is one year of imprisonment, while the minimum punishment is not even determined.

In spite of proscribed penalties, there is no obstacle for conduction of special investigatory actions for crimes proscribed in Chapter XVII, given that Croatian criminal law does not limit the possibility of use of that measures by proscribed minimum penalty.

Relating to this we explain that Article 334 of the Criminal Procedure Act contains a catalogue of criminal offences for which special investigatory actions can be conducted, dividing them into three groups according to their severity. First group covers the most serious crimes, which, pursuant to the understanding of the Constitutional Court, represent the threat to the society, including here criminal offences of sexual abuse of child under the age of fifteen, pandering and abuse of children for pornography, in which cases investigatory measures can be applied against the perpetrator for the period of eighteen months. In second group are offences for which investigatory actions can be ordered for the period of twelve months, including here criminal offences of sexual abuse of a child over the age of fifteen, grooming and abuse of children for the participation in pornographic performances. Third group includes some minor criminal offences for which investigation can be conducted only with significant difficulties, among them those committed by the use of information and communication technologies.[28,29]

To conclude, Croatian law enables use of intrusive investigation powers in cases concerning webcam sexual abuse of children, which substantially increases the chances of success in criminal proceedings.

[28] Constitutional Court expressed an opinion that legitimate criteria for inclusion of specific criminal offence into the catalogue of Article 334 which proscribe offences for which special investigatory actions can be determined, besides their severity is also possible difficulties concerning their investigation and collection of evidence, for example criminal offences committed by the use of computer systems or networks (U-I-448/09).

[29] Pavlović 2014.

6.2.3 *Possible Obstacles in Substantive Law Concerning Sweetie*

Considering the above cited provisions of the Croatian Criminal Code which incriminate sexual abuse and exploitation of children, the main obstacle for their application in the Sweetie case is the fact that Sweetie is not a real child, therefore does not meet the standard of a child in terms of relevant provisions of the Criminal Code which proscribe that a child is a person under the age of eighteen.

Criminal offence that explicitly mentions virtual children, more specifically incriminates realistic presentation of non-existent children, is criminal offence of exploitation of children for pornography proscribed by Article 163 of the Criminal Code, which defines child pornography as

> a material that visually or otherwise displays a real child or a real illustrated non-existent child or a person who looks like a child, in actual or simulated sexually explicit conduct or that displays sexual organs of children in sexual purposes. Materials which have artistic, medical or scientific importance are not considered pornography.

Therefore, generally speaking, under Croatian law, use of virtual character of a child for sexual purposes can be punished only as a crime of abuse of children for pornography, of course under the condition that other requirements proscribed by Article 163 are met.

Nevertheless, there is a possibility that the Sweetie case could be prosecuted as an attempt of criminal offenses proscribed by Articles 158–166 of the Criminal Code, including here criminal offences of satisfaction of lust in the presence of a child under the age of fifteen, enticement of children for satisfaction of sexual needs, pandering, exploitation of children for pornographic performances and introducing pornography to children, depending on the circumstances of each specific case and under the condition that all relevant elements of attempt are realized.

In regard to this we explain that in Croatian Criminal Law attempt is not punishable for all criminal offences, but only for those criminal offences for which proscribed punishment is five or more years of imprisonment, or for the attempt of another criminal offence when the law expressly proscribes punishment for an attempt. An attempt of some criminal offences from Chapter XVII of the Criminal Code is punishable under the general criteria of the minimum prescribed punishment, while for the other offences law expressly proscribes punishment of attempt.

Croatian criminal law has adopted the "mixed" theory of attempt according to which the essence of attempt is in a criminal will of the perpetrator which reflected in action. Thus, the intention is necessary subjective presumption of attempt and has to correspond to intent in completed criminal offence.[30]

However, the intention is not sufficient. It is necessary that the perpetrator gets out of the phase of preparatory actions and enters into the phase of attempt which

[30] For that reason undercover investigators cannot be punished because the subjective element intent does not exist.

presupposes fulfillment of the objective condition, that is initiation of the realization of the crime. In this regard, the Croatian Criminal Law has adopted the stand point of individual-objective theory according to which the crime is initiated as soon as the perpetrator realizes his decision to commit criminal offence by doing action that immediately precedes the commitment of criminal offence. Therefore, attempt exists when perpetrator commits action that precede the act that represents a criminal offence.[31]

Also, Croatian law recognizes the concept of an inadequate attempt that exists when the perpetrator tries to commit a criminal offense by inappropriate means or against an inappropriate object. However, the law proscribes punishment[32]of such attempts. Thus, inadequate attempt is punishable just as ordinary attempt,[33] but unlike ordinary attempt, where legislator has proscribed option of reduction of sentence, in the situation of inadequate attempt possibility of exemption of punishment exists. In relation to that, it should be noted that although Croatian criminal law does not distinguish absolutely and relatively inadequate attempt, accepting in that way the position that the criminal will is the same in both cases, para 3 stipulates the following:

> The perpetrator who attempts to commit a criminal offense by inappropriate means or against an inappropriate object, due to the *rough* irrationality, may be exempted from punishment.[34]

From the interpretation of this provision two different conclusions can derive. First, that inadequate attempt exists only in the situations where the perpetrator, due to the rough irrationality, tried to commit a criminal offence by inappropriate means or against an inappropriate object, in which cases he can be exempted from punishment. Second, that inadequate attempt also exists in the situation where the perpetrator tried to commit a criminal offence by inappropriate means or against inappropriate object, however did not act in that way due to the rough irrationality, in which cases his punishment could be reduced but he could not be exempted from punishment. Therefore, this kind of attempt, despite of the inadequacy of the mean or the object would in fact represent an ordinary attempt.

It should be noted that Croatian law does not contain a definition of the term "rough irrationality" nor is that term defined by the case law. So, it represents unspecified legal standard, the interpretation of which is left to the courts and will probably depend on the particular circumstances of each case.

Therefore, taking into account all of the circumstances of the Sweetic case, the mentioned provisions that regulate attempt in Croatian criminal law could represent

---

[31] Novoselec 2009.

[32] In most cases inadequate attempt does not constitute such a decrease of guilt that would justify exemption or reduction of punishment. According to the theory, at sentencing inadequate attempt should be equalized with ordinary attempt and punish the same (Novoselec 2009).

[33] Novoselec 2009.

[34] The new regulation of inadequate attempt, according to which the exemption from punishment is possible only if the perpetrator acted due to the rough irrationality, places the perpetrator in a more difficult position.

several different options for the prosecution of perpetrator in the Sweetie case. If we agree that inadequate attempt exists in every case in which the perpetrator tries to commit an offence by inappropriate means or towards inappropriate object, regardless of the fact did he acted due to the rough irrationality or not, then the perpetrator in the Sweetie case could be prosecuted for an inadequate attempt. If his behavior was not a result of rough irrationality, his sentence could be reduced, but he could not be exempted of punishment in full.

On the other side, if we accept the position that inadequate attempt exists only in situations in which the perpetrator tried to commit criminal offence by inappropriate mean or against inappropriate object due to the rough irrationality, the problem occurs because in such situation the perpetrator could not be prosecuted for an inadequate nor for an ordinary attempt.

In the Sweetie case the existence of criminal intent of the perpetrator is undeniable, because the perpetrator is convinced that he is communicating with a real child and despite that consciously performs certain sexual acts which constitute criminal offence, however, Sweetie is not a real child so in some way she represents inappropriate object of criminal offence. It is highly *disputable what would be considered as a rough irrationality of the perpetrator in the Sweetie case.*

Question remains whether it would be possible to prosecute the perpetrator for the completed criminal offence?

In this regard we explain that in the Sweetie case the key condition that constitutes a criminal offence is missing and that is that the criminal offence has to be committed against a child, therefore a person, which condition is not fulfilled in the Sweetie case. For example, Article 158 of the Criminal Code incriminates incitement of a child to perform sexual intercourse or equivalent sexual act with a third person or on himself. In general, we can say that this provision can be applied to the Sweetie case, but we cannot ignore that one of the elements of the offence is missing- and that is a child, because virtual child in terms of criminal law is not considered a child. The issue here is whether it would be nevertheless possible to prosecute the perpetrator in such cases, as well as the question would that kind of behavior constitute an inadequate attempt or maybe even completed criminal offence.

Although Article 163 of the Criminal Code contains the definition of child pornography which is defined as material that visually or otherwise displays a real child or non-existent child or a person who looks like a child in a real or simulated sexually explicit position, it does not contain the definition of virtual child, nor does Sweetie, who only represents as ten year old girl, undress or perform any sexually explicit conduct. In the cases in which the criminal proceeding would despite that be initiated, the lack of definition of a virtual child could pose a problem in terms of the legality principle.

Since the relevant case law for the situations which include virtual children does not yet exist[35] it would certainly be interesting to see which position in such

[35] Op.a. According to unofficial information obtained from the Ministry of Interior and the Zagreb County Court.

situations courts will take. The best solution would definitely be if the legislator would anticipate possible obstacles with which courts and other competent authorities will be faced when dealing with such cases and extends descriptions of existent criminal offences in a way that all of them would be punished as well if committed against a virtual child, before which it will be necessary to establish comprehensive definition of virtual child.

Finally, it should be noted that the fact that in the Sweetie case there is no sexual behavior on the side of the victim, we do not consider relevant for the application of the above mentioned provisions of the Criminal Code, since none of the offenses applicable to the Sweetie case proscribes sexual behavior on the victim's side as a mandatory requirement for the existence of a criminal offense.

6.3 Analysis of Criminal Procedure Law

6.3.1 General Description of Legal Framework

The criminal proceeding in Croatia is regulated by the Criminal Procedure Act which is divided into three parts: general provisions, the course of proceeding and special proceedings.

The Criminal procedure consists of several different phases—pre-trial phase, including here pre-investigation proceedings, investigation, investigatory actions and indictment, the trial phase with the pronunciation of judgment and finally the phase following an appeal against a judgment and execution of the judgment.

The State Attorney's Office is the competent authority which institutes prosecution for the offences subject to public prosecution. According to the Constitution[36] the State Attorney's Office is an autonomous and independent body within the justice system responsible for prosecution of perpetrators of criminal offences and other punishable acts, conduction of legal actions aimed at protection of state property and application of legal remedies to protect Constitution and laws made by the Parliament.

Article 2 of the Criminal Procedure Act proscribes that criminal proceedings are instituted and conducted only upon the request of the authorized prosecutor. For certain criminal offences, when proscribed by law, the State Attorney can institute criminal proceedings only upon the motion of the injured person. In cases involving offences subject to public prosecution, the authorized prosecutor is the State Attorney, while in cases involving offences subject to private charge, the prosecution is instituted by private prosecutor. The State Attorney is bound to institute prosecution when there is a reasonable suspicion that a certain person committed an offence which is subject to public prosecution and when there are no legal obstacles for the prosecution of that person. If the State Attorney determines that there are no

[36] Article 125 of the Constitution of Republic of Croatia.

grounds for the institution or conduction of criminal proceedings his role may be assumed by the injured person acting as a subsidiary prosecutor.[37]

The pre-trial phase of criminal proceedings has a rather inquisitorial character and consists of two sub-phases: pre-investigatory proceedings and the investigation.[38] During the pre-investigatory proceedings, when grounds for suspicion seem to exist that criminal offence subject to public prosecution has been committed, the police authorities are bound to take necessary measures to discover the perpetrator and gather evidence and information that could be useful for successful conduct of criminal proceedings.

In this phase of criminal proceeding the central role is appointed to the State Attorney whose main function is to conduct investigation and perform investigative activities when there is a reasonable suspicion that a certain person has committed a criminal offence. The purpose of investigation is to collect evidence and information necessary for a decision on whether to refer an indictment or to discontinue proceedings. While conducting investigation the State Attorney has to independently and impartially clarify the suspicion of criminal offence for which prosecution is carried ex officio and with equal attention collect information on the guilt as well as the innocence of the defendant.[39] Evidence is obtained through investigatory actions which are proscribed by the law.[40] State Attorney is authorized to order the conduct of investigatory action to the investigator and for that he does not need the consent of the investigative judge. In the pre-trial phase he acts independently.

The State Attorney will terminate the investigation if the offence with which a/ the defendant is charged is not a criminal offence subject to public prosecution, if there are circumstances that exclude the guilt of the defendant, if the statute of limitation has occurred or if other circumstances exclude criminal prosecution, and finally if there is no evidence that the defendant committed criminal offence he is charged with.[41] In that case the injured party can take over the criminal prosecution and file a motion to the investigative judge to conduct an investigation.

After completion of the investigation the State Attorney will decide to bring up an indictment or to discontinue proceedings. If he decides to continue the prosecution the investigative judge will held evidence hearing where all relevant evidence that will be used during the trial will be presented.[42] The defendant has the right to file a written response to the indictment but if the Prosecution Council confirms the charge the procedure enters into trial phase.

---

[37] Articles 55–59 of the Criminal Procedure Act.

[38] Horvatić and Derečinović 2002.

[39] Pavlović 2014.

[40] Chapter XVIII of the Criminal Procedure Act.

[41] Article 224 of the Criminal Procedure Act.

[42] Articles 235–238 of the Criminal Procedure Act.

6.3.2 Investigatory Powers

Investigatory actions are procedural actions aimed at collection of information and evidence relevant to the criminal proceedings. The State Attorney leads an investigation and is authorized to order conduction of investigatory actions in order to collect evidence necessary for the decision on the indictment or discontinuation of prosecution. The Criminal Procedure Act in Chapter XVIII proscribes the possibility of conducting the following investigatory actions: search of dwellings and persons, temporary seizure of objects, interrogation of defendant, interrogation of witnesses, recognition, judicial view, taking fingerprints and prints of other body parts, expert witness testimony, evidence by document, evidence by record and electronic (digital) evidence.

Article 332 provides the possibility of conduction of special evidentiary actions in cases where the investigation of criminal offences could not be carried out in any other way or investigation would be possible to carry out only with disproportionate difficulties. Upon a written reasoned request of the State Attorney, the investigative judge can issue an order against a person who is under reasonable suspicion of having committed a criminal offence proscribed by Article 334, for conduction of special investigatory measures that temporarily restrict constitutional rights of citizens. Article 332 regulates following special investigatory actions: surveillance and recording of telephone conversations or means of remote technical communication; interception, collection and recording of computer data; entry on premises for the purpose of conducting surveillance and technical recording of the premises; secret following and technical recording of individuals and objects; use of undercover investigators and informants; simulated purchase of certain objects, simulated bribe giving and simulated bribe taking; offering simulated business services or closing simulated legal business; controlled transport and delivery of objects from offences, Article 339 regulates temporary seizure of postal deliveries, Article 339a establishment of telecommunications contact and Article 340 comparison of computer data.

Conduction of special investigatory actions has to be approved by the investigative judge and is carried out by the police. Special conditions related to the conduction of certain actions are proscribed by law as well as the action of the police during their execution. Results of special investigatory actions are used as evidence for the purpose of criminal proceedings.

Succinct Overview of Investigatory Powers

See Table 6.2.

Table 6.2 Succinct overview of investigatory powers [*Source* The authors]

Council of Europe Convention on Cybercrime	Criminal Procedure Act, Croatia
Article 16. Expedited preservation of stored computer data	Article 257 para 2
Article 17. Expedited preservation and partial disclosure of traffic data	Article 257 paras 1 and 2

(continued)

Table 6.2 (continued)

Council of Europe Convention on Cybercrime	Criminal Procedure Act, Croatia
Article 18. Production order	Article 263 para 2
Article 19. Search and seizure of stored computer data	Article 257 para 1
Article 20. Real-time collection of traffic data	Article 263 para 3
Article 21. Interception of content data	Article 332 para 1 point 2
[Other (special)investigatory powers, not covered by the Cybercrime Convention, such as undercover operations]	Article 332 special investigatory measures: Surveillance and recording of telephone conversations or means of remote technical communication; interception, collection and recording of computer data; entry on premises for the purpose of conducting surveillance and technical recording of the premises; secret following and technical recording of individuals and objects; use of undercover investigators and informants; simulated purchase of certain objects, simulated bribe giving and simulated bribe taking; offering simulated business services or closing simulated legal business; controlled transport and delivery of objects from offences; establishment of telecommunications contact; comparison of computer data

Human Rights

Special investigatory actions which temporarily restrict constitutional rights and freedoms of citizens in order that evidence for the criminal proceedings can be collected are proscribed in Article 332 of the Criminal Procedure Act. Special investigatory measures in most cases represent infringement upon the right to privacy and as such may lead to violation of the personal and family life of the perpetrator and people close to him since he is a subject of secret investigation. By limiting the guaranteed constitutional rights of citizens, the State creates conditions for combat against serious crimes, at the same time taking the obligation to protect the rights of defendants in criminal proceedings, especially the constitutionally guaranteed right to personal and family life and freedom and secrecy of correspondence and all other forms of communication. The European Court of Human Rights has expressed the view that the government's duty to protect the free democratic constitutional order from the imminent danger, represents a legitimate objective of the State which is pursuant to Article 8 para 2 of the Convention "necessary in a democratic society in the interests of national security and for the prevention of disorder and crime".[43]

General conditions for the application of special investigatory actions are based onto subsidiarity and proportionality principles, which are crucial in criminal

[43] *Klass and others v. Germany*, Application no. 5029/71.

proceedings whenever fundamental rights and freedoms of citizens are brought into question. Their use is possible only in cases where the investigation cannot be conducted in any other way or where that would be possible only with disproportionate difficulties. Also, special investigative actions can be used only against certain individuals who are under reasonable suspicion of having committed a crime. They can also be ordered only for the offences proscribed in Article 334 of the Criminal Procedure Act and only at the written and reasoned request of the State Attorney.

Therefore, specific requirements must be met in each specific case in order that special investigatory actions could be implemented, which includes the clear definition of the persons in respect of which these actions can be applied, court order for their conduction, judicial supervision of execution as well as clear defined periods of time during which the restriction of rights can last. The provisions of the procedural law in this regard has to be clear, precise and predictable so that each person could know what can happen when he violates rules of society.[44]

Article 332 para 2 prescribes that in exceptional cases, if there is a risk of delay and if the State Attorney has a reason to believe that he will not be able to obtain an order of the investigative judge to conduct evidentiary actions in time, State Attorney is allowed to issue an order for conduction of special investigative actions on his own, but only for a period of 24 h.[45]

This gives important discretionary powers to the State Attorney who can arbitrarily decide whether circumstances of the delay and conviction that he will not be able to obtain an order of the investigative judge in time exist, since the law does not provide specific rules on what this terms concretely mean.[46] However, para 3 states that a/the State Attorney cannot issue an order for the conducting of certain special investigatory measures, including in this case entering the premises for the interception, collection and recording of computer data and entering premises to carry out surveillance and technical recording of premises,[47] thereby guaranteeing the protection of the constitutional right to inviolability of communication[48] and the fundamental right to inviolability of personal and family life.[49] Therefore, the law does not allow the limitation of those rights to the State Attorney, but only to the Court which guarantees that the application of special evidentiary actions which limit the defendant's fundamental rights will be conducted in accordance with the principle of proportionality.[50]

---

[44] Gluščić 2012, pp. 555–573.

[45] Due to the vagueness and imprecised content as well as legal uncertainty and unpredictability the Constitutional Court repealed the provision of Article 332 para 2 (U-I-448/09), after which this provision has been modified in order to comply with the Constitution.

[46] Pavlović 2014.

[47] According to the opinion expressed by the Constitutional Court (U-I-448/09), home is under stronger protection of the Constitution and the Convention that other premises.

[48] Article 36 para 1 of the Constitution.

[49] Article 35 of the Constitution.

[50] Pavlović 2014.

In this regard, we should also point out to para 6 which proscribes that special evidentiary measures of entering the premises, in cases in which conduction of this measure requires entering into persons home, can be determined only by a court order. Hence, about the restriction of the right to inviolability of personal and family life and home can be decided only after the court determines whether a restriction is in accordance with the law, whether there is a legitimate objective and whether it is necessary in a democratic society.[51] The aim of the principle of proportionality is protection of the individual against arbitrary and unlawful interference of judicial authorities in his private life.

Entrapment

The Criminal Procedure Act contains provisions on the use of investigative methods of undercover investigators and informants, simulated sale and purchase of items and simulated bribe-giving and bribe-taking as well as offering simulated business services and concluding simulated legal business.[52] They differ from other special investigatory measures in that, that their result is not technically registered fact but observation of person, which is why it is necessary to establish specific procedural regulation for the use of such evidence in criminal proceedings.[53]

Constitutional Court expressed the view that

> …the legitimate power and duty of the State, especially in the early stages of criminal proceedings, is to use a variety of investigative methods, corresponding to the nature of the crimes that are being investigated…which also includes the use of undercover investigators and confidants … (Constitutional Court decision no. U-III-1383/2007 of June 2, 2010)

The Rulebook on the Manner of Conduction of Special Investigatory Actions[54] defines undercover investigator as a police officer who, upon the order of the investigative judge, carries out special investigatory actions, while a confidant is defined as a citizen who voluntarily or by order of the investigative judge and the instructions of the State Attorney or the police, carries out special investigatory actions.[55]

The primary purpose of undercover investigators is to obtain evidence by infiltrating into criminal organizations. For the purpose of hiding their identity undercover investigators are allowed to use measures of concealment. They are also allowed to use hidden audio and video devices and other technical means in order to obtain technical record of the investigatory action which they are conducting.[56,57]

---

[51] Article 8 para 2 of the European Convention.

[52] Article 332 para 1 point 5, 6, 7 of the Criminal Procedure Act.

[53] Karas 2012, pp. 127–160.

[54] Rulebook on the Manner of Conduction of Special Investigatory Actions, Official Gazette no. 109/09.

[55] Article 20 of the Rulebook on the Manner of Conduction of Special Investigatory Actions.

[56] Gluščić 2012, pp. 555–573.

[57] Article 337 para 5 of the Criminal Procedure Act proscribes that if, in addition to conditions under Article 332 para 1, reasonable suspicion exists that particularly serious criminal offences from Article 334 will be committed, or have already been committed, investigative judge can upon

Action of undercover investigators must be in accordance with the Convention, constitutional and procedural standards. Article 333 para 2 of the Criminal Code stipulates that undercover investigators and informants can be examined as witnesses[58] about the content of conversations with the persons against whom special investigatory action was ordered as well as with all other participants.

It is evident that this arises certain questions regarding the relation of criminal proceeding and the privilege of the suspect to not incriminate himself which means that the accusation must be conducted without relying exclusively on evidence obtained by coercion and requires that the will of the accused person must be respected.[59]

In establishing whether the investigative procedure violated the very essence of the privilege of self-incrimination the European Court of Human Rights uses the test of equivalence[60] stating that the fairness of the trial is not violated in situations in which information are collected by the use of undercover investigators, under the condition that the conversation between the undercover investigator and the perpetrator did not reach the level of formal interrogation of the perpetrator.[61]

Therefore, the possibility provided by the law which enables that undercover investigators can be examined as witnesses during the trial about the content of the conversation with the defendant, can be applied only to conversations which can be considered voluntary, not to the conversation which are by the intensity equivalent to the formal interrogation of the suspect.[62]

However, para 3 explicitly proscribes that the judgment cannot be based solely on the testimony of undercover investigators or informants, accepting in that way the position expressed through the case law of European Court of Human Rights.[63] In this regard the Constitutional Court took the following position:

> The following rules always have to be respected: the use of anonymous witnesses must be absolutely necessary, judicial authorities must compensate the fact that the defense did not know the identity of the witness and conviction cannot be exclusively or in a decisive part based on the testimony of anonymous witnesses.[64]

---

the motion of State Attorney allow that secret investigators use technical devices to record non-public conversations.

58 Undercover investigators and confident give their testimony in court as endangered or protected witness.

59 Karas 2012, pp. 127–160.

60 The equivalence test means that not every conversation between a secret investigator and the suspect needs to be considered as interrogation, but it is necessary to examine the circumstances of each particular case (*Allan v. UK*, Application no. 48539/99; *Khan v. UK*, Application no. 35394/97), ibid.

61 Karas 2012, pp. 127–160.

62 Ibid.

63 Case of *Doorson v. The Netherlands*, Application no. 20524/92; Case of *Van Mechelen v. The Netherlands*, Application no. 55/1996/674/861-864.

64 U-I-448/09 of 19 July 2012.

Also, special intention should be given to the fact that the incitement to commit criminal offence is forbidden so undercover investigators have to be careful not to cross the line of their authorities.[65] Incitement constitutes influence on the perpetrator in order to lead him to decide or strengthen his decision to commit a criminal offence.[66]

Domestic courts have taken the view that incitement exists only in cases in which undercover investigators persistently persuade perpetrator to make the decision of commission of a criminal offense, or to encourage him to do so in cases in which a/the perpetrator did not make a firm decision to commit a criminal offense.[67] Incitement implies encouraging of the perpetrator with the intention to provoke or strengthen his decision to commit a criminal offense. Thus, enticement exists in cases in which a person did not previously make a decision to commit a criminal offense or is insecure in that decision.[68]

The influence of undercover investigators on the perpetrator is possible in situations of repeated persuasion, threat or error, as well as in the cases in which the perpetrator is dependent of, or is subordinated to the person who persuades him to commit a criminal offence.[69]

6.3.3 Succinct Overview of Investigatory Powers in an Online Context

Criminal offences committed via information and communication technologies are specific in a way that they exclude the personal contact between the perpetrator and the victims and the crime itself occurs in a virtual space, therefore neither can the evidence of commission of such acts be obtained by traditional methods of investigation. It is necessary to use specific methods in order to collect evidence in these cases. Also, it is necessary to continuously track trends in the development of information and communication technologies.

Article 207 of the Criminal Procedure Act stipulates that if there are grounds for suspicion that a criminal offense subject to public prosecution has been committed, the police has the right and duty to take the necessary measures to identify the perpetrator, to ensure and detect traces of criminal act and objects that can be used to determine facts and to gather all information that might be useful for the successful conduct of criminal proceedings.

---

65 Kž-37/02-7 of 23 November 2005; Kž-Us 98/09-10.

66 Article 37 of the Criminal Code.

67 I-Kž-429/03, of 2 November 2003; I-Kž-37/02, of 23 November 2005, I-Kž-1255/04 of 16 February 2006.

68 Karas 2010, pp. 363–366.

69 Ibid.

Also, Article 212 of the Criminal Procedure Act proscribes that the police is authorized to conduct certain investigatory actions, before the commencement of criminal proceedings, if there is a risk of delay, but only for the offences punishable by up to five years imprisonment.

The powers of the police are determined by the Law on police duties and powers.[70] The provisions applicable to the collection of evidence in digital form regarding the crimes committed in an online context, refer to the powers of the police to check the establishment of electronic communications, the inspection of people, objects and means of transportation and especially the use of undercover operations.[71]

In this regard, we explain that for the purpose of prevention and detection of offenses which are prosecuted ex officio, prevention of violence, and the search of persons and objects, police is authorized to request from the providers of communication services to verify identity, frequency and duration of a communication with a specific electronic address.[72] This request may also include the determination of the position of telecommunication devices, the determination of the location where persons who use electronic communication are, as well as the determination of identification marks of the device.

For the conduction of these measures it is enough that requirements from Article 207 are fulfilled, hence that the ground of suspicion that a criminal offence for which criminal proceeding is initiated ex officio has been committed, since this power does not include the right of access to the content of the communication.[73]

Furthermore, the police are authorized to conduct the inspection of objects[74] when it is necessary to discover traces of the criminal offence or other evidence. However, if the police, during the conduction of this measure, reveals traces of a criminal offence prosecuted ex officio, the inspection should be suspended and procedure continued according to the provisions of the Criminal Procedure Act which regulate search. This measure also includes the inspection of computer and other devices which contain computer data and therefore is important for the investigation of the criminal offences committed via information and communication technologies.

During the investigation the police is authorized to conduct undercover operations,[75] under the condition that the goal of police action could not be realized by the use of other actions. Undercover police actions are watching, monitoring, traps and ambush. Actions of observation and monitoring cannot last longer than 30 days, while actions of traps and ambushes can last as long as there are reasons for their application.

[70] Law on Police Duties and Powers.

[71] Article 13 of Law on Police Duties and Powers.

[72] Article 68 of Law on Police Duties and Powers.

[73] Pavlović 2014.

[74] Article 75 Law on Police Duties and Powers.

[75] Article 80 Law on Police Duties and Powers.

The Criminal Procedure Act regulates several investigatory actions applicable specifically to the cases of criminal offences committed via information and communication technologies.

One of this measure is search of movable proscribed by Article 257, which include search of computer and other devices connected to computer, as well as other devices intended for collection, preservation and transmission of data.[76] Upon the request of the authority who is conducting the search, the person who uses the computer, as well as the provider of telecommunication services are obliged to enable access to computer and other devices and give relevant information necessary for the realization of the search.[77] Furthermore, upon the order of the authority who is conducting the search, the person who uses computer or has access to computer, as well as the provider of telecommunication services are obliged to take immediate measures in order to prevent the destruction or modification of data.

The Supreme Court expressed the view that the search of objects includes the authorization to determine its content:

> search of mobile phones includes authorization to identify information about conversations and text messages, but also determination of the content of text messages if the record of these messages remained in mobile phone
>
> (Supreme Court I Kz-686/06-3 of 8 August 2006).[78]

Furthermore, the investigatory action of temporary seizure of objects[79] should be mentioned as relevant, since it allows the seizure and storage of objects that can be used to determine facts in criminal proceedings. Temporary seizure applies to data stored in computers and other devices intended to collect, preserve and transmit data as well as to the information available to the service procedure. Such information must be handed over to the State Attorney in a legible and comprehensible form. When obtaining them, the State Attorney has to proceed pursuant to the regulations related to maintaining the confidentiality of certain data.

Special investigatory actions which enable determination of the facts and collection of evidence for crimes involving the use of information and communication technology, are also proscribed in the Criminal Procedure Act, with the difference that, in the comparison with ordinary investigatory actions, this measures significantly limit the constitutional rights and fundamental freedoms of persons against whom they are applied.

Special investigatory actions relevant for the criminal investigation in an online context are surveillance and recording of telephone conversations or means of remote technical communication, interception, collection and recording of computer data, entry on premises for the purpose of conducting surveillance and

---

[76] Article 257 para 1 of the Criminal Procedure Act.

[77] Ibid.

[78] Similar also I Kž-696/04-7, od 22 February 2004; I-Kž-537/05-3, od 21 June 2005; I-Kž-515/07-5 od 7 October 2008.

[79] Article 261 of the Criminal Procedure Act.

technical recording of the premises, secret following and technical recording of individuals and objects ant the use of undercover investigators and informants.[80]

The Law also regulates special investigatory action of establishing telecommunications contact[81] if there is a suspicion that the user of telecommunication device committed a criminal offense subject to public prosecution. Upon an order of the investigative judge, the police is authorized to request information about identification, duration and frequency of communication between certain electronic addresses, location of the device and location of the person who uses electronic communication, from the telecommunication provider.

Finally, it should be noted that the police is authorized to compare personal data of citizens kept in a database and other registers, with police data records, registers and automatic data processing bases, provided that there are grounds for suspicion that a criminal offence subject to public prosecution has been committed.[82]

6.3.4 Application of Relevant Investigatory Powers to the Sweetie Case

The Criminal Procedure Act does not regulate the use of artificial intelligence such as Sweetie in criminal investigations. It is possible that some of the above described investigatory actions and police powers could include the use of such technologies, since Croatian law does not contain direct provisions on the use of specific technologies for investigative purposes, nor either excludes their use. Given the fact that special investigatory measures of the use of undercover investigators allow operations in which undercover investigators set fake profiles on social networks and others internet sites in order to realize communication with the perpetrators, hiding their own identity, we consider that the use of Sweetie for the same purpose would be possible as well.

Therefore, it can be concluded that the use of artificial intelligence during the investigation would be possible, under the condition that it complies with all other requirements proscribed by the law regarding the use of investigatory action within which this kind of technologies would be used. However, the possible obstacle for the use of Sweetie for the purpose of criminal investigation is the fact that Sweetie proactively contacts the perpetrator in the chat room and solicits the suspect which, from the standpoint of Croatian criminal law, could represent incitement to commit criminal offence.

---

[80] Article 332 para 1 of the Criminal Procedure Act.

[81] According to the Law on Electronic Communications, Communication is every notice altered or transferred, between the final number of participants via a publicly available electronic communications service.

[82] Article 340 of the Criminal Procedure Act.

In this regard we explain that under Croatian law, incitement of the perpetrators by undercover investigators is forbidden because the aim of the investigation is not the capture of persons who did not intend to commit a crime. There is no incitement if a person has already preconceived decision to commit a criminal offence, in which situations active participation of undercover investigators is permitted. Therefore, their actions are not necessarily limited to passive waiting whether the suspect will propose to commit a criminal offence or not.[83]

In this regard, the Supreme Court took the following position:

> Appreciating whether there are elements of incitement … has to be taken into account that incitement…must be persistent, long-term and has to constitute a decisive factor in the creation of the will of the perpetrator to commit criminal act. Statement of the undercover investigator that he wanted to buy the narcotic drogue, even several times repeated, does not represent incitement, it is an integral part of drugs sale … (VSRH, I Kž-1255/04 of 16 February 2006)[84]

The request of completely passive activity of undercover investigator would be justified only in situations where there would be no previous suspicion that a person has committed a crime, but in such cases the use of undercover investigators under Croatian law would not be permitted.[85]

According to Article 332 para 1 of the Criminal Procedure Act special investigatory actions can be applied only against specific persons for whom there is reasonable suspicion of having committed a criminal offence under Article 334. Since Sweetie does not target a specific person, her use in the context of special investigatory actions of use of undercover investigators would not be allowed under Croatian law. However, if the undercover investigators used Sweetie during the investigation of specific person for whom reasonable doubt that he committed a crime exists, we believe that such use of Sweetie would be possible.

Here we consider important to point out to the difference between the legal regulation of special investigatory actions of the use of undercover investigators regulated by the Criminal Procedure Act and undercover police operations regulated by the Law on Police Duties and Powers. For the conduction of undercover police operation sit is enough that grounds for suspicion that criminal offence has been committed exist, while the use of undercover investigators as a special investigatory action requires that a reasonable suspicion exists that a particular person has committed a criminal offence from Article 334. Therefore, we consider that Sweetie could be used in the context of undercover police operations in order to find the perpetrator of criminal offence.

---

[83] Karas 2010, pp. 363–366.

[84] VSRH I Kž-429/03 of 2 September 2003; VSRH, I Kž-37/02 of 2 November 2005.

[85] Karas 2010, pp. 363–366.

6.3.5 *Relevant Aspects of Digital Forensic Evidence*

Criminal offences committed by the use of information and communication technology can be proved by electronic (digital) evidence, which include data generated, stored or transferred with electronic devices.[86] Therefore, computer systems are the main providers of electronic evidence, and the concept of a computer system for that purposes should be interpreted in the broadest sense, so that in addition to the classic computer, includes tablets, smart-phones and similar devices.

The Criminal Procedure Act defines electronic (digital) evidence as information that is as evidence in electronic (digital) form obtained according to the law.[87] However, although the law contains a specific provision on electronic evidence, such evidence does not have special evidentiary value nor priority compared to traditional types of evidence.[88] So, for electronic evidence the same principles apply as to the regular evidence, primary the principle of free evaluation of evidence proscribed by Article 9 of the Criminal Procedure Act, which means that the courts are not obliged or limited with special formal evidentiary rules when evaluating electronic evidence.

Given the fact that webcam sexual abuse includes offences committed via information and communication technology, evidences of such offences will always be preserved in digital form on computer such as photos, videos, messages etc.

Obtaining of digital evidence in real time can be realized by conduction of investigatory actions of search and inspection of computers and other devices and seizure of data stored in computer.[89] By its nature, electronic evidence is sensitive, easily erased and changed, and therefore special efforts has to be taken in order that their credibility and integrity would be preserved fort the court, so that they can be used as valid evidence in criminal proceedings.

6.4 Conclusions and Recommendations

Following all mentioned above we can conclude that Croatian legislation is at an adequate level regarding the issues of protection of children from sexual abuse on the internet. Substantive criminal law, proscribed in Chapter XVII of the Criminal Code, is harmonized with the international treaties, primarily the Lanzarote Convention and Directive, containing several possibilities for the punishment of webcam sexual abuse of children.

---

[86] Kokot 2015, pp. 231–259.

[87] Article 202 para 2 point 33 of the Criminal Procedure Act.

[88] Kokot 2015, pp. 231–259.

[89] Article 331 of the Criminal Procedure Act.

Procedural law, codified in the Criminal Procedure Act, contains all relevant procedural measures for the investigation of computer related offences regulated by the Cybercrime Convention and at the same time adequately protects fundamental rights and freedoms of persons against whom such measures are applied.

However, disputable issues do exist, especially from the perspective of criminal offences of sexual abuse committed against non-existent children, like it is in the Sweetie case. Since, under Croatian law, a child is defined as a person under the age of fifteen, while criminal offences of sexual exploitation and abuse of children differ two groups of children, under the age of fifteen, which category enjoys stricter level of protection, and between the age of fifteen and eighteen, the obligatory requirement for the existence of all relevant criminal offences is that they were committed against a child—therefore a real person.

The only provision that expressly mentions virtual children is criminal offence of abuse of children for pornography, but that provision either does not contain the definition of a virtual child.

The new regulations of criminal law on attempt cause some doubts regarding the meaning of inadequate attempt and the term of rough irrationality connected to it, for which reason it remains unclear whether it could be considered that the perpetrator in the Sweetie case could be prosecuted for an attempt, given the fact that Sweetie represents inadequate object of criminal offence.

In procedural aspect, the main issue is that procedural law allows conduction of special investigatory actions only against specific person for whom reasonable suspicion exist that she committed a criminal offence from the catalogue of criminal offences proscribed in the Criminal Procedure Act. Therefore, the use of Sweetie would not be legitimate if she would target unspecified persons, because that would be considered incitement, which is forbidden under Croatian Law.

Since the relevant case law for the situations which include virtual children does not yet exist it would certainly be interesting to see which position in such situations courts will take. The best solution would definitely be if the legislator would anticipate possible obstacles with which courts and other competent authorities will be faced when dealing with such cases and extended descriptions of existent criminal offences in a way that all of them would be punished as well if committed against a virtual child, before which it will be necessary to establish comprehensive definition of virtual child. It would also be useful to establish precise conditions under which attempt would be considered inadequate as well as to determine the meaning of the term "rough irrationality". Finally, regarding the procedural aspect of Sweetie case, more precisely questions about the possibilities of use of artificial intelligence like Sweetie as an investigatory measure, the legislator would have to precisely regulate conditions under which the use of Sweetie would be allowed and would not represent incitement.

Annex

Relevant Legal Provisions (in Original)

Kazneni zakon

Članak 158.—Spolna zlouporaba djeteta mlađeg od petanest godina

Tko izvrši spolni odnošaj ili s njim izjednačenu spolnu radnju s djetetom mlađim od petnaest godina, ili ga navede da izvrši spolni odnošaj ili s njime izjednačenu spolnu radnju s trećom osobom ili da nad samim sobom izvrši sa spolnim odnošajem izjednačenu spolnu radnju, kaznit će se kaznom zatvora od jedne do deset godina.

Tko nad djetetom mlađim od petnaest godina izvrši bludnu radnju, ili ga navede da izvrši bludnu radnju s drugom osobom ili da nad samim sobom izvrši bludnu radnju, kaznit će se kaznom zatvora od šest mjeseci do pet godina.

Nema kaznenog djela iz stavka 1. i 2. ovoga članka ako razlika u dobi između osoba koje vrše spolni odnošaj ili s njime izjednačenu spolnu radnju ili bludnu radnju nije veća od tri godine.

(…)

Članak 159.—Spolna zlouporaba djeteta starijeg od petnaest godina

Tko s djetetom koje je navršilo petnaest godina koje mu je povjereno radi odgoja, učenja, čuvanja, dušebrižništva ili njege izvrši spolni odnošaj ili s njime izjednačenu spolnu radnju, ili ga navede da s drugom osobom izvrši spolni odnošaj ili s njime izjednačenu spolnu radnju, ili da samo nad sobom izvrši sa spolnim odnošajem izjednačenu spolnu radnju, kaznit će se kaznom zatvora od šest mjeseci do pet godina.

(…)

Članak 160.—Zadovoljenje pohote pred djetetom mlađim od petnaest godina

Tko pred djetetom mlađim od petnaest godina čini spolne radnje namijenjene zadovoljavanju vlastite ili tuđe pohote, kaznit će se kaznom zatvora do jedne godine.

(…)

Članak 161.—Mamljenje djece za zadovoljenje spolnih potreba

Punoljetna osoba koja osobi mlađoj od petnaest godina, u namjeri da ona ili druga osoba nad njom počini kazneno djelo iz članka 158. ovoga Zakona, putem informacijsko komunikacijskih tehnologija ili na drugi način predloži susret s njom ili drugom osobom i koja poduzme mjere da do tog susreta dođe, kaznit će se kaznom zatvora do tri godine.

Tko prikuplja, daje ili prenosi podatke o osobi mlađoj od petnaest godina radi počinjenja kaznenog djela iz stavka 1. ovoga članka, kaznit će se kaznom zatvora do jedne godine.

Za pokušaj kaznenog djela iz stavka 1. ovoga članka počinitelj će se kazniti.

Članak 162.—Podvođenje djeteta

Tko radi zarade ili druge koristi dijete namamljuje, vrbuje ili potiče na pružanje spolnih usluga, ili organizira ili omogući pružanje spolnih usluga s djetetom, a znao je ili je morao i mogao znati da se radi o djetetu, kaznit će se kaznom zatvora od jedne do deset godina.

Tko koristi spolne usluge djeteta koje je navršilo petnaest godina uz davanje bilo kakve naknade ili protučinidbe, a znao je ili je morao i mogao znati da se radi o djetetu, kaznit će se kaznom zatvora od šest mjeseci do pet godina.

Tko osobu za koju je znao ili je morao i mogao znati da je dijete radi zarade silom ili prijetnjom, obmanom, prijevarom, zlouporabom ovlasti ili teškog položaja ili odnosa zavisnosti, prisili ili navede na pružanje spolnih usluga, ili tko koristi spolne usluge tog djeteta uz naplatu, a znao je ili je morao i mogao znati za navedene okolnosti, kaznit će se kaznom zatvora od tri do petnaest godina.

Tko oglašava iskorištavanje spolnih usluga djeteta, kaznit će se kaznom zatvora od šest mjeseci do pet godina.

Članak 163.—Iskorištavanje djece za pornografiju

Tko dijete namamljuje, vrbuje ili potiče na sudjelovanje u snimanju dječje pornografije ili tko organizira ili omogući njezino snimanje,

kaznit će se kaznom zatvora od jedne do osam godina.

Kaznom iz stavka 1. ovoga članka kaznit će se tko neovlašteno snima, proizvodi, nudi, čini dostupnim, distribuira, širi, uvozi, izvozi, pribavlja za sebe ili drugoga, prodaje, daje, prikazuje ili posjeduje dječju pornografiju ili joj svjesno pristupa putem informacijsko komunikacijskih tehnologija.

Tko dijete silom ili prijetnjom, obmanom, prijevarom, zlouporabom ovlasti ili teškog položaja ili odnosa ovisnosti, prisili ili navede na snimanje dječje pornografije, kaznit će se kaznom zatvora od tri do dvanaest godina.

Posebne naprave, sredstva, računalni programi ili podaci namijenjeni, prilagođeni ili uporabljeni za počinjenje ili olakšavanje počinjenja kaznenog djela iz stavka 1., 2. i 3. ovoga članka će se oduzeti, a pornografski materijal koji je nastao počinjenjem kaznenog djela iz stavka 1., 2. i 3. ovoga članka će se i uništiti.

Dijete se neće kazniti za proizvodnju i posjedovanje pornografskog materijala koji prikazuje njega samog ili njega i drugo dijete ako su oni sami taj materijal proizveli i posjeduju ga uz pristanak svakog od njih i isključivo za njihovu osobnu upotrebu.

Dječja pornografija je materijal koji vizualno ili na drugi način prikazuje pravo dijete ili realno prikazano nepostojeće dijete ili osobu koja izgleda kao dijete, u pravom ili simuliranom spolno eksplicitnom ponašanju ili koji prikazuje spolne organe djece u spolne svrhe. Materijali koji imaju umjetnički, medicinski, znanstveni, informativni ili sličan značaj ne smatraju se pornografijom u smislu ovoga članka.

Članak 164.—Iskorištavanje djece za pornografske predstave

(1) *Tko dijete namamljuje, vrbuje ili potiče na sudjelovanje u pornografskim predstavama, kaznit će se kaznom zatvora od jedne do osam godina.*
(2) *Tko zarađuje od pornografskih predstava u kojima sudjeluje dijete ili na drugi način iskorištava dijete za pornografske predstave,*
kaznit će se kaznom zatvora od jedne do deset godina.
(3) *Tko silom ili prijetnjom, obmanom, prijevarom, zlouporabom ovlasti ili teškog položaja ili odnosa zavisnosti, prisili ili navede dijete na sudjelovanje u pornografskoj predstavi,*
kaznit će se kaznom zatvora od tri do dvanaest godina.
(4) *Kaznom zatvora iz stavka 1. ovoga članka kaznit će se tko gleda pornografsku predstavu uživo ili putem komunikacijskih sredstava ako je znao ili je morao i mogao znati da u njoj sudjeluje dijete.*
(5) *Posebne naprave, sredstva, računalni programi ili podaci namijenjeni, prilagođeni ili uporabljeni za počinjenje ili olakšavanje počinjenja kaznenog djela iz stavka 1., 2. i 3. ovoga članka će se oduzeti, a pornografski materijal koji je nastao počinjenjem kaznenog djela iz stavka 1. i 2. ovoga članka će se i uništiti.*

Članak 165.—Upoznavanje djece s pornografijom

(1) *Tko djetetu mlađem od petnaest godina proda, pokloni, prikaže ili javnim izlaganjem, posredstvom računalnog sustava, mreže ili medija za pohranu računalnih podataka ili na drugi način učini pristupačnim spise, slike, audiovizualne sadržaje ili druge predmete pornografskog sadržaja ili mu prikaže pornografsku predstavu, kaznit će se kaznom zatvora do tri godine.*
(2) *Predmeti, posebne naprave, sredstva, računalni programi ili podaci namijenjeni, prilagođeni ili uporabljeni za počinjenje ili olakšavanje počinjenja kaznenog djela iz stavka 1. ovoga članka će se oduzeti, a pornografski materijal će se i uništiti.*
(3) *Pornografijom se u smislu ovoga članka smatra materijal koji vizualno ili na drugi način prikazuje osobu u pravom ili simuliranom spolno eksplicitnom ponašanju ili koji prikazuje spolne organe ljudi u spolne svrhe. Materijali koji imaju umjetnički, medicinski ili znanstveni značaj ne smatraju se pornografijom u smislu ovoga članka*

Zakon o kaznenom postupku

Članak 207

Ako postoje osnove sumnje da je počinjeno kazneno djelo zakoje se kazneni postupak pokreće po službenoj dužnosti, policijaima pravo i dužnost poduzeti potrebne mjere:

1) *da se pronađe počinitelj kaznenog djela, da se počinitelj ili sudionik ne sakrije ili ne pobjegne,*
2) *da se otkriju i osiguraju tragovi kaznenog djela i predmeti koji mogu poslužiti pri utvrđivanju činjenica te*
3) *da se prikupe sve obavijesti koje bi mogle biti od koristi za*
(...)

Članak 212.—Hitne dokazne radnje

Policija može, ako postoji opasnost od odgode, i prije započinjanja kaznenog postupka za kaznena djela za koja je propisana kazna zatvora do pet godina obaviti pretragu (članak246.), privremeno oduzimanje predmeta (članak 261.), očevid (članak 304.), uzimanje otisaka prstiju i drugih dijelova tijela (članak 211. i 307.).

Članak 240.—Pretraga—Zajedničke odredbe

Pretraga je istraživanje predmeta pretrage uporabom osjetila i njihovih pomagala, pod uvjetima i na način propisan ovim Zakonom i drugim propisima.

Pretraga doma i drugih prostora, sredstva prijevoza i druge pokretne stvari te osobe poduzima se radi pronalaženja počinitelja kaznenog djela, predmeta ili tragova važnih za kazneni postupak, kad je vjerojatno da se oni nalaze u određenom prostoru, kod određene osobe ili na njezinom tijelu.

Propisi o pretrazi ne odnose se na prirodne, javne te napuštene prostore.

Članak 257.—Pretraga pokretne stvari i bankovnog sefa

Pretraga pokretnih stvari obuhvaća i pretragu računala i s njim povezanih uređaja, drugih uređaja koji služe prikupljanju, pohranjivanju i prijenosu podataka, telefonskim, računalnim i

drugim komunikacijama i nositelja podataka. Na zahtjev tijela koje poduzima pretragu, osoba koja se koristi računalom ili ima pristup računalu ili drugom uređaju ili nositelju podataka, te davatelj telekomunikacijskih usluga, dužni su omogućiti pristup računalu, uređaju ili nositelju podataka, te dati potrebne obavijesti za nesmetanu uporabu i ostvarenje ciljeva pretrage.

Po nalogu tijela koje poduzima pretragu, osoba koja se koristi računalom ili ima pristup računalu i drugim uređajima iz stavka 1. ovog članka, te davatelj telekomunikacijskih usluga, dužni su odmah poduzeti mjere kojima se sprječava uništenje ili mijenjanje podataka. Tijelo koje poduzima pretragu, može provedbu tih mjera naložiti stručnom pomoćniku.

Osobu koja koristi računalo ili ima pristup računalu ili drugom uređaju ili nositelju podataka, te davatelj telekomunikacijskih usluga, a koji ne postupe prema stavku 1. i 2. ovog članka, premda za to ne postoje opravdani razlozi, sudac istrage može na prijedlog državnog odvjetnika kazniti prema odredbi članka 259. stavka 1. ovog Zakona. Odredba o kažnjavanju ne odnosi se na okrivljenika.

Članak 261.—Privremeno oduzimanje predmeta

Predmeti koji se imaju oduzeti prema kaznenom zakonu, ili koji mogu poslužiti pri utvrđivanju činjenica u postupku, privremeno će se oduzeti i osigurati njihovo čuvanje.

Tko drži takve predmete, dužan ih je predati na zahtjev državnog odvjetnika, istražitelja ili policije. Državni odvjetnik, istražitelj ili policija će državatelja predmeta upozoriti na posljedice koje proizlaze iz odbijanja postupanja po zahtjevu.

Osobu koja ne postupi prema zahtjevu za predaju, premda za to ne postoje opravdani razlozi, sudac istrage može na obrazloženi prijedlog državnog odvjetnika kazniti prema članku 259. stavku 1. ovog Zakona.

Mjere iz stavka 2. ovog članka, ne mogu se primijeniti prema okrivljeniku niti osobama koje su oslobođene dužnosti svjedočenja (članak 285.).

Članak 263.

Odredbe članka 261. ovog Zakona odnose se i na podatke pohranjene u računalima i s njim povezanim uređajima, te uređajima koji služe prikupljanju i prijenosu podataka, nositelje podataka i na pretplatničke informacije kojima raspolaže davatelj usluga, osim kad je prema članku 262. ovog Zakona, privremeno oduzimanje predmeta zabranjeno.

Podaci iz stavka 1. ovog članka, na pisani zahtjev državnog odvjetnika se moraju predati državnom odvjetniku u cjelovitom, izvornom, čitljivom i razumljivom obliku. Državni odvjetnik u zahtjevu određuje rok u kojemu se imaju predati podaci. U slučaju odbijanja predaje, može se postupiti prema članku 259. stavku 1 ovog Zakona.

Podatke iz stavka 1. ovog članka, snimit će u realnom vremenu tijelo koje provodi radnju. Pri pribavljanju, snimanju, zaštiti i čuvanju podataka posebno će se voditi računa o propisima koji se odnose na čuvanje tajnosti određenih podataka (članak 186. Do 188.). Prema okolnostima, podaci koji se ne odnose na kazneno djelo zbog kojega se postupa, a potrebni su osobi prema kojoj se provodi mjera, mogu se snimiti na odgovarajuće sredstvo i vratiti toj osobi i prije okončanja postupka.

Članak 331.—Elektronički digitalni dokaz

Ako drukčije nije propisano ovim Zakonom, elektronički dokaz pribavlja se primjenom odredaba članka 257., 262. i 263. ovog Zakona.

Članak 332.—Posebne dokazne radnje

Ako se izvidi kaznenih djela ne bi mogli provesti na drugi načinili bi to bilo moguće samo uz nerazmjerne teškoće, na pisani obrazloženi zahtjev državnog odvjetnika, sudac istrage može protiv osobe za koju postoje osnove sumnje da je sama počinila ili zajedno s drugim osobama sudjelovala u kaznenom djelu iz članka 334. ovog Zakona, pisanim, obrazloženim nalogom odrediti posebne dokazne radnje kojima se privremeno ograničavaju određena ustavna prava građana, i to:

1) *nadzor i tehničko snimanje telefonskih razgovora i drugih komunikacija na daljinu,*
2) *presretanje, prikupljanje i snimanje računalnih podataka,*
3) *ulazak u prostorije radi provođenja nadzora i tehničko snimanje prostorija,*
4) *tajno praćenje i tehničko snimanje osoba i predmeta,*
5) *uporabu prikrivenih istražitelja i pouzdanika,*
6) *simuliranu prodaju i otkup predmeta te simulirano davanje potkupnine i simulirano primanje potkupnine,*
7) *pružanje simuliranih poslovnih usluga ili sklapanje simuliranih pravnih poslova,*
8) *nadzirani prijevoz i isporuku predmeta kaznenog djela.*

Članak 333.

*Snimke, isprave i predmeti pribavljeni provedbom radnji izčlanka 332. stavka 1. točke 1. do 8. ovog Zakona, mogu se*upotrijebiti kao dokaz u postupku.

Prikriveni istražitelj i pouzdanik mogu se ispitati kao svjedoci o sadržaju razgovora koje su vodili s osobama prema kojima je određena radnja iz članka 332. stavka 1. točke 5. do 8. Ovog Zakona, kao i svim sudionicima kaznenog djela radi čijeg otkrivanja i dokazivanja je ta radnja bila određena, a njihovi iskazi se mogu upotrijebiti kao dokaz u postupku.

Presuda i ocjena o nezakonitosti dokaza ne može se temeljitiisključivo na iskazu svjedoka iz stavka 2. ovog članka.

Članak 334.

Posebne dokazne radnje iz članka 332. stavka 1. ovog Zakonamogu se odrediti za sljedeća kaznena djela iz Kaznenog zakona:

1) … spolne zlouporabe djeteta mlađeg od petnaest godina (članak 158.), podvođenja djeteta (članak 162.stavak 1. i 3.), iskorištavanja djece za pornogra ju (članak 163.stavak 2. i 3.), teških kaznenih djela spolnog zlostavljanja i iskorištavanja djeteta (članak 166.)…
2) … spolne zlouporabe djeteta starijeg od petnaest godina (članak 159.), mamljenja djece za zadovoljenje spolnih potreba (članak 161.), podvođenja djeteta (članak 162.), iskorištavanja djece za pornogra ju (članak 163.), iskorištavanja djece za pornografske predstave (članak 164.)…

(…)

Članak 339.a—Provjera uspostavljanja telekomunikacijskog kontakta

Ako postoji sumnja da je registrirani vlasnik ili korisnik telekomunikacijskog sredstva počinio kazneno djelo iz članka 334. ovog Zakona ili neko drugo kazneno djelo za koje je propisana kazna zatvora teža od pet godina policija može, na temelju naloga suca istrage, a radi prikupljanja dokaza, putem Operativno-tehničkog centra za nadzor telekomunikacija od operatora javnih komunikacijskih usluga zatražiti provjeru istovjetnosti, trajanja i učestalosti komunikacije s određenim elektroničkim komunikacijskim adresama, utvrđivanje položaja

komunikacijskog uređaja, kao i utvrđivanje mjesta na kojima se nalaze osobe koje uspostavljaju elektroničku komunikaciju, te identifikacijske oznake uređaja.

(…)

Članak 340.

Policija može osobne podatke građana, pohranjene u zbirkamai drugim registrima sravnjivati s policijskim evidencijama, registrima i zbirkama s automatskom obradom podataka akopostoje osnove sumnje da je počinjeno kazneno djelo za koje seprogoni po službenoj dužnosti. Tako prikupljene obavijesti će se, uz izvješće državnom odvjetniku, izbrisati iz navedenih evidencijačim prestanu biti potrebne za uspješno vođenje kaznenogpostupka, ali najkasnije u roku od dvanaest mjeseci od dananjihove pohrane. Taj rok može sudac istrage na prijedlog državnogodvjetnika iznimno produljiti za tri mjeseca ako je vjerojatno daće se na taj način uspješno okončati raspisana potraga zaodređenom osobom ili predmetima.

(…)

English Translation of Relevant Legal Provisions

Criminal Code

Article 158—Sexual abuse of a child under the age of fifteen:
Whoever performs sexual intercourse or an equivalent sexual act with a child under the age of fifteen, or induces a child to perform sexual intercourse or an equivalent sexual act with a third person, or to commit sexual act equivalent to sexual intercourse on himself, shall be punished by imprisonment from one to ten years.

Who performs a lewd act upon a child under the age of fifteen or induces a child to perform a lewd act with a third person or to commit a lewd act on himself, shall be punished by imprisonment from six months to five years.

(…)

Article 159—Sexual abuse of a child over the age of fifteen:
Whoever performs sexual intercourse or an equivalent sexual act with a child over the age of fifteen, who was entrusted to his care for the educational, religious or other purposes, or induces a child who was entrusted to his care to perform sexual intercourse or an equivalent sexual act with a third person, or to perform sexual act equivalent with sexual intercourse on himself, shall be punished by imprisonment from six months to five years.

(…)

Article 160—Satisfaction of Lust in front of a Child under the age of fifteen:
Whoever in the in front of a child under the age of fifteen performs acts aimed at satisfying his own lust or the lust of a third person shall be punished by imprisonment up to one year.

(…)

Article 161—Enticement of children for satisfaction of sexual needs:
An adult with an intention that he or another person commits criminal offence from art. 158. upon a person under the age of fifteen, via information and communication technologies, or otherwise proposes a meeting with her or another person, and who takes measures to realize that meeting, shall be punished up to three years imprisonment.

Whoever collects, gives or transmits information about the person under the age of fifteen with an intention to commit criminal offence from par. 1 shall be punished by imprisonment up to one year.

The perpetrator shall be punished for the attempted criminal offence from par. 1.

Article 162—Pandering
Whoever, for profit or other benefit, entices, recruits or induces a child to offer sexual services, or organizes or enables offering of sexual services with a child, and knew or should have known that it was a child, shall be punished by imprisonment from one to ten years.

Whoever uses sexual services of a child who has reached the age of fifteen years by giving compensation or other benefit, and knew or could have known that it was a child, shall be punished by imprisonment from six months to five years.

Whoever, by force, threat, deception, fraud, abuse of power or position of dependency forces or induces another person for whom he knew or should have known and could have known that it was a child to offer sexual services, or who uses sexual services of the child with charge, and he knew or should have and could have known for the above mentioned circumstances, shall be punished by imprisonment of three to fifteen years.

Whoever advertises exploitation of sexual services of a child, shall be punished by imprisonment for six months to five years.

Article 163—Abuse of children for pornography
Whoever entices, recruits or induces a child to take part in making of child pornography or who organizes or enables making of child pornography, shall be punished by imprisonment from one to eight years.

Whoever illegally records, produces, offers, makes available, distributes, exports, imports, obtains for himself or another, sells, gives, presents or possess child pornography or via information and communication technologies knowingly access to child pornography shall be punished with the punishment from par. 1.

Who forces or induces a child, by force or threat, deception, fraud, abuse of power or of a position of vulnerability or dependency, to the recording of child pornography shall be punished by imprisonment from three to twelve years.

Special devices, equipment, computer programs or data designed, adapted or used to commit or facilitate the commission of the offense shall be confiscated, and pornographic material which originated from the offense will be destroyed.

The child will not be punished for the production and possession of pornographic material that displays himself or him and another child if they produced that material and possess it with the consent of each of them, exclusively for their personal use.

Child pornography is a material that visually or otherwise displays a real child or a real illustrated non-existent child or a person who looks like a child, in actual or simulated sexually explicit conduct or that displays sexual organs of children in sexual purposes. Material which have artistic, medical or scientific importance are not considered pornography in terms of this Article.

Article 164—Abuse of children for the participation in pornographic performances

Whoever entices, recruits or induces a child to participate in pornographic performances shall be punished by imprisonment from one to eight years.

Whoever gains profit from pornographic performances which include children, or otherwise exploits children for pornographic performances, shall be punished by imprisonment from one to ten years.

Whoever, by force or threat, deception, fraud, abuse of power or of a position of vulnerability or dependency, forces or induces a child to participate in pornographic shows, shall be punished by imprisonment from three to twelve years.

Whoever watches pornographic performance, live or via communication technology shall be punished accordingly to par. 1 of this article, if he knew or should have known and could have known that a child is involved.

Special devices, equipment, computer programs or data designed, adapted or used to commit or facilitate the commission of the offense referred to in paragraphs. 1, 2 and 3 will be confiscated, and pornographic material which originated from the offense will be destroyed.

Article 165—Introducing Pornography to Children

Whoever sells, gives, presents or by public display via computer system, network or storage media of computer data or otherwise, makes available documents, images, audiovisual content or other pornographic material or shows a pornographic performance to a child under the age of fifteen, shall be punished by imprisonment up to three years.

Objects, special devices, equipment, computer programs or data designed, adapted or used to commit or facilitate the commission of this offence shall be confiscated and pornographic material originated from the offence shall be destroyed.

For the purpose of this article pornography is material that visually or otherwise shows a person in a real or simulated sexually explicit conduct or that displays sexual organs in sexual purposes. Materials that have artistic, medical or a scientific character are not considered pornography.

Criminal Procedure Act

Article 207

If there are grounds for suspicion that a criminal offence subject to public prosecution has been committed, the police shall have the right and duty:

1) to take necessary measures aimed at discovering the perpetrator of the criminal
offence, preventing the perpetrator or accomplice from fleeing or going into hiding;
2) to discover and secure traces of the offence and objects of evidentiary value, and
3) to gather all information which could be useful for successfully conducting criminal
Proceedings
(…)

Article 212 Urgent investigatory actions

If there is a risk of delay, police can before the commencement of criminal proceedings for criminal offences punishable by imprisonment up to five years, conduct search (Article 246), temporary seizure of objects (Article 261), inspection (Article 304), take fingerprints and other parts of the body (Articles 211 and 307).

Article 240—Search

The search represents the investigation of the searched object by means of senses and aids under conditions and in the manner stipulated in this Act and in other regulations.

A search of a dwelling, other premises, means of transportation, movable and a person shall be undertaken with the purpose of finding the perpetrator of a criminal offence, objects or traces important for the criminal procedure, when it is probable that these may be found in certain premises, with a certain person or on its body.

The regulations on search shall not apply to natural, public or abandoned premises.

Article 257—Search of Movable Property and Bank Safe

The search of movable property also includes a search of a computer and devices connected with the computer, other devices for collecting, saving and transfer of data, telephone, computer and other communications, as well as data carriers. Upon the request of the authority carrying out the search, the person using the computer or having access to the computer or data carrier or the telecommunications service provider shall provide access to the computer, device or data carrier and give necessary information for an undisturbed use and the fulfilment of search objectives.

Upon the order of the authority carrying out the search, the person using the computer or having access to the computer and other devices referred to in paragraph 1 of this Article or the telecommunications service provider shall immediately undertake measures for preventing the destruction or change of data. The authority carrying out the search may order a professional assistant to undertake such measures.

The person using the computer or having access to the computer or other device or data carriers or the telecommunications service provider, who fail to comply with paragraphs 1 and 2 of this Article, even though there are no justifiable causes whatsoever, may be penalized by the investigating judge upon the motion of the State Attorney in accordance with provisions of Article 259 paragraph 1 of this Act. The penalty clauses hall not apply to the defendant.

Article 261—Temporary seizure of objects

Objects which have to be seized pursuant to the Penal Code or which may be used to determine facts in proceedings shall be temporarily seized and deposited for safekeeping

Whoever is in possession of such objects shall be bound to surrender them upon the request of the State Attorney, the investigator or the police authorities. The State Attorney, the investigator or the police authorities shall instruct the holder of the objection consequences arising from denial to comply with the request.

A person who fails to comply with the request to surrender the objects, even though there are no justified causes, may be penalized by the investigating judge upon a motion with a statement of reasons of the State Attorney pursuant to Article 259 paragraph 1 of this Act.

The measures referred to in paragraph 2 of this Article shall not apply either to the defendant or persons who are exempted from the duty to testify (Article 285).

Article 263

The provisions of Article 261 of this Act also apply to data saved on the computer and devices connected thereto, as well as on devices used for collecting and transferring of data, data carriers and subscription information that are in possession of the service provider, except in case when temporary seizure is prohibited pursuant to Article 262 of this Act.

Data referred to in paragraph 1 of this Act must be handed over to the State Attorney upon his written request in an integral, original, legible and understandable format. The State Attorney shall stipulate a term for handing over of such data in his request. In case handing over is denied, it may be pursued in accordance with Article 259 paragraph 1 of this Act.

Data referred to in paragraph 1 of this Act shall be recorded in real time by the authority carrying out the action. Attention shall be paid to regulations regarding the obligation to observe confidentiality (Articles 186 to 188) during acquiring, recording, protecting and storing of data. In accordance with the circumstances, data not related to the criminal offence for which the action is taken, and are required by the person against which the measure is applied, may be recorded to appropriate device and be returned to this person even prior to the conclusion of the proceedings.

(…)

Article 331—Electronic (Digital) Evidence
Unless otherwise prescribed by this Act, electronic evidence shall be obtained by applying the provisions of Articles 257, 262 and 263 of this Act.

Article 332
If the investigation cannot be carried out in any other way or would be accompanied by great difficulties, the investigating judge may, upon the written request with a statement of reasons of the State attorney, order against the person against whom there are grounds for suspicion the he committed or has taken part in committing an offence referred to in Article 334 of this Act, measures which temporarily restrict certain constitutional rights of citizens as follows:

1) surveillance and interception of telephone conversations and other means of remote
technical communication;
2) interception, gathering and recording of electronic data;
3) entry on the premises for the purpose of conducting surveillance and technical
recording at the premises
4) covert following and technical recording of individuals and objects
5) use of undercover investigators and informants
6) simulated sales and purchase of certain objects, simulated bribe-giving and simulated bribe-taking
7) offering simulated business services or closing simulated legal business
8) controlled transport and delivery of objects from criminal offences

(…)

Article 333
Recordings, documents and objects obtained by the application of the measures referred to in Article 332 paragraph 1 item 1 to 8 of this Act may be used as evidence in criminal proceedings.

An undercover agent and an informant may be interrogated as witnesses on the content of discussions held with the persons against whom the measures referred to in Article 332 paragraph 1 items 5 to 8 of this Act are imposed, as well as all accomplices in the criminal offence for whose disclosure and evidence collecting the measure was imposed and their statements may be used as evidence in the proceedings.

A ruling and evaluation on inadmissibility of evidence may not be based exclusively on the witness testimony referred to in paragraph 2 of this Article.

Article 334
Special investigatory actions proscribed in art. 332 par. 1 can be determined for the following criminal offences:

1) …sexual abuse of a child under the age of fifteen (art. 158), pandering (art. 162 par. 1 and 2), abuse of children for pornography (art. 163 par. 2 and 3), severe criminal offences of sexual abuse and exploitation of children…
2) … sexual abuse of a child over the age of fifteen (art. 159), enticement of children for satisfaction of sexual needs (art. 161), pandering (art. 162), abuse of children for pornography (art. 163), abuse of children for participation in pornographic performances…
(…)

Article 339a—Establishment of telecommunication contact
If there is a suspicion that registered owner or user of telecommunication device committed criminal offence subject to public prosecution, the police shall, upon the order of investigative judge, from the telecommunication provider, request information about identification, duration and frequency of communication between certain electronic addresses, location of the device and location of the person who uses electronic communication.
(…)

Article 340
The police authorities may compare personal data of citizens kept in a database and other registers with police data records, registers and automatic data processing bases, provided that there are grounds for suspicion that a criminal offence subject to public prosecution has been committed. Information thus collected shall, along with a report on this to the State Attorney, be erased from the above mentioned records as soon as it ceases to be necessary for successfully conducting proceedings, but not later than twelvemonths from the day when they are stored. Upon the motion of the State Attorney the investigating judge may exceptionally prolong this term for three months if it is likely that in such a manner a search for a certain person or object may be successfully completed.

References

Gluščić S (2012) Posebne dokazne radnje [Special Evidentiary Actions]. Polic. Sigur. Zagreb, 21(2): 555–573

Horvatić Ž, Derečinović D (2002) Criminal justice system in Europe and North America. The European Institute for Crime Prevention and Control

Karas Ž (2010) Poticanje od strane prikrivenog istražitelja [Incitement by the Secret Investigators]. Polic. Sigur. Zagreb, 19(3): 363–366

Karas Ž (2012) Neka dokazna pitanja o razgovoru prikrivenog istražitelja s osumnjičenikom [Some evidential issues on covert questioning by undercover agents] Hrvatski ljetopis za kazneno pravo i praksu (Zagreb) 19(1): 127–160

Kokot I (2015) Temeljne odrednice kriminalističkog postupanja u vezi sa sadržajem spolnog iskorištavanja djece na računalnom sustavu ili mreži u Republici Hrvatskoj [Basic determinants of criminal prosecution related to the topic of sexual exploitation of children on computer systems and networks in the Republic of Croatia]. ZPR 4(1): 231–259

Novoselec P (2009) Opći dio kaznenog prava [General Part of Criminal Law]

Pavlović Š (2014) Zakon o kaznenom postupku [Criminal Procedure Act]

Škrtić D (2013) Mamljenjedjeteta za zadovoljenje spolnih potreba uporabom infomacijsko komunikacijskih tehnologija [Incitement of children for satisfaction of sexual needs by use of information and communication technology]. Zb. Prav. Fak. Sveuč. Rij. 34(2): 1139–1170

Turković K et al (2013) Komentar kaznenog zakona [Comment on the Criminal Code]

Legislation

Constitution of Republic of Croatia, Official Gazette no. 56/90, 135/97, 8/98, 113/00, 124/00, 28/01, 41/01, 55/01, 76/10, 85/10, 05/14

Criminal Code, Official Gazette no. 125/11, 144/12, 56/15, 61/15

Criminal Procedure Act, Official Gazette no. 152/08, 76/09, 80/11, 121/11, 91/12, 143/12, 56/13, 145/13, 152/14

European Convention for the protection of Human Rights and Fundamental Freedoms Official Gazzette no. 18/97, 6/99, 14/02, 13/03, 9/05, 1/06, 2/10

Law on Electronic Communications, Official Gazette no. 73/08, 90/11, 133/12, 80/13, 71/14

Law on Juvenile Courts, Official Gazette no. 84/11, 143/12, 148/13, 56/15

Law on police powers and duties, Official Gazette no. 76/09, 92/14

Misdemeanour Act, Official Gazette no. 107/07, 39/13, 157/13, 110/15

Rulebook on the Manner of Conduction of Special Investigatory Actions, Official Gazette no. 109/09

Case Law

Constitutional Court of the Republic of Croatia, no. U-III-279/1998 of 9 October 1998

Constitutional Court of the Republic of Croatia, no. U-I-448/09 of 19 July 2012

County Court Osijek, no. Kž-339/08 of 23 May 2008

European Court of Human Rights, Allan v. UK, Application no. 48539/99; Khan v. UK, Application no. 35394/97, Doorson v. The Netherlands, Application no. 20524/92; Van Mechelen v. The Netherlands, Application no. 55/1996/674/861-864

European Cross Boarder Justice, The Case Study of the EAW, 2010

Supreme Court of the Republic of Croatia, no. I Kž-429/03 of 2 September 2003

Supreme Court of the Republic of Croatia, no. I Kž-696/04-7 od 22 February 2004

Supreme Court of the Republic of Croatia, no. I-Kž-537/05-3 od 21 June 2005

Supreme Court of the Republic of Croatia, no. I Kž-37/02 od 23 November 2005

Supreme Court of the Republic of Croatia, no. I Kž-1255/04 of 16 February 2006

Supreme Court of the Republic of Croatia, no. I-Kž-515/07-5 of 7 October 2008

Supreme Court of the Republic of Croatia, no. Kž-Us 98/09-10 of 17 February 2010

Ines Bojić is a practicing lawyer registered by the Croatian Bar Association with the seat of her office in Zagreb, Croatia. She graduated from the Law Faculty, University of Zagreb in 2000 and passed the Bar exam in 2002. Her legal experience she started to develop as a law trainee in a law office in Zagreb and further as a legal adviser at the Constitutional Court of the Republic of Croatia and the European Court of Human Rights. She has been running her own legal practice since June

2006 and in her daily work is mostly orientated to human rights cases, mostly in the fields of Family and Criminal law. She is representative in numerous cases before domestic Courts and the European Court of Human Rights (such as Branko Tomašić, Alliance of Churches Riječ života and others v. Croatia, D. J. v. Croatia, Đorđević v. Croatia and many others). Some of her cases before the European Court of Human Rights, influenced domestic legal framework, such as Criminal and Family Law and influenced on the way of perceiving of rights of victims of human rights violations and discrimination in Croatia. She is author of the book "How to escape from domestic violence" (November 2011), coordinator of the legal team of one of the most visible human rights organisations in Croatia, Croatian info point for the Council of Europe Program HELP (Human Rights Education for Legal Professionals), external associate at the Family Law department of Law Faculty in Zagreb etc.

Zvjezdana Kuprešak is a law trainee at Ines Bojić's law office where she started her internship in 2014. She graduated from the Law Faculty, University in Zagreb in 2015 and passed the Bar exam in 2017. During her studies she volunteered at the Legal Clinic of the University of Zagreb where she was a mentor, the Croatian Parliament as representative's assistant, as well as in the legal advisory of one of the most visible human rights organizations in Croatia. As a law trainee she assists in proceedings before domestic courts in various fields of law as well as human rights cases before the European Court and has a great interest to pursue a legal career in this field in the future.

Chapter 7
Substantive and Procedural Legislation in England and Wales to Combat Webcam-Related Child Sexual Abuse

Alisdair A. Gillespie

Contents

A. A. Gillespie (✉)
Lancaster University, Lancaster, UK
e-mail: a.gillespie@lancaster.ac.uk

S. van der Hof et al. (eds.), *Sweetie 2.0*, Information Technology and Law Series 31,
https://doi.org/10.1007/978-94-6265-288-0_7

Abstract This chapter considers how the Law of England and Wales approaches webcam abuse. It considers both the substantive and procedural law. It takes as its focus operations using Sweetie 2.0. It considers what crimes would be committed by those who offend via webcam, particularly those who seek to solicit a child into meeting them, or engaging in sexual activity via webcams. It also considers whether, in cases such as Sweetie, it matters whether there was ever a real child. The second part of the chapter considers the procedural law, assessing potential evidential rules that may assist in prosecuting those who commit child sexual abuse via webcams.

Keywords England and Wales · Criminal Law · Criminal Procedure · Child Sexual Abuse · Internet · Sex Offenders

7.1 The Legal System in England and Wales

The United Kingdom encompasses three distinct legal systems:

1. Laws of England and Wales.
2. Laws of Scotland.
3. Laws of Northern Ireland.

This position exists because of the way that the kingdom was formed. Wales was annexed by England and English law took precedence, with Welsh law ending.[1] Scotland and Ireland however joined the Union through the merger of existing kingdoms.[2] In each case, the legal systems of the constituent countries (England and Wales, Scotland and Ireland) were left alone. When the Republic of Ireland was created,[3] then it became only Northern Ireland that formed part of the United Kingdom and which had its own laws, with the Republic of Ireland forming its own distinct legal system, albeit based on the previous laws and structures.[4]

As a result of this constitutional position, there is arguably no such thing as UK law, although some law does extend to all constituent countries. However, even in those instances, the laws are interpreted and applied by the individual legal systems. Most UK law is civil rather than criminal, and certainly there is no UK-wide law that would be of relevance to Sweetie 2.0. For this reason, only the law of England and Wales will be considered in this chapter.

---

[1] Law of Wales Act 1535.

[2] *Act of Union 1707* (Scotland) and *Act of Union 1801* (Ireland).

[3] The *Government of Ireland Act 1920* partitioned Ireland into Southern Ireland (20 counties) and Northern Ireland (6 counties). The partition never took effect under domestic law because the 20 counties declared independence (as the Irish Free State). Technically Ireland did not become an independent republic until 1948 (see *Republic of Ireland Act 1848* (Irish legislation) and *Ireland Act 1949* (UK legislation).

[4] For further discussion on this, see Gillespie and Weare 2017, Chapter 1.

English law is famously the originator of the common law. Whilst the common law continues to be an important principle of English law, most notably in terms of how laws are interpreted and applied, the reality is that most law is now created by statute. The United Kingdom has a bicameral legislature. The 'lower' house is the House of Commons and its members are democratically elected. The 'upper' house is known as the House of Lords and traditionally its members were a mixture of Church of England bishops[5] and hereditary peers. The *House of Lords Act 1999* altered this so that alongside the Lords Spiritual, the remaining members were to be largely appointed for life.[6] This has led to complaints that the House of Lords is one of the largest legislatures in the world, with the vast majority of peers not doing their work.[7] Whilst a UK Parliament, the Westminster Parliament passes criminal laws only for England and Wales. Criminal law is a devolved matter to both the Scottish Parliament and the Northern Ireland legislature.

Parliament is supreme in that the courts do not have the power to 'strike down' legislation (as, for example, the Federal courts of the USA are so empowered). Their power is restricted to interpreting laws save that under the *Human Rights Act 1998* (HRA 1998) if a court believes that an Act of Parliament is incompatible with the provisions of the *European Convention on Human Rights* it may make a declaration to this effect.[8] However the declaration does not affect the enforceability of the law and its purpose is simply to allow for expedited change of the primary legislation.[9]

7.1.1 Criminal Trials

There are two principal criminal courts in England and Wales. The first is the 'Magistrates' Court' and the second is the 'Crown Court'. The Magistrates' Court hears summary trials. Approximately 97% of all crime is dealt with by the magistrates' court. The powers of the court are limited to imposing a custodial sentence of no more than six months' imprisonment and thus it tends to be the less serious offences that are tried in the court. That said, a number of sexual offences can be

---

5 Known as the 'Lords Spiritual'. These are the Archbishops of Canterbury and York, the bishops of London, Durham and Winchester and the next most senior 21 bishops. However, provision has now been made to allow female bishops to sit irrespective of seniority. The Lords Spiritual sit in the House only so long as they hold their bishopric.

6 Known as the Lords Temporal. 92 hereditary peers remain eligible to sit in the House. An election is held when one of the 92 die amongst all of the hereditary peers and they elect a peer to represent them until their death or retirement.

7 There are currently 839 peers eligible to sit in the House of Lords. Combined with the 650 members of the House of Commons, it means the legislature for the UK comprises nearly 1,500 members.

8 Section 4, *Human Rights Act 1998.*

9 Gillespie and Weare 2017, p. 174.

heard in the magistrates' court, including some that will be discussed in this chapter.

There are two types of magistrates who sit in the court. The first, and most populous, is the lay magistrate. As their name suggests, these are people who have no formal legal qualification and sit as members of the community to judge. They are advised by a legally-qualified clerk but theoretically decisions of both law and fact are for the magistrates' alone.[10] Lay magistrates sit in benches of three and decide matters by majority although this is never publicly expressed. The second type of magistrate is a District Judge (Magistrates' Court).[11] These are professionally-qualified judges who sit alone and make decisions of both fact and law. Their power is restricted to the same as lay magistrates.

The Crown Court is the court that most are familiar with from television and films. It is where the formal trial is held before judge and jury. The barristers who appear in court are bewigged, as is the judge[12] and other members of the court. A professional judge presides over the trial and the case is heard by a jury of 12 persons randomly selected.[13] Decisions of fact are largely the province of the jury and ultimately only the jury can decide to convict someone.[14] A judge can, however, decide as a matter of law to acquit a person.[15] The Crown Court has much wider powers, the Act of Parliament creating the offence normally stating what the maximum sentence is.

Offences are broken into three categories:

(a) *Offences triable only summarily.* That is to say offences that can only take place in the Magistrates' Court.
(b) *Offences triable only on indictment.* That is to say offences that can only take place in the Crown Court.
(c) *Offences that are either-way.* That is to say offences that can be tried in *either* the Magistrates' Court or Crown Court.

The statute creating the criminal offence will state what category of offence it is. Where the offence is either-way then the prosecution and defence will argue which level of court the matter should be tried in. However apart from low-level shoplifting cases[16] a defendant cannot be tried in the magistrates' court on an

[10] Save that if a magistrate ignored pertinent legal advice then this decision could be challenged by administrative review and could render the magistrate personally liable for the costs of the litigation (*Jones v. Nick* [1977] RTR 72).

[11] Deputy District Judges also exist. District Judges are full-time judges and deputy judges are part-time judges who normally remain in independent practice as a lawyer.

[12] Wigs are no longer worn in civil cases in England and Wales but continue to be worn in criminal cases.

[13] There is no provision under English law to pick a jury through objections etc. (*R v. Smith* [2003] 1 WLR 2229).

[14] *R v Wang* [2005] UKHL 9.

[15] *R v Galbraith* (1981) *73 Cr App R 124.*

[16] *Criminal Justice and Courts Act 2015*, s.52.

either-way offence without his consent. Where the consent is withheld then the trial will be heard by the Crown Court.[17]

7.1.2 Relevant Treaties and Cybercrime Laws

Notwithstanding the point above concerning the differences in law between the different constituent parts, foreign policy is an issue reserved for the UK government. The UK has, to its great credit, played an important role in the development of international law relating to child sexual exploitation and cybercrime.

Global Treaties
The UK signed the UN Convention on the Rights of the Child on 19 April 1990 and ratified it on the 16th December 1991. Its Optional Protocol on the sale of children, child prostitution and child pornography was signed on the 7th September 2000 but it was not ratified until the 20th February 2009. This was partly because the UK government believed that there was no need to ratify the treaty because its domestic laws already catered for the provisions within it.

Regional Treaties
The UK has played an important role in the Council of Europe and is a party to all of the principal treaties in this area.

The UK signed the *Council of Europe Convention on Cybercrime* ('Budapest Convention') on 23rd November 2001 but did not ratify it until 25th May 2011. As with the *Optional Protocol* this was partly because there was a belief that it was not necessary to ratify the Convention because its terms were already to be found within UK law, although the government also conceded that some new legislation would be required for this.[18]

The UK signed the *Council of Europe Convention on the Protection of Children against Sexual Exploitation and Sexual Abuse* ('Lanzarote Convention') on 5th May 2008 but it has not yet ratified the Convention. Again, it has been questioned why the UK has not ratified this convention and the answer was that there was a belief that most UK law is compliant with it.[19]

Supranational Law
At the time of writing, the United Kingdom is a member of the EU although it is currently negotiating its departure. The UK had an opt-out for criminal justice matters as part of the Lisbon Treaty negotiations,[20] but it decided that it would opt into two measures. The first was *Directive 2011/92/EU of the European Parliament*

[17] For further information on this, see Gillespie and Weare 2017.

[18] See *Hansard, HC Deb*, vol 405, col 288, 14 May 2003.

[19] *Hansard, HL Deb, col WA134: 18 November 2013.*

[20] Arcarazo and Murphy 2014, p. 21.

and Council of 13 December 2011 on combating the sexual abuse and exploitation of children and child pornography and the second was *Directive 2013/40/EU of the European Parliament and Council of 12 August 2013 on attacks against information systems.*

It is not known what will happen as a result of leaving the EU but, of course, the Directives will cease to be binding on the UK after its departure. Presumably reliance will instead be pleased on the Council of Europe treaties and any treaty it negotiates with the EU in respect of these measures as part of its exit talks.

7.1.3 Jurisdiction

Before considering the criminal law and how this could apply to Sweetie 2.0, it is perhaps worth noting how England and Wales approach the issue of jurisdiction in criminal cases. It will be seen that in some substantive offences, jurisdiction is expressly mentioned, but generally English laws are silent as to jurisdiction. Instead, the law relies on its general approach to jurisdiction.

Until comparatively recently, the law in England and Wales operated on the basis of terminatory jurisdiction.[21] This meant that in cross-border crimes, the jurisdiction was based on where the crime is completed, i.e. where the last (criminal) act took place. This would potentially cause difficulties for cases of webcam abuse. For example, let us take a simple example:

> D tells V, aged 13, to masturbate in front of a webcam.

Where does this act take place? Let us presume that D is in England and V is in the Netherlands. It could be argued that it takes place where V masturbates, i.e. the Netherlands. However, V is the victim and masturbation is not the crime. As will be seen, it is a criminal offence for an adult to cause or encourage a child to engage in sexual activity,[22] and this would include telling someone to masturbate. Thus whilst the sexual activity takes place in the Netherlands, the more salient question is where does the causing or encouraging takes place? Is it where D resides (and from where he makes the comments) or where V resides (as she is encouraged)?

The increase in cross-border crimes, particularly as a result of digital communication technologies, led to the courts reconsidering the rules on jurisdiction. In *R v. Smith (Wallace Duncan)(No 4)*[23] the Court of Appeal moved away from the terminatory rule and said, 'when substantial activities constituting a crime takes

[21] Williams 1965.

[22] Section 10, *Sexual Offences Act 2003*, discussed in Sect. 7.2.1.

[23] [2004] QB 1418.

place in England, the court shall have jurisdiction'.[24] This is a major shift from the terminatory theory and means that the key question is no longer about where the last act took place but rather it is concerned with whether 'substantial' actions in respect of the crime took place in England. The court did not define 'substantial' but there was no suggestion that, for example, the majority of activities need to take place within the jurisdiction.

That the shift away from the terminatory theory of jurisdiction applies to the internet, was specifically confirmed in the case of *R v Sheppard and Whittle*.[25] This concerned the publication of race-hate material on the internet. A specific point of attack in this case was that *Smith* (*Wallace Duncan*) was wrong[26] and the Court of Appeal emphatically rejected this. They noted that the key question was whether sufficient activities took place within this jurisdiction to justify the courts accepting jurisdiction.

If we apply this to webcam abuse it is easy to see how this could be useful. Let us take the example above again:

> D, in England, chats to V, in the Netherlands, and asks her to masturbate in front of a webcam.

It would be difficult to argue that 'substantial activities' did not take place in England and Wales. Much of the conversations took place whilst D was in the jurisdiction, and D made the suggestion that V should be masturbate from England. Thus the courts would almost certainly accept jurisdiction, and this demonstrates the flexibility of the new rules of jurisdiction, something that is likely to be directly relevant in the context of Sweetie.

7.2 Sexual Offence Laws

There is no penal code in England and Wales, with sexual offence laws forming part of the ordinary laws of England and Wales. The principal statute in this area is the *Sexual Offences Act 2003* (SOA 2003), which was a major piece of legislation that repealed, replaced and updated most of the law relating to sexual offences. This Act included the first law in Europe to specifically tackle the so-called grooming of children, particularly over the internet, something discussed below.

Perhaps somewhat controversially, the SOA 2003 did not consolidate all existing sexual offence laws and therefore some forms of sexual exploitation remain outside

---

[24] *Ibid.*, at 1434.

[25] [2010] 1 WLR 2779.

[26] Gillespie 2012.

of this Act. In the context of internet related child abuse, the most notable example of this would be that related to 'child pornography'.[27] This is a complicated area of the law and is covered by no fewer than three statutes.[28]

Since the passing of the SOA 2003, it has been amended by a number of statutes. Indeed a criticism of English law in recent years is that too much legislation has passed, including introducing overlapping provisions. This will be outlined further below but since 2003 the SOA 2003 has been amended almost annually. There is a belief that passing legislation is 'taking action' irrespective of whether the law is necessary, desirable or helpful.

Age of 'a Child'

The age of majority in England and Wales is 18 but this is not the common age of consent. Under English law, the ordinary age of consent is 16 and since 2000 this has been true regardless of whether it relates to heterosexual or homosexual activities.[29] That said, there are some instances when the age of consent is 18. This is primarily in respect of offences relating to child pornography[30] or where the offence relates to familial sexual activity[31] or the offender is in a position of trust.[32]

Whilst 16 is the ordinary age of consent, English law draws a distinction in its criminal laws between those aged under 13 and those between 13 and 16. Sexual offences against children under 13 are considered more serious and carry higher penalties. More than this, however, the laws relating to sexual offences against children under 13 are crimes of strict liability, that is to say there is no mental fault requirement. So, for example, there are two offences of sexual activity with a child. Section 9, SOA 203 prohibits the sexual touching of a child under the age of 16, with the fault requirement that the offender does not reasonably believe that the child is aged 16 or over. However, section 7, SOA 2003 prohibits the sexual touching of a child under the age of 13. Here, there is no requirement that the offender has any reasonable belief as to the victim's age. The fact that the child was aged under 13 suffices. Whilst this may seem harsh it is designed to protect young children from sexual abuse and exploitation and perhaps reflects the fact that children aged under 13 do not generally look aged over 16 and so, in essence, it is a requirement for people to check the age of the person they are intending sexual activity with.[33]

---

[27] It is conceded that the term 'child pornography' is a term that is disliked by many within law enforcement and child protection but it is used here as it is a term that is commonly used within international instruments, including the Lanzarote Convention (discussed in Gillespie 2011).

[28] Discussed in Sect. 7.2.3 below.

[29] *Sexual Offences (Amendment) Act 2000.*

[30] s.7(6), *Protection of Children Act 1978.*

[31] Sections 25–29, SOA 2003.

[32] Sections 18–24, SOA 2003. Positions of trust include teacher/pupil; medical practitioner/patient; social workers, probation officers etc.

[33] As the age of consent is 16, it is not the case that an offender should be checking whether the child is aged over 13 but over 16 because any sexual activity with someone under the age of 16 is illegal.

Table 7.1 Mapping sexual offences to domestic law

Lanzarote treaty	England and Wales
Article 18. Sexual abuse	s.5, SOA 2003 (Rape of a child) s.6, SOA 2003 (Assault of a child under 13 by penetration) s.7, SOA 2003 (Sexual assault of a child under 13) s.8, SOA 2003 (causing or inciting a child to engage in sexual activity) s.9, SOA 2003 (sexual activity with a child) s.10, SOA 2003 (causing or inciting a child to engage in sexual activity) s.16, SOA 2003 (abuse of a position of trust: sexual activity with a child) s.17, SOA 2003 (abuse of a position of trust: causing or inciting a child to engage in sexual activity)
Article 19. Offences concerning child prostitution	s.47, SOA 2003 (sexual exploitation of a child) s.48, SOA 2003 (Causing or inciting sexual exploitation of a child) s.49, SOA 2003 (Controlling a child in relation to sexual exploitation) s.50, SOA 2003 (Arranging or facilitating sexual exploitation of a child)
Article 20. Offences concerning child pornography	s.1, Protection of Children Act 1978 (making, taking, distributing or showing an indecent photograph of a child) s.160, Criminal Justice Act 1988 (possession of an indecent photograph of a child) s.62, Coroners and Justice Act 2009 (possession of a prohibited image of a child). s.48, SOA 2003 (Causing or inciting the sexual exploitation of a child) s.49, SOA 2003 (Controlling a child involved in sexual exploitation) s.50, SOA 2003 (Arranging or facilitating sexual exploitation of a child)
Article 21. Offences concerning the participation of a child in pornographic performances	s.2, Theatres Act 1968 (prohibition of presentation of obscene performance of plays) s.11, SOA 2003 (engaging in sexual activity in the presence of a child) s.12, SOA 2003 (causing a child to watch a sexual act)
Article 22. Corruption of children	s.12, SOA 2003 (Causing a child to watch a sexual act) s.19, SOA 2003 (Abuse of position trust: causing a child to watch a sexual act)
Article 23. Solicitation of children for sexual purposes	s.14, SOA 2003 (Arranging or facilitating commission of a child sex offence) s.15, SOA 2003 (Meeting a child following sexual grooming etc.)

(continued)

Table 7.1 (continued)

Lanzarote treaty	England and Wales
[other offences, not covered by the Lanzarote Convention]	s.15A, SOA 2003 (sexual communication with a child) s.69, Serious Crime Act 2015 (possession of a paedophile manual) s.1, Computer Misuse Act 1990 (unauthorised access to computer material) s.2, Computer Misuse Act 1990 (unauthorised access with intent to commit or facilitate commission of further offences)

[*Source* The author]

Potentially relevant criminal offences

Table 7.1 lists the offences under English law that might be relevant to Sweetie 2.0. It adopts the approach of bunching them against the articles of the Lanzarote Treaty to facilitate a comparison to other jurisdictions.

Clearly there is a multitude of offences that are designed to protect children from abuse and indeed it could be argued that there are too many offences. Rather than try and go through all the offences listed above and how they apply in all circumstances, this section will only concentrate on those offences that could apply to webcam abuse.

7.2.1 Sexual Abuse

The largest grouping of offences relates to sexual abuse. However, this is partly due to the issue of age discussed at above. Many of the offences cover the same broad elements but differ as to age or the absence of a fault element.

Applied to webcam abuse, the principal offences are:

(a) s.8, SOA 2003 (causing or inciting a child under the age of 13 to engage in sexual activity).
(b) s.10, SOA 2003 (causing or inciting a child to engage in sexual activity).
(c) s.17, SOA 2003 (abuse of a position of trust: causing or inciting a child to engage in sexual activity).

Each of these relate to the causing or inciting of sexual abuse because it is taken for these purposes that there is no physical meeting where contact abuse takes place (given Sweetie is a virtual construct).[34]

The principal difference between these offences is that of age and a position of trust. Section 8 applies to children under the age of 13 and, as explained above, this

[34] Where physical (sexual) contact did take place (i.e. in respect of a real child) then Sections 5–9 and 16, SOA 2003 would apply.

Table 7.2 Descriptors of sexual abuse offences [*Source* The author]

Offence	Activity caused or incited	Mode of trial	Maximum sentence
Section 8	Penetrative activity[a]	Indictable only	Life imprisonment
	Non-penetrative activity	Either-way	14 years' imprisonment[b]
Section 10	Penetrative activity[c]	Indictable only	14 years' imprisonment
	Non-penetrative activity	Either-way	14 years' imprisonment[d]
Section 17	All activity.	Either-way	5 years' imprisonment[e]

[a]The penetration of the victim's anus or vagina or such penetration by the victim; the penetration of the victim's mouth by a penis or the penetration of another's mouth with the victim's penis (see s.8(2) SOA 2003)
[b]Six months' imprisonment when tried summarily
[c]Defined in the same way as with s.8 (see s.10(2) SOA 2003)
[d]Six months' imprisonment when tried summarily
[e]Six months' imprisonment when tried summarily

means that there is no fault element in respect of the age of the child. It is irrelevant what age the person thought the child was: if it was under 13, then belief as to age is irrelevant. Section 10 applies to children aged between 13 and 16 and does include a fault element in respect of age. The child must either be under 13 or, if the child is aged 13 or over, the defendant must reasonably believe that the victim was aged 16 or over. Section 17 applies to children between the ages of 16 and 18. Theoretically it applies to younger children too but it would not be charged as the penalty is lower and thus s.10 would be charged. 'Abuse of a position of trust' is defined in a particular way by the Act[35] but it will not be discussed here as it will not be relevant to Sweetie; the mode of trial and maximum sentence differs depending on the offence but can be summarised as is set out in Table 7.2.

It may seem peculiar that for s.10 the maximum sentence is the same irrespective of the type of activity, but the mode of trial differs. Thus for penetrative sexual activity, the matter can only be heard by the Crown Court, with the full sentencing powers being available to the judge.

For simplicity, reference will be made to Sections 8 and 10. The sections apply where D causes or incites a child to engage in a sexual activity. Thus the principal terms are 'cause' and 'incite'. Both of these are terms of ordinary usage in the law[36] but it is clear that the offence is covering both situations where the sexual conduct has occurred ('caused'—it must have already happened) and those where the defendant is hoping that it will occur ('incited'—the conduct need not have happened so long as the defendant has incited the child to engage in this conduct).

The application of this offence to webcam conduct is clear. For example, if D asks V to masturbate herself then this would amount to an offence under s.10. V will engage in a sexual activity and D has either caused this activity (where it has

[35] Section 21, SOA 2003. Positions of trust include teacher/pupil; medical practitioner/patient; social workers, probation officers etc.

[36] 'Caused' usually means is responsible for bringing about the result. 'Incite' in this context means urging or persuading particular conduct.

happened) or incited it (where D asks V to do it but it has not yet happened). That said, there must be a causal link. So if D and V were communicating sexually and V masturbated herself without prompting from D and then said "I've been masturbating whilst talking" then this would not would satisfy s.10 because D has not caused or incited that activity, V initiated it.

There is no requirement that only the offender and the victim are involved in the activity and thus it could apply to other situations. Let us take an example:

> D tells V, a 14-year-old boy, to perform oral sex on X, another 14-year-old and then email D to tell him how it felt.

This would satisfy the requirements of s.10. Clearly D is inciting V to undertake a sexual activity, in this case with a third-person, and as it is a penetrative activity it would be triable only on indictment.

Where the offences have proven particularly useful is as an alternative to offences relating to child pornography. Although the substantive offence will be discussed below, there can be a logical difficulty in using these offences. Let us take an example:

> D, an adult, persuades V, a child aged 14, to send a pornographic picture of herself to him.

As will be seen below, the distribution of an indecent photograph of a child is illegal.[37] Indeed in the example above the child would technically be acting illegally by sending the image, although clearly there would be no question of prosecuting her in these circumstances. It would be possible to charge D with encouraging V to distribute an indecent photograph of a child,[38] but logically this still labels V as an offender. The offence under the *Serious Crime Act 2007* is encouraging the commission of an offence and thus it is premised on the basis that V commits a criminal offence. It would be deeply unfortunate if V was labelled as the offender, and indeed doing so could cause trauma. The alternative therefore is to use s.10.

The taking of a photograph is undoubtedly 'an activity' and if it is an indecent photograph then that activity is likely to meet the test of 'sexual'.[39] Accordingly,

[37] Section 1, *Protection of Children Act 1978.*

[38] s.44, *Serious Crime Act 2007.*

[39] This is defined in s.78, SOA 2003 as a two-part test. The test applies where a reasonable person 'would consider that (a) whatever its circumstances or any person's purpose in relation to it, it is because of its nature sexual, or (b) because of its nature it may be sexual and because of its circumstances or the purpose of any person in relation to it (or both) it is sexual'.

encouraging a child to send an indecent picture of herself would amount to causing or inciting a child to engage in sexual activity.[40] Section 10 has also been used where a child is asked to procure a picture of another child.[41] Again, this is undoubtedly because it would be inappropriate to label the child who is being encouraged to procure the picture as an offender, not least because the child has almost certainly been 'groomed' to undertake such behaviour.

Realistically the breadth of Sections 8 and 10 mean that it is likely that these would be the most appropriate offences in connection with Sweetie. As Sweetie has the appearance of a child, and Sweetie 2.0 has a chat facility, it is likely that offenders will try to encourage Sweetie to act in a sexual way, thus bringing Sections 8 and 10 into play.

7.2.2 Sexual Exploitation ("Child Prostitution")

2015 quietly ushered in what is both a welcome and perhaps quite radical change. The law in England and Wales ceased to make reference to the term 'prostitute' in the context of children. The SOA 2003 introduced four offences that specifically related to the prostitution of children, three of which used the term 'prostitute' or 'prostitution'. This was problematic as the term 'prostitute' carries with it connotations of 'whore' and plays to a particular stereotype.[42] The term is problematic in relation to adults where it has been stated that some people—although by all means not all—may choose to engage in sex work, but the same argument cannot be used in respect of children. A child does not choose to become prostituted but is instead placed in a position where it is forced—either physically, economically or socially—to engage in such behaviour. The *Serious Crime Act 2015* altered the SOA 2003 to remove reference to 'prostitution' in respect of children, with the offence now referring to 'commercial child sexual exploitation'. This is to be welcomed.

There are three offences of relevance to webcams:

(a) s.47, SOA 2003 (sexual exploitation of a child).
(b) s.48, SOA 2003 (causing or inciting child sexual exploitation of a child).
(c) s.50, SOA 2003 (arranging or facilitating sexual exploitation of a child).

Section 47
This section reads:

(1) A person ([D]) commits an offence if—
(a) he intentionally obtains for himself the sexual services of another person (B),

---

[40] For an illustration of this, see *R v Burford* [2015] EWCA Crim 615.
[41] *R v Honey* [2015] EWCA Crim 371.
[42] Barrett 1997.

Table 7.3 Descriptors of sexual exploitation offences

Age of child	Activity	Sentence
<13	Any sexual activity	Life imprisonment
13 – <16	Penetrative activity	14 years' imprisonment (triable only on indictment)
	Non-penetrative activity	14 years' imprisonment (triable either-way)
16 – <18	Penetrative activity	7 years' imprisonment (triable only on indictment)
	Non-penetrative activity	7 years' imprisonment (triable either-way)

[*Source* The author]

(b) before obtaining these services, he has made or promised payment for those services to B, or a third person, or knows that another person has made or promised such payment, and
(c) either—
(i) B is under 18, and [D] does not reasonably believe that B is 18 or over, or
(ii) B is under 13.

The punishment for the offence differs depending on the age of the child and the activity that has been paid for, according to Table 7.3.

The main difference between penetrative and non-penetrative activity is not in respect of the punishment but in respect of the mode of trial. The expectation being that the Crown Court would impose a heavier sentence and thus making an offence triable only on indictment means that there is an expectation about greater punishment.

It is clear from the provision that they key aspects of the offence are as follows:

(a) D intentionally obtains,
(b) the sexual services of a child,
(c) before so obtaining the service, D pays or promises to pay B or another person.

There is no appellate case-law that deals with Section 47 but the offence is relatively simple. The grammar makes clear that 'intentional' relates to obtaining the sexual service, not to the sexual service of a child. A separate fault element applies where the child is aged over 13 (that D does not reasonably believe that B is 18 or over). Where the child is under 13 there is no fault requirement in respect of the age of a child. It should be noted that D must reasonably believe that the child is over 18 and not just reasonably believe that the child was a particular age.

For example:

D pays B to watch V (aged 14) masturbate.

In this instance for D to be acquitted he would need to show that he reasonably believed that V was aged 18 or over. It would not matter if, for example, he thought that he was aged 16 (the ordinary age of consent).

The term 'sexual services' is not defined within the statute although it is unlikely to be problematic. Whilst it was probably initially conceived as preventing contact offending there is nothing within the legislation that requires this so long as D obtains a sexual service from a child then the legislation applies. Accordingly D paying to watch V (aged under 18) engage in sexual activities in front of a webcam should suffice.

Section 48

Section 48 states:

(1) A person [D] commits an offence if-
(a) he intentionally causes or incited another person (B) to be sexually exploited, in any part of the world, and
(b) either
(i) B is under 18, and A does not reasonably believe that B is 18 or over, or
(ii) B is under 13.

This offence is triable either-way and punishable by a maximum sentence of 14 years' imprisonment.[43]

'Sexually exploited' is defined in s.51(2) as where a person:

(a) on at least one occasion and whether or not compelled to do so, B offers or provides sexual services to another person in return for payment or a promise of payment to B or a third person, or
(b) an indecent image of B is recorded.

'Payment' is defined in s.51(3) as

> 'any financial advantage, including the discharge of an obligation to pay or the provision of goods or services (including sexual services) gratuitously or at a discount'.

The terms 'cause' and 'incite' bear the same definition as in respect of Sections 8 and 10 (discussed above). Thus the essence of this offence is that D causes or incites a child to perform sexual services in respect of a payment (broadly defined). This could be of clear relevance to webcam abuse.

Let us take an example:

> D approaches V, aged 16, online and says that X is willing to pay £30 to watch V masturbating himself.

In the example above, D has clearly incited V to be sexually exploited and thus s.48 will apply. It is important to note that the wording of s.48(1)(a) specifically

[43] Six months' imprisonment if tried summarily.

states that the approach can be anywhere in the world, rendering jurisdiction irrelevant where D is in England.

So, for example:

> D, in England, is chatting with V, a 17-year-old Thai student. He states that X is willing to give £500 to V's family if V agrees to perform a sex act on herself.

This would breach s.48 irrespective of the fact that V, or indeed X, is based outside of the jurisdiction. Clearly this is an important feature in the offence and potentially widens the potential usefulness of it in the context of Sweetie.

Section 50

Section 50 may also be relevant to Sweetie. It provides:

(1) A person ([D]) commits an offence if
(a) he intentionally arranges or facilitates the sexual exploitation in any part of the world of another person (B), and
(b) either
(i) B is under 18, and D does not reasonably believe that B is 18 or over, or
(ii) B is under 13.

Section 50 is tried and punished in the same way as s.48 and the term 'sexual exploitation' is defined in the same way.

Whereas s.48 relates to those who cause or incite *the child* to be sexually exploited, s.50 can be considered wider. Whilst some arrangements may be involved directly with the child, it could be with other people and facilitation can include many other types of conduct. For example:

> D advertises that X and Y, two 15-year-old children, will perform a live sex show via a webcam at a particular site. D states that he will provide the website location for a fee of £25 per person.

In this instance D may not actually be in direct contact with the children, it is quite possible that someone is else in instigating all of this. It is unarguable that D is facilitating the sexual exploitation of a child by his conduct since it is this money that leads to the webcam abuse taking place.

As with s.48, the sexual exploitation that is to be arranged or facilitated can take place anywhere in the world, again providing protection beyond the borders of England.

7.2.3 *Offences Relating to Child Pornography*

A number of offences exist in respect of child pornography although the term 'child pornography' is not used anywhere in English legislation, partly due to the general dislike of the term.

English law differentiates between four categories of images:

(a) *Photographs* or equivalent, including moving images.
(b) *Pseudo-photographs*. These will be discussed below but are, in essence, derivatives of photographs but which have been manipulated.
(c) *Tracings*. These are images that have been produced by tracing the outline of a photograph.
(d) *Non-photographic images of children*. These are images produced by computer or other means that depict children but which are not photographs, pseudo-photographs or tracings.

The first three of these categories constitute what is commonly understood to be 'child pornography' but the fourth is not and this will be discussed separately.

Photograph-Based Images

For those offences relating to photographs there are three aspects that are relevant to the offences; age, type of material and the nature of the material. Each needs to be considered in turn.

As noted earlier in this chapter, the age of 'a child' differs depending on the type of offence. Offences relating to child pornography are an example of where the 'higher' age of consent is used and thus for the purposes of photograph-based child pornography the age of 18 is used.[44]

The type of material has been identified above. For it to be an indecent photograph of a child, it must be a photograph, pseudo-photograph or tracing.[45] 'Photograph' and 'tracing' are self-explanatory given the discussion above put 'pseudo-photograph' does require an explanation. According to s.7(7), *Protection of Children Act 1978* (PoCA 1978) this is 'an image, whether made by computer-graphics or otherwise howsoever, which appears to be a photograph'. Whilst there is reference to computer-generated images it is not immediately clear that this applies to so-called 'virtual child pornography' because of the final wording of the section; that it must appear to be a photograph. In *Goodland v DPP*[46] the defendant created an item that was a photograph of a young girl and a second photograph of the naked body of a young woman. The photographs were hinged together by sellotape, meaning that in certain instances it would "appear" that the image of the child would be 'on' the naked body of the young woman. The High Court ruled that this was not a pseudo-photograph as it did not appear to be a

---

[44] See s.7(6), *Protection of Children Act 1978*.

[45] s.7(4), 7(4A), PoCA 1978.

[46] [2000] 1 WLR 1427.

photograph, it was patently two separate photographs. That said, the court declined to comment on whether if the image was photocopied in the hinged position, it would then be a pseudo-photograph but opined that it would depend on whether the resultant image appeared to be a photograph. It is thus that part of the test that is most important. Where an image does not appear to be a photograph (because it looks to be a cartoon etc.) then the test is not satisfied. However, where the quality of the image is of photographic quality (which arguably the original Sweetie image was) then it could amount to a pseudo-photograph.

The more usual examples of pseudo-photographs are derivatives of photographs. So, for example, it may be that a photograph of a young adult is graphically manipulated to reduce the size of the breasts, airbrush out pubic hair, thin the hips etc. to leave the resultant image looking like someone under the age of 18. Such a photograph would constitute a pseudo-photograph. The other type is where images are spliced together. A good example of this would be *R v H*[47] where the offender was a teacher who legitimately took photographs of pupils in his school. However unbeknown to them, he would then digitally manipulate the images so that the heads of children would appear on nude images (making it appear the child was nude) or splice images of semen onto the images so as to make it look like the child had performed in sexual activity. These resultant images would be examples of pseudo-photographs.

The final element is the nature of material, i.e. what it is about child pornography that makes it illegal. As noted above, the term 'child pornography' is not used and instead the term 'indecent photograph[48] of a child' is used. Therefore, the nature of the material is that it must be 'indecent'. The term 'indecent' is not defined within the Act, perhaps because it was thought that it was a word of ordinary usage. It has been held that the tribunal of fact must adopt the ordinary standards of decency and that 'obscene' and 'indecent' are on opposite sides of the same scale.[49] Whilst context cannot be taken into account[50] the age of the child can.[51] The latter is important because it recognises that what may be not be indecent for an adult could be indecent for a child or indeed between children. So, for example, a topless picture of a 24-year-old may not be indecent, but it may if it depicts a 14-year-old. Similarly, the topless photograph of a 2-year-old may not be indecent but the depiction of a 14-year-old may.

In *R v. Smethurst*[52] the Court of Appeal rejected an argument that the definition of 'indecent' contravenes the *European Convention on Human Rights*[53] due to it

---

[47] [2005] EWCA Crim 3037.

[48] Or 'pseudo-photograph'.

[49] *R v Stanford* [1972] QB 391.

[50] *R v Graham-Kerr* [1988] 1 WLR 1098.

[51] *R v Owen* [1988] 1 WLR 134.

[52] [2001] EWCA Crim 772.

[53] The principal attack was in respect of Article 10 (Freedom of Expression) and Article 8 (Right to Respect for Private Life).

being uncertain. The Court of Appeal rejected the allegation that it was uncertain and instead noted that whilst it was based on the judgment of the individual juror this did not mean it was either uncertain or imprecise. This is true. Whilst it may be a standard that raises questions about its breadth,[54] it is clear that it is a test that has been used without problems for a number of years. In *O'Carroll v United Kingdom*[55] the European Court of Human Rights rejected a petition based on the fact that the term 'indecent' breached Article 7 of the ECHR due to it being too uncertain a concept.[56] The rejection by the Court was emphatic and ensures that the test remains valid.

Now that 'indecent photograph of a child' has been defined, it is possible to consider the offences. The offences are split between two pieces of legislation; PoCA 1978 and the *Criminal Justice Act 1988*.

Section 1, PoCA 1978 states the following:

> (1) Subject to sections 1A and 1B it is an offence for a person-
> (a) to take, or permit to be taken or to make, any indecent photograph or pseudo-photograph of a child; or
> (b) to distribute or show indecent photographs or pseudo-photographs, or
> (c) to have in his possession such indecent photographs or pseudo-photographs, with a view to their being distributed or shown by himself or others; or
> (d) to publish or cause to be published any advertisement likely to be understood as conveying that the advertiser distributes or shows such indecent photographs or pseudo-photographs, or intends to do so.

This creates up to twenty different offences[57] but not all are relevant to this discussion. The offences under s.1 are triable either-way, with a maximum sentence of ten years' imprisonment.[58] Section 1A and 1B create two defences, neither of which are relevant here.[59] There are also defences to (1)(b) and (1)(c), these are set out in s.1(4):

---

[54] Gillespie 2011.

[55] (2005) 41 EHRR SE1.

[56] Article 7 requires the law to be accessible and foreseeable (see, for example, *Cantoni v France* App 17862/91.

[57] Take, permit to be taken, make, distribute, show, possession with intent to distribute; possession with intent to show; publish an advertisement conveying the advertiser distributes, publish an advertisement conveying the advertiser shows; conveying the advertiser intends to distribute or conveying the advertiser intends to show. Each of these could be for a photograph or pseudo-photograph.

[58] s.6, *Protection of Children Act 1978*.

[59] Section 1A provides a defence to 16 and 17-year-olds who are married or living in an enduring family relationship. Section 1B provides a defence for the purposes of law enforcement or where the security and intelligence services are legitimately involved. For a discussion on this see Gillespie 2004.

(4) Where a person is charged with an offence under subsection (1)(b) or (c), it shall be a defence for him to prove-
(a) that he had a legitimate reason for distributing or showing the photographs or pseudo-photographs or (as the case may be) having them in his possession, or
(b) that he had not himself seen the photographs or pseudo-photographs and did not know, nor had any cause to suspect, them to be indecent.

It is clear from *R v. Collier*[60] that this is a legal burden of proof, i.e. the defendant must prove that the defence applies rather than the prosecution disproving them. That said, whenever the defence must prove something they only prove it to the civil standard (preponderance of probabilities) rather than the (higher) criminal standard that the prosecution must show.

Neither defence is likely to be of relevance to the topics under discussion although it is worth pausing over 'legitimate reason' in case the following example occurred:

D is detected distributing indecent photographs of a child to X, who he believes is a sex offender but who is, in fact, an undercover operative. When arrested, D claims that he was only distributing the photographs to provide evidence of X's criminality that he was then going to pass to the police.

Following *Atkins v. DPP*[61] the term 'legitimate reason' will be considered a matter of fact and therefore D would need to prove that this was truly his reason and not simply a story. This, coupled with the defendant having the burden of proof, means that an evidential basis would be required, making spurious claims less likely.

PoCA 1978 does not include reference to simple possession, something to be found in other legislation (s.160, *Criminal Justice Act 1988*[62]). It is unlikely that this will be particularly relevant to Sweetie however because in *R v. Bowden*[63] the High Court held that downloading an image from the internet constitutes 'making' rather than possession.[64] This means that it falls within PoCA 1978 rather than the 1988 Act. Whilst it would seem that this would lead to a potentially higher sentence, the judiciary have stated that making should be sentenced as though the offender was convicted of possession.[65] It does however mean that there are reduced opportunities for defences, something that remains controversial.[66]

The most likely offence in respect of webcam abuse is most likely to be the distribution of an indecent photograph of a child. It is not uncommon for offenders to provide other offenders or indeed children with child pornography and s.1 is designed to prohibit this.

The term 'distribute' is an ordinary term but it is given a wider definition in s.1 (2), *Protection of Children Act 1978* as 'a person is to be regarded as distributing an

[60] [2005] 1 WLR 843.
[61] [2000] 1 WLR 1427.
[62] Triable either-way and punishable by up to five years' imprisonment.
[63] [2000] QB 88.
[64] For a discussion on the reasoning behind this, see Gillespie 2011.
[65] See Sentencing Council 2013.
[66] Discussed in Akdeniz 2007.

indecent photograph if he parts with possession of it to, or exposes or offers it for acquisition by, another person'. This is notable because it states that distribution includes parting with possession of it. Thus where D emails a picture to X then this would constitute distribution irrespective of whether X actually receives it or looks at the image. By sending the image, or a copy of the image, it has left the control of the offender and thus distribution is satisfied.

What about situations where offenders use Peer-to-Peer (P2P) networks? It is not uncommon for offenders to share images through P2P and during chats to make their libraries available to others. Ostensibly there are two charges that could be used here. The first is possession of an indecent photograph with a view to distributing it,[67] and the second would be distribution.[68] In *R v Dooley*[69] the defendant had been identified by the police and when they examined his computer they found that he was a user of P2P and that some indecent photographs were in his 'my shared' folder. The defendant was charged under s.1(1)(c) but the Court of Appeal quashed his conviction, noting that in order to be convicted under s.1(1)(c) the prosecution needed to prove that at least one of his purposes of possessing the image was to allow others to access the material and in the absence of that evidence, a conviction could not be sustained.

Dooley is an unhelpful case because proving this extra intention could be difficult. In Scotland, they adopted a similar approach but held that knowledge of how P2P worked would help satisfy this requirement.[70] This is a more appropriate basis upon which to draw inferences. If a person knows that material contained within their shared folder is accessible by others and they choose not to hide or move material from that folder, it would seem appropriate to suggest that they are prepared to distribute it. Indeed a simpler way to escape the logic of *Dooley* would be to rely not on possession with intent/view to distribute but distribution itself. It will be remembered that s.1(2) defines distribution, *inter alia*, as exposing it for acquisition by another. That being the case, then material found within the shared folder must be considered exposed and liability will arise. In *R v. Price*[71] the Court of Appeal held that distribution was a crime of strict liability because statutory defences existed to excuse liability. Therefore if s.1(1)(b) rather than s.1(1)(c) were to be relied upon in P2P cases the onus will be on the defendant to show that he was not aware that he was distributing this material.

Non-Photographic Images of Children

The uncertainty as to the definition of pseudo-photograph and increased concern about computer-generated images and other types of drawings led the government

67 Section 1(1)(c), PoCA 1978.

68 Section 1(1)(b), PoCA 1978.

69 [2006] 1 WLR 775.

70 *Peebles v. HM Advocate* (2007) JC 93.

71 [2006] EWCA Crim 3363.

to introduce new legislation to tackle sexualised images of children that were not photographs. Section 62, *Coroners and Justice Act 2009* (CJA 2009) created an offence of possessing a prohibited image of a child. In this context, a prohibited image of a child meant an image (other than a photograph or pseudo-photograph) that displays prohibited conduct and is 'grossly offensive, disgusting, or otherwise of an obscene character'.[72] Unlike with photograph-based forms of child pornography, this is therefore a matter of obscenity rather than indecency. An image is prohibited if:

(a) It is an image which focuses solely, or principally, on a child's genitals or anal region or,
(b) It depicts:
(c) The performance by a person of an act of intercourse or oral sex with or in the presence of, a child;
(d) An act of masturbation by, of, involving or in the presence of a child;
(e) An act which involves the penetration of the vagina or anus of a child with a part of a person's body or with anything else;
(f) An act of penetration, in the presence of a child, of the vagina or anus of a person with a part of the person's body or with anything else;
(g) The performance by a child of an act of intercourse or oral sex with an animal (whether dead or alive or imaginary);
(h) The performance by a person of an act of intercourse or oral sex with an animal (whether dead or alive or imaginary) in the presence of a child.[73]
(i) It is clear that the child need not be real and indeed can be imaginary or a creature that has child-like features.[74] This makes the offence extremely wide and indeed some have criticised the offence as being over-broad.[75]

The offence is one of simple possession,[76] with the offence being either-way and punishable by a maximum sentence of three years' imprisonment.[77] As an offence of possession, it would not seem particularly relevant to webcam abuse. Some concern has been raised that this legislation is being used for distribution when it was designed for possession. However, it is undoubtedly easier to prove than using obscenity legislation. The latter has had a chequered past and which is considered somewhat difficult to prosecute.[78] Given that it is necessary to possess an image in order to distribute it, it can be seen why there is temptation to use the (simpler) CJA 2009. Whether that is appropriate is perhaps more questionable, and it would perhaps be wiser to introduce a distribution offence for non-photographic images of

---

[72] s.62(2), CJA 2009.

[73] Section 62(6), (7), CJA 2009.

[74] s.65(6), CJA 2009.

[75] Ost 2010.

[76] s.62(1), CJA 2009.

[77] s.66(2), CJA 2009. The sentence is six months' imprisonment when tried summarily.

[78] Edwards 1998.

children, modelled on the legislation for indecent photographs of children. This would send a key message that distributing material relating to children is inappropriate and is not the equivalent of transmitting adult pornography.[79]

7.2.4 *Offences Concerning the Participation of a Child in Pornographic Performances*

The offences in England and Wales that specifically relate to the participation of children in pornographic performances primarily relate to offline performances, not online. Thus the *Theatres Act 1968* would be of little use because it applies to theatres within England, and not the broadcasting of abuse.

Section 12 (causing a child to watch a sex act) is discussed below and this could be of relevance where, for example, the performance involves the child being forced to watch the sexual performance, including through participation or presence. Section 11, SOA 2003 (engaging in sexual activity in the presence of a child) is unlikely to be of any assistance. Section 11 requires one or more persons to engage in a sexual activity and for the purposes of obtaining sexual gratification he engages in it when a child is present or in a place where they can be observed, and D knows or believes that the child is aware of the sexual activity. However, it does not apply to webcams because it requires physical presence, and thus s.12 would be necessary.

Where the abuse of a child is live-streamed then the principal offences would be:

(a) Sexual abuse: those offences discussed above would apply as the performance involves the sexual abuse of a child. The person who streams it and 'directs' such action is likely to be an accomplice to such matters.[80]
(b) Obscenity: the *Obscene Publications Act 1959* (as amended) makes it an offence to publish an obscene article. Whilst the Act was clearly written before digital technology, it has been held to apply to such media. 'Publish' means to make it available and therefore the live streaming of abuse would meet this definition. Whilst 'obscene' is a higher threshold than 'indecent' (used for the purposes of child pornography) it is unlikely that this will cause too many problems where the child is involved in pornographic performances.

---

[79] The OPA 1959 can, and sometimes is, used to criminalise the distribution of material that is not in itself illegal to possess: see, for example, *R v Perrin* [2002] EWCA Crim 747.

[80] English law recognises the law of complicity (s.8, *Accessories and Abettors Act 1861*) whereby a secondary party can be liable *as a principal* if he aids, abets, counsels or procures the commission of a crime. Aiding means assisting; abetting means encouraging; counselling is providing assistance in the preparation of the offence and procurement means to bring about the offence. The broadcasting of abuse is likely to be considered either abetting or procurement.

7.2.5 Corruption of Children

The corruption of children relates to causing a child to watch sexual activities, even when they are not involved themselves. In England and Wales this is covered by two offences:

(a) s.12, SOA 2003 (causing a child to watch a sexual act).
(b) s.19, SOA 2003 (abuse of a position of trust: causing a child to watch a sexual act).

The principal difference between the two offences is the existence of the breach of a position of trust, something discussed already. The reality is that whilst this may be relevant to webcam abuse in some situations (e.g. a teacher causing a pupil to watch a sex act) it is unlikely to be particularly relevant to Sweetie and therefore this section will focus primarily on Section 12.

Section 12 states:

> (1) A person aged 18 or over ([D]) commits an offence if:
> (a) for the purpose of obtaining sexual gratification, he intentionally causes another person (B) to watch a third person engaging in sexual activity, or to look at an image of any person engaging in an activity, and
> (b) the activity is sexual, and
> (c) either-
> (i) B is under 16 and [D] does not reasonably believe that B is 16 or over, or
> (ii) B is under 13.

The offence is triable either-way and is punishable by a maximum sentence of 10 years' imprisonment.[81]

Whilst it may appear that there is a loophole within s.12 in D causing Vto watch himself (D) performing a sex act is not within the provisions of this offence (as it requires a third party), ut this is covered by another offence.[82] That offence only applies where the offender and child are physically proximate though and thus it is unlikely to be of relevance to webcam situations. However, s.12 remains relevant for webcam abuse because the provision expressly makes reference to causing a child to watch an 'image'. The term 'image' is defined in s.79(4) as meaning 'a moving or still image and includes an image produced by any means and, where the context permits, a three-dimensional image'. Setting aside three-dimensional images (which appears to be more about future-proofing) it is clear that footage on a webcam must constitute an image as it is a moving image and it is produced 'by any means'. Also, the offence states that it can be an image of 'any person' and thus this must include D. Therefore, where D exposes himself or performs a sex act in front of a webcam this would breach s.12.

[81] Six months' imprisonment when tried summarily.

[82] s.11, *Sexual Offences Act 2003.*

It is not enough that a person *intentionally* causes a child to watch a sex act or a sexual image, they must do so for the purposes of obtaining sexual gratification. In *R v. Abdullahi*[83] the Court of Appeal stated that the purpose of this fault element is to ensure that legitimate showings of sexual material, for example for the purposes of sex education, would not be captured.[84] The court rejected an assertion that immediate sexual gratification was required, stating:

> …we can find nothing in the section which suggests that sexual gratification must be taken immediately, or putting the same point the other way round, that it cannot extend to a longer term plan to obtain further or greater sexual gratification in the form of the eventual working out of a particular sexual fantasy or activity involving the child. The purpose [of the showing] may involve short-term and long-term sexual gratification; immediate and deferred gratification.[85]

This is an important statement although it broadens the offence considerably and arguably means that it overlaps significantly with other offences, most notably s.14, SOA 2003. That said, this could be useful in the context of webcam abuse. Let us take an example:

> D meets V, a child, in an online chatroom devoted to art. They begin to talk about nude photographs and D sends V some photographs of women posing nude or semi-nude. His intent is that V will send him some photographs of herself, which he will use for sexual gratification.

A number of offences could be used to tackle this example but s.12 is one of them and it would apply irrespective of whether V does, in fact, comply. D has clearly shown a sexual image to a child (V) and he has done so for the purposes of sexual gratification. It may be that the gratification will only arise later (when V has sent the images) but *Abdullahi* makes clear that this is irrelevant.

7.2.6 *Solicitation of a Minor*

England and Wales was the first country in Europe, and one of the first outside of the USA, to specifically introduce legislation to tackle the solicitation of minors via the Internet. Within England and Wales this behaviour has commonly been referred to as 'grooming' rather than solicitation although the latter is arguably more

---

83 [2006] EWCA Crim 2060.

84 *Ibid.*, at [16].

85 *Ibid.*, at [17].

accurate, not least because the law does not really tackle grooming *per se* but rather the results of grooming.[86]

Whilst a number of offences could cover this behaviour, the two that do so expressly are:

(a) s.14, SOA 2003 (arranging or facilitating the commission of a child sex offence.
(b) s.15, SOA 2003 (meeting a child following grooming etc).

Section 15

Section 15 will be examined first because it is the offence that was created specifically to deal with the issue of solicitation Whilst passed in 2003, it has undergone a number of amendments. The current text is:

> (1) A person aged 18 or over ([D]) commits an offence if:
> (a) [D] has met or communicated with another person (B) one one or more communications and subsequently:
> (i) [D] intentionally meets B.
> (ii) [D] travels with the intention of meeting B in any part of the world or arranges to meet B in any part of the world, or
> (iii) B travels with the intention of meeting [D] in any part of the world.
> (b) [D] intends to do anything in respect of B, during or after the meeting mentioned in paragraph (a)(i) to (iii) and in any part of the world, which if done will involve the commission of a relevant offence,[87]
> (c) B is under 16, and
> (d) [D] does not reasonably believe that B is 16 or over.

This offence is triable either-way and is punishable by a maximum of 10 years' imprisonment.[88]

Until 2015 the requirement in s.15(1)(a) was to have met or communicated with a child on at least two occasions but the *Criminal Justice and Courts Act 2015* amended this to remove the requirement for a second act. The reality is that nothing turns on this. The requirement for two acts was never fully articulated in the original drafting of the offence and was almost certainly down to the fact that two or more acts was requirement for the offence of harassment.[89] There was pressure to remove this requirement from child protection agencies but at no point was a single example produced of a case that had collapsed because a second act of meeting or communication could not be proven. Indeed it is highly unlikely that such proof could ever have been found. Given the nature of this type of offending it is highly unlikely that a person could persuade a child to meet him after only a single

---

[86] Gillespie 2004.

[87] A relevant offence in this context would include any offence within Part I of the *Sexual Offences Act 2003* or anything that is done outside of England and Wales that would constitute an offence if done in England and Wales (see s.15(2)(b)). For our purposes the most relevant offences would those set out as 'sexual abuse' in the table above.

[88] Six months' imprisonment if triable summarily.

[89] See *Protection from Harassment Act 1997*.

communication. Whilst this may be possible in the context of commercial sexual exploitation, it has already been noted that the law would tackle such behaviour. That said, whilst the reduction to one incident will not add anything to the offence, neither will it detract from it. Given the other elements of the offence, there is little danger someone innocently talking to a child on a single occasion could find them within this provision because it is necessary to prove the intent to sexually assault the child.

Section 15 was designed to tackle the *results* of grooming and therefore its focus is on the subsequent acts (discussed below) and the intention to commit a sex offence. During the drafting of the legislation it was noted that the literature surrounding grooming showed that the initial contacts were often innocuous and indeed the sexualisation of contact may not take place until the final meeting. Section 15 was specifically designed to ensure that this did not create a loophole and therefore nothing within the Act requires that initial contact or communication is sexual. Whilst invariably the communication will include sexual content, because that is frequently an essential part of the 'grooming cycle',[90] there is nothing that *requires* the communication to be sexual.[91]

Section 15 is, in essence, a preparatory offence and mere communication is not enough. There is a requirement that D does a further act. Whilst set out in three sub-paragraphs (s.15(a)(i)–(iii)) it is clear that there are, in fact, four separate acts contemplated:

(a) D meets B.
(b) D travels to meet B in any part of the world.
(c) D arranges to meet B in any part of the world.
(d) B travels to meet D in any part of the world.

The phrasing of #2 and #4 makes clear that there is no requirement that D should actually meet B (which would therefore come under #1) and instead the further act is satisfied by the travelling. Of course there is the evidential requirement to prove that D was intending to meet B (and subsequently the intention to commit a sex offence) but in many instances this would not be problematic. Let us take an example:

> D lives in Newcastle-upon-Tyne (in North-East England). He has been talking to B, a child, who lives in Bristol. They eventually agree to meet up and D intends to sexually assault B when they do. On the agreed day, D purchases a train ticket and sets off on his journey. The police have monitored the communications. They can arrest him on the train to Bristol, the train ticket stating Bristol making it somewhat obvious why he is travelling.

[90] Ost 2009.

[91] *R v G* [2010] EWCA Crim 1693.

The advantage of this approach is that it provides the police with more flexibility and they can arrest the suspect in a controlled environment (e.g. the train) rather than a public place where containing him may be more difficult.

When the offence was first passed #4 was not available but it was introduced by the *Criminal Justice and Immigration Act 2008*. This closed a potential loophole:

D lives in Coventry and he is talking to V, a child, who lives in Leicester. They agree to meet up and D sends V a train ticket for him to come to Coventry.

There was doubt whether under this original wording the offence would be satisfied. One argument is that by sending the ticket D has 'arranged to meet' V (which is within the scope of the offence) but it was thought safer to introduce specific wording for the offence. Again, this provides flexibility in terms of the point at which an offender can be arrested.

It is notable that s.15 specifically addresses the issue of jurisdiction, meaning that it is not necessary to rely on the ordinary rules of jurisdiction. Whilst D must be within England and Wales when he meets or communicates with B, it is clear that B can be anywhere in the world, as can the meeting. This was purposely introduced to prevent situations where D would seek to evade the ordinary rules of jurisdiction. For example:

D is speaking with V, a child. They agree to meet up and V mentions that she is going to a school trip to France next week. D says (falsely) that, by coincidence, he is going to be in the same village and they agree to meet.

The (subsequent) meeting or travelling to meet must be accompanied by an intention to commit a relevant sexual offence. Intent in this context means aim or purpose and therefore the prosecution must prove that at the subsequent meeting D had as his aim or purpose the commission of a child sex offence. Part 1 of the SOA 2003 covers most sex offences against children but it does not include the taking of an indecent photograph of a child which, as will be remembered, are to be found in separate legislation.[92] Ostensibly this would mean that if D's intention was to take an indecent photograph of a child then s.15 would not apply. However two arguments could be raised to counter this. The first is that the act of getting a child to pose for an indecent photograph of a child would be a 'sexual activity' within the meaning of s.9, *Sexual Offences Act 2003*. As that is within Part 1, the s.15 offence could still be satisfied. The second alternative would be to rely on the sexual exploitation offences discussed above. Section 48 prohibits the causing of a child to

[92] *Protection of Children Act 1978.*

be sexually exploited. Section 51(2)(b) states that a child is sexually exploited if, *inter alia*, an indecent photograph of the child is recorded. Accordingly, the taking of an indecent photograph could be fall within s.48, which is also within Part 1 of the Act meaning that s.15 would apply here too.

It should be noted that it does not matter where in the world D intends to commit the sex offence so long as the relevant act would constitute an offence if it took place in England and Wales. This is an important point given the fact that offenders will travel to abuse children they have met online (or indeed pay for children to travel) but also because the phraseology of the legislation differs from ordinary principles of jurisdiction. It is not uncommon for extraterritorial offences to impose a requirement of dual criminality, i.e. that an act should be illegal in both the prosecuting state and the state where the act took place.[93] However, this causes difficulty with the sexual abuse of children where it known that some countries still do not have adequate laws and therefore there has been calls for dual criminality to be abolished.[94] Section 15 does not require the act to be illegal in the country that D intends the act to take place: it must simply be illegal if it were to take place in England and Wales. This would mean that a person could not try to evade the law by, for example, targeting children in areas of the world that have poor child protection laws.

The final fault element relates to the age of a child. Unlike some of the other offences that have been discussed before now, it makes no difference if the child is aged under 13. The only fault element is that the child is under 16 and that B does not reasonably believe that B is aged 16 or over. The age of 16 is chosen as the age of consent and it will be ultimately a matter of fact for the jury as to whether this is satisfied.

Section 14

The second offence that could be used for solicitation is s.14. This is purposely written in much wider language and was designed to tackle not only those who solicited children for abuse, but also those who assisted them. The wording of the offence is:

(1) A person commits an offence if-
(a) he intentionally arranges or facilitates something that he intends to do, intends another person to do, or believes another person will do, in any part of the world, and
(b) doing it will involve the commission of an offence under any of sections 9 to 13.

The offence is triable either-way with a maximum sentence of fourteen years' imprisonment.[95] A statutory defence exists to this offence.

(2) A person does not commit an offence if-
(a) he arranges or facilitates something he believes another person will do but that he does not intend another person to do, and

[93] Fredette 2009.
[94] Discussed in Svensson 2006.
[95] Six months' imprisonment when tried summarily.

(b) any offence within subsection (1)(b) would be an offence against a child for whose protection he acts.
(3) For the purposes of subsection (2), a person acts for the protection of a child if he acts for the purpose of-
(a) protecting the child from sexually transmitted infection,
(b) protecting the physical safety of the child,
(c) preventing the child from becoming pregnant, or
(d) promoting the child's emotional well-being by the giving of advice.

and not for the purpose of obtaining sexual gratification or for the purpose of causing or encouraging the activity constituting the offence within subsection (1)(b) or the child's participation it.

Perhaps the best example of when this defence could apply would be in respect of the provision of contraceptives to a child under 16. A doctor or pharmacist who provides contraceptives could be said to be facilitating a child sex offence in such circumstances as it is illegal for a person to have sex with a child under 16. Providing contraceptives or advice on how to undertake safe sex could be considered to be facilitating this act. However, assuming the person who supplies the contraceptives is doing so for the best interests of the child (to prevent a sexually transmitted infection or pregnancy) then the defence would operate.

The terms of Section 14 are broad and six separate offences are created ('arranging' or 'facilitating' leading to him committing a sexual offence, intending that someone else commits the offence or believing that someone will commit the offence). 'Intending' and 'believing' are presented separately because they represent different types of offending. Intending implies that it is D's aim or purpose that the offence will take place, i.e. the person is actively seeking to bring about the commission of that offence. However belief implies knowledge that something is going to happen (or may happen) but not that it is within the control of the individual or that they desire it.

The terms 'arranging' or 'facilitating' are not defined but they are words of ordinary usage. 'Arranging' is clearly putting in place the steps required to commit the offence. This could include, for example, travel arrangements:

D is talking with V, a 14-year-old boy. They agree to meet up where D intends that he will have sex with V. He sends V a rail ticket and books a hotel room where he intends they will meet.

This clearly constitutes an arrangement. Whilst if V actually travelled it is possible that s.15 could be used: by using s.14 it would not be necessary to wait for V to travel: by sending the train ticket and booking the room, D is clearly arranging that meeting which, in turn, will lead to the sexual assault taking place.

'Facilitation' is obviously significantly wider than arranging and can include acting as an intermediary. An example of this is *R v. Eller*[96] where the offender was found to be using Skype to 'repeatedly direct…, to a paying audience, live broadcasts of a serious sexual abuse of very young girls in the Philippines. He had also been involved in transmitting payments to the abused children in the Phillipines and had directly received payments from people who were paying to view these broadcasts'.[97] Clearly, this is the facilitation of conduct that would amount to the commission of a sex offence, meaning s.14 applies. It is also an interesting example of the difference between 'intent' and 'belief'. Whilst in this case there was clear evidence that the offender did all of these activities for the purposes of sexual gratification, it does not follow that this would be true in each case. For some it may simply be a way to earn money. Thus they may not care whether a child is abused so long as they get their money. However, even if they do not intend abuse, they would undoubtedly *believe* that the abuse will take place and so a conviction could be sustained.

Section 14 has been used on a number of occasions, particularly where an offender has sought to procure a child for sexual purposes. In *R v Jordan*[98] and *R v Robson*[99] the defendants approached (adult) prostitutes and asked whether they could find young children (a 12-year-old in *Jordan* and a 13/14-year-old in *Robson*). In neither case did the prostitute do so, or attempt to do so, and instead contacted the police. In both cases they were convicted of an offence contrary to s.14.[100] This demonstrates the fact that an early intervention is possible. The logic is that in both cases an arrangement was made (or attempted), which would be the identification of a child, and the defendant intended to do something with that child that would involve the commission of a child sex offence. There is no reason why the same could not apply to virtual transactions, so where a person seeks to procure a child online, Section 14 should apply in exactly the same way as it would offline.

Where s.14 has proven particularly useful is in respect of proactive law enforcement agencies where there has never been a real child, i.e. the so-called 'sting' operations. Since the offence is phrased in such a way that the defendant must only intend that a child sex offence takes place, there is no requirement that a real child is involved so long as the offender believes or intends that there is. It is this which makes s.14 ideal in a Sweetie-type situation. A good example of this is *R v. Wright*[101] where an undercover police officer posed as the father of an 8-year-old boy and 10-year-old daughter whom he had regularly sexually abused. The police

---

96 [2014] EWCA Crim 2995.

97 *Ibid.*, at [10].

98 [2006] EWCA Crim 3311.

99 [2009] EWCA Crim 1472.

100 In *Robson* it constituted an attempt but in *Jordan* it was the substantive offence. Nothing turns on this point and it was just simply prosecutorial (and arguably judicial) preference as to how an indictment was drawn up and what constituted an 'arrangement'.

101 [2014] EWCA Crim 664.

officer and defendant met on the internet and exchanged sexually-explicit emails. They arranged to meet for coffee, and to then return to the police officer's house where the defendant would be allowed to rape the 8-year-old boy. The defendant turned up and was in possession of items that could have been used to facilitate the sexual abuse of the child. He was arrested, charged and convicted of an offence contrary to s.14.

7.2.7 *Sexual Communication*

An offence that is not contained within the Lanzarote Convention, which may be of relevance to Sweetie 2.0 is a new offence of sexual communications. The offence was introduced by s.67, *Serious Crime Act 2015* although it inserts a new s.15A into the SOA 2003. The offence is:

> (1) A person aged 18 or over ([D]) commits an offence if-
> (a) for the purpose of obtaining sexual gratification, [D] intentionally communicates with another person (B);
> (b) the communication is sexual or is intended to encourage B to make (whether to [D] or to another) a communication that is sexual, and
> (c) B is under 16 and [D] does not reasonably believe that B is 16 or over.
> (2) For the purposes of this section, a communication is sexual of-
> (a) any part of it relates to sexual activity, or
> (b) a reasonable person would, in all the circumstances but regardless of any person's purpose, consider any part of the communication to be sexual'
>
> and in paragraph (a) 'sexual activity' means an activity that a reasonable person would, in all the circumstances but regardless of any person's purpose, consider to be sexual.

The offence is triable either-way and punishable by a maximum sentence of two years' imprisonment (s.15A(3), SOA 2003).[102]

The essence of the offence is quite simple. D engages in a sexualised conversation with a child (B). This differs from the solicitation offences because there is nothing within the legislation that requires D to be inciting V to do anything. Where it did this then it would constitute one of the offences discussed above. D must be engaging in the conversation for the purposes of sexual gratification which may be relatively difficult to prove. Where, as in the other offences, D is trying to engage V in sexual activity, such proof is relatively easy to identify. It may not be when there is no incitement although obviously if, for example, D tells V that he is masturbating whilst they talk then that would suffice.

[102] Six months' imprisonment when tried summarily.

Section 15A was not implemented until recently,[103] which is somewhat unusual for a criminal offence. No reason for the delay was given although some believed it was because the offence could be considered incompatible with Article 10 of the ECHR,[104] although perhaps less so where it is for the purposes of sexual gratification.[105] The fact that D has to be at least 18 does minimise the potential for adolescents to be convicted for talking to each other about sex, but even still the offence could be harsh:

D, on his 18th birthday, is talking to V, aged 15 (but whose 16th birthday is tomorrow) online. The conversation turns sexual when they realise that they are attracted to each other.

This is not an unrealistic scenario and yet would seem to breach s.15A. Obviously one would hope that there would be no prosecution in such circumstances but, strictly speaking, it would constitute an offence under this provision.

Notwithstanding the difficulties of s.15A, it could potentially be useful in the context of Sweetie. It is likely that a suspect would try to engage Sweetie in sexualised chat and so long as this is for the purposes of sexual gratification, then liability would arise. It is inevitable that the decision in *R v. Abdullahi*[106] would apply equally to this offence as the terminology and intention is the same. Thus, it is not necessary to prove that D was chatting sexually to gain *immediate* sexual gratification, it would suffice if it could be shown that the sexual conversation was designed as part of the 'grooming' process, with the intention of D later receiving sexual gratification. This widens the offence considerably and means that s.15A could conceivably be an additional tool in combatting the solicitation of children.

7.2.8 The Law of Attempt

Whilst the previous section has identified offences that are relevant where a child is sexually abused or exploited, what is the position where there was never a child? Sweetie 2.0, it will be remembered, is not a child. It is a graphic representation of a child accompanied by an AI interface. So there was never a child that could be abused or exploited. Under the law of England and Wales that would not matter.

---

[103] *Serious Crime Act 2015 (Commencement No 6) Regulations 2017* (SI 2017/451).

[104] Freedom of Expression.

[105] The ECtHR has generally allowed a greater margin of appreciation in respect of cases which involve minors being exposed to sexually-explicit content. See, for example, *Handyside v UK* A 24 (1976) and *Müller v Switzerland* 13 EHRR 212.

[106] [2006] EWCA Crim 2060.

For some offences, most notably s.14 SOA 2003, there is no requirement for a real child to be involved due to the phrasing of the offence. Thus s.14 refers to an *intent* to commit a sexual offence and others talk about the *belief* as to whether a child sex offence will take place. In neither instance must the substantive offence take place, it suffices that there was the relevant intent or belief and thus the fact that there was never a child in the first place is irrelevant so long as the defendant believed that there was a (real) child.

With other offences there is a requirement for a child. So, for example, s.10 SOA 2003 (causing or inciting a child to engage in sexual activity) refers to the victim as necessarily being under the age of 16. That child must then be caused or incited into engaging sexually. Without a child therefore, it may seem that there can be no substantive offence. However, English law has quite a nuanced approach to the law of attempts, something now set out in s.1, *Criminal Attempts Act 1981*:

> (1) If, with intent to commit an offence to which this section applies, a person does an act which is more than merely preparatory to the commission of the offence, he is guilty of attempting to commit the offence.
> (2) A person may be guilty of attempting to commit an offence to which this section applies even though the facts are such that the commission of the offence is impossible.

All of the offences that are set out previously in this chapter would be an offence to which the *Criminal Attempts Act 1981* would apply.[107] This means that it does not matter that the victim is not a real child. A good example of this would be in respect of s.10, causing or inciting a child to engage in a sexual activity:

> D is talking to Sweetie and he asks her to remove her top and pose in front of the webcam for him so that he can see her topless.

As noted earlier, if Sweetie were real then this would amount to an offence under s.10, *Sexual Offences Act 2003*. Sweetie is not real and thus it is impossible for him to be inciting a child to engage in sexual activity because there is no child. However, s.1(2), *Criminal Attempts Act 1981* expressly states that it is irrelevant that something is impossible and so the test becomes (under s.1(1)) does D, with intent to commit an offence, do an act which is more than merely preparatory to the commission of an offence under s.10, SOA 2003 were it not impossible? Clearly the answer is 'yes' and therefore D would be guilty of the offence of attempting s.10, SOA 2003. An attempt is punishable in the same way as the substantive offence.[108]

There are a number of cases where a conviction has been sustained irrespective of the fact that a real child does not exist. In the majority of situations, the situation

---

[107] Section 1(4), *Criminal Attempts Act 1981* states that the section applies to all indictable offences (subject to some exceptions which do not apply in these circumstances).

[108] s.4(1), *Criminal Attempts Act 1981*.

arises due to proactive operations by law enforcement. A good example of this is to be found in *R v. Collins*[109] where the offender made contact with two women online and made arrangements for them all to have sex with an 8-year-old child known as Chloe.[110] Unfortunately for the offender, Chloe did not exist and the two women were undercover police officers working for the *Child Exploitation and Online Protection Centre* (CEOP), a national law-enforcement agency that leads on child sexual exploitation investigations. An extensive number of messages and communications were passed between the women and the offender and he eventually travelled to meet them and the child whereupon he was arrested and ultimately pleaded guilty to offences contrary to s.14, SOA 2003.

7.3 Criminal Procedure

Criminal procedure in England and Wales is based on the adversarial process. There is a clear distinction between the investigation, prosecution and trial stages. At pre-trial level there is a clear division between the investigators (law enforcement) and prosecutors (the Crown Prosecution Service).

Law Enforcement

The police in England and Wales are divided into 43 police forces, each headed by a chief constable[111] who has operational independence within their force boundary. Thus there is no 'head' police officer across all of England and Wales or indeed the United Kingdom.[112] Alongside the police exist national law enforcement and security agencies. Both the Security Service and Secret Intelligence Service[113] have a role in the fight against child sexual exploitation[114] although this is mainly in respect of international operations and thus outside of the remit of this report. There is also the *National Crime Agency* (NCA) which, as its name suggests, is a law enforcement agency that has national jurisdiction. That said, 'national' does not mean UK-wide. Whilst it has jurisdiction across all of England, Wales and Northern Ireland, it only has limited jurisdiction in Scotland.

[109] [2015] EWCA Crim 915.

[110] *Ibid.*, at [2].

[111] In the City of London and Metropolitan police the head of the police area is known as a Commissioner rather than chief constable.

[112] There is a single police force in Northern Ireland (*Police Service of Northern Ireland*) and a single force in Scotland (*Police Scotland*). Each has a chief constable in operational command.

[113] Aka MI5 and MI6 respectively.

[114] Recognised in statute by s.1B, PoCA 1978.

An important part of the NCA is the *Child Exploitation and Online Protection Centre* (CEOP). Established in 2006, CEOP was a standalone organisation until 2007 when it was subsumed within the NCA. Whilst this was somewhat controversial[115] there is undoubted synergies between the work of the NCA and CEOP and also the back-office could be shared. CEOP has quickly established itself as one of the leading law enforcement organisations in the world dedicated to tackling child sexual exploitation and this has included a number of successful proactive operations, including through being a founding member of the *Virtual Global Task Force.*

At local level, each police force organises themselves in their own way although some forces have combined resources to produce regional units. Some police forces have paedophile units, some have cybercrime units and others have established POLIT (Paedophile OnLine Investigation Team) units, which are designed to tackle online child sexual abuse and exploitation. Some police forces have no specialist units and instead leave these matters to be investigated by ordinary detectives.

The police and NCA have operational independence and thus nobody can tell the police how to investigate a matter. That said, the police and CPS have worked hard in recent years to ensure that they work together and this is particularly true in cases of child sexual exploitation. The police will often seek the advice of the CPS in respect of complicated cases, including identifying the relevant evidential base.

Prosecutors

The CPS is a statutory body established in 1986.[116] It is headed by the Director of Public Prosecutions, an ancient office which has been folded into the new organisation. It has headquarters in both London and York and this includes specialist teams. It is organised into 13 areas (which overlap across the 43 police forces) and also CPS Direct. The latter is a 24/hour, 365-day service that works closely with the police, in particular in relation to charging. There is also 'casework divisions' which relates to specialist work. Currently this includes Special Fraud Division, Organised Crime Division and the Special Crime and Counter Terrorism Division. None of these casework divisions however have any impact on the subject matter at hand.

Whilst the police have investigatory independence, it is the responsibility of the CPS to decide whether a person should be formally charged with a crime and, if so, what charge that will be. Where a quick decision is made (including where the defendant will be remanded into custody before trial) then the matter is dealt with by *CPS Direct.* Where the matter takes longer because the offender is on bail (something not unusual in cybercrime cases where the forensic examination of a number of computers may be required) then the matter will be taken by a CPS caseworker.

---

[115] The founding commander of CEOP, Jim Gamble, resigned in protest at the subsuming of CEOP into the NCA.

[116] It was established by the *Prosecution of Offenders Act 1985.*

Whilst there is not a specialist casework division to tackle cybercrime, some CPS lawyers do have specialist training on these issues. That said, however, the CPS is under significant pressure, with significant staffing reductions having been made.[117] Indeed concern has been raised as to whether the cuts have gone too far and are affecting the ability of the CPS to discharge their responsibilities.[118] This has certainly meant that specialist lawyers are being used on non-specialist cases and that some cases are being dealt with by non-specialists. This is a matter of some regret because there is something 'different' about cybercrime, including being able to understand the evidence and how this evidence is to be presented.

Criminal Procedure

As with the substantive law, there is no single statute that governs investigatory powers and instead the rules are spread across a number of statutes. The most notable piece of legislation is the *Police and Criminal Evidence Act 1984* (PACE 1984) which sets out the principal rules on the powers of the police to investigate crime and how evidence will be dealt with at court. The statute has been heavily amended over the years. The *Criminal Procedure and Investigation Act 1996* is, as its name suggests, a significant piece of procedural law. This Act primarily regulates the disclosure of material pre-trial (i.e. between parties) and also streamlined many pre-trial protocols. The *Criminal Justice Act 2003* also made significant changes to the laws of evidence.

In terms of procedure, perhaps the most significant and relevant piece of legislation is the *Investigatory Powers Act 2016*. This is the latest piece of legislation that seeks, in part, to regulate investigatory (surveillance) powers. It follows previous legislation, including the *Regulation of Investigatory Powers Act 2000* and the *Data Retention and Investigatory Powers Act 2014*, the latter of which was a piece of legislation that incorporated its own demise as the legislation stated that it would automatically be repealed on the 31 December 2016.[119] The DRIPA 2014 was, in essence, a 'bridging' piece of legislation that was designed to allow for legislative consultation to take the country from RIPA 2000 to the IPA 2016. As will be seen, it remains extremely controversial.

To aid our analysis, Table 7.4 presents the investigatory powers contained within the *Council of Europe Convention on Cybercrime*[120] and where these can (approximately) be found within the law of England and Wales.

[117] Over the past five years there has been a reduction in staffing by almost 2,400 people, including trial lawyers and caseworkers.

[118] See, for example, http://www.bbc.co.uk/news/uk-wales-35496012. Accessed 14 March 2016.

[119] Section 8(3), *Data Retention and Investigatory Powers Act 2014*.

[120] Also known as the 'Budapest Convention'.

Table 7.4 Investigatory powers relevant to webcam-based abuse

Council of Europe Convention on Cybercrime	England and Wales
Article 16. Expedited preservation of stored computer data	Investigatory Powers Act 2016, Part 4
Article 17. Expedited preservation and partial disclosure of traffic data	Investigatory Powers Act 2016, Part 2
Article 18. Production order	Investigatory Powers Act 2016
Article 19. Search and seizure of stored computer data	Police and Criminal Evidence Act 1984
Article 20. Real-time collection of traffic data	Investigatory Powers Act 2016, Part 2
Article 21. Interception of content data	Investigatory Powers Act 2016, Part 1
Undercover operations conducted on the internet	Regulation of Investigatory Powers Act 2000, Part II

[*Source* The author]

7.3.1 Preservation of Data

The preservation and disclosure of data (Articles 16 and 17) are topics of extreme controversy at the moment. The United Kingdom had previously enacted statutory instruments to comply with the EU Directive on Data Retention.[121] Following the striking down of this Directive by the Court of Justice of the EU,[122] it was thought that the legality of data retention was being called into question and therefore new legislation was required to clarify this. The *Data Retention and Investigatory Powers Act 2014* was ultimately passed but, as noted already, this predicted its own demise.

The DRIPA 2014 was subject to legal challenge. Unlike the position in many countries, the courts of England and Wales do not have the power to 'strike down' primary legislation.[123] However, its obligations under EU law mean that the courts have the power to disapply law that is incompatible with EU law. Rather ironically, given that he became Secretary of State for Exiting the European Union, David Davis sought a ruling from the courts that DRIPA 2014 was incompatible with EU law because it did not provide sufficient safeguards to overcome the ruling in the *Digital Rights Ireland* case. The High Court ruled in his favour,[124] although the

---

[121] Directive 2006/24/EC of the European Parliament and of the Council on the retention of data generated or processed in connection with the provision of publicly available electronic communications services or of public communications networks and amending Directive 2002/58/EC. (2006) OJ L105/54. The most notable UK instrument implementing this is the *Data Retention (EC Directive) Regulations 2009 (SI 2009/859)*.

[122] *Digital Rights Ireland v Minister for Communications, Marine and Natural Resources* C-293/12.

[123] Gillespie and Weare 2017.

[124] *R (on the application of Davis) v. Secretary of State for the Home Department* [2015] EWHC 2092 (Admin).

Court of Appeal suggested it was 'minded' to disagree with the decision,[125] but ultimately referred matters to the Court of Justice of the European Union before making its final ruling.

The IPA 2016 now covers most areas relating to communications data. Part IV deals with the retention of communications data and s.87 authorises the Home Secretary to issue a retention notice if (a) she believes that it is necessary and proportionate for one of a series of purposes listed in s.87(2), and (b) the judicial commissioner approves the notice.[126] The list of purposes includes 'for the purposes of preventing or detecting crime'[127] and thus it will be applicable for investigations relating to Sweetie.

For these purposes, communication data means:

> communication data which may be used to identify, or assist in identifying any of the following-
>
> the sender or recipient of a communication (whether or not a person),
>
> the time or duration of a communication,
>
> the type, method or patter, or fact, of communication,
>
> the telecommunication system (or part of it) from, to or through which, or by means of which, a communication is or may be transmitted, or
>
> the location of any such system,
>
> and this expression therefore includes, in particular, internet connection records.

The list is quite wide and would cover the 'who', 'to', 'when', 'where' and 'how' of communication data, although not its contents, which is dealt with separately as an interception.

A notice must be for no more than 12 months,[128] and it is thought that most notices will be for a 12-month period. The principal purpose of this retention policy is that it allows a resource to be accessed when a crime is detected. It is not uncommon for there to be a delay in reporting a crime, and thus the communication data that could be useful in proving a crime or disproving an alibi (for example), would have gone. The retention order ensures that such data is accessible. Indeed, the statute itself states that requirements can be placed on the order that requires, for example, it to be stored in a way that facilitates efficient disclosure where appropriate.[129]

The 'judicial commissioner' is a judge that is given specific authority to consider these matters. Their role, appointment and tenure are set out below in the section on

[125] *Secretary of State for the Home Department v. Davis* [2015] EWCA Crim 1185.

[126] s.87(1), IPA 2016.

[127] s.61(7)(b), IPA 2016.

[128] s.87(3), IPA 2016.

[129] s.87(9), IPA 2016.

human rights[130] since they are, in essence, a key plank of the protection of human rights in this field.

7.3.2 *Disclosure of Data (Articles 17 and 18)*

Part 4 of IPA 2016 deals with the retention of the data but no more than that. It does not provide that the data can be used. The rules for this are located in Part 3 of the Act.

Unlike the issuing of a retention notice, which requires judicial authorisation, access to the data is controversially an administrative issue. A 'designated senior officer' can authorise an officer of a relevant public authority to obtain communications data[131] if he thinks:

> it is necessary to obtain communications data for a specified purpose,
>
> that it is necessary to obtain the data, inter alia, for the purposes of a specific investigation or a specific operation, and
>
> that it is proportionate to do so.[132]

For our purposes, the relevant public authority will usually be the police. The designated senior officer for the police will ordinarily be a superintendent, although it could be an inspector for 'entity data'. The latter is part of a distinction that is drawn between communication data, that is between entity data and event data.

'Entity data' is data that concerns entities (meaning an individual or corporation) or links between them, but does not include information about events.[133] 'Event data' is that which 'describes events taking place on a telecommunication system or other device which consists of one or more entity engaging in an activity at a specific point, or points, in time and space'.[134] The *Explanatory Notes* which accompany the Act[135] provide two examples to assist in understanding this (see Table 7.5).

It was noted above that the senior officer must consider that it is necessary to obtain information for a specified purpose, and this includes for the prevention and detection of crime, which would be obviously relevant for Sweetie.[136] The requirement that it must be necessary and proportionate should mean that it is not considered routinely, and that the data is only accessed when there is no other approach. However, considerable doubt exists as to whether that actually happens

[130] See Sect. 7.4.1 below.

[131] s.61(1),(2), IPA 2016.

[132] s.61(1), IPA 2016.

[133] s.261(3), *Investigatory Powers Act 2016*.

[134] Paragraph 726, *Explanatory Notes* explaining s.261(4), *Investigatory Powers Act 2016*.

[135] For an explanation of explanatory notes, see Gillespie and Weare 2017 p. 53.

[136] See s.61(7)(b), IPA 2016.

Table 7.5 Distinction between entity and events data [*Source* The author]

Entity data	Events data
Phone numbers or other identifiers linked to communication devices; addresses provided to a communications service provider; or IP address allocated to an individual by an internet access provider	The fact that someone has sent or received an email, phone call, text or social media message; the location of a person when they made a mobile telephone call or the Wi-Fi hotspot that their phone connected to; or the destination IP address that an individual has connected to online[a]

[a]Paragraph 727, *Explanatory Notes*

or whether necessity and proportionate are now more 'tick-box', with communication data routinely featuring in numerous investigations.

The authorisation is subject to some conditions and restrictions. Most pertinently for our purposes are those that relate to internet connection records. A request for such records requires one of three conditions to be satisfied.[137] The three conditions (A–C) are summarised as follows:

(a) The data is necessary for, inter alia, the detection or prevention of a crime, to identify the sender of an online communication. This could be to resolve an IP address resolution, i.e. identifying who an individual is from an IP address.[138]
(b) Condition B is not relevant for our purposes.
(c) Condition C applies to the prevention and detection of *serious* crime and communication data is necessary to identify:

> which communication services a person has been using, for example determining whether they are communication through apps on their phone;
>
> Where a person has accessed illegal content, for example an internet service hosting child abuse imagery; or
>
> Which internet service is being used and when it is being used.[139]

'Serious crime' is defined, inter alia, as an offence which a person aged 18 or over is capable of being sentenced to a term of imprisonment for a term of at least 12 months'.[140] All of the offences that were discussed in Section 2 of this chapter are offences that meet this threshold and thus obtaining connection records for these purposes should not be troublesome.

Requests for communication data will normally be routed through a Single Point of Contact (SPoC), and that person should ordinarily be consulted before an authorisation is approved.[141] The SPoC process has been used for several years,[142]

[137] s.62(2), IPA 2016.

[138] See para 180, *Explanatory Notes*.

[139] Paragraphs 184 and 181, *Explanatory Notes*.

[140] s.62(6), *Explanatory Notes*.

[141] s.76, IPA 2016.

[142] For a discussion, see McKay 2015.

and is designed to ensure that there is consistency as to the type of information sought, a realisation of what can, and cannot be provided, and also that the data is understandable. SPoCs are given specific training but also develop a good working relationship with the communication providers, meaning that it is often possible to ensure that mistakes or over-reach is identified and weeded out.

7.3.3 Search and Seizure of Stored Computer Data (Article 19)

Article 19 requires states to ensure that the police have the correct competences to search and seize computer equipment. The *Police and Criminal Evidence Act 1984* (PACE 1984) governs the search and seizure of property. Searches are authorised by s.8 and requires the police to seek judicial permission. The police must seek the permission of a justice of the peace[143] who will decide whether there are reasonable grounds to believe, *inter alia*, that there is evidence relevant to a criminal offence on the premises given. It is not uncommon for businesses to require a search warrant before allowing the search of premises, so as to assure customers that they *had* to permit the search rather than just allowed law enforcement a fishing expedition.

Seizure is dealt with under s.19. A constable has a general power to seize anything that he has reasonable grounds to believe has been obtained as a result of the commission of a crime or that it contains evidence about a crime and that it is necessary to seize it to prevent it being concealed, lost, abandoned or altered (s.19 (2), (3), PACE 1984). In terms of computer data, s.19(4) states:

> (4) The constable may require information which is stored in any electronic form and is accessible from the premises to be produced in a form in which it can be taken away and which it is visible and legible or from which it can readily be produced in a visible and legible form if he has reasonable grounds for believing-
> (a) that-
> (i) it is evidence in relation to an offence which he is investigating or any other offence, or
> (ii) it has been obtained in consequence of the commission of an offence, and
> (b) that it is necessary to do so in order to prevent it being concealed, lost, tampered or destroyed.

Whilst s.19(4) is useful it generally applies to third-party evidence since where it was the computer itself that was to be seized then this would be done under s.19(3), discussed above. However it could be useful and, for example, s.19(4) could be used to gain evidence of who logged onto a computer in a cybercafé, what credit card payments were made at a particular time or who stayed in a hotel at a particular time.

---

143 In England and Wales justices of the peace (also known as magistrates) do not generally hold any legal qualifications (and are known as 'lay magistrates') although they are obviously given training. Some justices do hold legal qualifications and they are known as District Judges (Magistrates' Court).

However s.19(4) could also be used in more broader contexts. For example, if it was alleged that D had accessed child pornography from his office computer, it is quite possible that the terminal is a dumb-terminal without any proper storage. The files downloaded may be stored in his folder within the larger networked file-store. It is unlikely to be practical (or proportionate) for the police to seize the whole mainframe, but they could request that the contents were produced in a format that allowed them to take them away and analyse them (e.g. the files are duplicated onto a portable storage device).

Encryption

The fact that a computer can be seized does not, by itself, help law enforcement. Increasingly law enforcement identifies material that is found to be encrypted. This means that even if the computer is seized it is of no real use to law enforcement because they are not able to access the contents. England and Wales is one of a small number of countries that has a law specifically to tackle this.

The rules are set out in Part III of RIPA 2000. The power applies when data lawfully comes into the possession of the police which contains protected data (for our purposes, encrypted data). If the relevant police officer believes on reasonable grounds that:

(a) the key to the protected data is in the possession of a person;
(b) that it is necessary to gain access to the data for the purposes of, *inter alia*, the prevention or detection of crime;
(c) that it is proportionate to require access to the data; and
(d) it is not reasonably practicable to gain access to the information by another way.

then he may apply to a circuit judge for permission to issue a notice requiring either the key to the disclosure to be supplied or for the data to be rendered into readable data.[144]

A person who knowingly fails to comply with an order commits an offence that is triable either-way.[145] The sentence depends on what it is believed is being protected by the encryption. Where it is believed to be information relating to terrorism or indecent photographs of children then the punishment is five years' imprisonment[146] otherwise it is two years' imprisonment.[147]

The offence is somewhat controversial,[148] not least because there is concern as to whether this requirement infringes the privilege against self-incrimination, something that is protected both by the common law and Article 6 of the ECHR.[149] In *R v S(F)*[150] the Court of Appeal was called upon to specifically consider whether the

144 s.49, RIPA 2000.

145 s.53, RIPA 2000.

146 s.53(5A)(a), RIPA 2000.

147 s.53(5A)(b), RIPA 2000.

148 Chatterjee 2011.

149 Choo 2014.

150 [2009] 1 WLR 1489.

encryption rules breaches the privilege against self-incrimination. The court noted that the privilege was not absolute, either in European or domestic contexts, and could be subject to limitations where they are in the public interest. The Lord Chief Justice,[151] giving the judgment of the court, held that the key was separate to the will of the individual because it does exist distinctly from the mind: the police could identify the key, it is just very difficult to do so. Whilst this is technically correct it fails to recognise that it is *de facto* a product of the mind since without the defendant testifying it is highly unlikely that the key would be identified.[152]

In *Greater Manchester Police v Andrews*[153] the High Court was perhaps more realistic. The offender was suspected of using encryption to hide child abuse images from the police. The court held that forcing the offender to disclose the key did trigger the privilege against self-incrimination but they held that it was engaged only to a limited extent and that where there was unfairness this could be remedied through the provisions of s.78, PACE 1984 (the general power a judge has to exclude prosecution evidence). Whether, in reality, this means anything is perhaps more open to question. Section 78 normally guards against improper gathering and unfair evidence, indeed an informal paraphrasing of it is that it operates where the prejudicial effect of the evidence outweighs its probative value. It is not clear that in cases of encryption that—save where it is truly unfair—that there is any prejudicial effect. More likely, is the courts will consider that the probative value of the indecent photographs (and the forensic evidence surrounding that) would be such that the privilege could be qualified.

The rules on encryption are undoubtedly controversial but they are undoubtedly of assistance to law enforcement and, in particular, where an offender seeks to hide their criminal actions behind encryption.

7.3.4 Interception of Content Data (Article 21)

Communication data has so far been discussed but it will be remembered that this did not contain the 'what' of the communication. If the content of data is to be identified then this requires an interception. This was previously dealt with by the *Regulation of Investigatory Powers Act 2000* but, in common with communication data, has now been superseded by the IPA 2016.

An interception is considered to be extremely serious, and an unauthorised interception amounts to a criminal offence, punishable by up to two years' imprisonment.[154] An authorised interception is most commonly through a warrant. Section 15, IPA 2016 creates a warrant that, inter alia, allows for the 'interception,

151 The most senior judge in England and Wales.

152 For a further discussion on this, see Mason and Gillespie 2017.

153 [2011] EWHC 1966 (Admin).

154 Section 3, IPA 2016.

in the course of their transmission by means of a… telecommunication system, of communications described in the warrant'.[155]

A warrant must be applied for by the chief constable of a police service,[156] or the Director-General of the National Crime Agency.[157] Whilst obviously such a person will not be the investigator, it demonstrates the seriousness of warrants. Whilst there will be internal scrutiny prior to the point at which it will be laid before the chief constable, he or she will carefully scrutinise the application and probably ask questions of the (internal) applicant before signing the application. The decision-maker for a warrant is the Home Secretary,[158] and again this is something that would ordinarily be done personally by the Secretary of State, with it being delegated to a junior minister only at times when she is unavailable. A warrant requires the approval of a judicial commissioner,[159] and again this is a scrutinised application. The commissioners operate a 'duty' system so that for urgent matters there will be a designated judge who will act as the on-call commissioner, and who can consider the matter through secure telephony and/or secure messaging. The judicial commissioner must apply the same principles as judicial review,[160] which is an interesting phrase. Presumably this means that the judge should consider whether the application is lawful, has been made in a procedurally proper manner and is not one which is perverse or objectively unreasonable.[161]

It is theoretically possible to make a warrant without judicial approval in urgent cases.[162] In such cases, the judicial commissioner must be notified as soon as practicable thereafter, including identifying why there was not time to contact the judicial commissioner. Where, in such a case, a judicial commissioner disagrees with the granting of a warrant, he can order the destruction of some or all of the data gathered, or restrict its use.[163] Given the use of technology, it will be extremely rare that a warrant would be given without judicial authority, and it would ordinarily be a matter of life or death. It is unlikely that such a circumstance could arise in respect of Sweetie.

[155] Section 15(2)(a), IPA 2016.

[156] England and Wales has 43 police forces, each of which is headed by a(n) (independent) chief constable.

[157] Section 18(1), IPA 2016.

[158] Section 19, IPA 2016.

[159] Section 23, IPA 2016.

[160] Section 23(2), IPA 2016.

[161] For a summary of judicial review grounds, see Barnett 2017, p. 615.

[162] Section 24, IPA 2016.

[163] Section 25(3), IPA 2016.

7.4 The Protection of Privacy

Unusually for an Act of Parliament, and perhaps a reflection of the controversy that surrounded the Act, the IPA 2016 has specific reference to the protection of rights, in particular privacy. Section 1(5) states that protection can be found in the oversight provisions of IPA 2016 and:

(a) by virtue of the Human Rights Act 1998,
(b) in Section 55 of the Data Protection Act 1998.
(c) in Section 48 of the Wireless Telegraphy Act 2006
(d) in Sections 1 to 3A of the Computer Misuse Act 1990
(e) in the common law offence of misconduct in public office, and
(f) elsewhere in the law.[164]

The legislation mentioned in (ii) to (iv) encompass a number of criminal offences that relate to inappropriate access or disclosure of data or technology. Paragraphs (i) and (vi) echo the principal protection for human rights, that of the common law and the *Human Rights Act 1998*.

The first is through the common law. One of the essential features of the common law is that it is designed to ensure that trials and processes around the investigation of crime are fair. The second, and arguably more important in recent years, is through the *Human Rights Act 1998* (HRA 1998). This Act is sometimes misleadingly said to have incorporated the *European Convention on Human Rights* into domestic law. This is not quite true for if this were true it would apply to private disputes when it does not. What the Act does is to place an obligation on public bodies—including the police and courts—to act in a way compatible with the ECHR.[165] In the context of Sweetie this means ensuring that any investigation does not infringe either Articles 6 or 8 of the ECHR. A breach of the ECHR by a public authority can be used either as a defence in legal proceedings[166] or be used as a cause of action against a public authority.[167] That said, there are restrictions on litigation when surveillance powers are involved, and that is something that will be discussed further below.

RIPA 2000 and subsequent legislation has been passed as a direct consequence of human rights, particularly the Human Rights Act 1998. Whilst covert surveillance, and indeed interceptions of communication, has occurred for many years, the common law adopts a different approach to that of the *European Court of Human Rights* (ECtHR). The common law operates on the basis that there is a presumption of legality unless the contrary is shown. Thus as there was no instrument stating that a police officer could not conduct covert surveillance, it was lawful. The ECHR operates on a different premise. It is clear that for the purposes of Article 8, there

164 Section 1(5)(b), IPA 2016.

165 s.6, *Human Rights Act 1998*.

166 s.7(1)(b), *Human Rights Act 1998*.

167 s.7(1)(a), *Human Rights Act 1998*.

must be a legal basis upon which one can act.[168] Whilst the ECtHR has accepted the common law in respect of some issues, there must be a clear rule that allows for everyone to appreciate the extent of the law. This is generally lacking in respect of surveillance, particularly involving communication technologies, in part because surveillance is, by its very nature, covert.

RIPA 2000 and now the IPA 2016 provides the legality for the purposes of a justification under Article 8(2). Both RIPA 2000 and IPA 2016 refer to the requirement to act in pursuit of a legitimate aim, in a way that is necessary and proportionate, thus satisfying the other aspects of Article 8(2).

7.4.1 Investigatory Powers Commissioners

It will be remembered that several provisions of the IPA 2016, most notably s.87 that permits the retention of communications data, mentions the Investigatory Powers Commissioner (IP Commissioner) and judicial commissioners. These are creatures of the IPA 2016, although it mirrors an earlier approach that was adopted in the *Regulation of Investigatory Powers Act 2000*. Section 227, IPA 2016 creates two offices: the IP Commissioner and judicial commissioners. The IP Commissioner is the senior of all the judicial commissioners and must ‘keep under review (including by way of audit, inspection and investigation) the exercise by public authorities [of their powers]’.[169] The reference to audit, inspection and investigation demonstrates that the Commissioner does not only consider applications, but will also visit relevant public authorities (with a team of inspectors) to audit how they are exercising their powers, including how they are recording matters. A good example of this will be requests for disclosure of communications data. It will be remembered that where the police make this request there is no need for judicial authorisation. However, the power is subject to the scrutiny of the IP Commissioner and his inspectors (usually accompanied by a Commissioner) will perform an audit on checks to ensure the powers were used appropriately.

The IP Commissioner will publish an annual report, which will be addressed to the Prime Minister,[170] which will contain opinions on how the powers are being used, including statistical data and qualitative statements about their use. The report must be published and laid before Parliament.[171]

The IP Commissioner, and other judicial commissioners, must be either serving or former senior members of the judiciary.[172] For our purposes, this means a judge of at least the rank of high court judge. The first (and current) IP Commissioner is

---

168 *Malone v UK* (1991) 13 EHRR 448.

169 s.229(1), IPA 2016.

170 s.234, IPA 2016.

171 s.234(6), IPA 2016.

172 s.227(2), IPA 2016.

the Rt. Hon. Lord Justice Fulford, a serving Lord Justice of Appeal, which is a judge of the Court of Appeal. The Commissioners can only be appointed if they are recommended jointly by:

(a) the Lord Chancellor,
(b) the Lord Chief Justice of England and Wales,
(c) the Lord President of the Court of Session,
(d) the Lord Chief Justice of Northern Ireland.[173]

The Lord Chancellor is also the minister of justice for England and Wales and has special responsibility to uphold the independence of the judiciary.[174] Whilst the ministers of justice in Scotland and Northern Ireland are not included, the Scottish ministers must be consulted when making the recommendation.[175] The judges listed in (a)–(d) are the most senior judges in each of the three judicial systems in England and Wales. Where it is a judicial commissioner being appointed, the IP Commissioner must also agree.

A commissioner will hold office for three years and can only be dismissed in two ways:

(a) By an address of both Houses of Parliament.[176] This is the equivalent of an Act of Parliament and is the same certainty of office as a superior judge of England and Wales holds.[177]
(b) Where the commissioner is declared bankrupt, disqualified from working as a director or other financial order, or is convicted in the United Kingdom of an offence for which he has is sentenced to a term of imprisonment (suspended or otherwise).

The manner in which the commissioners are appointed and hold office are designed to ensure that they are treated the same as the judiciary and should therefore be classed as independent. Of course, there will be some who will question whether the judiciary are part of the establishment and therefore whether this is the establishment picking people to ensure that other parts of the establishment can conduct surveillance on its citizens. However, the reality is that it is impossible to guarantee judicial independence in a way that theoretically eradicates all influence.[178] The commissioners act in a judicial capacity, have security of independence and are appointed with the consent of the judiciary. Therefore they are as independent as they can realistically be for these purposes.

---

173 s.227(4), IPA 2016.

174 s.3, *Constitutional Reform Act 2005*.

175 s.227(5), IPA 2016.

176 s.228(4), IPA 2016.

177 See Gillespie and Weare 2017, p 298.

178 On this see Bingham 2011, a former Lord Chief Justice of England and Wales, who discusses the concept of judicial independence.

Similar commissioners exist to regulate other forms of surveillance (for example, the 'interception of communications commissioner'[179]). Their duties are set out in the *Regulation of Investigatory Powers Act 2000* and they discharge very similar roles. Again, they hold judicial office and are designed to ensure that public authorities are held to account for the powers they exercise.[180]

7.4.2 Investigatory Powers Tribunal

Cases in England and Wales are usually heard in public and so there is a reluctance to discuss covert techniques in open court because it could allow criminals to understand their extent and limitations. Within criminal trials this is largely unproblematic. In most cases the evidence is obvious and there is no need to discuss the source in open court: the discussion being limited as to whether it is admissible. However in civil trials it is more likely that the source or technique would be relevant. A civil court will be asked to consider whether the right to respect for private life has been infringed by the police conducting surveillance, including through the interception of communications or the acquisition of communications data.

It was thought that, for the reasons noted above, it would not be appropriate to litigate surveillance powers in the ordinary civil courts. Thus RIPA 2000 included an exclusionary power to prevent actions under the HRA 1998 concerning investigative powers to be litigated in the ordinary courts.[181] Instead it created the *Investigatory Powers Tribunal*[182] which has the same standing and powers of the High Court.[183] The Tribunal will usually sit in secret and (controversially) it could choose not to hear both sides of the case together.[184] The argument in support of this is that it allows the tribunal to ask law enforcement (or the security services) detailed questions about the techniques or case without any danger that the offender may identify the strength of the case against them. There are currently 10 members of the Tribunal, including four serving judges,[185] with the remaining six members being of Queen's Counsel[186] or a retired high court judge. Clearly therefore the

179 s.57, RIPA 2000.

180 See McKay 2015 for an overview.

181 s.65(2)(a), RIPA 2000.

182 s.65(1), RIPA 2000.

183 Other than it does not have any inherent jurisdiction: cf the powers of the High Court of Justice. In essence, this means it has no common-law powers as its jurisdiction is conferred upon it by statute.

184 See Reg 9, *Investigatory Powers Tribunal Rules 2000* (SI 2000/2665).

185 Three of High Court rank and one of circuit rank.

186 This is the most senior of counsel in England and Wales.

Table 7.6 Determination of cases before the IPT

	2010	2011	2012	2013	2014	2015
Cases decided	210	194	191	161	201	219
Cases where complaint upheld	6	2	0	0	0	8

[*Source* The author]

tribunal sees itself as being the equivalent of a court. The jurisdiction of the IPT was extended to encompass powers within IPA 2016.[187]

Whilst the government intended that its activities would be secret, the tribunal itself has said that it will try to be open when it can. However, the reality is that most cases are going to be heard in private. The latest available figures suggest that in 2015, the tribunal sat in open session on only 15 occasions.[188] In that year, the tribunal dealt with 219 cases, demonstrating that closed hearings are certainly the norm.

Restricting public access to the tribunal is perhaps understandable, albeit controversial, but of more concern is the success rate. The principal statistics available are for the time period 2010–2015 (see Table 7.6).

This is a remarkable statistic. Out of 1,178 cases, only 12 were upheld (1.4%). The Tribunal notes that it can only adjudicate those cases before it, which is correct but it seems remarkable that in almost 99% of cases there has been absolutely no breaches of the rules. Either the police and security services are extremely adapt at dealing with investigatory powers, and legislation has provided adequate safeguards, or there is a problem with the tribunal. Certainly a 'success' rate of approximately 1% is somewhat suspicious but it is difficult to identify the reasons for this.[189]

7.5 Entrapment

Theoretically this should be a short section. Under English law there is no defence of entrapment.[190] Therefore, should this not be the end of the matter? No. Whilst entrapment is not a defence, it has been recognised by the courts that entrapment can be relevant in two circumstances; the admissibility of evidence and whether a prosecution amounts to an abuse of process, i.e. the police and/or prosecutors have acted in such a way as to bring the criminal justice system into disrepute. If they have, then the prosecution should be stopped.

---

[187] s.243, IPA 2016.

[188] http://www.ipt-uk.com/content.asp?id=30. Accessed 31 August 2017.

[189] Of course one difficulty is that if the surveillance is done correctly, the person who was the target will not know that they were under surveillance and cannot therefore complain about it.

[190] *R v Sang* [1980] AC 402 and *R v Looseley* [2001] UKHL 53.

The first aspect—exclusion of evidence—arises from the general rule that a criminal trial should be fair. English law provides a judge with a general exclusionary rule for any prosecution evidence where the prejudicial effect of it overweighs its probative value.[191] It is clear that a judge can take all of the circumstances of a case into concern and, since the introduction of the HRA 1998, this includes considering whether the exclusion of evidence is necessary to ensure that an offender has a fair trial. Section 78 is not used to mark disapproval but rather to deal with prejudicial evidence and this could include situations where it was thought that an individual had been led into committing a crime rather than being allowed to commit it.

The second, and arguably more common in the context of entrapment is abuse of process. This is a common-law doctrine that allows the courts to stay a prosecution due to the improper actions of the police, prosecutors or indeed someone else. This can, but need not, serve as a mark of disapproval to conduct.[192]

What is the test for entrapment? The leading authority is *R v. Looseley*[193] where it was said that the essence of the test could be summarised thus:

> On this a useful guide is to consider whether the police did no more than present the defendant with an unexceptional opportunity to commit a crime. I emphasise the word 'unexceptional'. The yardstick for the purpose of this test is, in general, whether the police conduct preceding the commission of the offence was no more than might have been expected from others in the circumstances.[194]

Whilst the test is perhaps somewhat oblique, it is well-understood within England and Wales and it means that there is no barrier in, for example, pretending to be a child. This by itself would not be considered entrapment so long as the police officer does not go beyond what a child would ordinarily do. It is providing an opportunity to the offender to commit an offence and thus is a legitimate law enforcement tactic. A good example of this in context is *R v. Jones*[195] where the defendant wrote graffiti on a train and toilets intimating that he was seeking girls aged 8 to 13 for sex. A journalist noticed the message and tried to get in contact, posing as a child. However she quickly contacted the police and a police officer posed as 12-year-old girl and exchanged messages with the defendant. The defendant sent a series of messages that made clear that he wanted her to perform sexual acts on himself (and herself) and ultimately arranged a meeting. When he turned up, he was arrested and charged with attempting to cause or incite a child under the age of 13 to engage in sexual activity. The defendant claimed entrapment but the judge rejected this, with the Court of Appeal upholding this ruling, making it clear that the police had merely provided an opportunity for the defendant to commit an offence. There had been no pressure or inducement to do so, the

[191] Section 78, *Police and Criminal Evidence Act 1984*.

[192] See the comments of Lord Nicholls in *R v Looseley* [2001] UKHL 53 at [17].

[193] [2001] UKHL 53.

[194] *Ibid.*, at [23] per Lord Nicholls of Birkenhead.

[195] [2007] EWCA Crim 1118.

defendant had freely communicated with 'the child' and had suggested sexual activity.[196]

It should be noted that the rules relating to entrapment are not just based on police activity but also to private citizens.[197] That said, the courts have tended to be slower to accept arguments about abuse of process, in part because the decision in *Looseley* was premised on the basis of the state forcing someone to commit a crime and where the police are not involved, there is arguably doubt as to where the state pressure has come from. However this is perhaps not strictly appropriate as the abuse of process doctrine is supposed to protect people from improper prosecutions and it should not matter where the pressure comes from; if a person is coerced into committing a crime they would not ordinarily commit that is unfair. That said, there is some evidence this is beginning to change,[198] perhaps because of concerns that citizens are trying to trap people who think they are going to meet with a child. The evidential integrity of these investigations is often weaker than a police inquiry and thus the courts are beginning to question whether the defence needs to be extended to take account of those who are leading people into disclosing potential criminal actions.

It is unlikely that the laws in England and Wales will cause any problems for Sweetie. Assuming that Sweetie responds as a child does, and allows a person to approach them and to suggest the sexual activity, rather than coercing them into doing so, then entrapment is unlikely to be an issue. That said, the chain of evidence will be important and thus it is important that Sweetie records the whole conversation in a transparent way, and one that demonstrates the actions of both Sweetie, those who control Sweetie and the potential defendant.

7.6 Conclusion

Sweetie is a challenging concept in some countries but it should not prove challenging in England and Wales. The English laws on sexual offences are constantly under review to ensure that they are fit for contemporary understandings of online abuse. If anything, English law can be criticised for being over-broad: there are multiple offences that could deal with the same conduct in different ways. However, the government has been careful to ensure that online abuse can be tackled. The law also appreciates the importance of proactive operations in this context and this means allowing criminalisation even where there is no real child.

The substantive criminal law under the *Sexual Offences Act 2003* allows for the substantive offence to be committed irrespective of whether a real child is involved. So, for example, where the offence is inciting a child to engage in sexual activity,[199]

---

[196] *Ibid*., at [23].

[197] See the comments of the Court of Appeal in *R v Shannon* [2001] 1 WLR 51.

[198] Dyer 2015.

[199] s.9, SOA 2003.

it is irrelevant whether there is a real child so long as the defendant *believes* that he is communicating with a child. Where the substantive offence cannot take place without a real child, the law of attempts ensures that liability remains. The *Criminal Attempts Act 1981* is a powerful and well-developed piece of legislation that has been used successfully in respect of child abuse.

In terms of procedure, the police have sufficient powers to identify offenders, including through the use of communications data. The police in England and Wales are now used to dealing with international operations and identifying offenders through traffic data. Internet Service and Internet Content Providers have developed a good relationship with law enforcement. Whilst the providers will wish to ensure that the appropriate authorisations are in place, when required to provide the data they are used to doing so and are able to act quickly in situations where this is important. Once the suspect has been identified then the police have sufficient powers to seize relevant devices and have the expertise to forensically examine them.

English law does not recognise a defence of entrapment *per se*. Instead it allows evidence to be excluded or prosecutions to be stopped for misconduct. The definition of entrapment is clear and is clearly based on impropriety. Where a person—be that a law enforcement agent or a private individual—acts in a way similar to how a real person (in this case, a child) would react then it is unlikely that a defendant would be considered to have been entrapped.

In conclusion therefore, English law should be able to react well to Sweetie 2.0 and it will be interesting to see the extent to which this is carried forward and cases are identified and prosecuted.

References

Akdeniz Y (2007) Possession and dispossession: a critical assessment of defences in possession of indecent photographs of children cases. Criminal Law Review 274–288

Arcarazo D A, Murphy C (2014) EU security and justice law after Lisbon and Stockholm. Hart Publishing, Oxford

Barnett H (2017) Constitutional and Administrative Law, 12th edn. Routledge, Oxford

Barrett D (1997) Child Prostitution in Britain: Dilemmas and Practical Responses. The Children's Society, London

Bingham T (2011) The business of judging: Selected essays and speeches. Oxford University Press, Oxford

Chatterjee B (2011) New but not improved: a critical examination of revisions of the Regulation of Investigatory Powers Act 2000 encryption provisions. International Journal of Law and Information Technology (19): 264–284

Choo A L T (2014) The privilege against self-incrimination and criminal justice. Hart Publishing, Oxford

Dyer A (2015) The problem of media entrapment. Criminal Law Review 311–331

Edwards S S M (1998) The contemporary application of the Obscene Publications Act 1959. Criminal Law Review 843–853

Fredette K (2009) International legislative efforts to combat child sex tourism: evaluating the Council of Europe Convention on Commercial Child Sexual Exploitation. Boston College International and Comparative Law Review (32): 1–43
Gillespie A A (2004) Tinkering with child pornography. Criminal Law Review 361–368
Gillespie A A (2011) Child pornography: law and policy. Routledge, London
Gillespie AA (2012) Jurisdictional issues concerning child pornography. International Journal of Law and Information Technology (20): 151–177
Gillespie A A, Weare S (2017) The English legal system, 6th edn. Oxford University Press, Oxford
Mason S, Gillespie A A (2017) Encrypted data. In: Mason S (ed) Electronic evidence. Society of Advanced Legal Studies, London, pp 261–284
McKay S (2015) Covert policing: law and practice, 2nd edn. Oxford University Press, Oxford
Ost S (2009) Child pornography and sexual grooming: legal and societal responses. Cambridge University Press, Cambridge
Ost S (2010) Criminalising fabricated images of child pornography: a matter of harm or morality? Legal Studies (30): 230–256
Svensson N L (2006) Extraterritorial accountability: an assessment of the effectiveness of child sex tourism laws. Loyola of Los Angeles International and Comparative Law Review (28): 641–664
Williams G (1965) Venue and the ambit of the criminal law (part 3). Law Quarterly Review (81): 518–530

Alisdair A. Gillespie is Professor of Criminal Law and Justice, and Head of the Law School, at Lancaster University. After qualifying as a barrister (Middle Temple), Alisdair returned to academia and has taught at the Universities of Durham, Teesside, De Montfort and Lancaster. His research expertise is in cybercrime and, in particular, the law relating to child sexual abuse and exploitation when facilitated by information and communication technologies. Alisdair has worked with numerous NGOs, public bodies, the police and the Crown Prosecution Service. He has also undertaken work for the EU, Council of Europe, G8 and the United Nations. He has published six books and over 50 articles in leading law journals.

Chapter 8
Substantive and Procedural Legislation in Estonia to Combat Webcam-Related Child Sexual Abuse

Kaspar Kala

Contents

Abstract Estonia is a civil law country influenced by the German legal doctrine. The virtual character of Sweetie (a neutral digital avatar of a 10-year-old girl) poses several challenges for Estonian material and procedural criminal law if used in

K. Kala (✉)
Proud Engineers, Tartu, Estonia
e-mail: kasparkala@ut.ee

S. van der Hof et al. (eds.), *Sweetie 2.0*, Information Technology and Law Series 31,
https://doi.org/10.1007/978-94-6265-288-0_8

practice. This chapter answers the question whether the use of Sweetie would be permitted under current material and criminal procedural law of Estonia. Under material criminal law, the act of requesting access to a webcam session with Sweetie is not criminalized, as Sweetie would not be considered 'child pornography'. However, in the context of sexual enticement, law enforcement bodies have used police agents depicted as children and courts have declared such acts punishable as impossible attempts. Theoretically, enticing a digital avatar could be punishable under the same logic. In the case of making arrangements for meeting a child for sexual purposes, the law implies that the victim must be a 'person'. Considering that this criminal act is a delict of abstract danger and that Sweetie is an ultra-real digital persona, one could argue that trying to meet with Sweetie for sexual purposes and making a preparatory act for this purpose also could be punishable as an impossible attempt under Estonian law. Although material criminal law provides room for interpretation, criminal procedure law sets limits to the use of a digital avatar for the apprehension of digital offenders. For Sweetie to be a tool for catching sex offenders online, the Estonian Code of Criminal Procedure would need to explicitly permit the use of digital avatars as their use could be categorized as a surveillance activity and would need a clear legal base. This is a debate Estonia is still to have.

Keywords Sexual Abuse of a Child · Digital Avatar · Online Criminal Offence · Sexual Self-Determination · Estonian Criminal Law · Surveillance Activity

8.1 Introduction—Legislation in Estonia

8.1.1 General Description of the Legal Framework

The Legal System and Substantive Criminal Law

Estonia is a democratic republic with a civil law tradition. The laws are passed by the unicameral Parliament (*Riigikogu*), consisting of 101 members. Members of the *Riigikogu* are elected in free elections according to the principle of proportional representation. Elections are general, uniform and direct.[1]

Estonian law is heavily influenced by the German legal doctrine. Criminal law is no exception. The criminal acts are codified in the Penal Code ('PC') which was adopted in 6 June 2001.[2] The PC went through reform (2011–2015) and the 'new'

[1] Section 60 of the Constitution of the Republic of Estonia.

[2] The PC entered into force on 1 September 2002. Prior to this a modified Soviet-era Criminal Code from 1961 was in force.

PC entered into force in 1 January 2015.[3] As a result, 247 offences were declared obsolete and 400 criminal and misdemeanour[4] offences went through modification.[5]

Most Important Principles of Criminal Law

The most important principles of criminal law in Estonia are: (i) no punishment without law (*nullum crimen, nulla poena, sine lege scripta, stricta, praevia,* (*'nulla poena sine lege'*)); (ii) no retroactive effect of a law exacerbating the situation of a person; (iii) *ne bis in idem*; (c) no analogy; (iv) *nulla poena sine culpa*; and (v) *ultima ratio.*

Nulla poena sine lege principle is set out in the Constitution of the Republic of Estonia[6] ('Constitution') and in the PC.[7] In the case law of the Supreme Court of Estonia ('Supreme Court'), a principle of the *exact definition* of the offence is derived from the *nulla poena sine lege* principle. This implies that the law setting out the punishment for an act must be clearly defined in order for every person to foresee what kind of acts are prohibited and criminalised in the society and what kind of punishments are set. This enables a person to adjust its behaviour accordingly.[8] The same principle obliges the state to describe offences in the law in a clear and unambiguous manner.[9]

The principle of *no retroactive effect of a law exacerbating the situation of a person* is set out in Subsection 23(2) of the Constitution and it stipulates that no one may be sentenced to a penalty that is more severe than the one that was applicable at the time the offence was committed. If, subsequent to the commission of the offence, the law makes provision for a lighter penalty, the lighter penalty applies.[10]

The *ne bis in idem* principle means that no-one may be prosecuted or sentenced for a second time for an act in respect of which he or she has been the subject of a final conviction or acquittal pursuant to the law.[11] According to settled case law of the European Court of Human Rights and the Supreme Court, the principle of *ne bis*

[3] The reason for reform was over criminalization. PC was reformed due to over-criminalization. In 2010, the Chief Justice of the Supreme Court of Estonia brought out the fact that over-criminalization has led to the situation where 55.6% of the labour force aged between 15 and 74 have a registered criminal or misdemeanour offence in the criminal records database and the *ultima ratio* principle of criminal justice has, therefore, failed. Accessed 24 February 2016.

[4] A misdemeanour is an offence which is provided for in the Penal Code or in another legal act (criminal acts are only in the PC) and the principal punishment prescribed for which is a fine (up to 1200 euros), detention (up to 30 days) or deprivation of driving privileges (up to two years).

[5] Siitam-Nyiri 2014, p. 578.

[6] Subsection 23(1) of the Constitution stipulates that 'No one may be convicted of an act which did not constitute a criminal offence under the law in force at the time the act was committed'.

[7] Subsection 2(1) of the PC.

[8] Judgment of the Supreme Court of 23 November 2012, case number 3-1-1-103-12, para 9.

[9] Judgment of the Supreme Court of 5 October, case number 3-1-1-66-12, para 13.

[10] This principle is further elaborated in Subsections 5(2) and 5(3) of the PC.

[11] The principle of *ne bis in idem* is set out in Subsection 23(3) of the Constitution and Subsection 2(3) of the PC.

in idem also prohibits carrying out a second procedure if a case concerns identical or basically identical factual circumstances (ECtHR *Zolotukhin v. Russia*).[12]

The *no analogy* principle means no act may be declared to be an offence by analogy in law.[13]

The *nulla poena sine culpa* principle guarantees the respect to human dignity[14] and is set out in Subsection 32(1) of the PC. The principle of guilt means that a person shall be punished for an unlawful act only if the person is guilty of the commissioning of the act.

The *ultima ratio* principle is not, however, set out in black and white in the PC.[15] Nevertheless, it is a principle distinguished in the case law of the Supreme Court. The Supreme Court has highlighted that under of law, punishing must be an *ultima ratio* measure and the state may exercise punitive measures only when other less intrusive measures do not protect the legal rights sufficiently.[16]

Regulation Concerning Age

A 'minor' is any person less than 18 years old. However, the PC makes further differentiations of age. A person less than 14 year of age is considered a 'child'.[17] In Estonia, fourteen is considered the 'sexual self-determination age'.[18] This implies that a child less than 14 years of age is not considered as mature to understand the meaning of sexual intercourse with an adult and the concomitant consequences. It must be noted that the sexual self-determination limit does not criminalise consensual sexual intercourse between 14 and 17-year olds and consensual sexual intercourse between children from 10 to 17. The consent of a 11 to 13-year-old to engage in sexual intercourse with an adult is not legally valid.[19]

Section 147 of the PC provides that a child less than 10 years old is considered to be 'a person incapable of comprehending' under the division of offences against sexual self-determination.[20] This implies that sexual intercourse with a child less than 10 years old is always considered rape in Estonia.[21]

---

12 Accessed 30 December 2015.

13 Subsection 2(4) of the PC.

14 Rosin 2016, p. 660.

15 It is, however, set out in the CCP in the section about surveillance activities. See '*The Code of Criminal Procedure*' in 8.3.3.

16 Judgment of the Supreme Court of 3 December 2014, case number 3-1-1-81-14, para 8.

17 See Section 145 of the PC.

18 Tamm and Ploom 2011, p. 20.

19 Ibid., p. 4.

20 Division 7 of the PC, Section 147.

21 See '*Rape*' below.

Attempt

Attempt is criminalised. According to the PC, attempt is an intentional act which is aimed to commit an offence and an attempt is deemed to have commenced from the moment the perpetrator, according to his or her own understanding of the criminal act, directly starts with the commission of the offence.[22] In case of attempt, the court may reduce the punishment by a third of the maximum.[23]

The PC also recognises 'impossible attempt' which is an attempt to commit a crime which cannot be completed due to the unsuitability of the object or subject of the offence or due to the unsuitability of the object or method used to commit the offence.[24] In other words, impossible attempt is an attempt which cannot be objectively completed with the chosen object (e.g. trying to kill somebody with Vitamin-C), method or by the subject. Impossible attempt is also punishable. However, if the person does not understand that the attempt is impossible due to his or her mental infirmity, the court may release a person from punishment or reduce the punishment by a third.[25]

To exemplify, in the context of the criminal act of enticement of children, where the perpetrator according to his knowledge entices a child (e.g. talks about sexual topics and asks the child to undress and touch his or her genitals),[26] but the 'child' is in fact an undercover police agent, the perpetrator has committed an impossible attempt as the actual enticement could not be completed, as the object of the act is not in fact a child.[27] However, according to the perpetrator's best knowledge and will, he was enticing a child.[28]

Criminal Procedure

Estonian criminal procedure is regulated in the Code of Criminal Procedure ('CCP') which has its roots in the German law doctrine.

The most important principles of criminal procedure are the principle of: (i) legality (Section 6 of the CCP); (ii) principle of state jurisdiction (Section 5 of the CCP) in regard to the principles of commencement of criminal procedure. In regard to principles on progression of the criminal procedure, the following principles need highlighting: (i) fair trial;[29] (ii) principle of no self-incrimination (Subsection 22(2)

---

[22] Subsections 25(1) and 25(2) of the PC.

[23] Subsection 25(6) of the PC.

[24] Subsection 26(1) of the PC. The best example here is that one cannot kill a person who is already dead. Therefore, if a person attempts such an offence, it is considered punishable as an impossible attempt because the intent was to kill a person.

[25] Subsection 26(2) of the PC.

[26] Facts taken from Judgment of Harju County Court of 5 August 2013, case number 1-13-6110.

[27] See Judgment of the Harju County Court of 20 January 2015, case number 1-15-82.

[28] See case law in '*Sexual enticement of children*' and '*Entrapment*'.

[29] An identical right to the principle of fair trial as set out in Article 6 of the ECHR is not set out in the Constitution or in the CCP, however, the rights comprising the righto of fair trial as the ECHR describes it are all found in different legal acts: (i) the right to impartial courts (Section 146 of the Constitution); (ii) public announcement of judgments (Subsection 24(4) of the Constitution); (iii) presumption of innocence (Subsection 22(1) of the Constitution and Subsection 7(1) of the

of the Constitution and 7(2) of the CCP); (iii) the principle of *in dubio pro reo* (Subsection 7(3) of the CCP); (iv) right of silence (*nemo tenetur se ipsum accusare;* Subsection 22(3) of the Constitution and Subsection 34(1) and Section 71 of the Code of Criminal Proceedings); (v) *ne bis in idem*;[30] (vi) right to be present at trial (Subsection 24(2) of the Constitution); (vii) right to appeal (Subsection 24(5) of the Constitution); (viii) right to impartial courts (Section 156 of the Constitution); (ix) principle of adversarial court procedure (Subsection 14(1) of the CCP); (x) safeguarding of personal liberty and respect for human dignity (Section 9 of CCP); (xi) principle of direct hearing (Section 15 of the CCP); and (xii) principle of free assessment of evidence (Subsection 61(1) of the CCP). The organisation of criminal procedure is based on the following principles: (i) publicity of the trial (Subsection 24(3) of the Constitution); and (ii) the principle of national language (Subsection 24(4) of the Constitution).

8.1.2 Relevant Treaties and Cybercrime Laws

Estonia has ratified the UN Convention on the Rights of the Child (on 20 November 1991), the Optional Protocol on the sale of children, child prostitution and child pornography (on 3 September 2004) and the Cybercrime Convention (on 1 July 2007). Also, Estonia has ratified the Lanzarote Convention (on 1 November 2016).

In regard to EU law in the field relevant for this report, Estonia has transposed the rules of Directive 2011/93/EU on combating the sexual abuse and sexual exploitation of children and child pornography[31] into the PC.[32]

At a national level, all criminal offences in regard to sexual abuse of minors are codified into the PC.

CCP); (iv) right to receive information in a language understandable to the accused (Subsection 9 (2) of the CCP); (v) right to receive adequate amount of time and facilities for the preparation of his/her defence and the right to receive legal assistance (set out in Section 8 of the CCP); and (vi) the right of the assistance of an interpreter is set out in Subsections 10(2), 10(21) and 34(1)(21) of the CCP.

[30] Subsection 199(1)(5) of the CCP states that criminal proceedings shall not be commenced if a decision or a ruling on termination of criminal proceedings has entered into force in respect of a person in the same charges.

[31] OJ L 335 of 17.12.2011.

[32] The changes to the PC are the most relevant taking into account the scope of the report. In total, the requirement to impose the Directive 2011/93/EU brought about the need to change twelve legal acts.

8.2 Analysis of Substantive Criminal Law

Crimes involving the sexual abuse of children are all found under different chapters of the PC.[33] The PC criminalises the following acts against minors which will be further analysed in Sect. 8.2.1.

8.2.1 *Relevant Criminal Offences*

Human Trafficking in Order to Take Advantage of Minors[34]

Human trafficking is an offence against the liberty of a person. The PC criminalises the act of human trafficking in order to take advantage of a minor (i.e. persons less than 18 years old). Subsection 175(1) of the PC criminalises the act of influencing (e.g. by way of persuasion or offering money) a minor to (i) commence or continue commission of a criminal offence; (ii) beg; (iii) engage in prostitution; (iv) work under unusual conditions; (v) marry against his or her will; or to (vi) appear as a model or actor in the manufacture of a pornographic or erotic performance or work. Such an act is punishable by two to ten years' imprisonment.

If the perpetrator uses force (i.e. violence, deceit, threats to cause damage to the minor or another person) to commit the acts set out in (i) to (v) above, the crime is punishable from three to fifteen years' imprisonment.[35]

According to the Act to Regulate Dissemination of Works which Contain Pornography or Promote Violence or Cruelty, 'pornography' means a manner of representation in which sexual acts are brought to the foreground in a vulgar and intrusive manner and other human relations are disregarded or relegated to the background.[36] What constitutes an 'erotic' performance is not defined in the law. The Supreme Court has stated that deciding whether a work is 'erotic' or 'pornographic' is a legal question and it is in the competence of the court to decide.[37]

'Pornographic or erotic work' could be, for example, a drawing, a painting, a sculpture, a photo, a phonogram, a movie, a play or a dance performance. Also, the work could be on a paper carrier or in the digital form.[38]

[33] Offences with sexual nature against minors are set out in the chapter of Offences against Liberty (chapter 9 division 6 of the PC), chapter of Offences against Sexual Self-determination (chapter 9 division 7 of the PC) and under the specific chapter of Offences Against Minors (chapter 11 division 2 of the PC).

[34] Section 175 of the PC.

[35] Subsection 133(2)(2) of the PC.

[36] Subsection 1(2)(3) of the Act to Regulate Dissemination of Works which Contain Pornography or Promote Violence or Cruelty.

[37] Judgment of the Supreme Court of 16 March 2006, case number 3-1-1-146-05, para 8.

[38] Pikamäe and Sootak 2015, p. 469.

A 'performance' is a work that is intended to be monitored live, e.g. striptease or a play including a sexual act.[39]

It must be highlighted that 'aiders' in the context of this offence could be all persons who participate in the creation of a pornographic or erotic work. These could be, but are not limited to, a photographer, a director, an operator, a screenwriter, an artist, the author of the text or an artist.[40] Aiding means also providing any physical, mental or material (e.g. funding) support for committing the crime.

According to official statistics, there were 15 registered cases of infringement of Section 175 of the PC in 2014, 63 in 2015 and 59 in 2016.[41]

Offline Criminal Offences against Sexual Self-Determination

As opposed to online offences, this section provides a list of crimes that are criminalised under the PC and which protect the sexual self-determination of minors.

Rape[42]

Rape is defined in the PC as sexual intercourse or commission of another act of sexual nature with a person against his or her will by using force or taking advantage of a situation in which the person was incapable of initiating resistance or comprehending the situation. Rape is punishable from one to five years' imprisonment.

'Sexual intercourse' means that the genitals of either the perpetrator or the victim are committed in the act. The Supreme Court has found that introducing a plastic bottle into the anus of the victim without the victim's consent and using force did not constitute 'sexual intercourse' as no genitals were committed to the sexual act.[43]

However, this is unproblematic as the offence also criminalises any act of sexual nature committed against the will of the victim. This means any kind of human act, excluding sexual intercourse, which is aimed at bringing forth sexual arousal. In settled case law, this has meant touching the victim's breasts and rubbing the body of the victim without her authorisation and using violence,[44] forcing a minor to massage the offender who was naked during the massage and kissing a child all over the body and putting a hand into the pants of the victim and touching his genitals.[45]

---

[39] Idem, p. 470.

[40] Idem.

[41] Crime in Estonia 2015, Ministry of Justice, p. 56.

[42] Section 141 of the PC.

[43] Judgment of the Supreme Court of 26 November 2011, case number 3-1-1-59-07, para 14. In this case, the Supreme Court did not agree with the Circuit Court that inserting a bottle in the victim's anus constitutes rape and explained that under the PC, at least the following acts constitute 'sexual intercourse' under the rate offence—vaginal intercourse, oral sex, anal sex or inserting an object into the vagina.

[44] Judgment of the Supreme Court of 15 October 2007, case number 3-1-1-25-07, para 1.

[45] Judgment of Tartu County Court of 25 September 2014, case number 1-14-8008.

Rape's distinctive element is that it is committed against the will of the victim.[46] Victim's will could be breached by (i) using force (physical violence or threats);[47] or (ii) by taking advantage of a situation in which the person is not capable of initiating resistance or comprehending the situation.

Rape committed against a minor is an aggravated circumstance[48] and raping a minor is punishable by six to fifteen years' imprisonment.[49] Age is an objective element of the offence which means that the offender must know or at least tacitly accept that the victim is less than 18 years old. If the offender may reasonably assume that the victim is not a minor (e.g. the victim lies about his or her age), the offender may not be taken accountable for aggravated rape.[50]

According to Section 147 of the PC, a child less than 10 years of age is deemed to be a person who is incapable to comprehend. This means that sexual intercourse or an act of sexual nature against a child younger than 10 years old is always considered as rape.

In 2015, there were 161 registered cases of rape. Two thirds of the victims were minors.[51] In 2016, there were nine cases less (152). Similarly, two thirds of the victims were minors.[52]

Compelling a Person to Engage in Sexual Intercourse or Another Act of Sexual Nature[53]

This offence criminalises the sexual intercourse or commission of another act of sexual nature with a person against his or her will by taking advantage of the dependency of the victim on the offender but without using force or outside a situation where the person was not capable of initiating resistance or comprehending the situation (as provided for in rape). The act is punishable by up to three years' imprisonment.

This offence implies that the victim is dependent on the offender and this dependency is exploited[54] to engage in sexual intercourse or commit another act of sexual nature.[55]

---

46 Violence during sex is not rape if it is consensual.

47 Using force must be understood as using violence as described in Sections 120 to 121 of the PC. Section 120 criminalises the act of threatening. A threat to kill, cause health damage or cause significant damage to property, if there is reason to fear the realisation of such threat. Section 121 criminalises the act of physical abuse (causing damage to the health of another person and physical abuse which causes pain). Physical abuse is punishable.

48 Aggravated circumstances are set out in Section 58 of the PC.

49 Subsection 141(2)(1) of the PC.

50 Pikamäe and Sootak 2015, p. 325.

51 Crime in Estonia 2016, p. 45.

52 Crime in Estonia 2015, p. 49.

53 Section 143 of the PC is the equivalent of Article 3(5)(i) alt 2 of the Directive 2011/93/EU.

54 Judgment of the Supreme Court of 18 February 2011, case number 3-1-1-109-10, p. 17.

55 For description for 'sexual intercourse' or 'act of sexual nature' see '*Rape*' under Sect. 8.2.1.

'Dependency' means that the actions of the victim are the results of the influence coming from the offender.[56] Such dependency may be based on a legal relationship —the offender has legal power to influence the victim,[57] for example a parent and a child, an employer and an employee, teacher and a student or a doctor and a patient.[58] Such dependency may also be based on a non-legal relationship, for example where the offender has compromising material about the victim. The Supreme Court has identified a dependency situation where the offender had a video where the child was shown naked and the possibility to disclose this video to the public.[59]

In the case of rape, the will of the victim is broken (either through violence or by exploiting the situation where the victim was unable to resist or comprehend), but under this offence, however, the decision to engage in sexual intercourse or commit an act of sexual nature is made by the victim himself or herself. However, this consent is not deemed legally relevant as the will of the victim is influenced by the offender.[60] The actual will of the victim must be ascertained in the proceedings.

Sexual Intercourse or Other Act of Sexual Nature Using Influence against a Minor[61]

This offence criminalises the engagement in sexual intercourse or commission of another act of sexual nature by an adult with a minor by taking advantage of the dependency of the minor on the offender or with the abuse of influence or confidence but without using force or outside a situation where the minor was not capable of initiating resistance or comprehending the situation (as provided for in the case of rape). The act is punishable by two to eight years' imprisonment.

In regard to 'dependency', see '*Compelling a Person to Engage in Sexual Intercourse or Another Act of Sexual Nature*' (Sect. 8.2.1). However, in the context of this offence, it is difficult to see a major difference between 'influence' and 'dependency'. In the case law, the Supreme Court has defined 'dependency' through the use of influence over the victim. The explanatory memorandum to the draft Act changing the PC set out that an example where the offender exploits 'influence' is where the victim is not dependent of the offender (see examples set out in '*Compelling a Person to Engage in Sexual Intercourse or Another Act of Sexual Nature*' in Sect. 8.2.1), but the offender is an idol or a role model for the victim.[62]

[56] Pikamäe and Sootak 2015, p. 426.

[57] Judgment of the Supreme Court of 18 February 2011, case number 3-1-1-109-10, para 16.2.

[58] Examples by Pikamäe and Sootak 2015, p. 427.

[59] Judgment of the Supreme Court of 26 August 2011, case number 3-1-1-61-11, para 9.

[60] Judgment of the Supreme Court of 18 February 2011, case number 3-1-1-109-10, p. 16.1.

[61] Section 1432 of the PC is the equivalent of Article 3(5)(i) alt 1 of the Directive 2011/93/EU.

[62] Explanatory Memorandum of the Act changing the PC, 469 SE. http://www.riigikogu.ee/tegevus/eelnoud/eelnou/a8261a8f-3708-45bf-bafc-2234bbbf3aa2/. Accessed 6 February 2016.

In regard to 'confidence', the offender must exploit the trust the victim has vested in the offender and by exploiting this, commit the offence.

In 2015, there were 18 reported violations.[63] In 2016, there were 20.[64]

Sexual Intercourse with a Descendant[65]

This section criminalises the sexual intercourse or commission of another act of sexual nature by a parent, person holding parental rights or a grandparent with a child or grandchild. The act is punishable by two to eight years' imprisonment.

It must be noted that age is not an objective element of the offence. Therefore, all sexual acts committed by a parent, person holding parental rights or a grandparent against the descendant are criminalised irrespective of the age of the victim.[66]

If the act is committed against a person less than 10 years old, it qualifies as rape (see '*Rape*' above).

In 2015, there were 14 reported violations.[67] In 2016, there were 6 registered infringements.[68]

Sexual Intercourse or Other Act of Sexual Nature with a Child[69]

The engagement in sexual intercourse or commission of another act of sexual nature by an adult with a child is punishable by up to five years' imprisonment. If the act is committed against a child younger than 10 years old, it must be qualified as rape.

Notably, in 2015, there were 30 registered violations.[70] In 2016, the number rose to 56.[71]

Buying Sex from a Minor[72]

Engaging in sexual intercourse or committing another act of sexual nature with a minor for monetary payment or any other benefit is punishable by up to three years' imprisonment under the PC. If the same is committed against a child, then the act is punishable up to five years' imprisonment.

'Monetary payment' in the context of this offence is a payment of money in whatever currency and means of payment (cash, bank transfer, credit card payment or payment environments in the Internet (e.g. PayPal)). 'Any other benefit' should be seen as a physical object or a legal right that has value that is measurable in

63 Crime in Estonia 2016, p. 45.
64 Crime in Estonia 2015, p. 49.
65 Section 144 of the PC.
66 Pikamäe and Sootak 2015, p. 431.
67 Crime in Estonia 2016, p. 45.
68 Crime in Estonia 2015, p. 49.
69 Section 145 of the PC.
70 Crime in Estonia 2016, p. 45.
71 Crime in Estonia 2015, p. 49.
72 Section 1451 of the PC is the equivalent of Article 4(7) of the Directive 2011/93/EU.

money (e.g. a ticket to a concert). In this regard, it is important that the payment is made or benefit is given in return for the sexual act.[73]

In 2015, there were 57 registered violations.[74] This number decreased by 21 registered incidents in 2016.[75]

Online Criminal Offences against Sexual Self-Determination

The offences described in this section are found in Chapter 11 Division 2 of the PC (Offences Against Minors). It is important to note that all of the online offences against sexual self-determination described in this section are delicts of abstract danger. This means that these delicts are punishable on the basis of the danger they pose to the society—in the context of offences against minors—real (i.e. existing) minors.[76] From the moral legal rights point of view, the offences described below are aimed to protect the normal mental and sexual development of minors.[77]

Arguably, only real persons have moral legal rights that need protection and such rights are not eminent in case of virtual avatars such as Sweetie. Therefore, Sweetie cannot be a victim. However, acts directed against Sweetie may be considered infringing to real minors on the basis of the threat the perpetrator thereby poses to the society. Acts directed against Sweetie could also be considered as attempts. As set out in settled case law referred to in '*Sexual Enticement of Children*' and '*Entrapment*', perpetrators who according to their knowledge sexually entice children, but who were actually police agents depicting themselves as children and conducting surveillance activities, have been punished as committing impossible attempts.

Requesting Access and/or Watching Child Pornography[78]

The PC criminalises the act of knowingly requesting access to child pornography or knowingly watching a pornographic performance involving a minor or of a pornographic or erotic performance involving a child.

'Requesting access' should be understood as looking for an opportunity to look, read or listen to a work or performance containing minors in a pornographic scenes and children in erotic scenes. Under this provision, the streaming of works of child pornography (and erotic performances including children respectively), not downloading works, (see '*Manufacture and Storing of and Making Available Child Pornography*') is criminalised. A criminal offence is committed after typing an

---

[73] Pikamäe and Sootak 2015, p. 433.

[74] Crime in Estonia 2016, p. 45.

[75] Crime in Estonia 2015, p. 49.

[76] Laurits 2014, p. 401.

[77] Pikamäe and Sootak 2015, pp. 471–472, 474 and 476.

[78] Section 1751 of the PC is the equivalent of Articles 4(4) and 5(3) of the Directive 2011/93/EU. 'Child pornography' must be understood as defined in this sub-chapter and in '*Manufacture and storing of and making available child pornography*', i.e. pornographic works involving minors and children and erotic works depicting children.

infringing keyword to the search engine, the actual access to the illegal content is irrelevant[79] for punishment.

Another important facet of this offence is that requesting access must be made 'knowingly' which describes intent, specifically direct intent. Legally, this means that the perpetrator wants or at least tacitly accepts the creation of the circumstances which belong to the necessary elements of an offence.[80] Conversely, this means that if a person is directed to or accesses a website that contains pornographic materials depicting minors in pornographic scenes or children in erotic scenes by accident, he[81] is not punished.

The act is punishable by a pecuniary punishment or an imprisonment of up to two years. According to the statistics, there was one registered case in 2014. In 2015 and 2016, there were none.[82]

Manufacture and Storing of and Making Available Child Pornography[83]
With this offence, the PC criminalises the manufacture, acquisition or storing, handing over, displaying or making available to another person in any other manner of pictures, writings or other works or reproductions of works depicting a minor in a pornographic situation, or a child in a pornographic or erotic situation. Such an act is punishable by a pecuniary punishment or up to three years' imprisonment. This section criminalises downloading (i.e. obtaining possession) pornographic materials depicting minors and erotic scenes depicting children.

'Works depicting minors in a pornographic situation' are works that depict a person younger than 18 years old (including younger than 14). If a child (i.e. a person less than 14) is depicted in an erotic work, this is considered equal to works depicting minors (i.e. persons from 14 to 17) in a pornographic context.[84]

It is not necessary to ascertain the exact age of the minor or child depicted in a pornographic work. It is sufficient if an independent observant third party could determine that the person depicted in a pornographic work is minor or a child depicted in an erotic work is less than 14 years old.[85]

'Manufacture' refers to the creation of a work depending on the nature of the work (writing, drawing, filming etc.). It is important to note that if an actual minor or child is used for the manufacture of a pornographic or erotic work, the act is punishable under Section 175 of the PC (see '*Human Trafficking in Order to Take Advantage of Minors*' in Sect 8.2.1). If a pornographic or erotic work is

---

[79] Pikamäe and Sootak 2015, p. 471.

[80] Subsection 16(3) of the PC.

[81] However, women are included here, the statistics indicates that most of the offenders are men. See Kirwan and Power 2013, p. 127.

[82] Crime in Estonia 2015, p. 49.

[83] Section 178 of the PC is the equivalent of Article 5(2) of the Directive 2011/93/EU.

[84] Pikamäe and Sootak 2015, p. 473.

[85] Pikamäe and Sootak 2015, p. 473.

manufactured purely based on imagination (e.g. writing a story), the act is punishable under this section.

The pornographic works of minors or erotic works of children do not need to depict real (i.e. existing) persons[86] as the manufacturing, storing or making available of virtual pornography and erotica is also in the remit of this provision.[87] In this context, a case arose in Estonia where an published writer published a work named "Untitled 12" depicting violent sexual abuse of fictitious minors. The work was published via a website located in the United Kingdom. At the time of publication, the author was in the U.S. The Circuit Court held that the work is a transgressive literary work in the U.S and UK and as such is not punishable under UK and U.S law (under the Miller test). Consequently, Estonian courts did not have jurisdiction over the case as the creation and publishing of the work has taken place abroad.[88]

In addition to what was mentioned in the introduction to this sub-chapter (see '*Online Criminal Offences against Sexual Self-Determination*'), Pikamäe and Sootak stipulate that this offence is aimed at reducing the demand for child pornography by demotivating the manufacture and thereby using minors and children in these illegal works.[89]

Statistics from 2014 indicates that there were 68 registered violations of this provision. In 2015, the number has increased to 120.[90] In 2016, the number of violations has stayed almost the same, decreasing by two cases.[91]

Making a Proposal to Meet with a Minor with the Aim of Committing an Offence of Sexual Nature[92]

The PC criminalises the making of a proposal to meet (and making preparations thereof) with a child *or* a minor who was not capable of comprehending the situation[93] for sexual purposes. The PC also criminalises concluding such an agreement with a child or a minor who was not capable of comprehending the situation. The sexual purpose refers to the fact that the perpetrator wishes to commit an offence of sexual nature (i.e. committing an illegal act described in Sects. '*Human Trafficking in Order to Take Advantage of Minors*', '*Offline Criminal Offences Against Sexual Self-Determination*' and '*Online Criminal Offences*

86 Judgment of Tallinn Circuit Court of 20 June 2017, case number, 1-17-689, p. 8.2.

87 Laurits 2014, p. 397.

88 Judgment of Tallinn Circuit Court of 11 October 2017, case number, 1-15-11024.

89 Pikamäe and Sootak 2015, p. 472.

90 Crime in Estonia 2016, p. 47. It must be noted, that this is an overall number and as Section 178 criminalises many acts in relation to child pornography, the statistics does not differentiate the acts and whether the persons depicted were less than 18 or less than 14.

91 Crime in Estonia 2015, p. 49.

92 Section 1781 of the PC.

93 See '*Rape*' under Sect. 8.2.1.

Against Sexual Self-Determination'). The act is punishable by a pecuniary punishment[94] or up to three years' imprisonment.

This provision protects both children and minors (14 to 17-year olds) who usually have exceeded the sexual determination age, but who, however, are not capable to foresee the sexual intentions of the offender due to his or her lack of knowledge in regard to sexual life. This, in return, means that if the minor is aware that the meeting contains sexual intercourse or another act of sexual nature and, nevertheless, agrees to meet the adult, there is no criminal offence.[95]

A 'meeting' in the context of this provision means a physical meeting. A proposal to meet, therefore, means an invitation to a physical meeting. The invitation, however, could may be made by electronic means. The invitation to the meeting may be of general nature.[96]

'Concluding an agreement to meet' refers to the situation where the child or the minor who is not capable of comprehending the situation gives his or her accept to the invitation made by the offender.[97]

In addition to making the invitation to meet, the criminal proceedings must identify that the offender has made an act to prepare for the meeting (e.g. booked a hotel room or appeared at the agreed location).[98] This is to mean that the intention of the perpetrator is real as fantasises are not criminalised.

In 2014, there were seven registered violations. In 2015, the number has decreased by four.[99] Similarly, in 2016, there were three registered incidents.[100]

Sexual Enticement of Children[101]

The PC outlaws the handing over, displaying or making otherwise pornographic works[102] or reproductions thereof knowingly available to children, showing sexual abuse to children or engaging in sexual intercourse in the presence of a child or knowingly sexually enticing a child in any other manner. Enticement is punishable by a pecuniary punishment or up to three years' imprisonment.

Whereas the handing over or making pornographic works or reproductions thereof knowingly available and other acts described above are quite self-explanatory, the part which criminalises 'sexual enticement in any other manner' needs clarification. According to Pikamäe and Sootak, this means that the

[94] According to Subsections 44(1), 44(2) and 44(3), the court may impose a pecuniary punishment of 30 to 500 daily rates. The daily rate is calculated on the basis of the average daily income of the offender. The minimum daily rate is 10 euros.

[95] Pikamäe and Sootak 2015, p. 475.

[96] Ibid. The invitation does not need to include the date, time, location and other details and it can be of general nature.

[97] Ibid.

[98] Pikamäe and Sootak 2015, p. 475.

[99] Crime in Estonia 2016, p. 45.

[100] Crime in Estonia 2015, p. 49.

[101] Section 179 of the PC is the equivalent of Article 3(3) of the PC.

[102] See section titled '*Human Trafficking in Order to Take Advantage of Minors*' above.

act of masturbating in front of the webcam, talking on sexual topics with a child and inducing a child to masturbate or touch his or herself are criminalised with this provision.[103] This makes it a very important provision and taking into account that sexual enticement is also a delict of abstract danger, the question arises whether a person who masturbating when seeing Sweetie via webcam infringes the legal rights of other minors.

In a case of 2012, a perpetrator according to his knowledge sexually enticed an 11-year-old A (described oral sex and sent videos to A where the act was depicted). A was in fact a police agent conducting surveillance activities.[104] The perpetrator was found guilty for committing an impossible attempt to entice a child.

In practice, the following acts have been considered as sexual enticement and punished as impossible attempts: (i) perpetrator was to his knowledge describing sexual intercourse, sent pornographic materials and provided detailed descriptions how an 11-year-old X should masturbate;[105] (ii) perpetrator to his knowledge asked a 12 year old K to describe her breasts and the perpetrator described how he would like to touch K. Also, the perpetrator asked K to appear naked on webcam;[106] (iii) perpetrator to his knowledge offered 500€ to 12-year-old X if X masturbated in front of the webcam.[107]

Therefore, it could be derived that if Sweetie would be instead of A, X or K, the case would logically need to be solved the same way and be punishable as an impossible attempt, irrespective of the fact whether Sweetie would be operated by a police agent or whether no human intervention is necessary. Yet, this is a theoretical position because there is no case law to support this as virtual avatars such as Sweetie are not in use today in Estonia.

In 2015, there were 93 registered cases of sexual enticement of children. Compared to 2014, the number has increased by 44.[108] In 2016, there were 90 registered cases.[109]

Legal Persons as Perpetrators

As prescribed by Article 12 of the Directive 2011/93/EU on combating sexual abuse and sexual exploitation of children and child pornography,[110] the PC stipulates that if criminal acts set out in Sect. 8.2.1 (except "Sexual Intercourse with a

103 Pikamäe and Sootak 2015, p. 477.

104 Judgment of the Harju County Court of 17 December 2012, case number 1-12-11607, p. 1.

105 Judgment of Harju County Court of 20 January 2015, case number 1-15-82/15.

106 Judgment of Harju County Court of 28 August 2013, case number 1-13-6413.

107 Judgment of Harju County Court of 5 August 2013, case number 1-13-6110.

108 Crime in Estonia 2016, p. 47.

109 Crime in Estonia 2015, p. 49.

110 OJ L 335, 17.12.2011, pp. 1–14.

Table 8.1 Relevant Criminalised Acts Under the PC

Human trafficking in order to take advantage of minors [Section 175 of the PC]	Two to ten years' imprisonment
Rape [Section 141 of the PC]	One to five years' imprisonment, if the same is committed against a minor, the act is punishable by six to fifteen years' imprisonment
Compelling a person to engage in sexual intercourse or another act of sexual nature [Section 143 of the PC]	Up to three years' imprisonment
Sexual intercourse or other act of sexual nature using influence [Section 143^2 of the PC]	Two to eight years' imprisonment
Sexual intercourse with a descendant [Section 144 of the PC]	Two to eight years' imprisonment
Sexual intercourse or other act of sexual nature with a child [Section 145 of the PC]	Up to five years' imprisonment
Buying sex from a minor [Section 145^1 of the PC]	Up to three years' imprisonment; if the same is committed against a child, the act is punishable by up to five years' imprisonment
Requesting access to child pornography and watching thereof [Section 175^1 of the PC]	Pecuniary punishment or up to two years' imprisonment
Manufacture of works involving child pornography or making child pornography available [Section 178 of the PC]	Pecuniary punishment or up to three years' imprisonment
Agreement of sexual purpose for meeting with child [Section 178^1 of the PC]	Pecuniary punishment or up to three years' imprisonment
Sexual enticement of children [Section 179 of the PC]	Pecuniary punishment or up to three years' imprisonment

[*Source* The author]

Descendant") are committed by a legal person, the act is punishable by a pecuniary punishment.[111] Such punishment ranges from 4000 to 16,000,000 euros according to the PC.[112]

8.2.2 *Interim Conclusion*

The PC criminalises the following acts and foresees the following punishments (Table 8.1).

[111] See Subsections 133(3), 1331(3), 141(3), 143(3), 1432(3), 145(3), 1451(4), 175(2), 1751(3), 178(2), 1781(2) and 179(2) of the PC.

[112] Subsection 44(8) of the PC.

8.2.3 *Possible Obstacles in Substantive Law Concerning Sweetie*

Requesting Access to Child Pornography and Watching Thereof
Sweetie, being an avatar of a 10-year old Filipino girl,[113] does not constitute a pornographic or an erotic work itself. This means that a person watching Sweetie on webcam is not knowingly watching a pornographic performance involving a child or an erotic performance including a child or requesting access thereof which is criminalised under Section 175^1 of the PC (see '*Requesting Access and/or Watching Child Pornography*', Sect. 8.2.1). Therefore, knowingly watching Sweetie via webcam is not punishable *per se*.

Agreement of Sexual Purpose for Meeting with a Child
Under material criminal law, the fact that Sweetie is a virtual avatar does not pose too much of a problem in the context of this offence as it is meant to protect children for physical sexual abuse. Therefore, in this context, possible obstacles arise from the criminal procedure rules which are analysed in Sect. 8.3.

Assuming that Sweetie would be permitted as a procedural tool to detect and apprehend sexual abusers, the intent to sexually abuse children could be identified with Sweetie as well as with other procedural means described in Sect. 8.3. However, the perpetrator would need to fulfil an extra element under the provisions which is *making an actual preparatory act*, e.g. appearing at the agreed location or buying airplane tickets to meet Sweetie.

An attempt to meet a child for sexual purposes is also punishable. An attempt to meet with a child is deemed to have committed from the moment when the person, according to the person's understanding of the act, directly commences the commission of the offence.[114] Therefore, theoretically the act of making the proposal to meet Sweetie would already be considered an attempt, however, it is very unlikely to be ever proven within the criminal proceedings.

Sexual Enticement
Sexual enticement means that the offender shows pornographic works, sexual abuse or otherwise knowingly sexually entices child (e.g. masturbates in front of the webcam).[115] Here, multiple questions arise. As Sweetie is not a real person and the offence of sexual enticement explicitly states that the victim must be a 'person', at first glance, the problem arises that Sweetie is virtual. However, as sexual enticement is a delict of abstract danger, there should not be a difference from the material

[113] Being 10, Sweetie would be considered a child under Estonian law, i.e. a person less 14 years old. A person less than 10-years old is considered as incapable of comprehending the sexual nature of the sexual acts due to young age (see '*Rape*' under Sect. 8.2.1).

[114] Subsection 25(2) of the PC.

[115] See case law in the sections '*Sexual Enticement of Children*' and '*Entrapment*'.

criminal law standpoint whether a virtual avatar is used to identify the criminal acts or whether real persons are enticed because the crime is punishable even without having a specific victim, but on the basis of the threat it poses to the society. Therefore, if Sweetie could be used as a procedural tool, the intent to entice would be punishable.

8.3 Analysis of Criminal Procedure Law

The CCP divides the persons subject to criminal proceedings into two-bodies conducting the proceedings and participants.[116] The first are courts, Prosecutor's Offices and investigative bodies. The latter are the suspect or accused, his or her counsel, victim, civil defendant and third parties are the participants in a criminal proceeding,[117] experts, clerks of court sessions and interpreters and translators.[118] Without going in depth with the rights and obligations of the participants, the following sub-chapter briefly outlines the competence of the investigative bodies and their roles in pre-trial and court proceedings.

8.3.1 Pre-trial Procedure

The Prosecutor's Office

The Prosecutor's Office comprises of the Office of the Prosecutor General's Office and four District Prosecutor's Offices. The Prosecutor's Office is headed by Prosecutor General.

The Prosecutor's Office, represented in a procedure by a public prosecutor, directs pre-trial proceedings and ensures the legality and efficiency thereof.[119] This means that it is up to the Prosecutor's Office to give guidance to the investigative bodies who perform procedural acts and ensure the legality thereof. The Prosecutor's Office may also perform procedural acts itself if necessary.[120]

The Prosecutor's Office also represents the public prosecution in court.[121] This implies that there is no private prosecution in Estonia.

Besides the powers described above, the Prosecutor's Office is, in pre-trial proceedings, competent to: (1) be present at the performance of procedural acts and intervene in the course thereof; (2) terminate criminal proceedings; (3) demand that

[116] Subsection 16(1) of the Code of Criminal Proceedings.

[117] Subsection 16(2) of the Code of Criminal Proceedings.

[118] Section 59 of the Code of Criminal Proceedings.

[119] Subsection 30(1) of the CCP.

[120] Subsection 213(1)(1) of the CCP.

[121] Subsection 30(1) of the CCP.

the materials of a criminal file and other materials be submitted for examination and verification; (4) issue orders to investigative bodies; (5) annul and amend orders of investigative bodies; (6) remove an official of an investigative body from a criminal proceeding; (7) alter the investigative jurisdiction over a criminal matter; (8) declare a pre-trial proceeding completed; (9) demand that an official of an investigative body submit oral or written explanations concerning the circumstances relating to a proceeding; and (10) assign the head of the probation supervision department with the duty to appoint a probation officer.[122]

Investigative Bodies

According to Subsection 31(1) of the CCP, the Police and Border Guard Board ('Police'), the Estonian Internal Security Service, the Tax and Customs Board, the Competition Board, the Military Police, the Environmental Inspectorate and the Prisons Department of the Ministry of Justice are investigative bodies within the limits of their competence.[123]

Investigative bodies perform procedural acts independently unless the permission of a court or the permission or order of a Prosecutor's Office is necessary for the performance of an act.[124]

Court

In pre-trial proceedings, the court is represented by the preliminary investigation judge who is a County Court judge who, sitting alone, authorises surveillance activities.[125]

Specific Regulation Concerning Minors

Specific rules exist in the CCP in regard to hearing witnesses who are minors. In pre-trial procedure, Subsection 70(1) of the CCP sets out that a body conducting investigative proceedings may involve a child protection official, a social worker, a teacher or a psychologist in the hearing of a witness who is a minor. Involvement of the afore-mentioned expert is mandatory when the official conducting the proceedings has not received appropriate training and if the witness is less than ten years old or less than fourteen and the hearing is related to domestic violence or sexual abuse.[126] If the body conducting the proceedings is obliged to involve an expert, the hearing must be video recorded.[127] In other cases, hearing a minor is video-recorded when necessary.[128]

---

122 Subsection 213(1) of the CCP.

123 See Section 212 of the CCP.

124 Subsection 32(1) of the CCP.

125 This, of course, is not the only task, but most relevant in the context of this report. See Subsection 21(2) of the CCP.

126 Subsection 70(2) of the CCP.

127 Subsection 70(3) of the CCP.

128 Ibid.

8.3.2 Court Procedure

The Prosecutor's Office

If the Prosecutor's Office is convinced that the necessary evidence in the pre-trial phase of the criminal matter has been collected, the Prosecutor's Office prepares a statement of charges which shall be sent to the court.[129] The charge is represented in court by the Prosecutor's Office.[130]

Court

Estonia has a three-tier court system. Judgments of the County Courts may be appealed to circuit courts. Circuit courts' judgments may be appealed to the Supreme Court which may decide whether to hear a case or not.[131]

The County Court as a court of first instance hears the criminal case brought to it by the Prosecutor's Office. Criminal offences in the first degree[132] are heard by a court panel consisting of the presiding judge and two lay judges.[133] Matters concerning criminal offences in the second degree[134] and criminal matters in which simplified proceedings are applied shall be heard by a judge sitting alone.[135]

Specific Regulation Concerning Minors

In trial hearings, children are not cross-examined[136] and the judge allows the child to tell the court everything he or she knows about the case.[137] As in the pre-trial procedure, the court may involve an expert in the hearing of a witness under fourteen years of age who may question the witness with the permission of the judge.[138] In certain cases, a court may decide not to summon a minor at all.[139]

---

129 Subsections 226(1) and 226(3) of the CCP.

130 Subsection 30(1) of the CCP.

131 Subsection 349(1) of the CCP.

132 A criminal offence in the first degree is an offence the maximum punishment prescribed for which in the PC for a natural person is imprisonment for a term of more than five years or life imprisonment. An offence of a legal person is a criminal offence in the first degree if imprisonment for a term of more than five years or life imprisonment is prescribed for the same act as maximum punishment for a natural person.

133 Subsection 18(1) of the Code of Criminal Proceedings.

134 A criminal offence in the second degree is an offence the punishment prescribed for which in this Code is imprisonment for a term of up to five years or a pecuniary punishment.

135 Subsection 18(2) of the Code of Criminal Proceedings.

136 Usually, witnesses are cross-examined in trial hearings. See Subsection 290(1) of the CCP.

137 Subsection 290(3) of the CCP.

138 Subsection 290(2) of the CCP.

139 When the testimony of the child was video recorded, and the opposing counsel has had the opportunity to pose questions to the witness in pre-trial procedure about the facts relating to the subject of proof.

8.3.3 *Investigatory Powers*

The investigatory powers are set out in the CCP. The Electronic Communications Act ('ECA') is very important in the context of the apprehension of online sexual abusers as it obliges the electronic communication undertakings to collect and retain communication metadata.

The Electronic Communications Act

The ECA sets out the obligations for electronic communications undertakings ('communications undertakings')[140] to retain certain information. Specifically, the ECA obliges two types of service providers: (i) providers of telephone or mobile telephone services and telephone network and mobile telephone network services ('telephone service providers'); and (ii) providers of Internet access, electronic mail and Internet telephony services ('Internet access providers'). The ECA sets out the criteria and the competent authorities who may request data from these service providers.[141] Such authorities are, for example, the Prosecutor's Office, the courts and the Police.

Obligation to Retain Data

The ECA obliges all communications undertakings to preserve the data necessary for the performance of the following acts:

- monitoring and detection of the source of communication;
- identification of the endpoint of the communication;
- identification of the date, time and duration of the communication;
- identification of the type of the communications service;
- identification of the terminal equipment or presumable terminal equipment of the user of the communication service; and
- pinpointing of the location of the terminal equipment.[142]

Telephone Service Providers

Telephone service providers are required to preserve the following data for a period of one year from the date of the communication:[143] (1) the number of the caller and the subscriber's (i.e. the client of the telephone service provider) name and address; (2) the number of the recipient and the subscriber's name and address; (3) in the cases of call forwarding or call transfer, the number dialled and the subscriber's name and address; (4) the date and time of the beginning and end of the call; (5) the telephone or mobile telephone service used; (6) the international mobile subscriber

---

140 According to Subsection 2(5) of the ECA, an 'electronic communications undertaking' is a person who provides publicly available electronic communications services to the end-user or to another provider of publicly available electronic communications services.

141 Subsection 111^{1}(11).

142 Subsection 1111(1) of the ECA.

143 Retention date is set out in Subsection 1111(4) of the ECA. If the data is requested by a competent body for investigatory purposes, the data must be retained for two years.

identity (IMSI) of the caller and the recipient; (7) the international mobile equipment identity (IMEI) of the caller and the recipient; (8) the cell ID at the time of setting up the call; (9) the data identifying the geographic location of the cell by reference to its cell ID during the period for which data are preserved; and (10) in the case of anonymous pre-paid mobile telephone services, the date and time of initial activation of the service and the cell ID from which the service was activated ('telephone communication metadata').[144]

The obligation to preserve the data by telephone service providers also applies to unsuccessful calls if those data are generated or processed upon providing telephone or mobile telephone services or telephone network or mobile telephone network services.[145]

Internet Access Providers

Providers of Internet access, electronic mail and Internet telephony services are required to preserve the following data for a period of one year from the date of communication: (1) the user IDs allocated by the communications undertaking; (2) the user ID and telephone number of any incoming communication in the telephone or mobile telephone network; (3) the name and address of the subscriber (i.e. client of the Internet access provider) to whom an Internet Protocol (IP) address, user ID or telephone number was allocated at the time of the communication; (4) the user ID or telephone number of the intended recipient of an Internet telephony call; (5) the name, address and user ID of the subscriber who is the intended recipient in the case of electronic mail and Internet telephony services; (6) the date and time of beginning and end of the Internet session, based on a given time zone, together with the IP address allocated to the user by the Internet service provider and the user ID; (7) the date and time of the log-in and log-off of the electronic mail service or Internet telephony service, based on a given time zone; (8) the Internet service used in the case of electronic mail and Internet telephony services; (9) the number of the caller in the case of dial-up Internet access; and (10) the digital subscriber line (DSL) or other end point of the originator of the communication ('Internet communication metadata').[146]

Inquiries from Competent Authorities

If the competent authority[147] submits a request to the communications undertaking, the communications undertaking is obliged by law to provide the data requested within ten working days.[148]

144 Subsection 1111(2) of the ECA.

145 Subsection $111^1(8)$ of the ECA.

146 Subsection 1111(3) of the ECA.

147 List set out in Subsection 1111(11) of the ECA.

148 Subsection 112(1) of the ECA. In case of an urgent request, the communications undertaking must provide the information within 10 hours.

Upon request, telephone service providers must be able to provide a surveillance agency and security authority and the Police real time identification of the location of the terminal equipment used in the telephone network.[149]

The Code of Criminal Procedure

Requests to Communications Undertakings for Information

Subsection $90^1(1)$ of the CCP provides a legal ground for a body conducting the proceedings to make inquiries to communications undertakings. The body conducting the proceedings may inquire data for the identification of an end-user related to the identification tokens used in the public electronic communications network (i.e. the so called call log).[150] Such a request does not cover the data relating to the fact of communication of messages.[151]

With the permission of the Prosecutor's Office, the investigative body may make inquiries in pre-trial procedure[152] to communications undertakings about the telephone communication metadata (see '*Telephone Service Providers*') and Internet communication metadata (see '*Internet Access Providers*') that is not covered by the call log.[153]

The CCP specifically states that inquiries described herein may only be made if this unavoidably necessary for the achievement of the objectives of criminal proceedings.[154] In other words, the *ultima ratio* principle must be complied with whenever the body conducting the proceedings requests communication metadata from telephone service providers and Internet access providers.

Surveillance Activities

The procedures regarding surveillance activities are set out in the CCP. According to the definition provided therein, 'surveillance activities' denote the processing of personal data for the performance of a duty provided by law with the objective of hiding the fact and content of data processing from the data subject.[155] The CCP provides a closed list of surveillance activities. Surveillance activities require either the warrant from the Prosecutor's Office or the court.

Surveillance agencies[156] may conduct surveillance activities on the following basis:

---

[149] Subsection 112(3) of the ECA.

[150] Subsection $90^1(1)$ of the CCP.

[151] Ibid.

[152] In court proceedings, the permission of the court is required for the inquiry to the communications undertaking.

[153] Subsection 901(2) of the CCP.

[154] Subsection 901(3) of the CCP.

[155] Subsection 1262(1) of the CCP.

[156] The Police, Estonian Internal Security Service, Tax and Customs Board, Military Police and the Prisons Department of the Ministry of Justice and prisons.

- the need to collect information about the preparation of a criminal offence for the purpose of detection and prevention thereof[157] ('Base 1');
- the execution of a ruling on declaring a person a fugitive[158] ('Base 2');
- the need to collect information in confiscation proceedings[159] ('Base 3'); and
- the need to collect information in a criminal proceeding about a criminal offence[160] ('Base 4').

For the purposes of Base 1 and Base 4, surveillance activities may be conducted for the prevention of and collecting information about the offences described in the following sections of this chapter: '*Human Trafficking in Order to Take Advantage of Minors*', '*Offline Criminal Offences Against Sexual Self-Determination*' and '*Online Criminal Offences Against Sexual Self-Determination*' above.

Covert Watching, Taking Comparative Samples and Entering into a Computer System

For the purposes of collecting information about the preparation of a criminal offence or for the purposes of detection and prevention thereof (i.e. Base 1), surveillance agencies may covertly watch a person, thing or an area, covertly take comparative samples and perform initial examinations, covertly examine a thing and covertly replace it.[161] Such surveillance activities require a permission from the Prosecutor's Office. The permission may be granted for up to two months which may be renewed for another two months.[162] If it is unavoidably necessary for the achievement of the objectives of the surveillance activities, the security agency may covertly enter into a building, vehicle, enclosed area or a computer system. However, in this case, a permission from the preliminary investigation judge must be obtained.[163]

157 In respect of the person in the case of whom there are serious reasons to believe that he or she commits the criminal offence referred to in Subsection 1262(2) of the CCP.

158 In respect of the person declared a fugitive.

159 In respect of the person who owns or possesses the assets which are the object of confiscation proceedings.

160 In respect of the person who is a suspect in a criminal proceeding or with respect to whom there is justified reason to believe that he or she has committed or commits the specified criminal offence.

161 Subsection 1263(1) of the CCP.

162 Subsection 1265(1) of the CCP.

163 Subsection 1264(5) of the CCP.

Observing a Postal Item, Wire-Tapping and Using a Police Agent

For the purposes of: (i) detection and prevention of a criminal offence set out in Sections 244,[164] 246,[165] 255,[166] 256[167] and Subsections 266(2) and 266(3);[168] (ii) gathering information in confiscation proceedings (i.e. Base 3) and for the purposes of collecting information in criminal proceedings about a criminal offence (i.e. Base 4), the Police and the Estonian Internal Security Service may covertly: (i) examine a postal item; (ii) intercept information; and (iii) use a police agent.[169]

The covert examination of a postal item requires a permission from the preliminary investigation judge—The permission may be granted for up to two months and ay be extended.[170]

Interception of messages requires a permission from the preliminary investigation judge. The permission may be granted for up to two months and the deadline may be extended by up to two months.[171]

Section 113 of the ECA provides that a communication undertaking must grant an authorised surveillance agency or security authority the access to its communication network.[172] The communications undertaking must enable the surveillance agency or security authority to select messages and ensure their transmission to a central or portable surveillance device of the surveillance agency or security authority in an unchanged form and in real time. Also, the communications undertaking must ensure the quality of message transmission which must be equivalent to the quality of the regular services provided by the communications undertaking and ensure the protection of the messages and of the data related to their transmission.[173]

The permission to use a police agent (see also '*Entrapment*' below) is granted by the Prosecutor's Office for up to six months. The deadline may be extended by six months at a time.[174]

Staging a Criminal Offence

For the purposes of detection of a criminal offence or detention of a criminal (i.e. Base 4), the Police and the Internal Security Service may stage a criminal

---

164 Section 244 of the PC criminalises the attack against the life or health of higher state public servants.

165 Section 246 of the PC criminalises the attack against the life or health of persons enjoying international immunity.

166 Section 255 of the PC criminalises the membership of criminal organisation.

167 Section 256 of the PC criminalises the formation of a criminal organisation.

168 Section 266 of the PC criminalises the illegal entry and failure to comply with the demand to leave.

169 Subsection 1263(2) of the CCP.

170 Subsection 1266(5) of the CCP.

171 Subsection 1267(3) of the CCP.

172 Subsection 113(1) of the ECA.

173 Subsection 113(3) of the ECA.

174 Subsection 1269(2) of the CCP.

offence.[175] Staging of a criminal offence is the commission of an act with the elements of a criminal offence with the permission of a court.[176] In case of staging a criminal offence, objective elements of a crime exist, however, the act is not punishable as there is no intent because the crime is authorised by the court and conducted for the purpose of apprehending a criminal or detecting a criminal offence.[177]

Staging may be performed by a police agent (see '*Observing a Postal Item, Wire-Tapping and Using a Police Agent*' and '*Entrapment*') or another person who covertly cooperates with a security agency.[178]

The CCP does not set out the limits when staging escalates to illegal incitement. The case law, however, explains that the police agent who is authorised by the court to stage a criminal offence may not incite illegal actions on persons who did not have a slightest intent to commit an offence[179] (i.e. the crime would not have been committed without the intervention of the *agent provocateur*[180]). According to the Supreme Court, illegal provocation exists where the police agent staging a criminal offence incites a person who had no intent (not even the slightest) to commit a crime prior to the actions of the police agent staging the offence and who had not been targeted by the Police or the Internal Security Service within the criminal procedure.[181]

This implies that staging may only be performed against specific persons whose criminal intent had been previously ascertained within the criminal procedure and that the intent must not be provoked by the police agent, but the intent had to be pre-existing (i.e. acts of the police agent staging a criminal offence must not outweigh the acts of the person being provoked).[182] Also, as an underlying principle for any surveillance activity, staging may only be used as an *ultima ratio* measure.[183]

Staging a criminal offence may only be performed with the permission of the preliminary investigation judge. The permission may be granted for up to two months which may be extended by up to two months.[184]

---

175 Subsection 1263(3) of the CCP.

176 Subsection 1268(1) of the CCP.

177 Lõhmus 2012, p. 119.

178 Kergandberg Pikamäe 2012, p. 324.

179 Judgment of the Supreme Court of 2 December 2004, case number 3-1-1-110-04, para 11.3.

180 Judgment of the European Court of Human Rights of 5 February 2008, case *Ramanauskas v. Lithuania,* para 39.

181 Judgment of the Supreme Court of 2 December 2004, case number 3-1-1-110-04, para 11.3.

182 Lõhmus 2012, p. 120. It must be noted that Lõhmus is very critical about the interpretation given by the Supreme Court stating that this does not give any clear guidelines what is illegal incitement and what is not.

183 Subsection 1261(2) of the CCP.

184 Subsection 1268(3) of the CCP.

Table 8.2 Transposition of the Council of Europe Convention on Cybercrime to Estonian Law [*Source* The author]

Council of Europe Convention on Cybercrime	Estonia
Article 16. Expedited preservation of stored computer data	Section 111^1 of the ECA
Article 17. Expedited preservation and partial disclosure of traffic data	Section 111^1 of the ECA
Article 18. Production order	Section 112 of the ECA; Section 90^1 of the Code of Criminal Proceedings
Article 19. Search and seizure of stored computer data	Section 113 of the ECA
Article 20. Real-time collection of traffic data	Subsections 112(3) and 113(3) of the ECA
Article 21. Interception of content data	Section 113 of the ECA; Section 126^7 of the CCP

Interim Conclusion
In total, the rights and obligations set out in the Budapest Convention may be brought into Table 8.2.

8.3.4 *Human Rights*

Right to Inviolability of Private and Family Life
The collection of telephone communication metadata and Internet communication metadata raise the question of conformity with the constitutional right to inviolability of private and family life enshrined in Section 26 of the Constitution. Subsection 26(2) of the Constitution provides that the right to private and family life is not an absolute right as it may be circumscribed by government agencies, local authorities, and their officials pursuant to a procedure provided by law to protect public health, public morality, public order or the rights and freedoms of others, to prevent a criminal offence, or to apprehend the offender.

The legality and conformity of the obligation to collect and store telephone and Internet communication metadata has been analysed by the Chancellor of Justice[185] and the Supreme Court in the context of criminal proceedings.

[185] The institution of the Chancellor of Justice in Estonia is not part of the legislative, executive or judicial powers, it is not a political or a law enforcement body. The institution of the Chancellor of Justice is established by the Constitution and the Chancellor only observes the Constitution and his conscience. The Chancellor of Justice is appointed by the Parliament on the proposal of the President of the Republic for a term of seven years. Once a year the Chancellor of Justice submits to the Parliament a report with an overview of their activities. The Chancellor of Justice in Estonia

The Chancellor of Justice analysed the constitutionality of Section 111^1 of the ECA (see '*The Electronic Communications Act*' above) when a citizen filed an application with the Chancellor for an opinion on such obligation by the communication undertakings taking into account that the European Court of Justice had declared the data retention directive (Directive 2006/24/EC) invalid.[186] The Chancellor highlighted that the judgment by the European Court of Justice does not have immediate effect on Estonia and the law of Estonia cannot be declared non-constitutional on that basis alone.[187] However, the Chancellor of Justice highlighted that the Judgment of the European Court of Justice may be used as a means to interpret the Constitution of Estonia.[188]

The Chancellor of Justice found that for the purposes of detection and apprehension of criminals, the right of inviolability of private and family life could be circumscribed and the collection of telephone and Internet communication metadata is a very effective tool for such purposes.[189] The Chancellor of Justice conducted the test of proportionality[190] and concluded that Section 111^1 of the ECA does not infringe any rights set out in the Constitution when obliging the electronic communication undertakings to collect and retain communications' metadata.[191]

The Supreme Court was faced with the question of constitutionality of the metadata collection regulation when an offender's guilt was proven on the basis of information received from a telephone service provider. At the time of proceedings,

combines the function of the general body of petition and the guardian of constitutionality. Such a combined competence is unique internationally

[186] Judgment of the European Court of Justice of 8 April 2014, joined cases C-293/12 and C-594/12, para 73.

[187] Chancellor of Justice's Position on the Constitutionality of Subsection 1111 of the ECA, p. 2. http://oiguskantsler.ee/sites/default/files/field_document2/6iguskantsleri_seisukoht_vastuolu_mittetuvastamise_kohta_elektroonilise_side_andmete_kogumine_sideettevotete_poolt.pdf. Accessed 23 February 2016. The reasoning arose from the fact that Member States had some discretion for the implementation of the directive into national law.

[188] Ibid.

[189] Ibid., p. 4.

[190] The proportionality test undertaken by the Chancellor of Justice is stipulated in Section 11 of the Constitution which states that rights and freedoms may only be circumscribed in accordance with the Constitution. Such circumscription must be necessary in a democratic society and may not distort the nature of the rights and freedoms circumscribed. The proportionality test comprises of: (1) an analysis of whether the interference with the right to inviolability of private and family life had a legitimate aim; (2) an analysis whether the interference is permitted by law (see Subsection 26(2) of the Constitution); and (3) an analysis whether the interference was suitable (does it have effect), necessary for achieving the aim (there is no less onerous measure) and reasonable considering the competing interests.

[191] Chancellor of Justice's Position on the Constitutionality of Subsection 1111 of the ECA, p. 11. It must be noted that the Chancellor of Justice highlighted that executive authorities (e.g. Data Protection Inspectorate) must further analyse Subsection 1111(9) of the ECA which describes the security and data protection requirements the communications undertakings must ensure in order to specify the obligations set out therein and to limit possible misuse.

requiring telephone communication metadata was a surveillance activity[192] and required a warrant. The accused claimed that the surveillance protocol was inadmissible as the data retention directive (Directive 2006/24/EC) had been declared invalid and as a result, the evidence was automatically inadmissible.[193] The Supreme Court did not agree with this. The Supreme Court noted that the invalidity of an EU directive does not automatically mean that national law is also invalid, as Member States have some discretion taking into account the objectives of the directive.[194] The Supreme Court accepted that the collection and retention of communication metadata prejudiced the rights of inviolability of private life of the accused. However, the Supreme Court stated that the circumscription of the right to private life was legitimate as inquiring telephone communication metadata helped to detect and apprehend a criminal and as such, it was an appropriate measure. Also, the Supreme Court highlighted that it was a necessary measure due its effectiveness.[195] When analysing the proportionality of the surveillance activity *in strictu sensu*, the Supreme Court stipulated that the retention period of one year is not an 'excessively long term'[196] and as the procedures of obtaining the warrant were followed correctly, the Supreme Court concluded that the CCP, when permitting requesting of communication metadata in the context of surveillance activities was in conformity with the Constitution.[197]

Applicable Checks and Balances

Inquiring Telephone and Internet Communication Metadata

Investigative bodies should only make inquiries to telephone service providers and Internet access providers if it is unavoidably necessary for the achievement of the objectives of criminal proceedings according to Subsection $90^1(3)$ of the CCP. Although there is no publicly available statistics on the number of requests made under Subsection $90^1(1)$ of the CCP to receive call logs or requests made under Subsection $90^1(2)$ of the CCP inquiring telephone and Internet communication metadata not covered with the call log on the basis of the permission of the Prosecutor's Office, it is questionable that the principle of *ultima ratio* is always followed. Moreover, as the Supreme Court has seen collection and retention of communication data as a legitimate mean to accomplish the goals of a criminal procedure, the question arises how much space for *ultima ratio* is actually left.

---

192 At the moment, it is not. See Subsection 901 of the CCP.

193 Judgment of the Supreme Court of 23 February 2015, case number 3-1-1-51-14, para 19.

194 Ibid., para 21.

195 Ibid., para 22.

196 Ibid., para 22.3.

197 Ibid., para 22.4. In their dissenting opinion, two Supreme Court Justices found that the Supreme Court was false to analyse the constitutionality of the then-in-effect provision from the CCP and should have focussed its analysis on the provisions of the ECA. Instead, the Supreme Court focussed its analysis on Section 117 of the CCP effect at the time of procedure which enabled making inquiries to communication authorities on the metadata retained by them.

Especially in the case of inquiring call logs which does not require the permission from the court nor the Prosecutor's Office under *de lege lata.*[198] Therefore, further clarification would be required as inquiring communication metadata prejudices the right to private and family life. The CCP should set out specific crimes for which such a measure may be used and the investigative bodies should be forced to give written reasons for the necessity to inquire call logs and metadata and such reasons should be added to the criminal file.[199]

Surveillance Activities

The CCP highlights that surveillance activities are permitted only on the bases provided for in the CCP itself and only if the collection of data by other activities or taking of evidence by other procedural acts is impossible (*ultima ratio*).[200] The activities may not endanger the life or health of the persons or cause unjustified property and environment damage or unjustified infringement of other personality rights.[201] Information obtained by surveillance activities constitutes evidence in criminal proceedings only if application for and grant of authorisation for surveillance activities and the conduct of surveillance activities is in compliance with the requirements of the law.[202]

Entrapment

The CCP provides two possible means of entrapment—staging a criminal offence (see '*Staging a criminal offence*') and using a police agent (see '*Observing a postal item, wire-tapping and using a police agent*'). However, staging is not used for the detection and apprehension of criminals relevant for this research. Therefore, this sub-chapter only focuses on the use of police agents for the identification and apprehension of online criminals.

A 'police agent' is a *person* who collects evidence in a criminal proceeding by using a false identity.[203] The statements of a police agent are used as evidence pursuant to the provisions of the CCP concerning witnesses.[204]

Using a police agent is very relevant in the detection and detention online offenders. This has been proved in the case law in the context of sexual enticement of children and concluding agreements to meet with minors for sexual purposes.

[198] As of 23 February 2016, there is no case law concerning Subsection $90^1(1)$ of the CCP.

[199] Antonova 2015, p. 51. Statistics from 2012 (when inquiring call logs or full set of metadata was a surveillance activity) indicates that there were 4060 (58% of all surveillance activities) permissions for obtaining call logs and 1444 (21% of all surveillance activities) permissions to receive a larger set of metadata. See Crime in Estonia 2012, Tallinn 2013, p. 34.

[200] Subsection 1262(2) of the CCP.

[201] Subsection 1262(3) of the CCP.

[202] Subsection 1262(4) of the CCP.

[203] Subsection 1269(1) of the CCP.

[204] Subsection 1269(4) of the CCP.

The case law indicates that the police has impersonated an 11-year-old Melissa,[205,206] an 11-year-old A[207] and 12-year-old K.[208] This means that police agents impersonate children with whom perpetrators speak via online channels. According to their best knowledge, the perpetrators are speaking to a child, when in fact they are speaking to a police officer conducting surveillance activities. The acts of the perpetrators have been qualified as impossible attempts to entice or commit an act of sexual nature with a child (see '*Attempt*' and '*Sexual Enticement of Children*' above).

8.3.5 Succinct Overview of Investigatory Powers in an Online Context

To sum up Sects. 9.0 and 9.0, in an online context, the investigative bodies may:

- request data from communications undertakings to trace and identify the source, date, time duration of the communication and the terminal equipment and destination of the terminal equipment of the user;
- In the context of surveillance activities, surveillance agencies may:
- enter into a computer system;
- wire-tap or intercept information;
- use entrapment
  - by using a police agent; and
  - by staging a criminal offence.

---

[205] Judgment of the Harju County Court of 20 December 2011, case number 1-11-13285, p. 3. According to the facts of the case, the offender used a social network site to make a proposal to meet what the offender thought would be an 11-year-old girl named Melissa in a gas station with the purpose of going into the woods and masturbate and have Melissa watch. Melissa was in fact a police agent conducting surveillance activities. Thereby, the offender committed an impossible attempt to meet with a child for sexual purpose (Section 178^1 of the PC) and was charged with a one-year imprisonment.

[206] Judgment of the Harju County Court of 27 September 2012, case number 1-12-6471, p. 1. In that case, the offender sexually enticed an 11-year-old M who was actually a police agent conducting surveillance activities and proposed to meet with M at a gas station and thereafter commit an act of sexual nature with M. Thus, the offender committed an illegal attempt of meeting a child for sexual purposes. The offender was punished by 9 months' imprisonment.

[207] Judgment of the Harju County Court of 17 December 2012, case number 1-12-11607, p. 1. The offender sexually enticed (described oral sex and sent videos to A where the act was depicted an 11-year-old A. The same offender sexually enticed and proposed to meet a 13-year-old 'Kity13' for sexual purposes. A was a police agent conducting surveillance activities and Kity13 was a journalist conducting a journalistic experiment. The offender was punished by a 18 months' imprisonment.

[208] Judgment of the Harju County Court of 19 April 2013, case number 1-13-3413, pp. 3–4. The offender sexually enticed a 12-year-old K and proposed to meet at a gas station and to engage in sexual intercourse during the meeting. K was a police agent conducting surveillance activities. The offender was punished by a pecuniary punishment of 400 daily units.

8.3.6 Application of Relevant Investigatory Powers to the Sweetie Case

Due to the novelty of the approach introduced by Sweetie, there has not been neither political nor academic discussions about the inclusion of virtual avatars such as Sweetie in the fight with online sexual abuse against minors and children. Although Sweetie's success of identifying more than a thousand paedophiles was covered across the news portals of Estonia,[209] no serious debate followed. Nevertheless, experts in the child protection area in the Police and at the Prosecutor's Office are very well aware of Sweetie.[210]

Under *de lege lata*, the Police uses police agents in the context of surveillance activities to investigate sexual crimes against minors and children.[211] Using police agents requires a permission of the Prosecutor's Office and may be used only for collecting information in a criminal proceeding about a criminal offence (i.e. Base 4).[212]

The definition of a 'police agent' refers to the fact that the agent is in fact a human being who uses a false identity.[213] Usually, it is an official of the Police. Therefore, the fact that Sweetie is not a real person deprives it the possibility to be used as a 'police agent'. However, if Sweetie or virtual avatars such as Sweetie could be operated by the Police, it could be argued that it is unproblematic as the police agent would swap its identity with the imaginary identity of Sweetie and it would not differ much of the situation today where police agents depict imaginary children. Nevertheless, the law today is created without considering the application of virtual software for the apprehension of online sexual abusers. Consequently, the word-by-word reading of the CCP does not permit the use of Sweetie. However, if the CCP is modified to explicitly permit virtual avatars, this would be unproblematic from the material criminal law point of view as already today, the Police punishes perpetrators by using police agents to identify their intent to commit sexual abuse of children (see case law in '*Sexual Enticement of Children*' and '*Entrapment*'). Therefore, if the CCP would permit this, the crimes of sexual enticement and arrangements to meet children for sexual purposes that are qualified today as

---

[209] See Postimees from 4 November 2013 '10-aastase filipiinlanna mängimine aitas tuvastada üle tuhande pedofiili'. http://maailm.postimees.ee/2585494/10-aastase-filipiinlanna-mangimine-aitas-tuvastada-ule-tuhande-pedofiili. Accessed 24 February 2016. See Postimees from 21 April 2014 'Tarkvaraline avatar jahib veebis pedofiile'. http://majandus24.postimees.ee/2768310/tarkvaraline-avatar-jahib-veebis-pedofiile. Accessed 24 February 2016

[210] See Laurits, E., Seksuaalne väärkohtlemine läbi internetikeskkonna. https://www.youtube.com/watch?v=uj-1FLfS6YAwww.youtube.com/watch?v=uj-1FLfS6YA. Accessed 24 February 2016.

[211] See case law in '*Sexual Enticement of Children*' and '*Entrapment*'.

[212] Subsection 1263(2) of the CCP.

[213] Subsection 1268(1) of the CCP.

impossible attempts[214] could also be impossible attempts committed against Sweetie. However, such changes in the CCP cannot be made easily and a public debate should precede.

8.3.7 Relevant Aspects of Digital Forensic Evidence

'Evidence' under the CCP means the statements of a suspect, accused, victim, the testimony of a witness, an expert's report, the statements given by an expert upon provision of explanations concerning the expert's report, physical evidence, reports on investigative activities, minutes of court sessions and reports on surveillance activities, and other documents, photographs, films or other data recordings.[215] However, this is not a closed list and evidence not listed above may also be used in order to prove the facts relating to a criminal proceeding, except in the case the evidence has been obtained by a criminal offence or violation of a fundamental right.[216]

The CCP sets out that surveillance activities should be recorded whenever possible. Subsection $126^5(2)$ of the CCP stipulates that the covert surveillance, collection of comparative samples and the conduct of initial examinations shall be video recorded, photographed or copied or recorded in another way if necessary (e.g. audio recorded). The same is set out for the covert examination of postal items[217] and staging a criminal offence.[218]

In case of wire-tapping and interception of information, recording of the information obtained by wire-tapping or interception of messages or other information transmitted by the public electronic communications is compulsory.[219]

The rules on documentation of surveillance activities set out that if necessary, the photographs, films, audio and video recordings and other data recordings made in the course of surveillance activities shall be appended to the surveillance report.[220]

The Forensic Examination Act sets out that it is possible to inquire from the Estonian Forensic Science Institute an 'information technology examination' which is a type of forensic evidence examination.[221] The Institute has a separate

---

214 See '*Attempt*'.

215 Subsection 63(1) of the CCP.

216 Subsection 62(2) of the CCP.

217 Subsection 1266(3) of the CCP.

218 Subsection 1268(2) of the CCP.

219 Subsection 1267(1) of the CCP.

220 Subsection $126^{10}(2)$ of the CCP.

221 Subsection 277(1) of the Forensic Examination Act.

department of information technology examination services.[222] A special subset of information technology examination is an information technology examination of materials relating to sexual abuse of a minor.[223]

8.4 Evaluation

8.4.1 Substantive Criminal Law

Substantive criminal law punishes intent to commit sexual crimes against children and minors. This intent does not need to materialise as attempt is also punishable.

Today, the Police uses surveillance activities to apprehend persons who sexually entice or wish to meet with children. Entrapment entails using police agents who, like Sweetie, are not who they claim to be. Police agents are not in fact minors or children and Sweetie is not real. Situations where a person has sexually enticed or wished to meet with a minor or a child who was actually a police agent conducting surveillance activities are qualified as impossible attempts (as the object of the crime is not in fact the minor or child anticipated by the perpetrator) and punishable because the intent of the perpetrator is clearly identified.

The fact that Sweetie is neutral means that the Police officer who would control Sweetie would not be posed with any moral or legal constraints (e.g. making child pornography available). However, the fact that Sweetie is neutral, means that requesting access to and watching Sweetie via webcam is not punishable *per se* under Estonian law.

However, if a person sexually entices Sweetie via webcam (e.g. masturbates), the question arises whether such an act poses an abstract threat to the society or not. On the one hand, it could be argued that no threat is posed by enticing a virtual avatar lack of personality. On the other hand, it could be said that if the person thinks Sweetie is real and intents to entice Sweetie, this should be punishable. This is a clash of the *ultima ratio* principle and the expectations of the society to define which acts are considered criminal. However, if case law has deemed enticing police agents who are not in fact children punishable, the situation should not differ in case of Sweetie.

In case of the offence of meeting with a minor for sexual purposes, the word-by-word reading of the provision implies that the victim must be a person. Further, the illegal act is committed only when the perpetrator makes a preparatory act (e.g. appears at the agreed location). If police agents are used, they are still persons. However, not who they claim to be as they use a false identity. Thus, in

[222] Subsection 11(3) of the Regulation on the European Fund for Strategic Investments, the European Investment Advisory Hub and the European Investment Project Portal and amending Regulations (EU) No 1291/2013 and (EU) No 1316/2013 (EFSI Regulation).

[223] Regulation on the list of examinations conducted in EFSI, Subsection 5(3)(1).

case of Sweetie, the first obstacle is its virtual character. However, taking into account that Sweetie is ultra-real, the person might not even make the difference. Thus, if a person wishes to meet with Sweetie and makes the necessary preparatory act, logically, such an act should also be punished as impossible attempt under Estonian law taking into account that this criminal act is a delict of abstract danger.

8.4.2 Substantive Criminal Procedure Law

Criminal procedure law today uses entrapment to apprehend criminals who commit sexual crimes against children and minors via computers. In multiple cases, police agents, who use a false identity (impersonating a child) have proved their worth. However, the same logic cannot be extended automatically to Sweetie. Although in Sects. 8.3.6 and 8.4.1, hypothetical constructions on the use of Sweetie have been made, Sweetie is not legal today. The CCP would need to explicitly permit Sweetie and using Sweetie as a police agent or as a means to identify and apprehend sexual offenders online would be most probably seen as a surveillance activity as the opposite could make Sweetie an illegal *agent provocateur*. Furthermore, a thorough debate would need to precede to the acceptance of Sweetie as it could lead to possible uptake of other virtual detection and apprehension tools in the future.

8.5 Summary and Conclusions

In summary, the substantive legal framework for the protection of minors against webcam sex in Estonia could be considered sufficient. In co-effect with the procedural rules of CCP and ECA, the Police and the Prosecutor's Office have a multitude of procedural means to identify and apprehend offenders in the online context and the substantive criminal is there to support this. Whether the adoption of Sweetie by the Police would enable the Police to identify more crimes related to sexual abuse of minors and children is unknown, however, probable.

However, the use of Sweetie as a procedural tool cannot be decided without a debate on both the expert level and on a larger societal scale. Taking into account the public opinion that sexual abuse of children and minors in the Internet is a problem, the author hopes that Sweetie would get positive feedback. However, it remains unanswered at the moment how much gain it would actually provide to the Police.

In order to adopt Sweetie as tool for the Police officers who identify and apprehend sexual enticers and paedophiles, the Code of Criminal Procedure of Estonia needs to be modified and a further analysis needs to be conducted which caveats lie ahead in the context of the Penal Code. Sweetie requires a firm basis in the law, especially when it would be considered a surveillance activity.

References

Act to Regulate Dissemination of Works which Contain Pornography or Promote Violence or Cruelty. https://www.riigiteataja.ee/en/eli/520012015009/consolide. Accessed 23 January 2018.

Antonova J (2015) Isikuandmtete kaitse kohtueelses kriminaalmenetluses. Eestis. Tartu. http://dspace.ut.ee/bitstream/handle/10062/47778/antonova_julia.pdf. Accessed 29 February 2016.

Chancellor of Justice's Position on the Constitutionality of Subsection 111 of the ECA. http://oiguskantsler.ee/en/estonian-model-of-the-institution-of-the-chancellor-of-justice-0. Accessed 23 February 2016.

Constitution of the Republic of Estonia. https://www.riigiteataja.ee/akt/115052015002. Accessed 23 January 2018.

Crime in Estonia (2012) Ministry of Justice, Tallinn 2013. www.kriminaalpoliitika.ee/sites/krimipoliitika/files/elfinder/dokumendid/17._kuritegevus_eestis_2012_0.pdf. Accessed 23 January 2018.

Crime in Estonia (2015) Ministry of Justice, Tallinn, 2016. www.kriminaalpoliitika.ee/sites/www.kriminaalpoliitika.ee/files/elfinder/dokumendid/kuritegevus_eestis_2015.pdf. Accessed 10 May 2015.

Crime in Estonia (2016) Ministry of Justice, Tallinn 2017. www.kriminaalpoliitika.ee/sites/krimipoliitika/files/elfinder/dokumendid/kuritegevus_eestis_est_web_0.pdf. Accessed 23 January 2018.

Cybercrime Convention.

Directive 2011/92/EU of the European Parliament and of the Council of 13 December 2011 on combating the sexual abuse and sexual exploitation of children and child pornography, and replacing Council Framework Decision 2004/68/JHA.

European Convention on Human Rights.

Forensic Examination Act. https://www.riigiteataja.ee/en/eli/ee/530102013102/consolide/current. Accessed 23 January 2018.

Kergandberg E, Pikamäe P (2012) Kriminaalmenetluse seadustik: kommenteeritud väljaanne. Tallinn.

Kirwan G, Powers A (2013) Cybercrime: the psychology of online offenders. Cambridge University Press.

Laurits E (2014) Virtuaalse isiku kujutamise probleemid karistusseadustiku § 178 kontekstis. Juridica 5: 395–405.

Law ratifying the Lanzarote Convention. https://www.riigiteataja.ee/akt/203112016001. Accessed 23 January 2018.

Lõhmus U (2012) Põhiõigustest kriminaalmenetluses. Tallinn, Juura.

Optional Protocol on the sale of children, child prostitution and child pornography.

Penal Code. https://www.riigiteataja.ee/en/eli/509012018005/consolide. Accessed 21 January 2018.

Pikamae P, Sootak J (2015) Karistusseadustik. Kommenteeritud väljaanne. Tallinn, Juura.

Ploom T, Tamm K (2011) Seksuaalse enesmääramise eapiiri muutmise analüüs. Ministry of Justice.

Regulation (EU) 2015/1017 on the European Fund for Strategic Investments, the European Investment Advisory Hub and the European Investment Project Portal and amending Regulations (EU) No 1291/2013 and (EU) No 1316/2013.

Rosin K (2016) Euroopa Liidu kriminaalõiguse areng Lissaboni leppe jõustumise järel. Juridica 9: 659–668.

Security Authorities Act. https://www.riigiteataja.ee/en/eli/521062017015/consolide. Accessed 20 January 2018.

Siitam-Nyiri K (2014) Karistusõiguse areng revisjonist Euroopa Liidu ühtse karistusõiguseni. Juridica 8: 576–581.

The Code of Criminal Procedure. https://www.riigiteataja.ee/en/eli/512122017005/consolide. Accessed 21 January 2018.
The Electronic Communications Act. https://www.riigiteataja.ee/en/eli/521082017008/consolide. Accessed 23 January 2018.
UN Convention on the Rights of the Child.

Kaspar Kala is currently project manager and legal expert at Proud Engineers. Previously, he worked as the data protection officer and legal advisor of the E-services Development and Innovation Department at the Ministry of Social Affairs of Estonia. Also, he has worked in a law firm dealing with intellectual property and data protection issues and with the Ministry of Economic Affairs and Communications working with e-government and data ownership issues. He has a BA degree from the University of Tartu and a LL.M. degree from the University of Tilburg. Currently he is in the process of obtaining an MSc in IT at the University of Tartu. Kaspar's interest in criminal matters arose during BA studies when he was a trainee at the Police and Border Guard Southern Prefecture's Criminal Bureau (division of criminal assaults) and at the Southern District Prosecutor's Office of Estonia. However, he has not worked as a criminal lawyer.

Chapter 9
Substantive and Procedural Legislation in Israel to Combat Webcam-Related Child Sexual Abuse

Asaf Harduf

Contents

Abstract The Sweetie project is designed to detect and collect evidence against online attempts made by adults to sexually interact with children who are victimized by sex abusers. This chapter analyses the compatibility of Israeli law with the utilization of the Sweetie software. The Israeli substantive law has undergone very few changes to accommodate the age of cyberspace. Most of the Israeli sex offenses

A. Harduf (✉)
Zefat Academic College, Zefat, Israel
e-mail: asafhardoof@gmail.com

S. van der Hof et al. (eds.), *Sweetie 2.0*, Information Technology and Law Series 31,
https://doi.org/10.1007/978-94-6265-288-0_9

have kept their physical-world version, including rape, forbidden intercourse by consent, sodomy and indecent act. By judicial broad interpretation, many of these offenses have been applied to online interaction between adults and minors online, including statutory rape and rape by fraud. Some of these offenses have been applied to interaction between adults and fake profiles of children. Offenses of obscenity, also not formally defined by Israeli law as sex offenses, have changed a little bit to accommodate for the computerized age. Today they are applied in Israel regarding any form of child nudity and child pornography. The Israeli law and jurisprudence do not define obscenity. Like the substantive law, Israeli procedural law is largely unimpressed by cyberspace. Some investigatory powers have not been set, such as expedited preservation of stored computer data. Other powers are very easy to use, such as seizure and search of computers in any offense suspected. There are no laws regarding entrapment. Undercover operations are not mandated by any law, but still they are far from rare. Israeli law of evidence is extremely thin in content and almost in all instances the court does not have to exclude evidence that was illegally obtained. The Israeli constitutional law has had very little effect on Israeli criminal law. In light of the above, and due to the easiness of operating agents online, the chapter suggests that possible consequences and implications must be contemplated before using a tool like Sweetie, including the effects on social media and of other fields of law, and after carefully analysing the destination of the initiative. Regarding online agent operation, especially paedophile operations, taking suspect human rights into account must be done in advance in a well-balanced and accountable procedure, since no remedy will be given later to anyone assumed to be a paedophile.

Keywords Sweetie · Paedophile · Undercover Operations · Criminal Law · Cyberspace · Virtual Child · Sexual Abuse · Sex Offenses · Online Agents · Online Enforcement · Human Rights · Accountability

9.1 Introduction: Legislation in Israel

9.1.1 General Description of the Legal Framework

The Israeli legal system has intricate and diverse roots. Israel is a young state, declared in 1948, a few years after the Holocaust. Prior to the declaration of independence, Britain had governed Palestine for many years, establishing a local Common Law system, but without jury. The Penal Code in Israel was similar to codes the British used for other colonies and territories. At the birth of Israeli legal framework, Israel has adopted most mandatory laws set by the British. The Israeli legal system is based in common law but also includes aspect of civil law.

At its first decades, the Israeli judicial system was significantly influenced by the British law and looked for British precedents to resolve legal debates. In time, the

Israeli courts had gained confidence and needed no longer to rely on foreign ruling in order to establish their verdicts. However, many of the laws set by the British sovereignty are still in effect even nowadays, as well as few of the Ottoman regime law which preceded the British rule.

While there was a methodical broad reform in the general part of the Israeli Penal Code in 1994, no such reform has ever been made in the general part of the code, i.e. the part of offenses. Much of the Israeli Penal Code is still based upon the British mandatory laws. For example, the Israeli definition of theft, section 383 of the Penal Code, is an exact translation of the British definition of theft as codified in 1916 Britain. More than an entire century later, while the British have codified a unified law of theft to suit the modern world of 1968, the Israeli legal definition of theft remains completely unchanged. This phenomenon is far from rare in Israel.

Many criminal proscriptions are found outside the Penal Code, scattered around the Israeli books of law. Some of these laws are criminally oriented; others combine civil and criminal provisions. There is no system of criminalization in Israel and finding all Israeli criminal prohibitions is a task that has never been done.

As for interpretation of criminal prohibitions, section 34U of the Penal Code requires the court to interpret in favour of the defendant in cases of unclear prohibitions. However, this section is seldom relied upon. Broad interpretation that relies on the hypothesized purpose of the law according to the court, designated "purposive interpretation", is the dominant form of interpretation in Israel in all legal fields, including criminal law.

Israel does not have a formal constitution. Some legislative acts, designated "Basic Laws", had been recognized by the Supreme Court as part of a constitution in the making. This informal constitution has had some influences over the criminal law, mainly over procedural law and a little bit over evidence law. However, it had nearly no effect whatsoever over substantive criminal law. In over twenty years of the Israeli so called "constitutional revolution" that followed two major basic laws established by the Israeli legislator ("the Knesset") in 1992, not once did the Supreme Court strike down a prohibition for being unconstitutional. In fact, only one offense was constitutionally scrutinized by the Supreme Court in Israel: organizing a prohibited game, regarding a bingo club. The offense was found constitutionally valid and all three judges reprimanded the petitioners for having brought up a constitutional petition.

The Israeli system includes diverse tribunals. Some specialize in one legal field, such as family law, labour law, military law and so forth. The general courts are of three kinds. The basic instance is the Magistrate court. It is authorized to judge offenses that include punishment of no more than seven years imprisonment (with few exceptions). A typical case is tried by one judge only. The judge is in charge of implementing the procedural, the evidential and the substantial law, as well as for fact finding. In criminal law, as opposed to civil law, there is no way to appeal most rulings before the entire case is complete. For these and other reasons, the Israeli judge is very powerful, mainly in criminal cases.

The District court is a higher instance. The six District courts in Israel are authorized to judge offenses that the Magistrate courts are unauthorized to judge, as well as appeals over them. Typical cases are tried by a panel of three judges who decide by virtue of majority. The most severe penalty they are authorized to inflict is death; however, other than two Nazi cases in all the history of Israel (one of which repealed), it is mainly a theoretical penalty. The most severe punishment in practice is imprisonment for life; it is inflicted mainly and regularly in homicide cases.

The highest instance in Israel is the Supreme Court. It deals with criminal and civil appeals over the District Courts; and also resides as the High Court of Justice, which rules in administrative and constitutive affairs. The Supreme Court's verdicts are usually final; however, sometimes a "further hearing" takes place, in which the Supreme Court re-evaluates its own judgment, due to inconsistent ruling or an important, difficult or novel ruling. The verdicts of the Supreme Court are publicized and updated daily in the judicial system's website. Most cases heard and decided in length in the Supreme Court are criminal law cases.

The conviction rate (at least a partial conviction) in Israel is near a hundred percent. Most cases end either by a plea-bar or by an admission of guilt and asking for leniency. When defendants plead "not-guilty" and evidence is heard, typical trials focus on factual issues of evidence (rather than laws of evidence), sometimes on procedure and seldom on substantive law. In other words, most cases discussed in criminal courts centre their attention on "who did what" questions. This makes the substantive criminal law under-analysed. Add to that the Israeli academy tendency (not to say obsession) to focus on foreign and mainly American scholarship and you get a highly undeveloped body of law.

9.1.2 Relevant Treaties and Cybercrime Laws

Israel has asked to join the Cybercrime Convention and is currently advancing towards joining, including taking steps to line up with the requirements of the convention.

Israel has signed and ratified the UN Convention on the Rights of the Child 1989 in 1991. Israel is also a party of the Protocol on the sale of children, child prostitution and child pornography. This fact has been stressed in the recent DOJ's bill to update the prohibition on obscenity.

Israel has not joined the Lanzarote Convention.

The Israeli Penal Code includes various provisions setting jurisdiction for offenses committed outside Israel. Two of these provisions relate to international law: section 16, "Offenses against international law", and section 17, "Vicarious applicability", regarding conventions. Although these provisions exist over forty years, to my knowledge they have never been applied.

9.2 Analysis of Substantive Criminal Law

9.2.1 Introduction

Crimes involving sexual abuse of minors are found mainly in the Israeli Penal Code. Section 34X of the General Part of the Penal Code maintains that a "minor" is a person under the age of eighteen.

The serious sex crimes are found in Chapter ten, Article five: "Sex Offenses". Most of them contain both general and minor-related aspects. I will hereby address only the minor-related aspects.

The first sex offense, *Rape*, is defined in section 345 of the Penal Code. Its first minor-related aspect is the statutory rape, defined in Section 345(a)(3). It proscribes having intercourse with a woman below the age of fourteen, even with her consent. According to Section 345(c), one "has intercourse" when introducing any part of the body or any object into the woman's sex organ.

The maximum penalty for each of the five forms of non-aggravated rape is 16 years imprisonment.

The second minor-related aspect of the offense is the *Aggravated Rape*. Section 345(b)(1) maintains that four types of rape are aggravated when the victim is under the age of 16:

- intercourse with a woman without her free consent, combined with Section 345 (a)(1);
- intercourse with a woman with her consent, "obtained by deceit in respect of the identity of the person or the nature of the act", combined with Section 345(a)(2);
- intercourse with a woman "by exploiting the woman's state of unconsciousness or other condition that prevents her from giving her free consent", combined with Section 345(a)(4);
- intercourse with a woman "by exploiting the fact that she is mentally ill or deficient, if—because of her illness or mental deficiency—her consent to intercourse did not constitute free consent", combined with Section 345(a)(5).

The maximum penalty for each of the above is 20 years imprisonment. The second sex offense, forbidden intercourse by consent, is set in Section 346. Although it also has a general aspect, regardless of age, it focuses on the age of the victim. Having intercourse with a minor between the ages 14 and 16 without marriage is one form of the offense. Another form is when the minor is between the ages 16 and 18 and the perpetrator exploits a relationship of dependence, authority, education or supervision, or makes a false promise of marriage. The maximum penalty is 5 years imprisonment.

Rape and forbidden intercourse by consent are gender-specific and apply to women victims only. The following offenses are general and not gender specific.

The offense of sodomy is set in Section 347. In some senses it reflects and completes the former two offenses, although it is not gender-specific. "Sodomy" is defined as an introduction of a bodily organ or an object into a person's anus, or introduction of a sex organ into a person's mouth. When done either on a minor between the ages 14 and 16, or on a minor between the ages 16 and 18 by exploiting relations of dependence, authority, education or supervision, it is parallel to forbidden intercourse by consent and the perpetrator is liable to five years imprisonment maximum.

When the act is done in one of the circumstances specified in Section 345 (rape), the perpetrator is liable to the penalties of a rapist.

Section 353 restricts criminal responsibility of forbidden intercourse by consent and sodomy.

The next sex offenses are Indecent Act (Section 348) and Indecent Act in Public (Section 349). Indecent act is "an act made for sexual arousal, satisfaction or abasement".

The first form of indecent act is set upon scenarios described in the basic offense of rape, when no intercourse took place. An act made for sexual arousal, satisfaction or abasement (and does not constitute intercourse or sodomy), when committed on a minor under the age of 14, sets liability of up to seven years imprisonment, even with the minor's consent. The second form of indecent act is set upon scenarios described in the aggravated offense of rape when once again no intercourse took place. Regarding minors, an indecent act that combines a minor-victim under the age of 16 with other aspects of the basic rape, such as achieving the victim's consent "by deceit in respect of the identity of the person or the nature of the act" (Section 345(a)(2)), may bring up to ten years imprisonment. A third form of indecent act relating to minors is 348(d)(1): committing an indecent act on a minor who is over 14, by exploiting a relationship of dependence, authority, education, supervision, employment or service, is a crime punishable by four years imprisonment.

Indecent act in public is defined in Section 349. Its first aspect sets a one year imprisonment liability. It focuses on the non-consent of the victim and does not relate to the age of the victim. The second aspect relates to minors under the age of 16: committing an indecent act in any place whatsoever before such a person is an offense punishable by three years imprisonment.

Another important section does not relate directly to minors; nevertheless, in practice it is implemented mainly regarding cases of minor-victims, as will be explained later: Section 350, Responsibility. This section proclaims that for purposes of sex offenses, it is immaterial whether a person performed an act or caused an act to be performed on him or on another person.

Section 351 focuses on minors: Sex Offenses within the Family and by Persons Responsible for Helpless Persons. When the above sex offenses are committed on a person who is a minor and is related to the perpetrator, then the initial punishment becomes more severe.

Three other forms of offenses are not legally defined as "sex crimes" but must be mentioned. Two of them are defined in Chapter eight, Article Ten: Prostitution and Obscenity.

As for Prostitution, Section 203B codifies aggravating circumstances for the general offenses of inducement to act of prostitution (Section 201) and inducement to engage in prostitution (Section 202), when the person induced is a minor. Section 203C criminalizes the act of consuming from a minor, whereas consuming prostitution from adults is not criminalized. The publication of a minor's prostitution is prohibited under Section 205A, even when the prostitution service is provided abroad. Section 208 criminalizes the act of permitting a minor to reside in brothel.

As for Obscenity, Section 214 (b) prohibits the publication of obscenity material that include minors, even representation or a drawing of minors. Section 214(b1) criminalizes the utilization of minors in obscenity. This act is aggravated if the person who utilizes them is the one responsible to their wellbeing. Section 214(b3) prohibits the possession or consumption of an obscene publication that includes the likeness of a minor, excluding those who possess or consume incidentally and in good faith.

In Israel there are no direct laws regarding child pornography and the "child pornography" does not appear in any prohibition; therefore the law of obscenity is currently practiced in Israel to deal with cases of sexual materials of minors. The law does not define the term "obscenity" and no criminal court in Israel has ever tried defining it. Although the prohibitions also cover materials that do not involve minors, the law as practiced in the last couple of decades focuses exclusively on materials that include minors.

The last type of sexual offense I will mention is the Law for Prevention of Sexual Harassment, 1998. It criminalizes sexual harassment and sets out various alternatives of the offense. One of which specifically deals with minors: Section 3(a)(6)(a) maintains that sexually oriented offers or references made towards a minor are forbidden even if the victim has not demonstrated reluctance, as long as the perpetrator exploits a relationship of dependence, authority, education or supervision. If the minor is under the age of 15, the offense is completed even without the above exploitation.

9.2.2 Possibly Relevant Criminal Offences

Succinct Overview of Sexual Offences Involving Minors

The following table lists the possibly relevant provisions of criminal law in Israel, grouped together by the provisions of the Lanzarote Convention, which gives the most comprehensive catalogue of sexual child-abuse offences available (Table 9.1).

Overview of Sexual Offences Related to Webcam Child Sexual Abuse

To begin with, one must point to a key aspect of the Israeli Penal Code: the law of attempts, which is set in the Introductory Part, Chapter 5, Article 1.

Section 25 of the Penal Code defines (in a very vague way) what constitutes an attempt: committing an act that constitutes more than preparation but does not complete the offense, with intent to commit the complete offense. Section 26

Table 9.1 Overview of sexual offences involving minors

Lanzarote treaty	Israel
Article 18. Sexual abuse	The Penal Code, Sections 345–351
Article 19. Offences concerning child prostitution	The Penal Code, Sections 203B, 203C, 205A
Article 20. Offences concerning child pornography	The Penal Code, Section 214 (as practiced)
Article 21. Offences concerning the participation of a child in pornographic performances	The Penal Code, Section 214 (b1)
Article 22. Corruption of children	The Penal Code, Section 208
Article 23. Solicitation of children for sexual purposes	The Penal Code, Section 203B
[other offences, not covered by the Lanzarote Convention]	The Law for Prevention of Sexual Harassment

[*Source* The author]

maintains that it is immaterial that commission of the offense was impossible. Section 28 sets an exemption because of remorse, a provision that to my knowledge has never been applied in court but has been informally applied by the prosecution in its decision to avoid indictment in certain offenses.

In scenarios in which there are no actual minors involved, all offenses related to webcam sexual child-abuse require the law of impossible attempt; otherwise, no offense is made at all. One cannot complete an offense which is set to protect minors where there are no minors at all; no offense in Israel protects avatars. The following discussion of various potentially applicable offenses to different forms of webcam sexual child-abuse assumes the law of attempts: none of the offenses applies on its own merit. Later I will address case-law regarding both actual minors victims and virtual ones. For the time being and in the spirit of the Sweetie Project, I will focus on scenarios in which there are only fictional minors.

Offenses of attempted sexual abuse might apply, due to Section 350, Responsibility. Historically this section was codified in anticipation of an amendment to the definition of "indecent act". According to the DOJ bill from 1986, a new definition was supposed to narrow down the latter term. Towards such change, the DOJ sought addressing a scenario in which the perpetrator, instead of the typical performance the indecent act on the victim, causes the victim to perform the act on the perpetrator; and so, Section 350 was born in 1988. Ironically, the definition of "indecent act" was not eventually amended, whereas Section 350 somehow found its way into the Penal Code, for no good reason. This section is currently used in order to apply offenses of sexual abuse to scenarios the legislator had not imagined, almost solely regarding offenses of minor-victims. Combining this section with the law of impossible attempt, the prosecution accuses people of attempted rape by fraud, according to Section 345(a)(2); attempted statutory rape, according to Section 345(a)(3); attempted aggravated rape, according to the combination of the previous two, Section 345(b)(1); attempted sodomy, according to Section 347(a), or attempted sodomy set on rape scenarios, according to Section 347(b); and eventually attempted indecent act, according to Section 348(a) or 348(b), and indecent act in public, Section 349(b).

One offense of child prostitution might also apply: Section 203B, attempted inducement to prostitution.

Offenses of child pornography are not likely to apply to webcam sexual child-abuse in cases of virtual children. However, they might apply in cases of real children: the utilization of minors in obscenity, Section 214(b1).

Finally, offenses of attempted sexual harassment under the Law for Prevention of Sexual Harassment might also apply to webcam sexual child-abuse in cases of virtual victims.

Fictional-Minors' Case Law

The most important and noted relevant Israeli case-law to date is LCrimA 1201/12 Ktiei vs. Israel (decided 9.1.2014). This is the first case-law in history of pedophile-entrapment set by the Israeli Supreme Court. It deals with a local Dateline-inspired television show produced by Channel 10, one of the only two commercial networks in Israel. Adult reporters went online and posed as 13 year old girls. Adult reporters went online and posed as 13 year old girls. They were approached by adult males and started conversing. If the conversation remained non-sexual, it was terminated; if it became sexual, the reporters' goal was bringing the "suspects" to meet the "child". The meeting took place at the child's home, while her parents are abroad. An eighteen years old actress portrayed the child and the meeting was nor more than a minute; the child would step out of the living room and in came the television host, beginning a process of televised shaming. Once the suspect would realize he was filmed he stepped out; there awaited police officers, arresting him.

Ktiei was the only defendant in this story who has reached the Supreme Court and maintained that he must be acquitted. He was also the only one acquitted by the Magistrate Court in this media legal story. However, the prosecution had won the appeal in the District Court. The District Court reprimanded the Magistrate Judge for not following an old precedent set by The Supreme Court in 1978 regarding the lack of legal consequences to entrapment. Eventually, in a third-instance hearing, the Supreme Court had affirmed the District Court conviction in two offenses: attempted indecency act and attempted sexual harassment.

The above Channel 10 story led to approximately twenty indictments, most of which had been concluded by admission of guilt without critical judicial review. However, some of them gave rise to diverse legal questions. One case focused on the defendant's plea against indicting him in with attempted rape instead of attempted indecency act, like the rest of defendants who had arrived to meet the "child". The District court denied the plea: Crim 1137/07 Israel vs. Ben-Guy (decided 21.10.2009). Eventually, the case was resolved by a plea-bar: the prosecution has changed the indictment into attempted indecency act and the defendant admitted guilt. Another defendant was charged with attempted indecency act in public, for having masturbated in front of a web camera in front of a "child". He claimed that the above fact does not amount to an offense. Nevertheless he was convicted and his appeal to the District Court was denied: CrimA 7476/09 Haim (decided 8.2.2000). Another related case has ended in the Magistrate court, convicting the defendant of attempted sexual harassment: Crim 2507/08 Israel vs. Rotem (decided 28.3.2011).

Exactly three years after Ktiei, the second and so far the last case-law in history of pedophile-entrapment set by the Israeli Supreme Court was decided: LCrimA 4275/16 Doe vs. Israel (decided 9.1.2017). The story began with a private citizen, self-declared as "The Paedophile Hunter", represented himself as a minor and chatted with the suspect. After the citizen involved The police, the suspect was interrogated and admitted to sexually assault his neighbor twenty years ago and his stepdaughter ten years ago, when she was 5 years old. While the neighbor confirmed his story, his spouse and daughter coherently denied that he ever assaulted the daughter. He was charged with assaulting the daughter and also charged with attempted sexual harassment, for his online chat with the citizen. The District Court convicted him in all charges. However, The Supreme Court repealed and acquitted him from the sexual assault. Now appeal was made on the attempted sexual harassment charge. He was sentenced to six months community sentence.

Another case worth mentioning regarding fictional minors is Adler. The defendant chatted with an adult woman. At some point she felt disturbed and asked her older sister to intervene. The older sister chatted with him for a while and later posed as a 14 year old girl, having watched a documentary on pedophile catching online. At some point he asked and received for her permission to turn on his webcam and touch himself. In the following days this occurred once again. The woman went to the police and pressed charges. Adler was charged with attempted indecency act in public; the law of attempt was required because there was no actual minor, on the one hand, and the sexual communication was consensual, on the other hand. The judge maintained the offense is applicable to cyberspace; however, many questions have surfaced, as the woman was unable to explain why she thought Adler was a pedophile to begin with and why some of the evidence was lost or even who helped her copy the evidence. A lingering doubt remained regarding Adler's mens rea and he was acquitted: Crim 13384-02-09 (decided 23.6.2011).

A recent case has shown a challenge to the prosecution's thesis of substantial law regarding the police pedophile-entrapment. Sexual chats took place between the suspect and a fictional 13 years old, operated by a police agent. He was charged with aggravated attempted indecency acts and with attempted sexual harassment. The defendant admitted the facts but maintained that they do not constitute as aggravated attempted indecency acts, but merely as attempted sexual harassment. The Magistrate Judge briefly denied the defense's claims and convicted the defendant on all charges: Crim 20996-08-15 *Israel v. Goldstein* (decided 17.7.2017). The defense intends to appeal.

Actual-Minors' Case Law

Other cases have included webcam sexual abuse of actual minors. In one case, the defendant posed as a fashion photographer and asked a 13 year old girl to pose nude in front of her webcam. She agreed. He pled guilty and was convicted of aggravated indecency act: Crim 2225/07 Israel vs. Lior (decided 25.6.2008). In a similar case, the defendant asked a 14 year old girl to undress and dance in front of her webcam; she complied. He recorded it without her knowledge and later uploaded the clip online, including the girl's name. Kids from her school found it online. He pled guilty to charges of publication of obscenity material that includes a minor and also to charges of violation of privacy: Crim 3990/07 Israel vs. Mashiach (decided 13.9.2010). Another defendant posed as a teenage boy although he was over sixty. He asked and received nude photos of teenaged girls. He pled guilty to various offenses, including attempted indecency act, possession of publication of obscenity material that include a minor, and obtaining anything by deceit: CrimA 6703/13 Cohen vs. Israel (decided 16.1.2014).

The most severe charges and punishment were inflicted on defendants who utilized cyberspace to develop relations with actual minors and later causing them to commit sexual acts. Most of them do not directly relate to webcam child-sexual abuse, but they are important and may be relevant in certain scenarios.

At the end of 2011, the Israeli prosecution has indicted a person for what can be described "cyber rape", for the first time in Israel history. The defendant was a fifty years old man. He posed as a teenage boy with various fraudulent characteristics and contacted minor girls, conversing with them in sexual manners. In one time he staged an event of self-infliction, as if cutting his genitals; in response, the minor who was in contact with him inserted scissors into her vagina. Eventually the prosecution dropped the rape charges and the defendant pled guilty to aggravated indecent acts: CrimA 538/13 Sabach vs. Israel (decided 26.12.2013). Other defendants also conversed with minors, always by posing as a young boy or girl, making sexual conversations with mutual masturbation. The prosecution claimed that causing a minor to penetrate herself is "cause of rape", either because of her age (less than 14) or because of the fraud. For example: CrimA 2656/13 Doe vs. Israel (decided 21.1.2014), a case that also involved charges of extortion and obscenity. Another defendant pled the District Court to dismiss the rape indictment, claiming his actions, even if proved, do not constitute "rape". His plea was denied: Crim 41309-12-14 Israel vs. Morobaty (decided 16.3.2015). Eventually the prosecution dropped the rape charges and settled for indecent acts, to which the defendant pled guilty. One defendant pled guilty to one charge of causing rape by fraud, after the prosecution dropped other charges, and was convicted by the District Court: CrimA 707/14 Doe vs. Israel (decided 6.7.2015). Another pled guilty to charges of causing sodomy: CrimA 512/13 Doe vs. Israel (decided 14.12.2013). In 2016, two defendants pled guilty and were therefore convicted of "cause of rape" in multiple cases: Crim 9232-07-15 Israel vs. Timsut (decided, 28.3.2016); and Crim 34838-04-15 Israel vs. Gavrilov (decided, 28.3.2016). The former was sentenced to 6.5 years imprisonment; the latter was convicted of almost fifty charges against tens of minors and was sentenced to 18 years imprisonment. Later that year the prosecution's rape thesis was applied to adult victim for the first time in Israel. A female Israeli soldier was extorted online to penetrate herself and take pictures. The defendant pled guilty to cause of rape and was sentenced by The Supreme Court to 5 years imprisonment: LCrimA 8720/15 Israel vs. Pinto (decided 11.9.2016).

Conclusion

- If the perpetrator induces or forces the minor to display breasts or genitals or to perform sexual activities (e.g., masturbate) in front of the webcam, this may constitute:
  - *Cause of Rape* (still not critically decided). Must include self-penetration by the minors. In addition, the victim's age must be under 14; or the consent must be obtained "by deceit in respect of the identity of the person or the nature of the act".
  - *Cause of Sodomy* (still not critically decided). Must include introduction of a bodily organ or an object into a person's anus by the minor. In addition, the victim's age must be under 14; or the consent must be obtained "by deceit in respect of the identity of the person or the nature of the act".
  - *Aggravated indecency act* (still not critically decided by high courts). Must include the above terms, save the penetration or introduction of a bodily organ or an object into a person's anus.
  - *Sexual Harassment* (still not critically decided by high courts). The victim must be under 15 and the perpetrator over 18 (unless the victim has expressed distress).
- If the perpetrator shows his genitals or masturbates in front of the webcam, this may constitute:
  - *Aggravated indecency act* (still not critically decided by high courts). The act must be without consent. Otherwise, the victim's age must be under 14; or the consent must be obtained "by deceit in respect of the identity of the person or the nature of the act".
  - *Indecency act in public.* The victim's age must be under 16. Although the offense is so titled, when it comes to minors the perpetration need not be "public"; and due to recent case law, that interpret the general offense of indecency act, the prosecutorial usage of indecency act in public, the lesser offense, has narrowed down almost completely.
  - *Sexual Harassment* (still not critically decided by high courts). The victim must be under 15 and the perpetrator over 18 (unless the victim has expressed distress).

There is no special limit regarding the intrusive investigation powers.

9.2.3 Possible Obstacles in Substantive Law Concerning Sweetie

The above provisions can be applied to the Sweetie case. The criminal attempt must be applied for criminal charges of any kind. If the avatar does not undress then attempted aggravated indecency act is the most serious crime.

9.3 Analysis of Criminal Procedure Law

9.3.1 General Description of Legal Framework

The Israeli criminal procedure involves many assorted laws. Some of them are general and include provisions on diverse issues, while others methodically focus on concrete topics. The latter include, for instance, a law on criminal procedure of pre-trial arrest and during-trial detention; a law on bodily searches; a law on conduction of investigation; and so forth. General procedure law are scarce.

The most important general Israeli law on the matter is the Law of Criminal Procedure [consolidated version, 1982]. Section 58 maintains that any person is allowed to file a complaint to the police. Section 59 maintains that when the police learn that an offense has been made, they must investigate. However, if the offense is a not a felony (a felony is an offense which includes over three years imprisonment), a police captain may rule that there is no public interest in the investigation. This decision may be repealed by a higher authority if the person who has filed the complaint appeals. The highest appeal instance is the Israeli Attorney-General. The only way to repeal the Attorney-General's decision by an administrative petition to the High Court of Justice; almost all of those are denied.

Israeli law does not specify how an investigation must or must not be made. No laws specifically deal with investigation, although one specific law deals with interrogation: the Criminal Procedure Law (suspect interrogation), 2002.

There is no formal law that sets out the rights of suspects. The Criminal Procedure Law (Powers of Enforcement—Arrests) 1996 sets out the right of a detainee to counsel; but there is no law that sets out such right for suspects who are not detained. An old ordinance (a law set prior to the independence of Israel) from 1927 sets out the privilege of self-incrimination; yet there is no formal law that sets out a right to remain silent during interrogation. On the one hand, the Israeli Supreme Court has recognized the above two as case-law rights; on the other hand, the police usually do not suffer any consequences for the violation thereof. There is no law that forbids the police from using means of pressure against suspects. The only potential remedy for such suspects is a possible exclusion of their admission of guilt; this remedy is very rare.

To sum up, the police have a very wide discretion on many of their tasks and authorities. The public prosecution is generally and typically uninvolved in police investigation, unless legal counsel is needed. When it comes to criminal law, there are no investigative judges in Israel. The involvement of the court is minimal and is usually limited to giving warrants. When the suspect is under arrest and the judge is asked to extend detention, the police inform the judge of the planned moves of investigation and the judge decides how many days detention are needed. Even in such cases the judge is typically uninvolved in the investigation and neither limits nor refers to its moves. The defence can inform the judge of supposedly illegal investigative procedures taken and ask for their immediate cease. However, the defence's ability to oversee the investigation in real time is very limited. The attorney cannot be present during investigation and is uninformed of the evidence gathered and the investigation planned; therefore, the attorney's only source of knowledge is the suspect. In conclusion, to say that the police have the upper hand at this point is a major understatement.

Whenever investigations do not lead to indictment, the evidence gathered is typically not exposed to the suspect or anyone else besides the police and prosecution. Therefore, this evidence, along with the police methods of obtaining it, is left unquestioned, uncriticised and unhindered. At the current time in Israel there is no authority that reviews the actions of the prosecution made out of litigation context; the issue is under discussion nowadays and the prosecution obviously objects any external review.

I should also emphasize that generally speaking, criminal litigation in Israel is publicly open. Investigations are sometimes even covered by the media. Names and photographs of suspects are revealed to any interested eye two days after the beginning of investigation. In some cases, the judge bans all publication, for example when the police need the investigation to remain secret or when the suspect's identity might reveal the victim's identity (mainly in sex offenses). The Supreme Court's verdicts are publicly accessible and are easily retrievable. In the age of cyberspace this sometimes means that long after the Israeli Statute of Limitations applies and the criminal record is cleansed, the verdict is still publicly visible.

9.3.2 Investigatory Powers

Succinct Overview of Investigatory Powers

See Table. 9.2.

Human Rights

Human rights standards set theoretical limitations on the use of investigation powers in Israel. According to Basic Law: Human Dignity and Liberty, which had been recognized by the Supreme Court as a part of the constitution in the making, no one may enter the premises of a person and no search can be made without consent. However, those rights may be violated by law, as long as this law befits the values of the State of Israel, was enacted for a proper purpose, and to an extent no greater than required, or by regulation enacted by express authorization in such law.

In effect, the constitutional limitations of investigation powers are very weak. The defence attorney and the suspect are obviously not present to object police requests for search warrants: such procedures are non-adversarial. The Magistrate

Table 9.2 Overview of investigatory powers

Council of Europe Convention on Cybercrime	Israel
Article 16. Expedited preservation of stored computer data	No such formal power in existence
Article 17. Expedited preservation and partial disclosure of traffic data	No such formal power in existence
Article 18. Production order	Obtained through the Law of Criminal Procedure (Enforcement Authorities—Communication Data) 2007, Sections 3–4 Prior to 2007: obtained by the Ordinance of the Criminal Procedure (Arrest and Search) (new version), 1969, Section 43
Article 19. Search and seizure of stored computer data	Obtained by the Ordinance of the Criminal Procedure (Arrest and Search) (new version), 1969, Section 23A
Article 20. Real-time collection of traffic data	Obtained by the Law of Criminal Procedure (enforcement authorities—communication data, 2007), Sections 3–4
Article 21. Interception of content data	Obtained by the Wiretap Law 1979, Section 6
Other (special) investigatory powers, not covered by the Cybercrime Convention, such as undercover operations	No such formal power in existence

[*Source* The author]

judge is in charge of balancing public interest with the rights of the suspect. Magistrate judges, asked to issue search warrants, seldom or even never refuse the police's requests; they are expected to issue hundreds of warrants in short time and cannot reflect on each. Appeal is impossible, both theoretically and practically: there is no legal right to appeal and no legal discussion is made proximately after the search. If no incriminating evidence is found, criminal procedure will not take place in court. If incriminating evidence is found and the search was illegal, the only balance for the suspect is the possible exclusion of evidence. However, this balance is somewhat theoretical too: evidence is seldom excluded in Israel, especially with regards to violation of privacy. The Supreme Court has never excluded key evidence (i.e. material evidence in cases there was no alternative evidence) in serious crimes, let alone evidence obtained by violation of privacy. Violation of the right to counsel has brought upon the exclusion of evidence by the Supreme Court in rare cases; and this is the right whose violation thereof has brought upon the most cases of evidence exclusion. Exclusion is more common in cases of small offenses or when other key evidence is found. When the court finds that the police violated human rights and even the formal law, there are usually no negative consequences for the state, perhaps except for the reprimand of the court.

Entrapment
There are no formal laws regarding entrapment in Israel. There is also no judicial precedent made by the Supreme Court that either forbids entrapment or sets limitations upon the police in this context. In one case the Supreme Court has mentioned that entrapment set against a person who had not been previously suspected of a serious crime might nullify the indictment by a preliminary pleading of abuse of process or outrageous conduct; however, the defendant was suspect of a serious crime: CrimA 1224/07 Beladev vs. Israel (decided 10.2.2010). To date no indictment was nullified due to such defence. In the above-mentioned case of Ktiei the Supreme Court discussed the issue of entrapment regarding the Dateline-inspired commercial television show. The three judges agreed that the defendant had not been entrapped and therefore his claim must be rejected: he was the one who approached the "child" and started a conversation; later, he was the one initiating and incorporating sexual content, before she cooperated. One judge ruled, obiter dyctum, that in the future the Supreme Court may re-evaluate the need to recognize the defense of entrapment. The other two judges have not shared his suggestion.

9.3.3 Succinct Overview of Investigatory Powers in an Online Context

The main investigatory powers in criminal investigations in an online context are regulated by three key laws.

The interception of content data is considered the most intrusive practice with regards to human rights. This practice is set and limited by the Wiretap Law, 1979. Only a high-ranked police officer may request the interception, regarding felonies alone. Only a chief justice or a vice-chief justice of a District court may grant it. If the procedure is knowingly improperly-followed by the police, the wiretap evidence must be excluded. If the procedure is disregarded, but not knowingly, the evidence may be used nevertheless, regarding serious felonies, seven years imprisonment or more.

The interception of traffic data, identification data and location data is mandated under the Law of Criminal Procedure (Enforcement Authorities—Communication Data, 2007). A police officer at the rank of no less than an inspector may request the interception, regarding an offense that is not a transgression. Any Magistrate court judge may issue the warrant. In some cases, mainly of emergency, even a high-ranked police officer may issue a warrant, but only regarding felonies. No exclusion clause is set in this law. Before 2007, no law specifically addressed the interception of traffic data and the police used an old law allowing them to ask a Magistrate judge to issue a warrant of disclosure of evidence.

Finally, the search and seizure of stored computer data is the easiest legal practice for the police. This is set out in the Ordinance of the Criminal Procedure (Arrest and Search) (new version, 1969). Any policeman may request search and seizure regarding any offense. Any Magistrate judge may issue the warrant. No exclusion clause is set out in the law. The police rely on this law to issue warrants in numerous investigations, not only for computer offenses.

Undercover operations are not mandated by any law. Prior to cyberspace they usually occurred in investigations of complex and serious crimes, like illicit drugs trafficking or illicit weapons trade. Nowadays they are also practiced in cyberspace, mainly to spot and indict paedophiles.

Political discussions on the topic are seldom, if any. The DOJ has issued a bill to amend the law of searches in order to update and facilitate the investigative tools. The amendment is yet to be codified.

The academic discussion in the issue is almost non-existent in Israel. The only Israeli scholar that critically reviews the issue is the writer of this chapter. Dr. Wismonsky, the head of the new cyber unit of the DOJ also deals with the issue academically, much more sympathetically to the choices of the prosecution and the DOJ, obviously. Israeli legal scholars who research cyberspace deal almost solely with civil aspects, mainly with intellectual property law.

9.3.4 Application of Relevant Investigatory Powers to the Sweetie Case

The investigation powers described above can be applied to the Sweetie case without any special limitation. Since there is no law that directly authorizes the police to go undercover or operate undercover agents, it does not matter if this unauthorized power is wielded in the context of offline crime or online crime; nor does it matter if the police use a human agent or a computer agent. One may argue that the police are not authorized to operate any software that inflicts upon human rights; for the time being this argument seems weak and incoherent to police practices known by the judicial system, including the Supreme Court. There are also no guidelines regarding the law of evidence in this context, and the script is not limited by any formal law. The defence may claim the police practice is illegal for lack of authorization and demand to exclude all evidence obtained by it. However, there is no law requiring courts to exclude evidence obtained without authorization and criminal courts tend not to exclude key evidence in general, let alone regarding paedophile cases.

9.3.5 *Relevant Aspects of Digital Forensic Evidence*

There are no specific rules on evidence collection in an online context. Generally speaking, the Israeli formal law of evidence is short and laconic and does not cover most evidential issues, therefore leaving very wide judicial discretion. In addition, the Israeli legislator tends to overlook cyberspace and has left the old and general procedural, evidential and material law completely unchanged. To date, no law directly deals with online or with electronic evidence; and there is no case-law that sets any clear requirements in this context. So there are no technical requirements that need to be taken into account as well, nor are there any relevant standards for digital forensics that need to be met.

9.4 Miscellaneous

There are a few other relevant aspects of the criminal procedure in need of mentioning. The first regards the magnitude and commonness of online agent operations and their broad social and legal implications.

While operating agents offline is usually both expensive and dangerous in many cases, operating agents online is both cheap and safe. In terms of resource-allocation, cyberspace facilitates and therefore encourages agent operations. Obviously, the development of complex software such as Sweetie requires major resources. However, operating online agents is a task which is not only cheap and safe; it is also very simple, at least when done by a human, like in Israel. Any citizen can be self-appointed online agent and locate paedophiles, acting on the citizen's moral code. The Israeli police, as mentioned, are not authorized by law to operate agents and therefore they too act by their own code. The process is not contemplated and reflected upon by psychologists or legal scholars. It is just done. Therefore, in terms of legal policy, possible consequences and implications must be contemplated. Online operations may and probably will become much more common in other aspects as well as paedophile offenses. For example, in July 2015 the media published that the Israeli police operated agents in Facebook and caught forty drug dealers who sold illicit drugs, half of whom were minors. What is the future of online agents? Will they be used also for incriminating suspects of petty offenses, like misdemeanours and transgressions? In the absence of a formal law authorizing the police to operate agents, are there legal limits to such operations, or can the police do as they find fit? If online agent operation becomes common, will it create safer environments for children and other potential victims or will it simply push paedophile offenders and other criminals to become more cautious and sophisticated? Will an environment of common agent operations have negative effects on social networks and surfing experience?

Another related aspect of online agent operations is the issue of public enforcement vs. private enforcement. Whereas traditional agent operations are conducted by the police, sometimes with an on-going legal counsel provided by the prosecution, cyberspace has the potential of revolutionizing agent operations. As described above, the process gets easy, cheap and simple offline. This invites private enforcement. Commercial shows might benefit from so-called research and editorial choices become crucial for criminal law. These choices do not necessarily cohere with suspect human rights on the one hand and with public interest on the other hand, as opposed to public ratings; and the police's and prosecution's choice to cooperate with commercial channels brings about further difficulties.

Furthermore, a commercial channel is at least bound by certain legal and social aspects. The same cannot be said regarding private factors. The easiness of going undercover, combined with the positive social feedback for paedophile hunt, attract ordinary citizens to participate in self-initiated self-arranged undercover operation. Aside for the Channel 10 story described above, Adler was the first Israeli case of such private operation. At that time, it was one of a kind case. Later, another citizen had famously declared himself to be "The Paedophile Hunter". He opened a Facebook page and published some of his operations and pictures of suspects. Sometimes the police cooperated and arrested the people he implicated. It was later published that he filed over a thousand complaints to the police, among which only a handful have ended with indictments. Counter-complaints have started piling up against him and eventually he left the country.

During 2015, a famous Israeli singer posted a story in his Facebook page of catching a paedophile, stressing the online dangers to children. The media covered the story the following days. Not long after, the Israeli Second Authority for Television and Radio, set and empowered by law, disqualified a so-called documentary movie about the story, due to hidden commercial aspects. It turns out the singer had signed a contract with "Tom", an internet security company.

In the age of "likes" and the commercial and social importance of popularity in social networks, the danger of vigilantism is inherent and perhaps radical.

A third related aspect is the possible effect of undercover operations on civil law. The easiness described above suggests that the judicial system will meet evidence achieved by undercover operations made by private parties not only for the benefit of paedophile criminal prosecution, but also for civil cases. For example, in Israel there are two parallel tribunals authorized to adjudicate family law: religious and secular. According to Jewish religious law, adultery is relevant for property distribution in divorce cases. The cyberspace environment offers private parties an attractive way of implicating the other party. Should the legal system encourage a cyberspace in which lies and deception are the tools of the righteous?

9.5 Conclusions and Recommendations

An important question that must be asked is the object of online undercover operations in general and automated operation specifically. Is it putting paedophiles in jail? Protecting children from online dangers? Diverting paedophile attention from actual children to fictional ones? Educating the public or calming it down? Letting surfers know that cyberspace is not the lawless arena some claim or desire it to be? Illustrating technology's power and designers' or policy makers' ingenuity? Some of these aims contradict others. The object of the whole initiative must be carefully examined and explicitly stated, foreseeing criminal litigation ahead.

In Israel, so far, the easiness of operating online agents is disturbing and there are no checks and balances at all; simply a paedophiles hunt. Add to that the potential automation of the process and the easiness becomes total; whereas accountability fades. An agent may be called to the stand and answer hard questions; this process may cause a somewhat chilling effect on undercover operations. Software cannot be interrogated in court and software developers are not likely to testify, especially regarding international operated software.

Israel is on the hunt for paedophiles. Add to the above the need to show results, i.e. criminal indictments, and you get a danger of over-use and misuse. If software agent, or any human agent to that matter, does not provide incriminating evidence, the person in charge of the operation, be it a police officer or a television editor, will not be pleased and may suggest resource reallocation. This means one way or the other, suspects must and will be found. This incentive is dangerous and promotes the danger of pushing users into crimes; and also promotes apathy towards human rights of suspects and dismissing the presumption of innocence. A person who contacts the agent might be considered a paedophile till proven otherwise. In a society and in a legal system that demonize paedophiles, this sets a major problem. Add to that the Israeli lack of law authorizing agents (and agent software), and the potential of misuse is significant. Add to that the tendency of such defendants to draw minimal judicial attention, in order to narrow their public exposure, and the legal limitations on the process become non-existent.

The lack of formal authorization also means that if criminal offenses are made by the agent during the operation, such as obtaining anything by deceit or violation of privacy etc., the agent is in theoretical danger of being indicted for the agent's crimes. No criminal defences for agents exist in the formal law, clearly, as the law does not even recognize the idea of agents and undercover operations. Of course, no one is expected to investigate and prosecute agents who are operated by the police; however, self-appointed vigilant agents may be exposed to legal sanctions, both criminal and civil. As for public agents, their work may theoretically bring about the exclusion of evidence, due to a judicial ruling that the agents acted illegally. There is also a policy issue: is it both right and legal to commit crimes in order to catch criminals? For the time being, the Israeli debate is almost non-existent. No one is outraged by acts of those who try catching and punishing paedophiles.

One need also give careful thought to the issue of automated investigation. The repercussions of moving police investigations from police hands to private hands are anything but light; the repercussions of moving police investigations from human hands to machines is nearly revolutionary. Can software engineers design tools sensitive enough to imitate a child? Can any adult imitate how a real child will act? How can engineers avoid creating emotional attachment between people and stimulating sexual desires where no such feelings have been there before? Can they accommodate the public interest with suspect human rights and other legal requirements? Can they do it not only more efficient than police agents, but perhaps also morally and legally better? If they can, do they want to? Whose values will their technology reflect? These difficult questions require elaborate philosophical and legal analysis. Robocop is alive and he's kicking paedophiles. What crimes and criminals are next?

All of the above bring me to the main conclusion. Regarding online agent operation, especially paedophile operations, taking suspect human rights into account must be done in advance in a well-balanced and accountable procedure. No remedy will be given later to anyone assumed paedophile.

Currently, the lack of formal law and judicial guidelines to the usage of agents and its limits, I find that the Israeli legal framework is normatively inadequate to welcome offenders investigating and incriminating software.

However, since the Israeli legislator appears to be generally indifferent to the cyber world on the one hand, and the DOJ is certainly untroubled by the lack of any formal limits to operation of police agents, one need not expect any change soon; and I believe any tool devised to combat paedophiles will be welcome in Israel, regardless of its intrusiveness and normative problems. Perhaps it will even present an improvement towards human rights, as the software was designed in a much more sensitive environment.

Annex

Relevant Legal Provisions (in Original)

מתוך חוק העונשין

עבירות נגד משפט העמים

16. (א) דיני העונשין של ישראל יחולו על עבירות-חוץ אשר מדינת ישראל התחייבה, באמנות בינלאומיות רב צדדיות ופתוחות להצטרפות, להעניש עליהן; והוא, אף אם נעברו בידי מי שאינו אזרח ישראלי או תושב ישראל, ויהא מקום עשיית העבירה אשר יהא.

(ב) הסייגים האמורים בסעיף 14 (ב)(2) ו-(3), ו-(ג) יחולו לגבי תחולת דיני העונשין של ישראל גם לפי סעיף זה.

תחולה שילוחית

17. (א) מדינת ישראל רשאית להתחייב באמנה בינלאומית להחיל את דיני העונשין שלה על עבירת-חוץ או להחיל את הוראות סעיף 10, לבקשת מדינה זרה ועל בסיס של הדדיות, גם במקרים אחרים מאלה האמורים בסעיפים 13 עד 16, ובלבד שנתקיימו כל אלה: (1) על העבירה חלים דיני העונשין של המדינה המבקשת; (2) העבירה נעברה בידי אדם הנמצא בתוך שטח ישראל והוא תושב ישראל, בין אם הוא אזרח ישראלי ובין אם לאו; (3) בכפוף למיצוי הדין בישראל כלפי האדם, תוותר המדינה המבקשת, בבקשתה, על תחולת דיניה היא בעניין הנדון.

(ב) לא יוטל בישראל בשל העבירה עונש חמור מזה שניתן היה להטיל לפי דיני המדינה המבקשת.

(ג) כל שאר התנאים ייקבעו באמנה.

סיווג עבירות

24. אלה סוגי העבירות לפי חומרתן:

(1) "פשע" – עבירה שנקבע לה עונש חמור ממאסר לתקופה של שלוש שנים;

(2) "עוון" – עבירה שנקבע לה עונש מאסר לתקופה העולה על שלושה חודשים ושאינה עולה על שלוש שנים; ואם העונש הוא קנס בלבד – קנס העולה על שיעור הקנס שניתן להטיל בשל עבירה שעונשה הוא קנס שלא נקבע לו סכום;

(3) "חטא" – עבירה שנקבע לה עונש מאסר לתקופה שאינה עולה על שלושה חודשים, ואם העונש הוא קנס בלבד – קנס שאינו עולה על שיעור הקנס שניתן להטיל בשל עבירה שעונשה הוא קנס שלא נקבע לו סכום.

ניסיון מהו

25. אדם מנסה לעבור עבירה אם, במטרה לבצעה, עשה מעשה שאין בו הכנה בלבד והעבירה לא הושלמה.

חוסר אפשרות לעשיית העבירה

26. לעניין ניסיון, אין נפקה מינה אם עשיית העבירה לא הייתה אפשרית מחמת מצב דברים שהמנסה לא היה מודע לו או טעה לגביו.

פטור עקב חרטה

28. מי שניסה לעבור עבירה, לא יישא באחריות פלילית לניסיון, אם הוכיח שמחפץ נפשו בלבד ומתוך חרטה, חדל מהשלמת המעשה או תרם תרומה של ממש למניעת התוצאות שבהן מותנית השלמת העבירה; ואולם, אין באמור כדי לגרוע מאחריותו הפלילית בשל עבירה מושלמת אחרת שבמעשה.

פרשנות

34כא. ניתן דין לפירושים סבירים אחדים לפי תכליתו, יוכרע העניין לפי הפירוש המקל ביותר עם מי שאמור לשאת באחריות פלילית לפי אותו דין.

הגדרות

34כד. לעניין עבירה –

"החזקה" – שליטתו של אדם בדבר המצוי בידו, בידו של אחר או בכל מקום שהוא, בין שהמקום שייך לו ובין אם לאו; ודבר המצוי בידם או בהחזקתם של אחד או כמה מבני חבורה בידיעתם ובהסכמתם של השאר יראו כמצוי בידם ובהחזקתם של כל אחד מהם ושל כולם כאחד;

"פומבי", לעניין מעשה – (1) מקום ציבורי, כשאדם יכול לראות את המעשה מכל מקום שהוא; (2) מקום שאינו ציבורי, ובלבד שאדם המצוי במקום ציבורי יכול לראות את המעשה;

"פרסום" – כתב, דבר דפוס, חומר מחשב, או כל מוצג חזותי אחר וכן כל אמצעי שמיעתי העשויים להעלות מלים או רעיונות, בין לבדם ובין בעזרת אמצעי כלשהו;

"פרסם" – (1) בדברים שבעל פה – להשמיע מלים בפה או באמצעים אחרים, בהתקהלות ציבורית או במקום ציבורי או באופן שאנשים הנמצאים במקום ציבורי יכולים לשמוע אותם, או להשמיען בשידורי רדיו או טלויזיה הניתנים לציבור או להפיצן באמצעות מחשב בדרך הזמינה לציבור, או להציען לציבור באמצעות מחשב; (2) בפרסום שאינו דברים שבעל פה – להפיצו בקרב אנשים או להציגו באופן שאנשים במקום ציבורי יכולים לראותו, או למכרו או להציעו למכירה בכל מקום שהוא, או להפיצו בשידורי טלויזיה הניתנים לציבור, או להפיצו לציבור באמצעות מחשב בדרך הזמינה לציבור, או להציעו לציבור באמצעות מחשב.

"ציבור" – לרבות כל חלק ממנו העלול להיפגע מהתנהגות שעליה מדובר בהקשרו של מונח זה.

"קטין" – אדם שטרם מלאו לו 18 שנים.

הבאת אדם לידי מעשה זנות

201. המביא אדם לידי מעשה זנות עם אדם אחר, דינו – מאסר חמש שנים.

נסיבות מחמירות

203. (א) נעברה עבירה לפי סעיפים 201 או 202 תוך ניצול יחסי מרות, תלות, חינוך או השגחה, או תוך ניצול מצוקה כלכלית או נפשית של האדם שהובא לידי מעשה זנות או לידי עיסוק בזנות, דינו של עובר העבירה – מאסר עשר שנים.

(ב) נעברה עבירה לפי סעיפים 201 או 202 באחת מנסיבות אלה, דינו של עובר העבירה – מאסר שש עשרה שנים: (1) תוך שימוש בכוח, או הפעלת אמצעי לחץ אחרים, או תוך איום באחד מאלה, ואחת היא אם נעשו אלה כלפי האדם שהובא לידי מעשה זנות או לידי עיסוק בזנות, או כלפי אדם אחר; (2) תוך ניצול מצב המונע את התנגדותו של האדם שהובא לידי מעשה זנות או לידי עיסוק בזנות, או תוך ניצול היותו חולה נפש או לקוי בשכלו; (3) בהסכמה שהושגה במרמה, של האדם שהובא לידי מעשה זנות או לידי עיסוק בזנות.

ניצול קטינים לזנות

203ב. (א) נעברה עבירה לפי סעיפים 199, 201, 202 או 203 בקטין שמלאו לו ארבע עשרה שנים, דינו של עובר העבירה – (1) אם נקבע לעבירה מאסר חמש שנים – מאסר שבע שנים; (2) אם נקבע לעבירה מאסר שבע שנים – מאסר עשר שנים; (3) אם נקבע לעבירה מאסר עשר שנים – מאסר חמש עשרה שנים; (4) אם נקבע לעבירה מאסר שש עשרה שנים – מאסר עשרים שנה.

(ב) נעברה עבירה לפי סעיפים 199, 201, 202 או 203 בקטין שטרם מלאו לו ארבע עשרה שנים, או שמלאו לו ארבע עשרה שנים ועובר העבירה אחראי על הקטין, דינו של עובר העבירה – כפל העונש שנקבע לעבירה אך לא יותר מעשרים שנים.

(ג) בסעיף זה, "אחראי על קטין" – כהגדרתו בסעיף 368א.

דין לקוחו של קטין

203ג.המקבל שירות של מעשה זנות של קטין, דינו – מאסר שלוש שנים.

איסור פרסום ומסירת מידע בדבר זנות של קטין

205א. המוסר מידע או המפרסם פרסום על מתן שירות של מעשה זנות, כשנותן השירות הוא קטין, דינו – מאסר חמש שנים; לעניין עבירה לפי סעיף זה, אחת היא אם שירות הזנות ניתן בישראל או מחוץ לישראל, אם המידע מתייחס לקטין מסוים אם לאו, או אם הפרסום מציין שנותן השירות הוא קטין אם לאו.

איסור ציון קטינות בפרסום שירותי זנות

205ב. המפרסם פרסום המציין שנותן שירותי זנות הוא קטין, כשנותן השירות אינו קטין, דינו – מאסר שישה חודשים.

המניח לקטין לדור בבית זנות

208. המניח לקטין בן שתים עד שבע-עשרה שנים הנתון למשמורתו או להשגחתו, שידור בבית זנות או שיבקר בו תכופות, דינו – מאסר שלוש שנים.

פרסום והצגת תועבה

214. (א) העושה אחת מאלה, דינו – מאסר שלוש שנים: (1) מפרסם פרסום תועבה או מכינו לצורכי פרסום; (2) מציג, מארגן או מפיק הצגת תועבה – (א) במקום ציבורי; (ב) במקום שאינו ציבורי – אלא אם כן הוא מקום המשמש למגורים או המשמש חבר בני אדם שהחברות בו היא למי שמלאו לו שמונה עשרה שנים ולתקופה רצופה.

(ב) המפרסם פרסום תועבה ובו דמותו של קטין, לרבות הדמיית קטין או ציור של קטין, דינו – מאסר חמש שנים.

(ב1) המשתמש בגופו של קטין לעשיית פרסום תועבה, או המשתמש בקטין בהצגת תועבה, דינו – מאסר שבע שנים.

(ב2) נעברה העבירה לפי ס"ק (ב) או (ב1) בידי האחראי על הקטין כהגדרתו בסעיף 368א, או בהסכמתו של אחראי כאמור, דינו של האחראי – מאסר עשר שנים.

(ב3) המחזיק ברשותו או הצורך פרסום תועבה ובו דמותו של קטין, דינו – מאסר שנה; לעניין ס"ק זה, "מחזיק" ו"צורך" – למעט המחזיק וצורך באקראי ובתום לב.

(ג) ביהמ"ש הדן בעבירה לפי סעיף זה שנעברה בידי בעל עסק במהלך עסקיו, רשאי להפעיל גם את הסמכויות לפי סעיפים 16 ו-17 לחוק רישוי עסקים, התשכ"ח-1968, ובלבד שלא ישתמש ביהמ"ש בסמכותו לפי סעיף 17 אלא אם כן השתכנע שיש ראיות לכאורה לביצוע העבירה ושהפעלת סמכותו דרושה לטובת הציבור.

(ד) לא יוגש כתב אישום – (1) לפי ס"ק (א) – אלא בתוך שנתיים מיום ביצוע העבירה, ובידי פרקליט מחוז או בהסכמתו בכתב; (2) לפי ס"ק (ב) עד (ב3) – אלא בידי פרקליט מחוז או בהסכמתו בכתב.

הגנות

214ב. לא יראו אדם כעובר עבירה לפי סעיפים 205א עד 205ג ו-214, אם מסירת המידע, הפרסום, ההחזקה או הצריכה נעשו למטרה כשרה, לרבות לשם דיווח נכון והוגן בעניין שסימן זה דן בו, ובלבד שמסירת המידע, הפרסום, ההחזקה או הצריכה אינם אסורים לפי דין אחר ולא נעשו כדי לעודד מעשים אסורים לפי סימן זה.

אינוס

345. (א) הבועל אשה – (1) שלא בהסכמתה החופשית; (2) בהסכמת האשה, שהושגה במרמה לגבי מיהות העושה או מהות המעשה; (3) כשהאשה היא קטינה שטרם מלאו לה ארבע עשרה שנים, אף בהסכמתה; או (4) תוך ניצול מצב של חוסר הכרה בו שרויה האשה או מצב אחר המונע ממנה לתת הסכמה חופשית; (5) תוך ניצול היותה חולת נפש או לקויה בשכלה, אם בשל מחלתה או

בשל הליקוי בשכלה לא הייתה הסכמתה לבעילה הסכמה חופשית; הרי הוא אונס ודינו – מאסר שש עשרה שנים.

(ב) על אף האמור בס"ק (א), דין האונס – מאסר עשרים שנים אם האינוס נעשה באחת מנסיבות אלה: (1) בקטינה שטרם מלאו לה שש עשרה שנים ובנסיבות האמורות בס"ק (א)(1), (2), (4) או (5); (2) באיום בנשק חם או קר; (3) תוך גרימת חבלה גופנית או נפשית או הריון; (4) תוך התעללות באשה לפני המעשה, בזמן המעשה או אחריו; (5) בנוכחות אחר או אחרים שחברו יחד עמו לביצוע האינוס בידי אחד או אחדים מהם.

(ג) בסימן זה – "בועל" – המחדיר איבר מאיברי הגוף או חפץ לאיבר המין של האישה.

בעילה אסורה בהסכמה

346. (א) (1) הבועל קטינה שמלאו לה ארבע עשרה שנים וטרם מלאו לה שש עשרה שנים, והיא אינה נשואה לו, או הבועל קטינה שמלאו לה שש עשרה שנים וטרם מלאו לה שמונה עשרה שנים, תוך ניצול יחסי תלות, מרות, חינוך או השגחה, או תוך הבטחת שווא לנישואין, דינו – מאסר חמש שנים; (2) לעניין ס"ק זה, יראו מטפל נפשי שבעל קטינה שמלאו לה שש עשרה שנים וטרם מלאו לה שמונה עשרה שנים, במהלך התקופה שבה ניתן לקטינה טיפול נפשי על ידו, כאילו עשה את המעשה האמור תוך ניצול יחסי תלות; חזקה זו לא תחול אם מעשים כאמור החלו לפני תחילתו של הטיפול הנפשי במסגרת קשר זוגי.

(ב) הבועל אשה שמלאו לה שמונה עשרה שנים תוך ניצול מרות ביחסי עבודה או בשירות או עקב הבטחת שווא לנישואין תוך התחזות כפנוי למרות היותו נשוי, דינו – מאסר שלוש שנים.

מעשה סדום

347. (א) (1) העושה מעשה סדום באדם שמלאו לו ארבע עשרה שנים וטרם מלאו לו שש עשרה שנים, או העושה מעשה סדום באדם שמלאו לו שש עשרה שנים וטרם מלאו לו שמונה עשרה שנים, תוך ניצול יחסי תלות, מרות, חינוך או השגחה, דינו – מאסר חמש שנים; (2) לעניין ס"ק זה, יראו מטפל נפשי שעשה מעשה סדום באדם שמלאו לו שש עשרה שנים וטרם מלאו לו שמונה עשרה שנים, במהלך התקופה שבה ניתן לאותו אדם טיפול נפשי על ידו, כאילו עשה את המעשה תוך ניצול יחסי תלות, חזקה זו לא תחול אם מעשים כאמור החלו לפני תחילתו של הטיפול הנפשי במסגרת קשר זוגי.

(א1) העושה מעשה סדום באדם שמלאו לו שמונה עשרה שנים תוך ניצול מרות ביחסי עבודה או בשירות, דינו – מאסר שלוש שנים.

(ב) העושה מעשה סדום באדם באחת הנסיבות המנויות בסעיף 345, בשינויים המחויבים, דינו כדין אונס.

(ג) לעניין סימן זה, "מעשה סדום" – החדרת איבר מאברי הגוף או חפץ לפי הטבעת של אדם או החדרת איבר מין לפיו של אדם.

מעשה מגונה

348. (א) העושה מעשה מגונה באדם באחת הנסיבות המנויות בסעיף 345(א)(2) עד (5), בשינויים המחויבים, דינו – מאסר שבע שנים.

(ב) העושה מעשה מגונה באדם באחת הנסיבות המנויות בסעיף 345(ב)(1) עד (5), בשינויים המחויבים, דינו – מאסר עשר שנים.

(ג) העושה מעשה מגונה באדם בלא הסכמתו אך שלא בנסיבות כאמור בס"ק (א), (ב) או (ג1), דינו – מאסר שלוש שנים.

(ג1) נעברה עבירה לפי ס"ק (ג) תוך שימוש בכוח או הפעלת אמצעי לחץ אחרים, או תוך איום באחד מאלה, כלפי האדם או כלפי זולתו, דינו של עובר העבירה – מאסר שבע שנים.

(ד) (1) העושה מעשה מגונה באדם שהוא קטין שמלאו לו ארבע עשרה שנים תוך ניצול יחסי תלות, מרות, חינוך, השגחה, עבודה או שירות, דינו – מאסר ארבע שנים; (2) לעניין ס"ק זה יראו מטפל נפשי שעשה מעשה מגונה באדם שמלאו לו ארבע עשרה שנים וטרם מלאו לו שמונה עשרה

שנים, במהלך התקופה שבה ניתן לאותו אדם טיפול נפשי על ידו, כאילו עשה את המעשה תוך ניצול יחסי תלות; חזקה זו לא תחול אם מלאו לאדם שש עשרה שנים, והמעשים החלו לפני תחילתו של הטיפול הנפשי במסגרת קשר זוגי.

(ד1) מטפל נפשי העושה באדם שמלאו לו שמונה עשרה שנים מעשה מגונה בנסיבות המפורטות בסעיף 347א(ב), דינו – מאסר שלוש שנים.

(ה) העושה מעשה מגונה באדם שמלאו לו שמונה עשרה שנים תוך ניצול מרות ביחסי עבודה או בשירות, דינו – מאסר שנתיים.

(ו) בסימן זה, "מעשה מגונה" – מעשה לשם גירוי, סיפוק או ביזוי מיניים.

מעשה מגונה בפומבי

349. (א) העושה מעשה מגונה בפומבי בפני אדם אחר, ללא הסכמתו, או העושה מעשה כאמור בכל מקום שהוא תוך ניצול יחסי תלות, מרות, חינוך, השגחה, עבודה או שירות, דינו – מאסר שנה.

(ב) העושה, בכל מקום שהוא, מעשה מגונה בפני אדם שטרם מלאו לו שש עשרה שנים, דינו – מאסר שלוש שנים.

גרם מעשה

350. לעניין עבירה לפי סימן זה, אחת היא אם העושה עשה את המעשה או גרם שהמעשה ייעשה בו או באדם אחר.

עבירות מין במשפחה ובידי אחראי על חסר ישע

351. (א) העובר עבירה של אינוס לפי סעיף 345(א) או של מעשה סדום לפי סעיף 347(ב) באדם שהוא קטין, והוא בן משפחתו או באדם שהוא חסר ישע והוא אחראי עליו, דינו – מאסר עשרים שנים.

(ב) הבועל אישה שמלאו לה ארבע עשרה שנים וטרם מלאו לה עשרים ואחת שנים או העושה מעשה סדום באדם שמלאו לו ארבע עשרה שנים וטרם מלאו לו עשרים ואחת שנים, והוא בן משפחתם, דינו – מאסר שש עשרה שנים.

(ג) העושה מעשה מגונה באדם שהוא קטין והוא בן משפחתו, דינו – (1) בעבירה לפי סעיף 348(א) או (ג1) – מאסר עשר שנים; (2) בעבירה לפי סעיף 348(ב) – מאסר חמש עשרה שנים; (3) בכל מקרה שאינו בין המנויים בפיסקאות (1) ו-(2) – מאסר חמש שנים.

(ד) העושה מעשה מגונה בפני אדם שהוא קטין, בכל מקום שהוא, והוא בן משפחתו, דינו מאסר ארבע שנים.

(ד1) אחראי על חסר ישע העובר עבירה לפי סעיף 349(א) בחסר הישע, דינו – מאסר שנתיים.

(ה) לעניין סעיף זה – "אומן" – אחד מאלה: (1) אב או אם במשפחת אומנה שאישר משרד הרווחה; (2) אחראי על קטין לפי פסקה (3) להגדרה "אחראי על קטין או חסר ישע" בסעיף 368א; "אח או אחות חורגים" – בן או בת של בן זוג של הורה; "אחראי על חסר ישע" – כהגדרה "אחראי על קטין או חסר ישע" בסעיף 368א; "בן משפחה" – (1) הורה; בן זוגו של הורה אף אם אינו נשוי לו; סב או סבתא; (2) מי שמלאו לו חמש עשרה שנים והוא אחד מאלה: אח או אחות; אח או אחות חורגים; דוד או דודה; גיס או גיסה. ואולם לעניין עבירה של בעילה אסורה לפי ס"ק (ב) או של מעשה מגונה לפי ס"ק (ג)(3) שנעשו במי שמלאו לו שש עשרה שנים, לא ייכללו דוד או דודה, גיס או גיסה בהגדרת "בן משפחה". (3) אומן, בן זוגו של אומן אף אם אינו נשוי לו; אביו או אמו של אומן; (4) מי שמלאו לו חמש עשרה שנים והוא אחד מאלה: בנו או בתו של אומן ובן זוגו של כל אחד מאלה; אחיו או אחותו של אומן ובן זוגו של כל אחד מאלה; ואולם לעניין עבירה של בעילה אסורה לפי ס"ק (ב) או של מעשה מגונה לפי ס"ק (ג)(3) שנעשו במי שלמלאו לו שש עשרה שנים, לא ייכלל בן זוג של בנו או בתו של אומן, אחיו או אחותו של אומן ובן זוגו של כל אחד מאלה בהגדרה "בן משפחה". "חסר ישע" – כהגדרתו בסעיף 368א.

איסור פרסום

352. (א) המפרסם ברבים שמו של אדם או של כל דבר שיש בו כדי לזהות אדם כמי שנפגע בעבירה או כמי שהתלונן כי הוא נפגע בעבירה לפי סימן זה, דינו – מאסר שנה.

(ב) לא ישא אדם באחריות פלילית לפי ס"ק (א) אם האדם ששמו או זהותו פורסמו כאמור נתן את הסכמתו לפרסום, בפני בימ"ש, או אם בימ"ש התיר את הפרסום מטעמים מיוחדים שיירשמו.

סייג לאחריות פלילית

353. באישום בשל עבירה לפי סעיפים 346(א) או 347(א) תהיה זו הגנה לנאשם שהבדל הגילים בינו לבין הקטין אינו עולה על שלוש שנים, אם הקטין הסכים למעשה ואם המעשה נעשה במהלך יחסי רעות רגילים וללא ניצול מעמדו של הנאשם.

מתוך חוק סדר הדין הפלילי [נוסח משולב], התשמ"ב-1982

תלונה

58. כל אדם רשאי להגיש תלונה למשטרה על שבוצעה עבירה.

חקירת המשטרה

59. נודע למשטרה על ביצוע עבירה, אם על פי תלונה ואם בכל דרך אחרת, תפתח בחקירה; אולם בעבירה שאינה פשע רשאי קצין משטרה בדרגת פקד ומעלה להורות שלא לחקור אם היה סבור שאין בדבר ענין לציבור או אם היתה רשות אחרת מוסמכת על פי דין לחקור בעבירה.

טענות מקדמיות

149. לאחר תחילת המשפט רשאי הנאשם לטעון טענות מקדמיות, ובהן – ... (10) הגשת כתב האישום או ניהול ההליך הפלילי עומדים בסתירה מהותית לעקרונות של צדק והגינות משפטית.

פקודת סדר הדין הפלילי (מעצר וחיפוש) [נוסח חדש], התשכ"ט-1969

חדירה לחומר מחשב

23א. (א) חדירה לחומר מחשב וכן הפקת פלט תוך חדירה כאמור, יראו אותן כחיפוש וייעשו על-ידי בעל תפקיד המיומן לביצוע פעולות כאמור; לענין זה, "חדירה לחומר מחשב" – כמשמעותה בסעיף 4 לחוק המחשבים, התשנ"ה-1995.

(ב) על אף הוראות פרק זה, לא ייערך חיפוש כאמור בס"ק (א), אלא על-פי צו של שופט לפי סעיף 23, המציין במפורש את ההיתר לחדור לחומר מחשב או להפיק פלט, לפי הענין, והמפרט את מטרות החיפוש ותנאיו שייקבעו באופן שלא יפגעו בפרטיותו של אדם מעבר לנדרש.

מתוך חוק האזנת סתר, התשל"ט-1979

האזנת סתר למניעת עבירות

6. (א) נשיא בית משפט מחוזי, או סגן הנשיא שהסמיכו הנשיא לענין זה רשאי, לפי בקשת קצין משטרה מוסמך, להתיר בצו האזנת סתר אם שוכנע, לאחר ששקל את מידת הפגיעה בפרטיות, שהדבר דרוש לגילוי, לחקירה או למניעה של עבירות מסוג פשע, או לגילוי או לתפיסה של עבריינים שעברו עבירות כאמור, או לחקירה לצרכי חילוט רכוש הקשור בעבירה שהיא פשע.

(ב) הבקשה תהיה על פי טופס שייקבע והיא תידון במעמד צד אחד בלבד, ומטעם המבקש יתייצב קצין בדרגת סגן ניצב ומעלה.

(ג) סירב השופט להעניק היתר כמבוקש, רשאי היועץ המשפטי לממשלה או נציגו לערער על ההחלטה לפני נשיא בית המשפט העליון או שופט של בית המשפט העליון שנשיאו מינהו לכך.

(ד) בהיתר לפי סעיף זה יתוארו זהות האדם אשר האזנה לשיחותיו הותרה או זהות הקו או המיתקן המשמשים או המיועדים לשמש לקליטה, להעברה או לשידור של בזק, ואשר האזנה אליהם הותרה, ומקום השיחות או סוגן, הכל אם הם ידועים מראש; כן יפורטו דרכי ההאזנה שהותרו.

(ה) בהיתר תפורש תקופת תקפו; התקופה לא תעלה על שלושה חדשים מיום נתינת ההיתר; ההיתר ניתן לחידוש מפעם לפעם.

(ו) המפקח הכללי של המשטרה יגיש, מדי חודש, דין וחשבון ליועץ המשפטי לממשלה על היתרים שניתנו לפי פרק זה, סעיף 9א(א)(2) וסעיף 2א לחוק חסינות חברי הכנסת, ועל תנאיהם.

(ז) שר המשטרה ימסור, מדי שנה, דין וחשבון לועדת החוקה חוק ומשפט של הכנסת, שיכלול את מספר הבקשות שהוגשו ואת מספר ההיתרים שניתנו לפי פרק זה, בציון מספר האנשים, וכמות קווי הבזק ומתקני הבזק, אשר האזנה אליהם הותרה.

ראיות

13. (א) דברים שנקלטו בדרך של האזנת סתר בניגוד להוראות חוק זה או להוראת סעיף 2א לחוק חסינות חברי הכנסת, לא יהיו קבילים כראיה בבית משפט, אלא באחד משני אלה: (1) בהליך פלילי בשל עבירה לפי חוק זה; (2) בהליך פלילי בשל פשע חמור, אם בית משפט הורה על קבילותה לאחר ששוכנע, מטעמים מיוחדים שיפרט, כי בנסיבות העניין הצורך להגיע לחקר האמת עדיף על הצורך להגן על הפרטיות. האזנת סתר שנעשתה שלא כדין בידי מי שרשאי לקבל היתר להאזנת סתר, לא תהיה קבילה כראיה לפי פסקה זו, אלא אם כן נעשתה בטעות בתום לב, תוך שימוש מדומה בהרשאה חוקית.

(א1) בקשה לקבילות ראיה לפי סעיף קטן (א) תהיה באישור היועץ המשפטי לממשלה, פרקליט המדינה או הפרקליט הצבאי הראשי בענין שבתחום סמכותו, והדיון בה יהיה, בשינויים המחוייבים, לפי סעיף 46 לפקודת הראיות, התשל"א-1971.

(א2) כדי להחליט בדבר קבילות כאמור בסעיף קטן (א), רשאי בית המשפט להקשיב לדברים או לעיין בהם; לענין פסקה זו סמכות להקשיב כמשמעותה בסעיף 2א(ד).

(א3) בית משפט שהחליט על קבילותה של ראיה לפי סעיף קטן (א) רשאי לשמוע אותה בדלתיים סגורות.

(ב) דברים שנקלטו בהאזנה כדין לפי סעיף 7 או בהאזנה כדין לשיחה חסויה לפי סעיף 5 לא יהיו קבילים כראיה אם לא אושרה ההאזנה כאמור באותם סעיפים לפי הענין.

(ג) דברים שנקלטו כדין בדרך האזנת סתר לא יהיו קבילים כראיה אלא בהליך פלילי שאינו על פי קובלנה.

(ג1) דברים שנקלטו כדין בדרך האזנת סתר יהיו קבילים כראיה בהליך פלילי להוכחת כל עבירה; לענין סעיף קטן זה, "הליך פלילי" – לרבות הליך אחר לחילוט רכוש הקשור בעבירה שהיא פשע.

(ד) אין באמור בסעיף זה כדי לגרוע מטענה בדבר קבילות ראיה גם אם נקלטה בדרך האזנת סתר בהתאם להוראות חוק זה.

מתוך חוק סדר הדין הפלילי (סמכויות אכיפה – נתוני תקשורת), התשס"ח-2007

צו לקבלת נתוני תקשורת ממאגר מידע של בעל רישיון בזק

3. (א) בית המשפט רשאי, על פי בקשה של קצין משטרה שהסמיך לעניין זה המפקח הכללי או של נציג רשות חוקרת אחרת (בסעיף זה – הבקשה), להתיר, בצו, למשטרה או לרשות החוקרת האחרת, קבלת נתוני תקשורת ממאגר מידע של בעל רישיון בזק, בדרך שיקבע בצו, אם שוכנע שהדבר נדרש למטרה מהמטרות המפורטות להלן, ובלבד שאין בקבלת נתוני התקשורת כאמור כדי לפגוע, במידה העולה על הנדרש, בפרטיותו של אדם: (1) הצלת חיי אדם או הגנה עליהם; (2) גילוי עבירות, חקירתן או מניעתן; (3) גילוי עבריינים והעמדתם לדין; (4) חילוט רכוש על פי דין.

(ב) היה המנוי שלגביו הוגשה הבקשה בעל מקצוע, לא יתיר בית המשפט קבלת נתוני תקשורת כאמור בסעיף קטן (א), אלא אם כן שוכנע בהסתמך על פירוט ברור לכך במסגרת הבקשה, שיש יסוד לחשד שבעל המקצוע מעורב בעבירה שבקשר אליה הוגשה הבקשה.

(ג) הבקשה תוגש בכתב ותיתמך בהצהרה שלאחר אזהרה או בתצהיר.

(ד) בבקשה יצוינו, בין היתר, כל אלה: (1) העובדות המקנות סמכות לבית המשפט; (2) פרטי זהותו ותיאור תפקידו של מגיש הבקשה ומקור סמכותו לבקש צו לפי סעיף זה; (3) תמצית העובדות והמידע שעליהם מבסס המבקש את הבקשה; (4) המטרות שלשמן נדרשים נתוני התקשורת; (5) נתוני התקשורת המבוקשים; (6) פרק הזמן שלגביו מתבקשים נתוני התקשורת, בהתייחס לתקופה שלפני מתן הצו, וכן בכפוף להוראות הסיפה של סעיף קטן (ז) בהתייחס לתקופה שלאחר מתן הצו (בסעיף זה – נתוני תקשורת עתידיים); (7) פרטי הזיהוי של המנוי או מיתקן הבזק שנתוני התקשורת מתבקשים לגביהם, אם הם ידועים מראש, לרבות היות המנוי האמור מי שחל לגביו חיסיון מקצועי לפי כל דין (בחוק זה – בעל מקצוע); בפסקה זו, "דין" – לרבות הלכה פסוקה; (8) פרטים בדבר בקשות לקבלת נתוני תקשורת שהתבקשו בעבר בקשר לאותו אדם באותו תיק חקירה (בסעיף זה – בקשות קודמות).

(ה) חומר חסוי, שעליו מבוסס המידע המפורט בסעיף קטן (ד)(3) ו–(4), יועמד לעיון בית המשפט בלבד; החומר יסומן ויוחזר למבקש לאחר העיון.

(ו) (1) לבקשה יצורפו – (א) החלטות בית המשפט שדן בבקשות קודמות; (ב) העתקים מבקשות קודמות ופרוטוקולים של דיוני בית המשפט בבקשות הקודמות, ככל שאלה נדונו בפני בית משפט אחר. (2) על אף הוראות פסקה (1), בית המשפט רשאי לדון בבקשה דחופה גם בלא המסמכים שבאותה פסקה, אם נחה דעתו שיש ברשותו המידע הנדרש למתן ההחלטה בבקשה, ומטעמים מיוחדים שיירשמו.

(ז) בהחלטה בבקשה ובקביעת פרק הזמן שלגביו יועברו נתוני תקשורת, יתחשב בית המשפט, בין היתר, בצורך בצו למימוש המטרות המנויות בסעיף קטן (א), במידת הפגיעה בפרטיותו של אדם, בחומרת העבירה, בהיות המנוי בעל מקצוע ובסוג נתוני התקשורת שהתיר לקבלם על פי הצו; בית המשפט רשאי לקבוע תקופות שונות לפרק הזמן לקבלת נתוני תקשורת לפי סוג נתוני התקשורת שהתיר לקבלם, ובלבד שהתקופה המרבית לקבלת נתוני תקשורת עתידיים לא תעלה על שלושים ימים מיום מתן הצו.

(ח) בצו לפי סעיף זה יפורטו כל אלה: (1) הנימוקים למתן הצו, ולעניין צו המתייחס למנוי שהוא בעל מקצוע – נימוקים מפורטים בדבר מתן צו בנסיבות אלה; (2) נתוני התקשורת שמותר לקבלם על פי הצו; (3) פרטי הזיהוי של המנוי או מיתקן הבזק שנתוני התקשורת התבקשו לגביהם, אם הם ידועים מראש, (4) פרק הזמן שלגביו ניתן לקבל נתוני תקשורת על פי הצו; (5) מועד מתן הצו ומועד תום תוקפו.

(ט) הנימוקים למתן הצו כאמור בסעיף קטן (ח)(1), לא יועברו לבעל רישיון בזק שהצו נוגע אליו.

(י) צו שניתן לפי סעיף זה יהיה בתוקף שלושים ימים מיום שניתן. (יא) אין בהוראות סעיף זה כדי לגרוע מסמכות בית המשפט לתת צווים נוספים לגבי אותה חקירה.

היתר לקבלת נתוני תקשורת במקרים דחופים

4. (א) קצין מוסמך רשאי, על פי בקשה של שוטר או של שוטר צבאי, לפי העניין, להתיר קבלת נתוני תקשורת ממאגר מידע של בעל רישיון בזק, בלא צו של בית המשפט לפי סעיף 3, אם שוכנע כי לשם מניעת עבירה מסוג פשע או גילוי מבצעה או לשם הצלת חיי אדם יש צורך, שאינו סובל דיחוי, בקבלת נתוני תקשורת כאמור, וכי לא ניתן לקבל בעוד מועד צו לפי סעיף 3.

(ב) היתר לפי סעיף זה יינתן ככל הניתן בכתב, ויהיה לתקופה שלא תעלה על 24 שעות; ההיתר יכלול את פרטי זהותו ותיאור תפקידו של הקצין המוסמך, תמצית העובדות והמידע שעל יסודם ניתן ההיתר, המטרות שלשמן נדרשים נתוני התקשורת וכן פרטים כאמור בסעיף 3(ח)(2) עד (4).

(ג) ניתן היתר לפי סעיף זה, רשאי שוטר או שוטר צבאי, לפי העניין, לדרוש מבעל רישיון בזק שההיתר נוגע אליו, להעביר למשטרה או למשטרה הצבאית החוקרת, נתוני תקשורת בהתאם להיתר; בעל רישיון בזק יעביר את נתוני התקשורת, בהתאם לדרישה, בלא דיחוי.

(ד) קצין מוסמך שנתן היתר לפי סעיף זה ידווח על כך בכתב, בפירוט תמצית נימוקיו למתן ההיתר כאמור, בהקדם האפשרי, לקצין משטרה בדרגת ניצב משנה ומעלה שהסמיך לעניין זה ראש אגף החקירות והמודיעין של משטרת ישראל (בחוק זה- ראש אגף החקירות והמודיעין) או למפקד המשטרה הצבאית החוקרת, לפי העניין.

(ה) (1) ראש אגף החקירות והמודיעין יגיש דין וחשבון ליועץ המשפטי לממשלה, אחת לשלושה חודשים, על היתרים שניתנו לפי סעיף זה. (2) מפקד המשטרה הצבאית החוקרת יגיש דין וחשבון לפרקליט הצבאי הראשי, אחת לשלושה חודשים, על היתרים שניתנו לפי סעיף זה.

(ו) המפקח הכללי וקצין משטרה צבאית ראשי יקבעו, בפקודות משטרת ישראל או בנוהלי משטרת ישראל, או בהוראות קצין משטרה צבאית ראשי, לפי העניין, הוראות לעניין סעיף זה, לרבות לעניין אופן מתן היתר כאמור בסעיף קטן (א), אופן העברת הדרישה על פי ההיתר לבעל רישיון בזק לפי סעיף קטן (ג) ואופן הדיווח לפי סעיפים קטנים (ד) ו–(ה), ורשאים הם לקבוע הוראות שונות בהתאם לנימוקים למתן ההיתר ולנסיבות שבהן ניתן; בסעיף זה, "הוראות קצין משטרה צבאית ראשי" – פקודות כלליות אחרות, כמשמעותן בסעיף 2א(ג) לחוק השיפוטי הצבאי, התשט"ו-1955, שהוציא קצין משטרה צבאית ראשי.

מתוך חוק יסוד: כבוד האדם וחירותו

פרטיות וצנעת הפרט

7. (א) כל אדם זכאי לפרטיות ולצנעת חייו.

(ב) אין נכנסים לרשות היחיד של אדם שלא בהסכמתו.

(ג) אין עורכים חיפוש ברשות היחיד של אדם, על גופו, בגופו או בכליו.

(ד) אין פוגעים בסוד שיחו של אדם, בכתביו או ברשומותיו.

פגיעה בזכויות

8. אין פוגעים בזכויות שלפי חוק-יסוד זה אלא בחוק ההולם את ערכיה של מדינת ישראל, שנועד לתכלית ראויה, ובמידה שאינה עולה על הנדרש, או לפי חוק כאמור מכוח הסמכה מפורשת בו.

English Translation of Relevant Legal Provisions

The Penal Code

Offenses against international law

16. (a) Israel penal laws shall apply to foreign offenses, which the State of Israel undertook—under multilateral international conventions that are open to accession—to punish, and that even if they were committed by a person who is not an Israel citizen or an Israel resident and no matter where they were committed. (b) The restrictions said in section 14(b)(2) and (3) and (c) shall also apply to the applicability of Israel penal laws under this section.

Vicarious applicability

17. (a) The State of Israel may assume obligation, by international conventions, to apply its penal laws to foreign offenses or to apply the provisions of section 10—at the request of a foreign state and on a reciprocal basis—also to cases different from those enumerated in sections 13 to 16, on condition that all the following conditions are met: (1) the penal laws of the requesting state apply to

the offense; (2) the offense was committed by a person who is on Israel territory and who is an Israel resident, whether or not he is an Israel citizen; (3) in its request, the requesting state waived—subject to the full application of Israel Law against that person—the applicability of its law to the case at hand.

(b) In Israel no more severe penalty shall be imposed for the offense, than could have been imposed under the laws of the state that made the request.

(c) All other conditions shall be set in the convention.

Classification of offenses
24. The following are the categories of offense, by their severity:

(1) "Felony"—an offense, for which a penalty more severe than three years imprisonment; has been set;
(2) "Misdemeanour"—an offense, for which a penalty of imprisonment for more than three months, but not more than three years was set; and if the penalty only consists of a fine—if the fine exceeds the amount of fine that can be imposed for an offense, for which the penalty is a fine the amount of which has not been set;
(3) "Transgression"—an offense, for which a penalty of imprisonment for not more than three months was set; if the penalty only consists of a fine—if the fine does not exceed the amount of fine that can be imposed for an offense, for which the penalty is a fine the amount of which has not been set.

What constitutes an attempt
25. A person attempts to commit an offense, if he—with intent to commit it—commits an act that does not only constitute preparation, on condition that the offense was not completed.

Commission of offense is impossible
26. For purposes of attempt, it is immaterial that commission of the offense was impossible, because of circumstances of which the person who made the attempt was not aware or in respect of which he was mistaken.

Exemption because of remorse
28. If a person attempted to commit an offense, he shall not bear criminal liability therefor, if he proved that—of his own free will and out of remorse—he stopped in the commission of the act or substantively contributed to prevention of the results, on which the completion of the offense depends; however, the aforesaid does not derogate from his criminal liability for another completed offense connected to the act.

Interpretation
34U. If an enactment can be reasonably interpreted in several ways in respect of its purpose, then the matter shall be decided according to the interpretation that is most favourable for whoever is about to bear criminal liability under that enactment.

Definitions
34X. In relation to an offense—

…

"Possession"—a person's control of a thing in his custody or in the custody of another person or in any place, whether it belongs to him or not; a thing held by or in the possession of one or several of a group, with the knowledge and consent of the rest, is deemed to be in the custody and possession of each of them and of all together;

"Publication"—written matter, printed matter, computer material or any other visual presentation and any audio means capable of presenting words or ideas, whether alone or with the help of any medium;

"Publish"—(1) in respect of spoken matter—to utter words orally or by other means at a public gathering or in a public place or so that they may be heard by persons in a public place, or letting them be heard by radio or television broadcasts to the public, or distributing them by means of computers in a manner accessible to the public, or offering them to the public by means of computers; (2) in respect of matter other than spoken—its distribution among people or its presentation so that people in a public place can see it, selling it or offering it for sale in any place, broadcasting it by television broadcasts to the public, or distributing it by means of computers in a manner accessible to the public, or offering it to the public by means of computers;

"Public"—including any part of it likely to be adversely affected by the conduct referred to in the context of this term; "minor"—a person who has not reached age 18.

Procurement 199 (a) The following are liable to five years imprisonment: (1) a person who wholly or in part, permanently or for any period of time lives on the earnings of a person engaged in prostitution; (2) a person who knowingly receives something that was given for a person's act of prostitution, or a part of what was so given. (b) If a person committed an offense under this section in connection with his spouse, child or stepchild, or if he committed the offense by exploiting a relationship of authority, dependence, education or supervision, then he shall be liable to seven years imprisonment. (c) For purposes of this section, it is immaterial—(1) whether what the offender received was money, valuable consideration, a service or some other benefit; (2) whether he received it from a person who engages in prostitution or from some other person; (3) whether he receives what was given for an act of prostitution or a substitute for what was so given.

Inducement to act of prostitution
201. If a person induces another to perform an act of prostitution with another person, then he shall be liable to five years imprisonment.

Aggravating circumstances
203. (a) If an offense under sections 201 or 203 was committed by exploiting a relationship of authority, dependence, education or supervision, or by exploiting the economic or mental distress of the person who was induced to perform an act

of prostitution or to engage in prostitution, then the person guilty of the act shall be liable to ten years imprisonment.

(b) If an offense under sections 201 or 203 was committed under one of the following circumstances, then the person guilty of the offense shall be liable to sixteen years imprisonment: (1) by use of force or by use of other means of pressure, or by threat of one of these, and it is immaterial whether it was done against the person who was induced to commit an act of prostitution or to engage in prostitution or against some other person; (2) by exploiting a situation that prevents opposition by the person induced to commit an act of prostitution or to engage in prostitution, or by the exploiting the fact that he is mentally ill or mentally incompetent; (3) by agreement obtained be deception of the person induced to commit an act of prostitution or to engage in prostitution.

Exploitation of minors for prostitution

203B. (a) If an offense was committed under sections 199, 201, 202 or203 against a minor who has reached age 14, then the person who committed the offense—(1) if for that offense a penalty of five years was set—shall be liable to seven years imprisonment; (2) if for the offense a penalty of seven years was set —shall be liable to ten years imprisonment; (3) if for the offense a penalty of ten years was set—shall be liable to fifteen years imprisonment; (4) if for the offense a penalty of sixteen years was set—shall be liable to twenty years imprisonment;

(b) If an offense was committed under sections 199, 201, 202 or 203 against a minor who has not yet reached age 14, or if he reached age 14 and the person who committed the offense is responsible for the minor, then the person who committed the offense shall be liable to double the penalty set, but not more than twenty years.

(c) In this section, "responsible for the minor"—as defined in section 368A.

Liability of a minor's customer

203C. If a person accepts sexual services from a minor, then he is liable to three years imprisonment.

Must not publicize or inform of a minor's prostitution

205A. If person delivers information or publishes any publication about the provision of a service that is an act of prostitution, the person who provides the service being a minor, then he shall be liable to five years imprisonment; for purposes of an offense under this section it does not matter whether the prostitution service is provided in Israel or abroad, whether the information refers to a specific minor, or whether the publication states that the person who provides the service is a minor.

Permitting minor to reside in brothel

208. If a person permitted a minor between age two and age seventeen, of whom he has custody, to reside in a brothel or to visit it frequently, then he is liable to three years imprisonment.

Obscene publication and display

214. (a) If a person did one of the following, then he is liable to three years imprisonment: (1) he published an obscene publication or prepared it for publication; (2) he presented, organized or produced an obscene display—(a) in a public place; (b) in a place which is not public—unless it is used for residential purposes or is used by a body of persons, membership in which is restricted to persons aged eighteen and up and is for a continuous period.

(b) If a person published an obscene publication and it includes the likeness of a minor, including a representation or a drawing of a minor, then he shall be liable to five years imprisonment.

(b1) If a person utilized the body of a minor in order to advertise an obscenity, or used a minor in the presentation of an obscenity, then he shall be liable to seven years imprisonment.

(b2) If an offense under subsections (b) or (b1) was committed by a person responsible for a minor, as defined in section 368A, or with the consent of an aforesaid responsible person, then the responsible person shall be liable to ten years imprisonment.

(b3) If a person has in his possession an obscene publication that includes the likeness of a minor, then he shall be liable to one year imprisonment; for purposes of this section, "has in his possession"—exclusive of whoever has in his possession incidentally and in good faith.

(c) If the Court deals with an offense under this section, committed by a person in the course of his business, then it may also use its powers under sections 16 and 17 of the Licensing of Business Law 5728-1968, but it shall not use its authority under section 17, unless it is satisfied that there is prima facie evidence of the offense, and that the use of its authority is necessary for the public good.

(d) An indictment shall only be filed—(1) under subsection (a)—within two years after the day on which the offense was committed, and only by the District Attorney or with his written consent. (2) under subsections (b) and (b3)—by the District Attorney or with his written consent. (d) An indictment under this section shall only be filed within two years after the day on which the offense was committed, and only by the District Attorney or with his written consent.

Defences

214B. A person shall not be deemed to have committed an offense under sections 205A to 205C and 214, if provision of the information, the publication or the possession were for a legal purpose, including true and fair reporting on the subject with which this Article deals, on condition that the provision of information, the publication or the possession are not prohibited under any other enactment and were not carried out in order to encourage acts prohibited under this Article.

Rape

345. (a) If a person had intercourse with a woman—(1) without her freely given consent; (2) with the woman's consent, which was obtained by deceit in respect of the identity of the person or the nature of the act; (3) when the woman is a minor below age 14, even with her consent; (4) by exploiting the woman's state of unconsciousness or other condition that prevents her from giving her free consent; (5) by exploiting the fact that she is mentally ill or deficient, if—because of her illness or mental deficiency—her consent to intercourse did not constitute free consent—then he committed rape and is liable to sixteen years imprisonment.

(b) Notwithstanding the provisions of subsection (a), a rapist shall be liable to twenty years imprisonment, if the rape was committed under one of the following circumstances: (1) upon a minor under age 16, and under the circumstances said in subsection (a)(1), (2), (4) or (5); (2) while threatening with a firearm or other weapon; (3) while causing bodily or mental injury, or pregnancy; (4) together with the abuse of the woman before, during or after the act; (5) in the presence of one or several others, who joined together to commit rape by one or several of them.

(c) In this Article—"has intercourse"—introduces any part of the body or any object into the woman's sex organ.

Forbidden intercourse by consent

346. (a) (1) If a person had intercourse with a minor who has reached age 14, but has not yet reached age 16 and who is not married to him, or if a person has intercourse with a minor who has reached age 16, but has not yet reached age 18, by exploiting a relationship of dependence, authority, education or supervision, or by a false promise of marriage, then he is liable to five years imprisonment.

(2) For the purposes of this subsection, if a person who provides mental health treatment to a minor who has reached age 16, but has not yet reached age 18, had intercourse with her during the period in which he gave her mental health treatment, then he shall be deemed to have performed the said act by exploiting a relationship of dependence; the said presumption shall not apply if such relations began in a pair relationship before the mental health treatments began.

(b) If a person had intercourse with a woman has reached age 16, aged more than 18 by exploiting his authority in employment or service, or by false promises of marriage while pretending to be single in spite of being married, then he is liable to three years imprisonment.

Sodomy

347. (a) (1) If a person committed sodomy on a person who has reached age 14, but has not yet reached age 16, or if he committed sodomy on a person who has reached age 16, but has not yet reached age 18 by exploiting relations of dependence, authority, education or supervision, then he is liable to five years imprisonment.

(2) For the purposes of this subsection, if a person who provides mental health treatment to a minor who has reached age 16, but has not yet reached age 18, committed sodomy on that person during the period in which he gave him mental health treatment, then he shall be deemed to have performed the said act by exploiting a relationship of dependence; the said presumption shall not apply if such relations began in a pair relationship before the mental health treatments began.

(a1) If a person committed sodomy upon a person who has reached age 18 or more, by exploiting his authority in employment or service, then he is liable to three years imprisonment.

(b) If a person committed sodomy upon a person under one of the circumstances specified in section 345, mutatis mutandis, then he is liable to the penalties of a rapist.

(c) For purposes of this Article, "sodomy"—introduction of a bodily organ or an object into a person's anus, or introduction of a sex organ into a person's mouth.

Indecent act

348. (a) If a person committed an indecent act on a person under one of the circumstances enumerated in section 345(a)(2) to (5), mutatis mutandis, then he is liable to seven years imprisonment.

(b) If a person committed an indecent act on a person under one of the circumstances specified in section 345(b)(1) to (5), mutatis mutandis, then he is liable to ten years imprisonment.

(c) If a person committed an indecent act on a person without his consent, but not under the circumstances said in subsections (a), (b), or (c1), then he is liable to three years imprisonment.

(c1) If an offense under subsection (c) was committed by use of force or by the application of other means of pressure, or by the threat of one of them—whether toward the person or toward anybody else—then the person who committed the offense is liable to seven years imprisonment.

(d) (1) If a person committed an indecent act on person who is a minor who has reached age 14, by exploiting a relationship of dependence, authority, education, supervision, employment or service, then he is liable to four years imprisonment; (2) for the purposes of this subsection, if a person who provides mental health treatment and performed an indecent act on a minor who has reached age 14, but has not yet reached age 18, during the period in which he gave that person mental health treatment, then he shall be deemed to have performed the said act by exploiting a relationship of dependence; this presumption shall not apply if that person has reached age 16 and the acts began in a pair relationship before the mental health treatments began.

(d1) If a provider of mental health treatments committed an indecent act on a person who has reached age 18 under the circumstances said in section 347A(b), then he shall be liable to three years imprisonment.

(e) In this Article, "indecent act"—an act for sexual arousal, satisfaction or abasement.

Indecent act in public

349. (a) If a person committed an indecent act on a person in public before another person without his consent, or if a person committed a said act anywhere by exploiting a relationship of dependence, authority, education, supervision, employment or service, then he is liable to one year imprisonment.

(b) If a person committed an indecent act in any place whatsoever before a person who has not yet reached age 16, then he is liable to three years imprisonment.

Responsibility

350. For purposes of an offense under this Article, it is immaterial whether a person performed an act or caused an act to be performed on him or on another person.

Sex offenses within the family and by persons responsible for helpless persons

351. (a) If a person committed an offense of rape under section 345(a), or of sodomy under section 347(b), on a person who is a minor and his relative or on a helpless person for whom he is responsible, then he is liable to twenty years imprisonment.

(b) If a person had intercourse with a woman who has reached age 14, but has not yet reached age 21, or committed sodomy upon a person who reached age 14, but has not yet reached age 21 and is his relative, then he is liable to sixteen years imprisonment.

(c) If a person committed an indecent act upon a minor who is his relative or on a helpless person for whom he is responsible, then he is liable—(1) for an offense under section 348(a) or (c1)—to ten years imprisonment; (2) for an offense under section 348(b)—to fifteen years imprisonment; (3) in any case not specified in paragraphs (1) and (2)—to five years imprisonment. (d) If a person in any place committed an indecent act upon a minor who is a member of his family, then he is liable to four years imprisonment;

(d1) If a person responsible for a helpless person committed an offense under section 349(a) against the helpless person, then he is liable to two years imprisonment;

(e) For purposes of this section—"foster parent"—one of the following: (1) the father or mother of a foster family approved by the Ministry of Welfare;

(2) the person responsible for a minor under paragraph (3) of the definition of the definition of "guardian of minor or of a helpless person" in section 368A;

"Stepbrother" or "stepsister"—son or daughter of a parent's spouse;

"Person responsible for a helpless person"—like the definition of "guardian of a minor or of a helpless person" in section 368A;

"relative"—(1) parent; spouse of parent, even if not married to him; grandfather or grandmother; (2) a person who has reached age 15 and who is one of these: brother or sister, stepbrother or stepsister, uncle or aunt, brother-in-law or sister-in-law; however, for purposes of the offense of prohibited intercourse under subsection (b), or of an indecent act under subsection (c)(3), committed upon a person aged 16 or more, uncle and aunt, brother-in-law or sister-in-law shall not be included in the definition of "relative"; (3) a foster parent; the spouse of a foster parent, even if not married to him; the father or mother of a foster parent; (4) a person who has reached age fifteen and is one of the following: the son or daughter of a foster parent and the spouse of each of these; the brother or sister of a foster parent and the spouse of each of these; however, in respect of the offense of prohibited intercourse under subsection (b) and indecent act under section (c)(3), committed on a person who has reached age sixteen, the spouse of a foster parent's son or daughter, the foster parent's brother and sister and the spouse of any of these shall not be included in the definition of "relative";

"Helpless person"—as defined in section 368A.

Section 352 Publication prohibited

(a) If a person published a person's name or anything that can identify a person was injured by an offense or who complained that he was injured by an offense under this Article, then he is liable to one year imprisonment.

(b) A person shall not bear criminal responsibility under subsection (a), if the person whose name or identity were made public gave his consent to the publication before a Court or if a Court permitted publication for special reasons that shall be recorded.

Restriction on criminal responsibility

353. In an indictment for an offense under sections 346(a) or 347(a), it shall be a defence for the defendant that the difference in age between himself and the minor is not more than three years, if the minor consented to the act, and if the act was performed within ordinary friendly relations and without exploitation of the defendant's position.

Criminal Procedure Law [Consolidated Version], 1982

Complaint

58. Any person is entitled to complain to the police that an offence has been committed.

Police investigation
59. If the police, whether by a complaint or in any other manner, learns that an offence has been committed, it will open an investigation. However, in the case of an offence other than a felony, a police officer with the rank of captain or higher is entitled to direct that no investigation will be held if he is of the opinion that no public interest is involved or if another authority is legally competent to investigate the offence.

Preliminary pleadings
149. After the commencement of the trial, the defendant is entitled to make preliminary pleadings, including the following: … (10) The filing of an indictment or the conduct of criminal proceedings is in material contradiction to the principles of justice and fair trial.

Criminal Procedure (Arrest and Search) (new version) 1969

Accessing computerized material
23A. (a) Accessing computerized material is considered a search and will be executed by a skilled functionary.

(b) No search will be executed without a judicial warrant clearly stating the authorization to access computerized material and specifying the objects of the search and its terms in a manner suitable to befit the right of privacy and not violating it more than required.

Wiretap Law 1979

Wiretap to prevent crimes
6. (a) Chief justice or a vice-chief of the District court may issue a warrant of wiretap if a high ranked police officer has so requested and the Judge has found that it is needed to reveal, investigate or prevent felonies.

Evidence
13. (a) Evidence obtained by wiretapping conducted opposed to this law will be inadmissible in the court of law, except in: (1) criminal charges of wiretapping; (2) criminal charges of a major felony (over seven years penalty) if the court so ordered, by special rationalization that will be specified, that the interest of finding the truth exceeds the interest to protect privacy. Illegal wiretapping conducted by those who are allowed so request it will not be admissible under this paragraph unless the wiretap was conducted in good faith.

Law of Criminal Procedure (Enforcement Authorities—Communication Data) 2007

A warrant to receive communication data
3. (a) A magistrate judge may issue a warrant of wiretap if a the police have so requested and the Judge has found that it is needed for one of the following objects, as long as the judge finds it does not violate the right of privacy more than required: (1) saving or protecting human life; (2) revealing, investigating and preventing offenses; (3) revealing and prosecuting offenders; (4) confiscation.

A permission to receive communication data in emergency cases
4. (a) A high ranked police officer is authorized by request of a police man to grant the reception of communication data without a judicial warrant, as long as the police officer finds it is urgently needed in order to reveal a felony or a felon or to save human life and there is no opportunity to issue a warrant.

Basic Law: Human Dignity and Liberty
Section 7: (a) All persons have the right to privacy and to intimacy.

(b) There shall be no entry into the private premises of a person who has not consented thereto.

(c) No search shall be conducted on the private premises of a person, nor in the body or personal effects.

(d) There shall be no violation of the confidentiality of conversation, or of the writings or records of a person

Section 8 There shall be no violation of rights under this Basic Law except by a law befitting the values of the State of Israel, enacted for a proper purpose, and to an extent no greater than is required, or by regulation enacted by virtue of express authorization in such law.

International Workshop on Sweetie—June 30, 2016

Location: Kamerlingh Onnes Gebouw (KOG), room A0.28—Steenschuur 25, 2311 ES Leiden

Timeframe: Start 10 a.m.—lunch 12:30 p.m.—end (expected) 5 p.m.

Further Reading

Harduf A (2009) Possession of electronic child pornography: where does society prefer its pedophiles? The Criminal Defender 150: 4–9

Harduf A (2010) Cybercrime. Nevo Publishing, pp 1–430

Harduf A (2010) The harassing lightness of seduction. The Criminal Defender 160: 11–14

Harduf A (2013) Criminal law tangles in the web: the virtual actus reus. Hapraklit 52: 67–149

Harduf A (2014) Cloak of rights, substance of justice: rhetoric and realism of criminal adjudication. Haifa Law Review 8: 33–98

Harduf A (2015) In the name of the child: the constitutionality of criminalizing the accessing of child pornography. Mozney Mishpat 10: 209–251
Harduf A (2015) Intimate technology, suspect autonomy and legal opportunism: accessing smartphones by "authorization" of detained suspects. The Criminal Defender 224: 4–12
Harduf A (2016) Criminalization downloads evil: re-examining the approach to the electronic possession when child pornography goes international. B.U. Int. L.J. 34: 279–318
Harduf A (2016) Rape goes cyber: the offense of rape, its rationales and its limits. Aley-Mishpat 13: 65–134
Harduf A (2017) Encrypted files, encrypted law: the concept of obscenity and conviction without watching the obscene materials. Hamishpat Online 70
Harduf A. Judicial passivism and fictitious criminal law: non-elementary convictions and the problem of the legal lie. Hamishpat 24: 39–73
Harduf A (forthcoming) Non-legislation, over-jurisprudence and under-freedom: legislative passivism and litigatory-judicial activism, the boundaries of the law and the boundaries of non-law. Mozney Mishpat
Harduf A (forthcoming) The process is the punishment: the shaming of criminal procedure. Law and Business
Wismonsky H (2010) Revealing pedophiles by undercover agents operations on the internet. The Criminal Defender 160: 5–10
Wismonsky H (2014) Sex crimes in cyberspace. Mishpat Matte'ah 2, 32
Wismonsky H (2015) Criminal investigation in cyberspace. Nevo Publishing, pp 1–339
Wismonsky H (2015) Fighting paedophilia: in the wake of Penal Code (Amendment 118), 2014, and of the bill to limit the use of site to prevent offenses (Amendment 2). Mozney Mishpat 10: 181–207

Relevant Case Law

Crim 20996-08-15 Israel vs. Goldstein (decided 17.7.2017)
LCrimA 4275/16 Doe vs. Israel (decided 9.1.2017)
LCrimA 8720/15 Israel vs. Pinto (decided 11.9.2016)
CrimA 707/14 Doe vs. Israel (decided 6.7.2015).
Crim 41309-12-14 Israel vs. Morobaty (decided 16.3.2015).
CrimA 2656/13 Doe vs. Israel (decided 21.1.2014).
CrimA 6703/13 Cohen vs. Israel (decided 16.1.2014).
LCrimA 1201/12 Ktiei vs. Israel (decided 9.1.2014).
CrimA 538/13 Sabach vs. Israel (decided 26.12.2013).
CrimA 512/13 Doe vs. Israel (decided 14.12.2013).
Crim 13384-02-09 (decided 23.6.2011).
Crim 2507/08 Israel vs. Rotem (decided 28.3.2011).
Crim 3990/07 Israel vs. Mashiach (decided 13.9.2010).
CrimA 1224/07 Beladev vs. Israel (decided 10.2.2010).
CrimA 7476/09 Haim (decided 8.2.2000).
Crim 1137/07 Israel vs. Ben-Guy (decided 21.10.2009).
Crim 2225/07 Israel vs. Lior (decided 25.6.2008).

Asaf Harduf Ph.D., Haifa University, “Is Online Crime Inherently Different from Offline Crime?” A senior lecturer at Zefat Academic College. Formerly a prosecutor and a defender at the IDF Judge Advocate General. Teaches and researches all branches of criminal law: substantive criminal law, procedural criminal law and the law of evidence. Specializes in cybercrime and criminalization. Published three books and over twenty peer-reviewed articles, most of which analyze cybercrime law, including offenses like online rape, online indecent act, online theft, unauthorized computer access, online gambling and possession of child pornography. Giving lectures to the Israeli Bar, the public prosecution, the public defense and judges. Chosen outstanding lecturer in three law faculties.

Chapter 10
Substantive and Procedural Legislation in the Netherlands to Combat Webcam-Related Child Sexual Abuse

Bart W. Schermer, Bert-Jaap Koops and Simone van der Hof

Contents

B. W. Schermer (✉) · S. van der Hof
Center for Law and Digital Technologies, Leiden University, Leiden, The Netherlands
e-mail: b.w.schermer@law.leidenuniv.nl

S. van der Hof
e-mail: s.van.der.hof@law.leidenuniv.nl

B.-J. Koops
Tilburg Institute for Law, Technology, and Society, Tilburg University, Tilburg, The Netherlands
e-mail: e.j.koops@tilburguniversity.edu

S. van der Hof et al. (eds.), *Sweetie 2.0*, Information Technology and Law Series 31,
https://doi.org/10.1007/978-94-6265-288-0_10

Abstract This chapter deals with Dutch substantive criminal law and criminal procedure and the extent to which it can combat webcam-related child sexual abuse. In summary we can say that the substantive legal framework for the protection of minors against webcam sex in the Netherlands is generally adequate, particularly with the amendments enacted by the Computer Crime III Act in 2018. Criminal procedural law offers generally adequate investigation powers, but questions remain regarding the legality of using Sweetie as an investigative power given its hybrid character (both as lure and as undercover method), and regarding the risk of incitement or entrapment.

Keywords The Netherlands · Criminal Law · Criminal Procedure · Child Sexual Abuse · Internet · Virtual Crime · Entrapment

10.1 Introduction: Legislation in the Netherlands

10.1.1 General Description of the Legal Framework

The Legal System

The Netherlands is a constitutional monarchy with a civil law tradition. The primary form of legislation in the Netherlands is an Act of Parliament, enacted jointly by Government and Parliament (*Staten-Generaal*).[1] The Dutch Parliament consists of two houses: the Lower House (*Tweede Kamer der Staten-Generaal*) and the Upper House or Senate (*Eerste Kamer der Staten-Generaal*).

Substantive criminal law is codified in the Dutch Criminal Code (*Wetboek van Strafrecht*, hereafter: DCC). The Criminal Code distinguishes between crimes (*misdrijven*) and misdemeanours (*overtredingen*). Definitions of crimes may only be enacted via an Act of Parliament, whereas misdemeanours may also be enacted by provinces and municipalities.

Procedural criminal law is codified in the Code of Criminal Procedure (*Wetboek van Strafvordering*, hereafter DCCP). Changes to the Code of Criminal Procedure may only be made via an Act of Parliament. However, Parliament may decide to delegate the authority to issue elaborative rules of a procedural nature to the Government or individual minister.[2]

The judiciary in the field of criminal law is comprised of the courts and the Public Prosecution Service (*Openbaar Ministerie*). The highest court in the Netherlands is the Supreme Court (*Hoge Raad*). The Supreme Court is responsible for hearing appeals in cassation and for a number of specific tasks with which it is charged by law.

---

[1] See Article 81 Dutch Constitution.

[2] Tak 2008, p. 4.

The Public Prosecution Service is charged with the prosecution of criminal offences. The Public Prosecution service is headed by the Board of Procurators General (*College van Procureurs-Generaal*). In the pre-trial phase the investigation of crimes is overseen by the examining magistrate (*rechter-commissaris*), the trial phase is overseen by the trial judge.[3]

Human Rights are codified in the Constitution, but laws and treaties cannot be reviewed by courts based on this constitution (Article 120 Constitution (*Grondwet*)). However, Article 94 of the Dutch Constitution states that:

> Statutory regulations in force within the Kingdom shall not be applicable if such application is in conflict with provisions of treaties that are binding on all persons or of resolutions by international institutions.

Therefore, the Dutch judiciary can assess legislation on the basis of international treaties. The most relevant human rights treaty is the European Convention on Human Rights (ECHR). The ECHR creates both positive and negative human rights obligations for the signatories that are binding. The Netherlands is also a party to the International Covenant on Civil and Political Rights.

Relevant Aspects of Criminal Law and Criminal Procedure

A first aspect that is relevant from the perspective of the Sweetie case is that in Dutch criminal law there is no definition of a minor.[4] For the purpose of the articles concerned with the sexual abuse of minors and lewd acts (*ontuchtelijke handelingen*), a minor is a person that has not yet reached the age of 16 years. However, the legislator also recognises that a person between the age of 16 and 18 years is vulnerable and therefore there are several articles which also protect persons under the age of 18 years. In the overview provided in this chapter the relevant age of the victim is noted for each of the crimes.

An aspect of Dutch criminal law that is also of particular relevance to the Sweetie 2.0 case is the difference between a completed criminal act and that of an attempt (*poging*). In general, an attempt is punishable, though the maximum penalty is reduced by a third (Article 45 DCC).

An attempt requires the beginning of an act of execution of behaviour that fulfils the description in the offence. From a legal perspective, an attempt can fail either because the action is not completed through circumstances outside of the actor's will, or when despite being completed, it does not meet the requirements set forth in the description of the crime. In the latter case, the attempt is deemed inadequate (*ondeugdelijk*). An attempt can be relatively inadequate (*relatief ondeugdelijke poging*) or absolutely inadequate (*absoluut ondeugdelijke poging*). With a relatively inadequate attempt the attempt would normally succeed (and thus fulfil the description of the crime), although in the concrete case it has not succeeded. With

---

[3] It seems that in Dutch literature, examining magistrate, examining judge and investigative judge are used interchangeably to translate the Dutch rechter-commissaris.

[4] When discussing child webcam sex abuse and the ways of combatting it, the Sweetie case will be taken as the main example as a basis for discussion.

an absolutely inadequate attempt, the crime could never be completed, despite the intention of the perpetrator.[5]

An example of a relatively inadequate attempt is an attempt to steal money from an empty vault. Normally stealing from a vault will result in the crime of theft, but since the vault in this concrete case is empty, the attempt is deemed relatively inadequate. An example of an absolutely inadequate attempt is trying to murder a corpse. In this case the desired result (killing a person) can never occur because that person is already dead. A relatively inadequate attempt is punishable, while an absolutely inadequate attempt is not punishable, regardless of the criminal intent of the perpetrator.

A relevant aspect in criminal procedure law is that there is no obligation in Dutch law to prosecute all offences: the Public Prosecutor can use discretionary power (*opportuniteitsbeginsel*) to decide whether an offence merits prosecution (see Articles 167 en 242 DCCP). As a consequence, Dutch criminal provisions are sometimes broadly formulated, since the Prosecution can decide not to prosecute minor cases. At the same time, Dutch law has the legality principle of *nullum crimen sine lege previa* (Article 1 DCC), and legal certainty requires that criminal provisions are described in sufficiently precise language (*Bestimmtheitsgebot*), so that criminalisations cannot be formulated too broadly or vaguely.[6] Moreover, the interpretation of criminal law allows for an extensive interpretation (interpreting terms widely as long as they fit grammatically, historically, or teleologically within the meaning of the term), but not for an analogous interpretation.

Procedural law also has a legality principle (Article 1 DCCP), requiring that law-enforcement powers are laid down by law, and the formulation needs to be sufficiently precise (also in light of the 'foreseeability by law' requirement of Article 8 ECHR). Minor infringements of privacy can be based on the general task description of the police (Article 3 Police Act 2012 (*Politiewet 2012*)), but law-enforcement powers that intrude upon people's privacy in a more than minor manner need to be explicitly regulated in a specific provision in the Code of Criminal Procedure.

There are gradations of intrusiveness of investigatory powers. The most intrusive ones require a warrant by the examining magistrate, most others require an order by the Public Prosecutor; the least intrusive powers can be performed by police officers themselves. An important criterion is also the category of crimes for which investigatory powers can be used. The main threshold is that many of the more intrusive powers (including most cyber-investigation powers) can only be used for crimes for which pre-trial detention is allowed (Article 67 para 1 DCCP); generally, that concerns crimes carrying a maximum penalty of at least four years' imprisonment. However, some crimes with a lower penalty are specifically enumerated in Article 67 para 1 DCCP, and this includes most cybercrimes and some crimes

[5] See: Machielse 2015.

[6] See for instance: Nan 2012.

against decency (Articles 248d and 248e), so that (cyber-)investigatory powers can also be used to investigate these.

10.1.2 Relevant Treaties and Cybercrime Laws

The Netherlands is a party to various conventions; relevant for this report are the Cybercrime Convention, the Lanzarote Convention, and the UN Convention on the Rights of the Child 1989 (including the Optional Protocol on the sale of children, child prostitution and child pornography). These conventions have been implemented in the Criminal Code and Code of Criminal Procedure.

Dutch cybercrime and cyber-investigation law is not enacted in a stand-alone statute, but relevant provisions have been inserted or adapted in the Criminal Code and Code of Criminal Procedure, primarily through the Computer Crime Act (*Wet computercriminaliteit*, 1993),[7] the Computer Crime II Act (*Wet computercriminaliteit II*, 2006)[8] and the Computer Crime III Act (*Wet computercriminaliteit III*, 2018)[9] but also through thematic laws such as the Act Partially Adapting Decency Legislation (*Wet partiële wijziging zedelijkheidswetgeving*, 2002).[10] For investigation powers, particularly the Special Investigatory Powers Act (*Wet bijzondere opsporingsbevoegdheden*, in force since 2000)[11] is relevant for the purposes of this chapter, since this regulates, *inter alia*, undercover operations; although the legislator in the late 1990s occasionally referred to the Internet in the Explanatory Memorandum to the Special Investigatory Powers Act, most powers have been formulated with physical forms of investigation in mind, and it is sometimes unclear whether and to what extent the law should apply to online forms of investigation.[12]

10.2 Analysis of Substantive Criminal Law

10.2.1 Introduction

Crimes related to the abuse of minors are codified in the Second Book (Crimes), Title XIV of the DCC, which deals with crimes against decency (*misdrijven tegen de zeden*), running from Article 239 DCC up to Article 254a DCC. Several of these articles are specifically devoted to crimes involving minors. Some other offences

[7] *Staatsblad* [Dutch Official Journal] 1993, 33.

[8] *Staatsblad* [Dutch Official Journal] 2006, 300.

[9] *Staatsblad* [Dutch Official Journal] 2018, 322, entry into force 1 March 2019.

[10] *Staatsblad* [Dutch Official Journal] 2002, 388.

[11] *Staatsblad* [Dutch Official Journal] 1999, 245.

[12] See for instance Schermer 2012.

that may coincide with the abuse of a minor (such as kidnapping) are dealt with in other titles, but as they are not relevant to the subject matter of this chapter, they will not be discussed.

10.2.2 Possibly Relevant Criminal Offences

Succinct Overview of Sexual Offences Involving Minors

The following table lists the possible relevant provisions of the Dutch Criminal Code, grouped together by the provisions of the Lanzarote Convention, which gives the most comprehensive catalogue of sexual child-abuse offences available (Table 10.1).

Overview of Sexual Offences Related to Webcam Child Sexual Abuse

In this section, we briefly discuss the criminal offences that are potentially applicable to different forms of webcam sexual child-abuse, again structured by the relevant types of offences from the Lanzarote Convention.

Sexual Abuse (Article 18 Lanzarote Convention)

Articles 244 and 245 criminalise rape (*verkrachting*) of a minor; since this consists (at least partly) in sexual penetration, these are generally not applicable to webcam sexual abuse and not relevant for the Sweetie case.

Article 247 deals with assault (*aanranding*) of a minor below 16 that does not involve penetration of bodily orifices. The article reads:

> Any person who engages in lewd acts with a person whom he knows to be unconscious, to have diminished consciousness or to be physically unable to resist, or to be suffering from such a degree of mental disease or defect that such person is incapable or not sufficiently capable of exercising or expressing his will in the matter or of offering resistance, or who engages in lewd acts, out of wedlock, with a person under the age of sixteen years, or who entices the latter into engaging in or tolerating such acts, out of wedlock, with a third party, shall be liable to a term of imprisonment not exceeding six years or a fine of the fourth category.

In 2004 the Supreme Court decided that for sexual assault in the sense of Article 247 DCC, physical contact between the assailant and the victim is not necessary.[13] This means that the physical presence of the perpetrator is not by definition required to meet the requirements of Article 247 DCC. This means that when the perpetrator is participating in a webcam session with a person under the age of 16 and the victim performs sexual acts (such as performing sexual acts with themselves or a third person), the perpetrator can be held accountable for sexual assault.

---

[13] See: ECLI:NL:HR:2004:AQ0950.

Table 10.1 Implementation of Lanzarote Convention requirements in Dutch criminal law [*Source* The authors]

Lanzarote Convention	Dutch Criminal Code
Article 18. Sexual abuse	Article 244: sexual contact consisting (in whole or in part) of penetration, with a person under the age of 12 Article 245: sexual contact consisting (in whole or in part) of penetration, with a person under the age of 16 Article 246: sexual assault using force or threat Article 247: sexual assault of a person under the age of 16 Article 248a: Inducing a person under the age of 18 to commit lewd acts, by promising money or goods, by abuse of actual relationships or through deception Article 249: Sexual contact with a person under the age of 18 if this person has a special relation to the victim (e.g. parent, caretaker, teacher)
Article 19. Offences concerning child prostitution	Article 248b: Engaging in sexual contact with a prostitute that has reached the age of 16, but not the age of 18 Article 248f: Urging or forcing a person under the age of 18 to have sexual contact with a third person Article 250: Intentionally causing or promoting sexual contact of a third party with a person under the age of 18
Article 20. Offences concerning child pornography	Article 240b: Criminalises spreading, offering, openly showing, manufacturing, importing, exporting, acquiring and possessing child pornography or getting access to child pornography (by automated means)
Article 21. Offences concerning the participation of a child in pornographic performances	Article 248c: Being intentionally present at sexual activities performed by a person under the age of 18 years Article 248f: Urging or forcing a person under the age of 18 to have sexual contact with a third person
Article 22. Corruption of children	Article 239: Indecency in a public place accessible to minors under 16, or in a non-public place where someone is present against his will Article 240: Sending an indecent image to someone without requested Article 240a: Offering or showing to someone under the age of 16 an image that is to be considered harmful to a minor below 16 Article 248d: With licentious intent making a person under the age of 16 years witness sexual activities
Article 23. Solicitation of children for sexual purposes	Article 248e: Grooming: proposing to meet a minor with the intention of sexual abuse or making an image of sexual activity with that minor
[other offences, not covered by the Lanzarote Convention]	None

Another form of sexual assault (against both minors and adults) is criminalised in Article 246 DCC: forcing someone (through violence, threat or some other factuality) to perform lewd acts. The article reads:

> Any person who by an act of violence or any other act or by threat of violence or threat of any other act, compels another person to engage in or to tolerate lewd acts, shall be guilty of indecent assault and shall be liable to a term of imprisonment not exceeding eight years or a fine of the fifth category.

This is for instance the case if the perpetrator has threatened to publicise nude pictures of the victim if the victim does not perform sexual acts in front of the webcam (sextortion).[14] If the perpetrator does not force or coerce the victim but rather tricks the victim (for instance by pretending to be someone else), then Article 248a applies.[15] This article reads:

> Any person who, by means of gifts or promises of money or goods, by abuse of the authority arising from de facto relationships or by deception, intentionally induces a person who is under the age of eighteen years or a person who poses – whether or not using a technical device, including a virtual creation of a person under the age of eighteen years – as a person under the age of eighteen years to engage in lewd acts or to tolerate such acts performed by him, shall be liable to a term of imprisonment not exceeding four years or a fine of the fourth category.

When the victim has not yet reached the age of 18 years and the perpetrator solicits the victim to perform lewd acts (*ontuchtige handelingen*) via the webcam (e.g. by promising money or by tricking the victim), then Article 248a DCC could be applicable. The term lewd act is more specific than sexual acts: it refers to acts of a sexual character that violate social-ethical norms; we can assume that soliciting a 10- or 11-year-old girl to masturbate in front of a webcam violates the social-ethical norm in the Netherlands, so this would count as lewd. There does not have to be physical contact between the perpetrator and victim[16] —making someone take a sexually oriented pose also qualifies as soliciting a lewd act.[17] Although the legislation was presumably drafted with a view to cases involving proximity between perpetrator and victim, it has also been applied to online solicitation leading to lewd acts in front of the webcam, for example to a man who had, through his mental superiority as an adult over a minor, induced the victim to undress, touch her breasts, masturbate and put an object into her vagina in front of the webcam.[18]

Child Prostitution (Article 19 Lanzarote Convention)

According to Article 248b DCC, when the victim has reached the age of 16, but not yet the age of 18 years and offers to perform sexual acts for the perpetrator in front

---

[14] See for instance: ECLI:NL:RBHAA:2008:BD8449.

[15] ECLI:NL:HR:2014:3140.

[16] Cleiren and Nijboer 2000, comment 6a on Article 248a DCC.

[17] HR [Dutch Supreme Court] 20 January 1998, NJ 1998, 336, mentioned in Cleiren and Nijboer 2000, comment 6a on Article 248a DCC.

[18] ECLI:NL:GHDHA:2013:3706.

of the webcam (e.g., in exchange for money or goods), then this falls under the heading of child prostitution. The article reads:

> Any person who sexually abuses a person who makes himself available for the performance of sexual acts with a third party for remuneration and who has reached the age of sixteen years but is under the age of eighteen years, shall be liable to a term of imprisonment not exceeding four years or a fine of the fourth category.

Article 248f criminalises forced prostitution of a minor. This can be the case for instance if a pimp or loverboy prostitutes a minor in front of a webcam. The article reads:

> He who by force, violence or any other act or threats of violence or any other act intentionally causes and promotes the performance of lewd acts with a third person by a person whom he knows or should reasonably suspect that has not yet reached the age of eighteen years, shall be punished with imprisonment not exceeding ten years or a fine of the fifth category.

Article 250 DCC states that those who support or initiate lewd act with their child, pupil or someone entrusted to their care can be held liable for child prostitution. This can for instance be the case when a parent prostitutes their own child in front of a webcam. The article reads:

1. Any person who:

1°. Intentionally arranges or encourages the sexual abuse of his minor child, step-child or foster child, his ward, a minor with whose care, education or supervision he is entrusted or his employee or subordinate who is a minor by a third party, shall be liable to a term of imprisonment not exceeding four years or a fine of the fourth category.

2°. Intentionally arranges or encourages, other than in the cases referred to in 1°, the sexual abuse of a minor, whom he knows or has reasonable cause to suspect is a minor, by a third party, shall be liable to a term of imprisonment not exceeding three years or a fine of the fourth category.

2. If the offender makes a habit of committing the serious offence, the terms of imprisonment may be increased by one third.

Child Pornography (Article 20 Lanzarote Convention)

Child pornography is defined in Article 240b as any image of a person apparently under the age of 18 that is engaged, or seemingly engaged, in a sexual activity. Virtual child pornography, i.e., images that do not contain real minors, also falls under the scope of this article through the clause 'seemingly involving'. Possession of child pornography is criminalised, as well as intentionally accessing (via automated means), producing, distributing, trafficking and publishing child pornography. The article reads:

1. Any person who distributes, offers, publicly displays, produces, imports, conveys in transit, exports, obtains, possesses or accesses by means of a computerised device or system or by use of a communication service an image - or a data carrier that contains an image - of a sexual act involving or seemingly involving a person who is manifestly

> under the age of eighteen years, shall be liable to a term of imprisonment not exceeding four years or a fine of the fifth category.
> 2. Any person who makes a profession or habit of committing any of the serious offences defined in subsection (1), shall be liable to a term of imprisonment not exceeding eight years or a fine of the fifth category.

When a minor (in this case a person under the age of 18) exposes him/herself via a webcam and/or when he or she performs sexual acts via the webcam, this is considered child pornography. The perpetrator does not have to actually store the images on a data carrier, merely consuming the images is a criminal act. As such, not only downloading and storing child pornography is criminalised, but also intentionally watching (live) child pornography via a webcam or other streaming technology. In the case of webcam sex and streaming, there may, however, be evidentiary problems for the public prosecution.

Furthermore, and this may be relevant for the Sweetie project, in order to qualify as child pornography, the images need not be real. Virtual child pornography is also criminalised. The definition of child pornography in Article 240b DCC also covers virtual child pornography, if the virtual images are sufficiently realistic. An image that is clearly (at first sight) manipulated, it does not fall under the definition of Article 240b DCC. It is required that the image portrays seemingly realistic child pornography. This follows from Parliamentary comments with regard to the DCC.[19] In 2013, the Supreme Court has confirmed this view.[20] The boundary between realistic and non-realistic is still somewhat fuzzy, but for the purposes of this report we can at least say that avatars such as Sweetie are sufficiently realistic for the purposes of Article 240b DCC.

Pornographic Performances (Article 21 Lanzarote Convention)

Pornographic performances with minors are criminalised in Article 248c DCC. The article reads:

> Any person who is intentionally present at the performance of lewd acts by a person whom he knows or has reasonable cause to suspect has not yet reached the age of eighteen years or who is intentionally present at the display of images of such acts in an establishment designated for that purpose, shall be liable to a term of imprisonment not exceeding four years or a fine of the fourth category.

Unlike for instance Article 247 DCC, this article requires the physical presence of the suspect.[21] This makes this article difficult if not impossible to use for prosecuting webcam sex.

Those who prostitute the minor for the purpose of a pornographic performances may be tried for prostitution (Articles 248, 248f and 250 DCC).

[19] *Kamerstukken II* 2001–2002, 27 745, nr. 6, p. 11.

[20] ECLI:NL:HR:2013:BY9719.

[21] See: ECLI:NL:RBBRE:2006:AV1470.

Corruption of Minors (Article 22 Lanzarote Convention)

Article 240a DCC criminalises showing harmful materials to a minor. The article reads:

> Any person who supplies, offers or shows to a minor he knows or has serious cause to suspect is under the age of sixteen years, an image, an object or a data carrier that contains an image, which if displayed could be harmful to persons under the age of sixteen years, shall be liable to a term of imprisonment not exceeding one year or a fine of the fourth category.

Material of a sexual nature is covered by this article, the rationale being that the legislator means to protect persons under the age of 16 against undesirable influence that may result from the confrontation with such material. Webcam sex (e.g., showing bare genitals) is considered harmful according to the courts and thus webcam sex falls under the definition of Article 240a DCC.[22]

When the victim has not yet reached the age of 16, displaying sexual activities with lewd intent via a webcam falls under the heading of 248d DCC, which criminalises having a minor witness sexual activities with lewd intent. This article can be considered a *specialist* of Article 240a DCC and carries a higher maximum penalty (two years instead of one-year imprisonment). A court held that showing bare genitals in itself does not constitute a 'sexual activity'.[23]

Grooming (Article 23 Lanzarote Convention)

Grooming, penalised in Article 248e DCC, is the offence whereby a minor is 'groomed' by a perpetrator via for instance a webcam in order to meet in real life. The article reads:

> Any person who, by means of a computerised device or system or by making use of a communication service, proposes to meet a person under the age of sixteen years or a person who poses – whether or not using a technical device, including a virtual creation of a person under the age of sixteen years – as a person under the age of sixteen years, with the intention of engaging in lewd acts with a person under the age of sixteen years or of creating an image of a sexual act in which a person under the age of sixteen years is involved, shall, if he undertakes any action to bring about that meeting, be liable to a term of imprisonment not exceeding two years or a fine of the fourth category.

The object of the legislator is to prevent meetings between perpetrators and minors that could result in sexual abuse. Thus, there need to be concrete preparations in order to achieve the purpose of meeting the minor.[24] So, only when a webcam session (either in the chat or in the audio-visual stream) leads to the setup of a meeting, and there is an actual activity aimed at realising the meeting, will this be considered grooming. Since Article 248e criminalises a preparatory act, an

---

[22] See e.g. Rechtbank Leeuwarden, 23 April 2009, ECLI:NL:RBLEE:2009:BI2330, and Rechtbank Leeuwarden, 10 mei 2011, ECLI:NL:RBLEE:2011:BQ4176.

[23] ECLI:NL:RBROT:2014:8074.

[24] ECLI:NL:HR:2014:3140.

attempt at grooming (e.g. talking to a minor about meeting but not actually setting up the meeting) is not possible, as this would be an attempt to attempt a meeting.[25]

Aggravating Circumstances

Article 248 lists a number of aggravating circumstances that can increase the maximum sentence of different crimes in Title XIV by a third. Aggravating circumstances include perpetration of the crime by a group, and the situation where the perpetrator has a specific relation to the child (e.g. that of parent, caretaker or teacher).

Conclusion

From the above, we can conclude that sexual abuse of minors via webcams can be prosecuted on the basis of various provisions, depending on the exact form and circumstances of the act.

- If the perpetrator induces or forces the minor to display breasts or genitals or to perform sexual activities (e.g., masturbate) in front of the webcam, this may constitute:
  - sexual assault (Article 247 DCC for minors below 16; Article 246 DCC, if threat or force is used);
  - solicitation of sex (Article 248a DCC, if the perpetrator induces the victim through gifts or promises or misleads the victim (e.g., by lying about his age));
  - child prostitution (Article 248f DCC or Article 250 DCC if the minor is forced or induced to have sex with a third person in front of the webcam; Article 248b DCC if a 16- or 17-year-old minor performs sexual activities for payment);
  - child pornography (Article 240b: manufacturing or storing child pornography (if images are saved on the perpetrator's computer), or intentional access to child pornography).
- If the perpetrator shows his genitals or masturbates in front of the webcam, this may constitute:
  - indecency in a non-public place where someone else is present against her will (Article 239);
  - sending an indecent image unrequestedly (Article 240);
  - displaying to a minor below 16 an image that is to be deemed harmful when shown to minors (Article 240a);
  - inducing a minor below 16 to witness a sexual act with lewd intent (Article 248d).

[25] See for instance: ECLI:NL:RBOBR:2014:7494.

Altogether, this offers many possibilities for combatting webcam sex. However, several of these are relatively minor offences with lower penalties. The offences of Article 240b (child pornography), Articles 246 and 247 (sexual assault), and Article 248a (soliciting sexual acts of a minor), Articles 248b and 248f (child prostitution) carry a maximum punishment of at least four years' imprisonment, which means that the use of more intrusive investigation powers (which are needed for cyber-investigations) is allowed for these crimes. This would imply that cyber-investigation is only possible for cases in which the minor performed some kind of sexual activity, but not in cases in which the minor does not undress but in which the perpetrator sends sexual images of himself. However, as observed above, Article 248d (inducing a minor below 16 to witness sexual acts), although carrying a maximum of two years' imprisonment, is also a crime for which pre-trial detention is allowed (Article 67(1) DCCP), so that intrusive investigation powers are possible for cases in which the perpetrator but not the victim displays sexual behaviour in front of the webcam.

Whether these provisions can also be applied to persons who have had webcam contact with Sweetie, remains to be seen, however, given that Sweetie is a virtual person and does not actually engage in sexual activities.

10.2.3 Possible Obstacles in Substantive Law Concerning Sweetie

In this section, we investigate whether trying to engage in webcam sex with Sweetie is a punishable offence under Dutch criminal law. There are two main differences when applying the provisions discussed above in the context of Sweetie versus applying it in the context of offences with real minors: (a) the fact that Sweetie is virtual, and (b) the fact that Sweetie does not undress, so there is no sexual behaviour on the victim's side.

The first potential obstacle for the criminalisation of webcam sex with Sweetie is the fact that Sweetie is not a real person. The descriptions of crimes relating to the abuse of minors generally involve real persons as they are the object of protection. This is reflected in the crime descriptions through components such as 'a person' or 'someone' under the age of 16 or 18 years.

In Dutch criminal law the rule is that if the behaviour of the suspect does not fit all the components of the description, the crime cannot be proven. The result is an acquittal. Given that an avatar is not a natural person, the component "person under the age of 16 years" can never be proven. So that would mean that webcam sex (in whatever form) with virtual minors cannot be proven under Dutch law. An attempt to commit illegal webcam sex could then also not be construed. The reason for this is that the object (the virtual minor) is not a person, so any attempt will be considered absolutely inadequate and thus not criminal (see above, Sect. 10.1.1 on attempt). This position was confirmed by the Appeals Court in the Hague in 2013.

In this case the suspect tried to groom a minor that in reality turned out to be a law enforcement officer.[26] The Court held that the component 'person under the age of 16 years' could indeed not be proven as the law enforcement officer was an adult.

The verdict of the Appeals Court prompted the legislator to propose revising Articles 248a and 248e DCC so that they also cover persons the suspect thinks to be a minor. Mistakenly thinking that there is interaction with a person under the age of 16 years (248e DCC) or 18 years (248a DCC) will also lead to criminal liability.[27] This opens up the way for law enforcement officers to pose as minors and trap suspects.

The question is whether Sweetie, who is not a person at all, can also be used by law enforcement based on the same logic. During the legislative process of changing the Articles 248a and 248e DCC (part of the Computer Crime III Act), the Dutch Council of State advised the legislator to specifically include the possibility for the use of 'virtual child lures'.[28] In the Explanatory Memorandum to the Computer Crime III Act the legislator argued that virtual child lures were not used and that as such the advice of the Council of State was not followed.[29] However, partly following news about the Sweetie 2.0 project, Parliament adopted an amendment to include the possibility of using virtual child lures.[30] This led to including the clause "whether or not using a technical device, including a virtual creation of a person under the age of sixteen years" in the text of the Articles 248a and 248e DCC. Thus, if a webcam session with Sweetie involves inducement to commit lewd acts or grooming, the fact that Sweetie is virtual is no longer an obstacle for criminal liability under Article 248a or 248e DCC.

Also the second obstacle is not necessarily a problem for these offences. The fact that Sweetie does not undress is immaterial for the offence of grooming, which consists of arranging a meeting and, e.g., sending a map or travel plan as a step towards realising the meeting. For sexual inducement under Article 248a DCC, Sweetie will not be induced to actually perform lewd acts, but if the perpetrator performs lewd acts himself during a webcam session with Sweetie, this will be a criminal act under Article 248a DCC.

Apart from using the Articles 248a and 248e DCC, there is also the option to prosecute webcam-related child sexual abuse using the offence of child pornography (Article 240b Sr). A sexually oriented webcam session with a minor can be considered accessing child pornography. Recording and storing the webcam session can be construed as the production and possession of child pornography. The fact that the minor in question is a virtual child does not change the situation, since the inclusion of 'seemingly involved' in the provision means that virtual child pornography is also criminal.

---

[26] See: ECLI:NL:GHDHA:2013:2302.

[27] *Kamerstukken II* 2015–2016, 34 372, nr. 2.

[28] Advice of the Dutch Council of State, *Kamerstukken II* 2015–2016, 34 372, nr. 4.

[29] *Kamerstukken II* 2015–2016, 34 372, nr. 3, p. 70.

[30] *Kamerstukken II* 2016/17, 34 372, nr. 15.

With Sweetie, however, we then run into the second obstacle: because Sweetie does not actually undress, the crime of accessing, producing or owning child pornography cannot be proven. This leaves us with the attempt to access, produce or own child pornography. For instance, if a suspect asks Sweetie to remove all her clothes, and Sweetie would start to 'undress', then this could be construed as an attempt to access child pornography. However, one could argue that if Sweetie would never remove clothing, she could be considered an absolutely inadequate object for the purpose of attempting to create child pornography, and hence this would not constitute a criminal attempt. As such, it depends on the circumstances and future interpretation in case-law whether Article 240b DCC can be used to prosecute persons who have communicated with Sweetie and asked her to perform sexual activities in front of the webcam.

A final relevant aspect to take into account in the context of an attempt is the possibility for the suspect to voluntarily retreat from the criminal act (*vrijwillige terugtred*). While this is a substantive law topic, it is more relevant in the context of criminal procedure: should Sweetie give room for retreat (or encourage retreat) and how should the avatar react to a perpetrator signalling retreat? This will be covered in the sections discussing entrapment.

10.3 Analysis of Criminal Procedure Law

10.3.1 General Description of the Legal Framework

Dutch criminal procedure law is regulated in the Code of Criminal Procedure (*Wetboek van Strafvordering*, henceforth DCCP). The Dutch criminal procedure system can be characterised as being moderately accusatorial.[31] The main goal of criminal procedural law is to establish the truth in order to convict the guilty and to prevent the conviction of those who are innocent.[32]

In March 2015, the Minister of Security & Justice outlined plans to modernise the Code of Criminal Procedure.[33] A draft of a new Book 2 regulating the investigatory powers was launched for consultation in 2017, which included most of the current investigatory powers (in slightly revised form) and a few new powers, such as systematically copying data from open sources.[34] These proposed changes will not be discussed in this report, as they may change considerably in the process

[31] Tak 2008, p. 29.

[32] Hirsch Ballin 2012, p. 39.

[33] *Kamerstukken II* 2015–2016, 29 279, nr. 278.

[34] Voorstel van wet tot vaststelling van Boek 2 van het nieuwe Wetboek van Strafvordering (Het opsporingsonderzoek), February 2017, https://www.rijksoverheid.nl/onderwerpen/modernisering-wetboek-van-strafvordering/documenten/kamerstukken/2017/02/07/wetsvoorstel-tot-vaststelling-van-boek-2-van-het-nieuwe-wetboek-van-strafvordering.

leading up to the new Code, and they are not directly relevant to the main investigatory powers we discuss here.

Phases in Criminal Procedure

We can roughly distinguish three main phases in Dutch criminal procedure: the pre-trial phase, the trial phase and the execution phase. In this report, we limit ourselves to the pre-trial phase (*voorbereidend onderzoek*), because criminal investigation takes place in this phase.

Pre-trial investigations largely consists of investigation by the police under the direction of a public prosecutor, but the examining magistrate can also conduct a preliminary judicial investigation.[35]

Principle of Legality

Article 1 of the DCCP states that criminal procedure law can only take place in the manner provided by the law. The criminal procedural legality principle has a twofold purpose:

- To protect the fundamental rights and freedom of citizens;
- To ensure the integrity of the criminal investigation.

Criminal procedure starts with the investigation carried out by the police under the direction of the public prosecutor.[36] The examining magistrate's role in the criminal investigation is primarily that of warrant judge, considering that the ex ante authorisation of the examining magistrate is required for the use of some investigative powers.[37]

The aim of the criminal investigation is to gather information on the offence and suspect that can be used for prosecution. A suspect is defined in the DCCP as 'anyone who, based on facts or circumstances, may reasonably be suspected of having committed an offence.'[38]

In order to gather information and evidence, the police use a variety of investigative techniques and procedures, some of which may infringe on the rights and freedoms of citizens. In those cases, the investigatory activity must under the legality principle have a basis in the law. Coercive measures (e.g. stopping and searching) and special investigative powers (e.g. wiretapping, undercover operations) are all described in the DCCP. Those techniques and procedures that do not constitute an infringement of fundamental rights (such as patrolling the street while keeping eyes and ears open) are covered by the general task description of the police (Article 3 Police Act 2012) and do not need a specific basis in the law.

[35] Tak 2008, p. 82.

[36] See Article 141 DCCP.

[37] Hirsch Ballin 2012, p. 62.

[38] Article 27 DCCP.

Coercive Measures and Investigatory Powers in the Netherlands

In order to uncover evidence, it might be necessary for the police to use coercive measures and/or investigative powers that infringe upon the rights and freedoms of the suspect. The suspect cannot be forced to cooperate with the investigation (*privilege against self-incrimination*), but has to tolerate the application of coercive measures and special investigative powers. For the latter, this is less relevant, since the goal of the special investigative powers is that they are used without the suspect having knowledge of their use.

The coercive measures are listed in Title IV of the DCCP and include amongst others:

- Stopping a suspect (Article 52 DCCP)
- Arresting a suspect (Articles 53–54 DCCP)
- Placing and keeping a suspect into custody (Article 57 DCCP)
- Searches (such as a body search (Article 56 DCCP), or a home search (Articles 97, 110 DCCP))
- Seizures (Articles 94 et seq. DCCP)

These coercive measures may also be used for digital materials. Moreover, special powers exist for searches to copy data from data carriers (Articles 125i–125o DCCP).

Special Investigative Powers

Apart from the coercive measures mentioned above, the DCPP regulates certain 'special investigative powers'. During criminal investigations conducted in the early nineties, Dutch law enforcement officers used a range of investigative methods, such as wiretapping and running informants that had no clear basis in Dutch criminal procedure law. This situation ultimately led to a parliamentary inquiry into the criminal investigation methods employed by the Dutch police. The committee of inquiry concluded that they needed a clear basis in the Dutch law. As a result, the Special Powers of Investigation Act (*Wet Bijzondere Opsporingsbevoegdheden*) came into effect on 1 February 2000. The special powers include amongst others: visual observation, infiltration, pseudo purchase and service delivery, 'sneak and peek' operations, undercover surveillance, oral interception of confidential communications, and the investigation of telecommunications.[39]

10.3.2 Investigatory Powers and Human Rights

When it comes to coercive measures and investigatory powers, the fundamental rights and freedoms of the suspect are infringed upon. In the case of coercive measures, the (physical) freedom of the suspect is limited, whereas the investigatory

[39] Schermer 2007, p. 93.

powers mainly infringe upon the right to private life. For Dutch criminal procedure law, especially the ECHR forms a relevant framework.

Right to Privacy

With special investigative powers, the suspect will most often be unaware of the use of the investigatory power and the infringement on his rights (more in particular the right to private life). Nonetheless, these investigative powers may violate the private life of the suspect and possibly of others as well. In the Netherlands, the right to privacy is regulated in the Articles 10–13 of the Constitution. Moreover, the right to privacy is regulated in Article 8 ECHR and Article 17 of the ICCPR, both of which the Netherlands is party to. In practice, Article 8 ECHR has the strongest influence. Article 8 para 2 ECHR states that:

> There shall be no interference by a public authority with the exercise of this right except such as is in accordance with the law and is necessary in a democratic society in the interests of national security, public safety or the economic well-being of the country, for the prevention of disorder or crime, for the protection of health or morals, or for the protection of the rights and freedoms of others.

The component "necessary in a democratic society" is a test of proportionality and subsidiarity. The component "in accordance with the law" is a legality requirement, which requires that the use of an investigative power that infringes on the private life of a citizen has a proper basis in the member state's law.

The ECHR distinguishes between investigatory methods that have little to no effect on fundamental rights and freedoms and those that have a significant impact on fundamental rights and freedoms. If the investigative method has little or no impact, it is covered by the general task description of the police (Article 3 Police Act 2012), if not, a specific legal basis in the DCCP is necessary.

Entrapment

The DCCP contains provisions on the use of investigative methods that entail the buying and selling of illegal goods or services (*pseudokoop en-dienstverlening*) in order to collect evidence against suspects. Furthermore, police have used lures in untargeted attempts to trap perpetrators. Lures used include bikes, cars and even prostitutes. However, trapping suspects is considered at odds with the right to a fair trial in the Netherlands. There is a fine line between luring/trapping suspects (*lokken*) and entrapment (*uitlokking*). The former is generally allowed, the latter is not, and as such, an incitement/entrapment plea can be put before the courts.[40] The difference between trapping and entrapment has been established in case law.

The landmark case in the Netherlands in this area is the 1979 *Tallon* case.[41] In this case defendant T. bought drugs from two undercover agents. The Supreme Court held that T. was criminally liable for the purchase, despite the fact that the transaction was orchestrated by the police. The main argument for this position was

[40] The ECtHR uses the terms incitement and entrapment interchangeably in their case law.

[41] HR 4 December 1979, NJ 1980, 356 m.nt ThWvV.

that the actions of the police did not bring T. to other actions than those he already intended to take. In other words, the intent of T. to buy drugs was formed independently of the actions of the undercover police action. This has become known as the *Tallon criterion* in Dutch criminal procedural law and is used in all cases to determine whether a suspect has been entrapped.

Entrapment also features in the case law of the European Court of Human Rights (ECtHR). The Court acknowledges that there is a need for special investigative techniques and that undercover investigations as such do not infringe on the right to a fair trial. However, given the risk of incitement/entrapment, these techniques and methods must be kept within clear limits.[42] In the case of *Ramanauskas v. Lithuania* it was held that prosecution on the basis of incitement is incompatible with Article 6 ECHR, as it exposes the accused to the risk of being deprived of a fair trial from the outset.[43] To ascertain if the conduct of law enforcement authorities is incitement/entrapment, a substantive test of incitement has been developed by the Court in this case. According to the Court, incitement can be defined as:

> A situation where the officers involved - whether members of the security forces or persons acting on their instructions – do not confine themselves to investigating criminal activity in an essentially passive manner, but exert such an influence on the subject as to incite the commission of an offence that would otherwise not have been committed, in order to make it possible to establish the offence, that is to provide evidence and institute a prosecution.

It follows from this definition that passivity of the law enforcement authorities is the deciding factor in determining whether the actions of the law enforcement authorities have incited the suspect. In deciding whether the investigation was "*essentially passive*", the Court examines the reasons underlying the covert operation and the conduct of the authorities carrying it out.[44] Relevant elements include:

- Whether there are objective suspicions against the suspect that he/she had been involved in criminal activity or was predisposed to commit a criminal offence (*Bannikova v. Russia, Teixeira de Castro v. Portugal*).
- Familiarity with the criminal environment (e.g. knowing the price of drugs) (*Shannon v. The United Kingdom*).
- Failure to withdraw from the criminal transaction, despite the opportunity to do so (*Shannon v. The United Kingdom*).
- Whether the suspect was pressured in some way to commit the crime. Indicative of pressure can be pro-actively contacting the suspect (*Burak Hun v. Turkey*), reiterating the offer despite initial refusal by the suspect and/or insistent prompting (*Ramanauskas v. Lithuania*), making an offer very difficult to refuse, for instance by raising the price far above the average market value (*Malininas*

[42] Council of Europe (2013), p. 25.

[43] Council of Europe (2013), p. 25.

[44] Council of Europe (2013), p. 26.

v. Lithuania), or appealing to the suspect's compassion, for instance by mentioning withdrawal symptoms (*Vanyan v. Russia*).[45]

It must be noted that the case law of the ECtHR is focused primarily on 'buy and bust' entrapment. When we translate these requirements to the Sweetie case, possible indicators for entrapment could be for instance that Sweetie pro-actively contacts the suspect in a chatroom and/or solicits the suspect to introduce a sexual element in a webcam conversation.

Non-targeted Entrapment

The above case law dealt primarily with—more or less—targeted entrapment. Apart from targeted entrapment we can also distinguish non-targeted entrapment. In the case of non-targeted entrapment, bait is set while there is no suspicion aimed at a specific person. The reason for the entrapment in this case is that there is a general suspicion, for instance that many bicycles are stolen in a certain area. Non-targeted entrapment is not regulated by criminal procedure, but the Dutch Supreme Court has decided on it in two important cases in 2008 and 2009.

The first case was about a bicycle being left for bait. It was left unlocked and the person who got on the bike and drove away with it was arrested for bicycle theft.[46] The second case involved a car as bait.[47] In both cases the Supreme Court decided that the use of bait was allowed, since the suspect was not brought to actions other than those at which his intention was already aimed (the *Tallon* criterion). The fact that there was no specific legal basis for placing the bicycle or car as bait was not deemed a violation of the legality principle, given that the infringement on fundamental rights and freedoms was limited.

From these cases we may deduce that non-targeted entrapment is allowed, when the bait or lure does not substantially change the situation in the area where the lure is placed, and thus has no significant effect on the decision-making process of the perpetrator. It may be different if the introduction of the bait creates a situation that is (substantially) different from the original situation, as it may create circumstances that 'make the thief'. In his opinion in the bicycle case, Knigge, the Advocate-General with the Supreme Court argues this point.[48]

10.3.3 Succinct Overview of Investigatory Powers in an Online Context

Given that many of the articles in the DCCP dealing with investigatory powers are technology-neutral in their formulation, they can also be applied in an online

45 Council of Europe (2013), p. 26.

46 HR 28 oktober 2008, ECLI:NL:HR:2008:BE9817 (*Lokfiets*-arrest).

47 HR 6 oktober 2009, ECLI:NL:HR:2009:BI7084 (*Lokauto*-arrest).

48 ECLI:NL:PHR:2008:BE9817 (Conclusion of the AG in *Lokfiets*-arrest).

Table 10.2 Overview of investigatory powers in an online context [*Source* The authors]

Council of Europe Convention on Cybercrime	Netherlands
Article 16. Expedited preservation of stored computer data	Article 126ni DCCP
Article 17. Expedited preservation and partial disclosure of traffic data	Articles 126n–126na DCCP, Article 126ni DCCP
Article 18. Production order	Articles 126na, 126nc–126ni DCCP
Article 19. Search and seizure of stored computer data	Articles 125i–125o DCCP
Article 20. Real-time collection of traffic data	Article 126n DCCP
Article 21. Interception of content data	Article 126m DCCP
[Other (special) investigatory powers, not covered by the Cybercrime Convention, such as undercover operations.]	Article 126g DCCP—systematic observation Article 126h DCCP—infiltration Article 126i DCCP—Pseudo-purchase and pseudo-service delivery Article 126j DCCP—systematic gathering of information Article 126l DCCP—oral interception Article 126nba DCCP—remote (covert) access to a computer system

context. Above is a table summarising the most relevant investigatory powers in an online context (Table 10.2).

Powers and Measures Related to Online Investigations[49]

There are several coercive measures and investigative powers pertaining to investigations in an online context that have been codified in the DCPP. These include access to and searching of stored data (Articles 125i, 125j and 125k DCCP), freezing of data (Article 126ni DCCP) and making stored data inaccessible (Article 125o DCCP). We shall not discuss these further, as they have little bearing on the Sweetie project.

Below we shall discuss the relevant special investigative powers that can be used in an online environment.

Systematic Observation (Article 126g DCCP)

Article 126g of the DCCP regulates the use of visual surveillance (*stelselmatige observatie*) as an investigative method. Article 126g DCCP defines this as: ‘To systematically follow a person or systematically observe his movements or behaviour.’

Surveillance is deemed systematic when it enables a more or less complete picture to be gained of certain aspects of a person’s life. Different factors determine whether surveillance is systematic: the duration of the observation, the place, the

[49] Parts of this section have been derived from Schermer 2007.

intensity, the frequency of the observation, and whether a technical device is used that can do more than enhance the senses.

Systematically following or observing a person is only permitted in the case of a suspected felony and at the order of the public prosecutor. The duration of the surveillance is bound to a maximum of three months, a period that can be extended repeatedly by three months by order of the public prosecutor. Whether behaviour is observed offline or online does not matter; both the text of the article and the explanatory memorandum do not exclude observation of persons on the internet, therefore systematic observation can also be conducted in places such as chatrooms and massive multiplayer online role-playing games.

The place where the systematic observation takes place in an online context does matter from a procedural law and human rights perspective. Generally speaking, conduct that takes place on the public internet is considered less privacy-sensitive, and can be monitored on an incidental basis without requiring an order for systematic observation. Private chatrooms and one-on-one conversations are generally considered more privacy-sensitive and would require an order on the basis of Article 126g DCCP or another investigatory power.[50] There is still discussion in the Netherlands to what extent open source data (e.g. blogposts, public Facebook profiles, Twitter feeds) may be monitored on a more structural basis.[51]

Infiltration (Article 126h DCCP)

Article 126h of the DCCP regulates the investigative method of infiltration. Covert investigation or infiltration can be defined as "participating or co-operating with a group of people that is believed to be planning crimes or to have committed crimes."

In any covert investigation, there is a serious risk that the covert investigator will have to commit criminal offences, lest his cover be blown. While a covert investigator may commit criminal offences in order to stay undercover, any actions that could give rise to a criminal offence should be listed in the order issued by the public prosecutor.

Infiltration is only allowed when there is a reasonable suspicion of one of the criminal offences mentioned in Article 67 para 1 DCCP,[52] that forms a serious breach of law and order. Besides an order by the public prosecutor, also specific authorisation is required from the Public Prosecution Office's board following advice of a Central Assessment Committee (*Centrale Toetsingscommissie*).[53]

The article also explicitly prohibits incitement (Article 126h para 2), using the wording of the aforementioned Tallon criterion.

---

[50] See for instance: Oerlemans and Koops 2013.

[51] See for instance: Oerlemans and Koops 2013. See also Commissie modernisering opsporingsonderzoek in het digitale tijdperk 2018.

[52] That is, crimes for which pre-trial detention is allowed—generally, crimes carrying a maximum penalty of at least four years' imprisonment, but also including some specifically designated crimes, including most cybercrimes and Articles 248d and 248e DCC.

[53] Aanwijzing opsporingsbevoegdheden, *Staatscourant* 2014, 24442, Sect. 5.1.

Pseudo Purchase/Service Delivery (Article 126i DCCP)

Article 126i DCCP regulates the use of pseudo-purchase or pseudo-service delivery as an investigative method. Pseudo-purchase/service delivery can be described as "the purchase of goods from, or the supply of services to the suspect."

Since a criminal offence could result from the use of this investigative method, the Tallon criterion is also applicable to pseudo-purchase/service delivery. In the Special Powers of Investigation Act pseudo-purchase/service was limited to the purchase of physical goods. The Computer Crime II Act amended Article 126i DCCP to include the purchase of data.

Systematically Gathering Intelligence Undercover (Article 126j DCCP)

Article 126j DCCP regulates the investigative method of systematically gathering intelligence undercover, which is the case when a police officer takes active steps to collect information about the suspect by engaging with the suspect or people in his environment, without it being apparent that he is a law enforcement officer. A police officer could, for instance, gain intelligence on a suspect through undercover activities, such as visiting places frequented by the suspect. Systematically gathering intelligence under cover differs from infiltration because the investigating officer is not committing any criminal acts himself. This also means that the undercover work poses fewer risks to the integrity and security of the investigation. Therefore, this investigative method is bound by less stringent requirements than infiltration and pseudo-purchase/service delivery.

Systematically gathering intelligence undercover is only allowed when there is a reasonable suspicion of a crime mentioned in Article 67 para 1 DCCP, that forms a serious breach of law and order. The public prosecutor has got to give his explicit authorisation through an order.

Oral Interception of Confidential Communications (Article 126l DCCP)

Article 126l DCCP regulates the use of technical aids, such as bugs, to record confidential communications. It differs from wiretapping in that the interception takes place in the vicinity of the communication. Nevertheless, this article might cover online communications, such as emails and chats,[54] if they are recorded through, e.g., a keylogger in the computer user's keyboard. As such, this investigative method may also be used in the context of Sweetie, for instance to record the webcam stream, or the associated chats. However, placing a keylogger in someone's computer usually requires physical entry into the user's dwelling, which is only possible in cases of crimes carrying a maximum penalty of at least eight years' imprisonment (Article 126l para 2 DCCP). Because of the intrusiveness of recording covert communications, a warrant from the examining magistrate is required.

[54] Nieuwenhuis 2015, Article 126l (en 126s).

Intercepting Telecommunications (Article 126m DCCP)

Article 126m DCCP regulates the use of wiretapping for law enforcement purposes. When an offence mentioned in Article 67 para 1 DCCP poses a serious breach to law and order, a wiretap can be ordered when this is urgently required for the investigation, and if the examining magistrate has given a warrant. Usually, the wiretap will be facilitated by a telecommunications provider (who has the duty to comply), but the police can also install interception devices themselves. Wiretapping is possible not only for public telecommunications, but also for non-public telecommunications, such as a company's internal network.

Remote Access to a Computer (Article 126nba DCCP)

The Computer Crime III Act has introduced the new investigatory power of covert remote access of automated works, otherwise known as (legal) 'hacking'. When the investigation urgently requires it, the public prosecutor may instruct police officers to hack computer used by the suspect. In the context of webcam-related child sexual abuse this investigatory power may be used for instance to ascertain the location of the computer and the identity of its user, or to assist visual surveillance by turning on the webcam[55] or oral interception by turning on the microphone. Furthermore, it may be used to gather evidence by copying data from the hard disk (including data generated later, for an (extensible) period of four weeks) and even for remotely destroying data, such as recorded child webcam sex recordings (after securing a copy of those for evidential purposes). In light of the high intrusiveness of covertly hacking into someone's computer, the requirements are strict, including a warrant by the examining magistrate and permission by the Public Prosecutor Office's board through the Central Assessment Committee, and it is only possible for crimes mentioned in Article 67 para 1 DCCP that pose a serious breach to law and order (for ascertaining location or identity, or facilitating visual surveillance or communications interception) of crimes carrying a maximum penalty of at least eight years' imprisonment or certain specifically designated crimes (for searching and copying data from the hard disk or deleting data).

10.3.4 Application of Relevant Investigatory Powers to the Sweetie Case

There are no specific rules on the use of artificial intelligences such as Sweetie in criminal investigations in the DCCP. The descriptions in the DCCP of the investigative powers described above do not cover the use of specific technologies. As

[55] Note, however, that visual observation in the form of recording images is not allowed inside dwellings, so the police is not allowed to turn on a webcam if there is reason to believe that the computer is inside a dwelling. *Kamerstukken II* 2015–2016, 34 372, nr. 3, pp. 26–27 and *Kamerstukken II* 2016–2017, 34 372, no. 6, p. 50.

such, the use of artificial intelligence for investigative purposes is not excluded, as long as the application of AI stays within the bounds of the description of the investigative power.

Up until now, the use of AI for investigative purposes has been limited. The most advanced forms of AI currently employed are data mining and (web)crawling. The use of data mining is covered by the Police Data Act (*Wet politiegegevens*), the use of crawlers is not specifically addressed in the DCCP. Given that both applications have little bearing on the Sweetie project, they shall not be discussed further in this context.

Chatbots, such as Sweetie, have not been employed by the Dutch police for investigative purposes. This is mainly because the intelligence is not yet strong enough to use in complex, targeted investigations. Also, the risk of entrapment is considered a serious possible obstacle. In answer to questions by parliament members on (among other things) the Sweetie project, the Minister answered that the Public Prosecutor's Office had investigated some dozens of cases handed over to them by Terre des Hommes following the first Sweetie project, but that none of these cases led to a prosecution since each had involved entrapment. The Minister added that, in so far as moving animations were to be used in future, the Tallon criterion should be adhered to.[56]

Legality: Investigative Powers Applicable to the Use of Sweetie

Given the possible infringement on the right to privacy, it is important to qualify the use of Sweetie from an investigatory perspective and determine whether or not its use has a basis in the law.

As explained above, utilising a lure (such as a car or bicycle) does not in itself require an explicit basis in the law. As such, it can be argued that the use of Sweetie does not require a specific basis in the law. However, the use of Sweetie goes beyond acting as a mere lure. Unlike stealing a car or a bike, a suspect interacts with Sweetie. So, once the suspect has taken the bait, the use of Sweetie is more akin to an undercover investigation. Conceptually, the use of Sweetie by law enforcement can be divided into two phases: (1) the bait phase and (2) the interaction phase. The former may not yet require an order or warrant, but the latter does.

This 'hybrid' character of a Sweetie investigation poses a problem, because using powers related to undercover investigations requires a reasonable suspicion (Article 126h or 126j DCCP).[57] It is questionable whether the reasonable suspicion required for utilising Sweetie is already present in the stage where the bait has been set, but not yet taken. Most likely, this is not the case. This means that the next phase (the actual interaction with Sweetie) is not covered by the required order.

---

56 Aanhangsel *Handelingen II* 2016–2017, 948, p. 2.

57 The use of infiltration (Article 126h DCCP) may not be possible, given that infiltration is aimed at entering a criminal group, rather than having one-on-one contact with individual crime perpetrators. It might be used, however, in investigations of online platforms that are specifically used by an underground community dedicated to facilitating online sexual child abuse.

This problem has also been signalled in Dutch literature in relation to the use of a physical child lure (i.e. a law enforcement officer instead of an AI).[58] Practically, this issue could be solved by having a public prosecutor issue an order on the spot when the bait phase transitions into the interaction phase. However, whether this is practically feasible remains to be seen; in any case this approach would be pushing the boundaries of criminal procedural law. In order to create more legal certainty, it would be preferable to regulate the use of 'interactive lures' more specifically.

Incitement via Sweetie

If the use of Sweetie is possible under the existing legal framework or in an amended legal framework, then the use of Sweetie must pass the substantive incitement test. This means that the use of Sweetie must be essentially passive. Relevant elements to determine the passivity include (1) the location and (2) the behaviour of Sweetie.

Ad (1) Location

Following the rationale of the cases on the use of lures, it is relevant that the use of Sweetie does not significantly alter the existing situation. So, for instance, Sweetie being present in a chatroom for young boys or girls where there is sufficient evidence that this chatroom is frequently used by perpetrators to contact children for grooming or online webcam sexual abuse, would not significantly alter the situation. However, introducing Sweetie in a regular chatroom for children, which is not known for being used (also) for sexual purposes, might significantly alter the situation, although this will also depend on Sweetie's profile and actual behaviour (see under 2). Placing Sweetie in, for example, a self-help chatroom for people struggling with their paedophilic interests will be considered incitement, as the presence of a young girl significantly alters the environment.

Ad (2) Behaviour

In order to maintain an essentially passive demeanour, Sweetie's behaviour must not incite potential perpetrators. A first element of this is that Sweetie should have a neutral profile, which resembles those of other children in the platform where she is active. Second, she should not actively solicit potential suspects, e.g., by offering webcam sex upfront or otherwise demonstrate an interest in sexual activity. A third element is that once a suspect engages Sweetie in (one-on-one) conversation, Sweetie's behaviour must confirm to the requirements set forth by the ECtHR. This could, amongst other things, mean that:

- Sweetie is never the first to introduce sexual topics or to offer webcam sex when approached by a suspect;
- The chat script must leave room for the suspect to retreat (*vrijwillige terugtred*);
- The chat script must not reiterate the idea to have webcam sex if the suspect, after having initially suggesting it, has retreated from the proposal.

58 See for instance Ölçer 2014.

10.3.5 Relevant Aspects of Digital Forensic Evidence

Under Dutch law, a judge may only conclude his judgement on the basis of lawful evidence (Articles 338–339 DCCP). Article 339 gives the five lawful forms of evidence:

- The court's own observation (evidence presented during the trial hearing to the court);
- Statements made by the suspects;
- Statements made by witnesses;
- Testimony by appointed experts;
- Written documents.

Written documents are the most important category of evidence, as they contain the official reports of law enforcement officers (*proces-verbaal*). Digital materials are also accepted as evidence in court. Through the court's own perception, video, sound and other electronic data can directly reach the judge. Also, written reports by forensic evidence can serve as evidence (as written documents), if necessary supplemented by statements as expert witness during trial.[59]

Evidence Gathering by the Police

(Digital) evidence is gathered first and foremost by the police. The rules for gathering are linked to the use if coercive measures and (special) investigative powers. If the rules of the DCCP are not followed, the evidence is considered to be unlawfully obtained. This may have one of the following consequences described in Article 359a of the DCCP: reduction of the penalty, exclusion of evidence, or dismissal of the trial (*niet-ontvankelijkheid van het OM*). In practice, many irregularities do not lead to exclusion of evidence, and very seldom to dismissal of the trial; often, there is at most a reduction of penalty, or the irregularity is merely noted in the judgement.

Digital Evidence

Digital evidence is admissible in the Dutch courts. When it comes to digital evidence, the integrity and authenticity of the data are of specific concern. Therefore, law enforcement uses specific procedures to maintain the integrity and authenticity of digital evidence, such as maintaining the chain of custody for digital evidence. Law enforcement officers must also use copies (images) of the original data carrier when conducting research in order to avoid changing the evidence.

Given that webcam streams, unlike downloads, are not by definition stored on the computer of the victim or the suspect, there might be practical issues to prove webcam sexual abuse. However, apart from witness testimony, it is possible to gather other forms of evidence related to the webcam streams, such as for instance

59 Buruma and Koops 2004, p. 117.

chatlogs that have been saved. When Sweetie is used by law enforcement, then streams can also be recorded.

To meet the standards of evidence and to ensure the proper operation of Sweetie, the use of Sweetie should be governed by the Decree on technical aids in criminal procedure (*Besluit technische hulpmiddelen strafvordering*).[60] This decree sets forth technical requirements that technical aids must meet in order to be cleared for use in a law enforcement setting. However, the decree is primarily suited for physical devices, such as cameras, and less suitable to assess and prove the proper functioning of software.[61]

Evidence Provided by Third Parties (Silver Platter)

Law enforcement may obtain evidence from third parties. When law enforcement requires the disclosure of information by a third party (e.g. an internet platform or an ISP), they must use the relevant investigative power to order production of the evidence. However, actors in society may voluntarily hand over evidence to law enforcement authorities. This evidence may have been gathered illegally by the third party. When authorities have had no involvement in the irregularity and the unlawfully gathered evidence is handed to them on a 'silver platter', then it is considered admissible in court.

10.4 Evaluation

10.4.1 Substantive Criminal Law

When we examine the substantive criminal law framework in the Netherlands we may conclude that there is no gap in the protection of minors in the area of webcam sex. Webcam sex with minors can result in amongst others, the crimes of sexual assault, access and production of child pornography, and corrupting minors. As such, from a substantive criminal law perspective, potential victims of webcam sex are properly protected.

The criminal offences generally only apply to real minors as the victim of the crime. This implies that engaging in webcam sexual activity with Sweetie is not criminal as such. The offence of child pornography could apply, since this includes virtual child pornography, but since Sweetie does not address, it could at most constitute a criminal attempt (if Sweetie is considered a relatively inadequate object, which is contestable). However, with the changes of the Computer Crime III Act, two offences specifically include engaging with a virtual creation of a minor. Therefore, inducing Sweetie to witness lewd acts by the perpetrator, or grooming

60 *Staatsblad* [Dutch Official Journal] 2006, 524.

61 Cf Commissie modernisering opsporingsonderzoek in het digitale tijdperk 2018.

Sweetie, constitute criminal offences. These changes enable the use of Sweetie and similar virtual creations in criminal investigations into these crimes.

10.4.2 Criminal Procedure Law

The use of Sweetie raises two main questions in relation to criminal procedural law. The first question is that of *legality*. How can we qualify the use of Sweetie in a criminal investigation, in terms of the proper legal basis for such use? Given the 'hybrid' character of Sweetie as both a lure and undercover agent, different investigatory powers apply to the different stages (baiting and interacting) of its use. While it is possible to cover the use of Sweetie under existing investigatory powers, it would be preferable to regulate the use of 'interactive lures' more specifically.

The second question is that of *incitement/entrapment*. If the use of Sweetie is deemed to have a legal basis in Dutch criminal law, we still need to be mindful of the fact that incitement using Sweetie is not allowed. Therefore, we need to take into account the Tallon criterion (the Dutch prohibition of entrapment), including the specific factors necessary for complying with this criterion. Since most of the criteria for incitement have been developed in case law related to drugs transactions in an offline context, it is somewhat unclear how interaction with Sweetie should be structured to avoid incitement. The suggestions offered in Sect. 10.3.2 should provide a good starting point for that discussion.

10.4.3 Summary and Conclusions

In summary we can say that the substantive legal framework for the protection of minors against webcam sex in the Netherlands is generally adequate, particularly with the amendments enacted by the Computer Crime III Act in 2018.

Criminal procedural law offers generally adequate investigation powers to investigate webcam-related child sexual abuse. However, questions remain regarding the legality of using Sweetie as an investigative power, given its hybrid character (both as lure and as undercover method), and particularly regarding the risk of incitement or entrapment.

References

Buruma Y, Koops BJ (2004) Formeel strafrecht en ICT. In: Koops BJ (ed) Strafrecht en ICT. Sdu Uitgevers, The Hague, pp 79–119

Cleiren CPM, Nijboer JF (eds) (2000) Tekst & commentaar strafrecht. Kluwer, Deventer

Commissie modernisering opsporingsonderzoek in het digitale tijdperk (2018) Regulering van opsporingsbevoegdheden in een digitale omgeving (Commissie Koops), June 2018 (in Dutch)

European Court of Human Rights (2013) Guide on Article 6, right to a fair trial (criminal limb)

Hirsch Ballin MFH (2012) Anticipative criminal investigation. Theory and counterterrorism practice in the Netherlands and the United States. T.M.C. Asser Press, The Hague
Machielse AJM (2015) Artikel 45 Sr, aant. 2.6.2. In: Noyon TJ, Langemeijer GE, Remmelink J (eds) Het Wetboek van Strafrecht. Kluwer, Deventer
Nan JS (2012) Het lex certa-beginsel. Sdu Uitgevers, The Hague
Nieuwenhuis M (2015) Commentaar op Wetboek van Strafvordering Article 126l (en 126s). Tekst & commentaar strafvordering
Oerlemans JJ, Koops BJ (2013) Surveilleren en opsporen in een internetomgeving. Justitiële Verkenningen 38/5
Ölçer FP (2014) De lokmethode bij de opsporing van grooming. Computerrecht 2014/3
Schermer BW (2007) Software agents, surveillance, and the right to privacy: a legislative framework for agent-enabled surveillance. Leiden University Press, Leiden
Schermer BW (2012) Digitale IRT-affaire of nieuwe opsporing? http://webwereld.nl/beveiliging/59972-digitale-irt-affaire-of-nieuwe-opsporing-opinie. Accessed 7 October 2015
Tak PJP (2008) The Dutch criminal justice system. Wolf Legal Publishers, Nijmegen

Relevant Case Law

Substantive Criminal Law
ECLI:NL:HR:2013:BY9719
ECLI:NL:RBROT:2014:8074
ECLI:NL:GHARL:2013:CA3968
ECLI:NL:HR:2014:3140
ECLI:NL:RBHAA:2008:BD8449
ECLI:NL:HR:2004:AQ0950
ECLI:NL:GHDHA:2013:2302
ECLI:NL:GHDHA:2013:3706
ECLI:NL:RBBRE:2006:AV1470
ECLI:NL:RBLEE:2009:BI2330
ECLI:NL:RBLEE:2011:BQ4176

Criminal Procedure Law
ECLI:NL:HR:2008:BE9817 (*Lokfiets*-arrest)
ECLI:NL:HR:2009:BI7084 (*Lokauto*-arrest)
HR 4 december 1979, NJ 1980, 356 m.nt ThWvV (*Tallon*)
Bannikova v. Russia
Teixeira de Castro v. Portugal
Shannon v. The United Kingdom
Burak Hun v. Turkey
Ramanauskas v. Lithuania
Malininas v. Lithuania
Vanyan v. Russia

Law
European Convention of Human Rights and Fundamental Freedoms
Dutch Code of Criminal Procedure (*Wetboek van Strafvordering*)
Dutch Criminal Code (*Wetboek van Strafrecht*)

Chapter 11
Substantive and Procedural Legislation in the Philippines to Combat Webcam-Related Child Sexual Abuse

Michael Anthony C. Dizon

Contents

Abstract This chapter examines the criminal laws and procedural rules in the Philippines that apply to webcam or online child sexual abuse. It analyzes the relevant treaties, legislation and case law, particularly those relating to sexual offenses involving minors and cybercrime investigations. The chapter also tackles specific legal issues concerning the use of Sweetie 2.0 in combating webcam-related child sexual abuse including child pornography, cybersex, child sex tourism, entrapment, and criminal investigations in an online context. Based on the research, many of the sex-related crimes under Philippine law can be used to investigate and prosecute the sexual abuse and exploitation of real children and

M. A. C. Dizon (✉)
Te Piringa – Faculty of Law, University of Waikato, Hamilton, New Zealand
e-mail: michael.dizon@waikato.ac.nz

S. van der Hof et al. (eds.), *Sweetie 2.0*, Information Technology and Law Series 31,
https://doi.org/10.1007/978-94-6265-288-0_11

minors via webcam or other similar means or media. However, most of these crimes do not apply in the case of Sweetie 2.0 because they require as an essential element for their commission the involvement of a real child or minor victim. As such, the possible crimes that may be committed in relation to Sweetie 2.0 are limited to cybersex, attempted cybersex, grooming, and luring. Also, since the general rule is that criminal investigations can only conducted when a crime has been, is being or is about to be committed, Sweetie 2.0 can only be used in the first stance to investigate these four offenses.

Keywords Child Abuse · Cybersex · Grooming · Luring · Pornography · Entrapment

11.1 Introduction: Legislation in the Philippines

11.1.1 General Description of the Legal Framework

Due to its colonial histories with both Spain and the United States, the Philippines has a mixed legal system that primarily follows the civil law tradition, but also draws from the US common law system. Many Philippine laws have its origins in Spanish or continental European legal systems (e.g., the Civil Code and the Revised Penal Code). However, laws enacted since the early 20th century and most special laws are based on or inspired by US laws and policies. To illustrate, while the Revised Penal Code has its roots in Spanish criminal laws, the Philippine Rules of Criminal Procedure are similar to and borrow from US procedural rules. The structure and substance of the Philippine Constitution is also very much influenced by the US Constitution. Like the United States, the Philippines has three branches of government (the legislature, the executive, and the judiciary) and subscribes to the system of checks and balances.[1] The Bill of Rights in the Philippine Constitution is based on the Amendments to the US Constitution and landmark US court decisions (e.g., the right to privacy, Miranda rights, etc.). In many instances, the rights under the Philippine Constitution have the exact same wording as those in the United States. With regard to Philippine criminal laws and procedures, it is the legislature that enacts criminal laws, the executive investigates and prosecutes criminal offenses through law enforcement agencies, and the judiciary is the one that promulgates the rules of procedure, interprets laws, decides whether a crime has been committed and what punishment to mete out (which is then carried out by the executive).

Philippine criminal laws adhere to the general principles of generality, territoriality, non-retroactivity, and *in dubio pro reo* (resolving all doubts in favor of the accused). Following the principle of the general application of criminal laws, the

[1] See 1987 Constitution, Articles VI, VII and VIII.

law provides that "[p]enal laws and those of public security and safety shall be obligatory upon all who live or sojourn in the Philippine territory, subject to the principles of public international law and to treaty stipulations".[2] Furthermore, pursuant to the territoriality principle, criminal laws are enforced "within the Philippine Archipelago, including its atmosphere, its interior waters and maritime zone"[3] and, save for a handful of specific exceptions, do not have extra-territorial application.[4] Criminal laws and laws in general are also applied prospectively.[5] The Philippine Constitution expressly prohibits the enactment of an "*ex post facto* law or bill of attainder".[6] However, penal laws may be given "retroactive effect in so far as they favor the person guilty of a felony".[7] Moreover, the Philippines subscribes to the principle that "penal statutes are construed strictly against the State and liberally in favor of the accused".[8] As explained by the Supreme Court:

> the fundamental principle in applying and in interpreting criminal laws is to resolve all doubts in favor of the accused. *In dubio pro reo*. When in doubt, rule for the accused. This is in consonance with the constitutional guarantee that the accused shall be presumed innocent unless and until his guilt is established beyond reasonable doubt.

Intimately related to the *in dubio pro reo* principle is the rule of lenity. The rule applies when the court is faced with two possible interpretations of a penal statute, one that is prejudicial to the accused and another that is favorable to him. The rule calls for the adoption of an interpretation which is more lenient to the accused.[9]

It is worth noting that, with respect to criminal intent, there are two types of crimes in Philippine law: *mala in se* and *mala prohibita*. According to the Supreme Court,

> The law has long divided crimes into acts wrong in themselves called "acts *mala in se*," and acts which would not be wrong but for the fact that positive law forbids the, called "acts *mala prohibita*." This distinction is important with reference to the intent with which a wrongful act is done. The rule on the subject is that in acts *mala in se*, the intent governs, but in acts *mala prohibita*, the only inquiry is, has the law been violated? When an act is illegal, the intent of the offender is immaterial.[10]

The Supreme Court further explains the differences between these two types of crimes:

> Generally, *mala in se* felonies are defined and penalized in the Revised Penal Code. When the acts complained of are inherently immoral, they are deemed *mala in se*, even if they are

---

[2] Civil Code of the Philippines, Article 14.

[3] The Revised Penal Code (as amended), Article 2.

[4] The Revised Penal Code (as amended), Article 2.

[5] Civil Code of the Philippines, Article 4.

[6] 1987 Constitution, Article III. Sec 22.

[7] The Revised Penal Code (as amended), Article 22.

[8] *People v. Valdez*.

[9] *Intestate Estate of Vda. de Carungcong v. People*.

[10] *Dunlao, Sr. v. Court of Appeals*.

> punished by a special law. Accordingly, criminal intent must be clearly established with the other elements of the crime; otherwise, no crime is committed. On the other hand, in crimes that are *mala prohibita*, the criminal acts are not inherently immoral but become punishable only because the law says they are forbidden. With these crimes, the sole issue is whether the law has been violated. Criminal intent is not necessary where the acts are prohibited for reasons of public policy.[11]

Mala in se crimes are considered harder to prosecute since they require proof of criminal intent or malice on the part of the accused. A number of the sex-related crimes committed against children and minors are punished under special laws rather than the Revised Penal Code. This means that, in general, they are considered *mala prohibita* crimes, and the mere commission of these offenses is punishable regardless of whether the accused had criminal intent or not.

11.1.2 Relevant Treaties and Cybercrime Laws

The Philippines is a signatory or state party to the following international treaties and agreements relating to combating child sexual abuse:

(a) UN Convention on the Rights of the Child;
(b) Optional Protocol on the sale of children, child prostitution and child pornography; and
(c) Convention on the Suppression of Traffic in Persons and of the Exploitation of the Prostitution of Others.

The Philippine laws that are relevant to child sexual abuse are:

(d) Special Protection of Children Against Abuse, Exploitation and Discrimination Act;
(e) The Revised Penal Code (as amended);
(f) The Anti-Rape Law of 1997;
(g) Anti-Trafficking in Persons Act of 2003 (as amended); and
(h) Anti-Child Pornography Act of 2009.

With regard to cybercrime, the following national laws apply in the Philippines:

(i) Cybercrime Prevention Act of 2012;
(j) Anti-Photo and Video Voyeurism Act of 2009;
(k) Electronic Commerce Act;
(l) Access Devices Regulation Act of 1998; and
(m) Data Privacy Act of 2012.

---

[11] *Garcia v. Court of Appeals.*

11.2 Analysis of Substantive Criminal Law

11.2.1 Introduction

Sex-related offenses committed against children and minors are penalized under the Revised Penal Code as well as special laws such as the Special Protection of Children Against Abuse, Exploitation and Discrimination Act and the Anti-Trafficking in Persons Act of 2003 (as amended), the Anti-Rape Law of 1997, and the Anti-Child Pornography Act of 2009.

Under Philippine law, a child or minor is a "person below eighteen (18) years of age".[12] However, these terms may also cover "those over [eighteen years of age] but are unable to fully take care of themselves or protect themselves from abuse, neglect, cruelty, exploitation or discrimination because of a physical or mental disability or condition".[13] The Anti-Child Pornography Act of 2009 similarly defines a child as "a person below eighteen (18) years of age or over, but is unable to fully take care of himself/herself or protect himself/herself from abuse, neglect, cruelty, exploitation or discrimination because of a physical or mental disability or condition".[14]

11.2.2 Possibly Relevant Criminal Offenses

Succinct Overview of Sexual Offenses Involving Minors

The following table lists the possibly relevant criminal law provisions in the Philippines, grouped together by the provisions of the Lanzarote Convention, which gives the most comprehensive catalogue of sexual child abuse offenses available (Table 11.1).

Overview of Sexual Offenses Related to Webcam Child Sexual Abuse

Many of the crimes listed in the table above are relevant and may be applicable to webcam child sexual abuse involving real children and minors because these offenses can also be committed online or at a distance. While there is no specific case law confirming this position, this conclusion is based on a legal analysis of the elements of the relevant criminal offenses and a general review of Philippine law and jurisprudence.

---

[12] See Special Protection of Children Against Abuse, Exploitation and Discrimination Act, sec 3 (a); see Anti-Trafficking in Persons Act of 2003 (as amended), sec 3(b).

[13] Special Protection of Children Against Abuse, Exploitation and Discrimination Act, sec 3(a); see also Anti-Trafficking in Persons Act of 2003 (as amended), sec 3(b).

[14] Anti-Child Pornography Act of 2009, sec 3(a).

Table 11.1 Sexual offences against children under the Lanzarote Convention and their Philippine equivalents

Lanzarote Treaty	The Philippines
Article 18. Sexual abuse	Article 266-A of the Revised Penal Code (as amended by The Anti-Rape Law of 1997) Rape. Rape is Committed (1) By a man who shall have carnal knowledge of a woman under any of the following circumstances (a) Through force, threat, or intimidation (b) When the offended party is deprived of reason or otherwise unconscious (c) By means of fraudulent machination or grave abuse of authority; and (d) When the offended party is under twelve (12) years of age or is demented, even though none of the circumstances mentioned above be present Article 336 of the Revised Penal Code. Acts of Lasciviousness. Any person who shall commit any act of lasciviousness[a] upon other persons of either sex, under any of the circumstances mentioned in the preceding article, shall be punished by prisión correccional Article 338 of the Revised Penal Code. Simple Seduction. The seduction of a woman who is single or a widow of good reputation, over twelve but under eighteen years of age, committed by means of deceit, shall be punished by arresto mayor Article 337 of the Revised Penal Code. Qualified Seduction. The seduction of a virgin over twelve years and under eighteen years of age, committed by any person in public authority, priest, house-servant, domestic, guardian, teacher, or any person who, in any capacity, shall be entrusted with the education or custody of the woman seduced, shall be punished by prisión correccional in its minimum and medium periods The penalty next higher in degree shall be imposed upon any person who shall seduce his sister or descendant, whether or not she be a virgin or over eighteen years of age Under the provisions of this chapter, seduction is committed when the offender has carnal knowledge of any of the persons and under the circumstances described herein Article 342 of the Revised Penal Code. Forcible Abduction. The abduction of any woman against her will and with lewd designs shall be punished by reclusión temporal The same penalty shall be imposed in every case, if the female abducted be under twelve years of age Article 343 of the Revised Penal Code. Consented Abduction. The abduction of a virgin over twelve years and under eighteen years of age, carried out with her consent and with lewd designs, shall be punished by the penalty of prisión correccional in its minimum and medium periods Section 3(b) of the Special Protection of Children Against Abuse, Exploitation and Discrimination Act. "Child abuse" refers to the maltreatment, whether habitual or not, of the child which includes any of the following: (1) Psychological and physical abuse, neglect, cruelty, sexual abuse and emotional maltreatment;… Section 10 of the Special Protection of Children Against Abuse, Exploitation and Discrimination Act. Other Acts of Neglect, Abuse, Cruelty or Exploitation and Other Conditions Prejudicial to the Child's Development (a) Any person who shall commit any other acts of child abuse, cruelty or exploitation or be responsible for other conditions prejudicial to the child's development including those covered by Article 59 of Presidential Decree No. 603, as amended, but not covered by the Revised Penal Code, as amended, shall suffer the penalty of prision mayor in its minimum period (b) Any person who shall keep or have in his company a minor, twelve (12) years or under or who is ten (10) years or more his junior in any public or private place, hotel, motel, beer joint, discotheque, cabaret, pension house, sauna or massage parlor, beach and/or other tourist resort or similar places shall suffer the penalty of prision mayor in its maximum period and a fine of not less than Fifty thousand pesos (P50,000): Provided, That this provision shall not apply to any person who is related within the fourth degree of consanguinity or affinity or any bond recognized by law, local custom and tradition or acts in the performance of a social, moral or legal duty (c) Any person who shall induce, deliver or offer a minor to any one prohibited by this Act to keep or have in his company a minor as provided in the preceding paragraph shall suffer the penalty of prision mayor in its medium period and a fine of not less than Forty thousand pesos (P40,000); Provided, however, That should the perpetrator be an ascendant, stepparent or guardian of the minor, the penalty to be imposed shall be prision mayor in its maximum period, a fine of not less than Fifty thousand pesos (P50,000), and the loss of parental authority over the minor (d) Any person, owner, manager or one entrusted with the operation of any public or private place of accommodation, whether for occupancy, food, drink or otherwise, including residential places, who allows any person to take along with him to such place or places any minor herein described shall be imposed a penalty of prision mayor in its medium period and a fine of not less than Fifty thousand pesos (P50,000), and the loss of the license to operate such a place or establishment

(continued)

Table 11.1 (continued)

Lanzarote Treaty	The Philippines
Article 19. Offences concerning child prostitution	Section 5 of the Special Protection of Children Against Abuse Exploitation and Discrimination Act. Child Prostitution and Other Sexual Abuse. Children, whether male or female, who for money, profit, or any other consideration or due to the coercion or influence of any adult, syndicate or group, indulge in sexual intercourse or lascivious conduct, are deemed to be children exploited in prostitution and other sexual abuse The penalty of reclusion temporal in its medium period to reclusion perpetua shall be imposed upon the following (a) Those who engage in or promote, facilitate or induce child prostitution which include, but are not limited to, the following (1) Acting as a procurer of a child prostitute (2) Inducing a person to be a client of a child prostitute by means of written or oral advertisements or other similar means (3) Taking advantage of influence or relationship to procure a child as prostitute (4) Threatening or using violence towards a child to engage him as a prostitute; or (5) Giving monetary consideration, goods or other pecuniary benefit to a child with intent to engage such child in prostitution (b) Those who commit the act of sexual intercourse or lascivious conduct with a child exploited in prostitution or subjected to other sexual abuse… (c) Those who derive profit or advantage therefrom, whether as manager or owner of the establishment where the prostitution takes place, or of the sauna, disco, bar, resort, place of entertainment or establishment serving as a cover or which engages in prostitution in addition to the activity for which the license has been issued to said establishment Section 6 of the Special Protection of Children Against Abuse, Exploitation and Discrimination Act. Attempt to Commit Child Prostitution. There is an attempt to commit child prostitution under Section 5, para (a) hereof when any person who, not being a relative of a child, is found alone with the said child inside the room or cubicle of a house, an inn, hotel, motel, pension house, apartelle or other similar establishments, vessel, vehicle or any other hidden or secluded area under circumstances which would lead a reasonable person to believe that the child is about to be exploited in prostitution and other sexual abuse There is also an attempt to commit child prostitution, under para (b) of Section 5 hereof when any person is receiving services from a child in a sauna parlor or bath, massage clinic, health club and other similar establishments. A penalty lower by two (2) degrees than that prescribed for the consummated felony under Section 5 hereof shall be imposed upon the principals of the attempt to commit the crime of child prostitution under this Act, or, in the proper case, under the Revised Penal Code Section 3(c) of the Anti-Trafficking in Persons Act of 2003 (as amended). Prostitution—refers to any act, transaction, scheme or design involving the use of a person by another, for sexual intercourse or lascivious conduct in exchange for money, profit or any other consideration Section 3(h) of the Anti-Trafficking in Persons Act of 2003 (as amended). Sexual Exploitation—refers to participation by a person in prostitution, pornography or the production of pornography, in exchange for money, profit or any other consideration or where the participation is caused or facilitated by any means of intimidation or threat, use of force, or other forms of coercion, abduction, fraud, deception, debt bondage, abuse of power or of position or of legal process, taking advantage of the vulnerability of the person, or giving or receiving of payments or benefits to achieve the consent of a person having control over another person; or in sexual intercourse or lascivious conduct caused or facilitated by any means as provided in this Act Article 341 of the Revised Penal Code. White Slave Trade. The penalty of prisión correccional in its medium and maximum periods shall be imposed upon any person who, in any manner, or under any pretext, shall engage in the business or shall profit by prostitution or shall enlist the services of women for the purpose of prostitution
Article 20. Offences concerning child pornography	Section 3(b) of the Anti-Child Pornography Act of 2009. Child Pornography refers to any representation, whether visual, audio or written combination thereof, by electronic, mechanical, digital, optical, magnetic or any other means, of a child engaged or involved in real or simulated explicit sexual activities Section 4(c)(2) of the Cybercrime Prevention Act of 2012. Child Pornography. The unlawful or prohibited acts defined and punishable by Republic Act No. 9775 or the Anti-Child Pornography Act of 2009, committed through a computer system: Provided, That the penalty to be imposed shall be (1) one degree higher than that provided for in Republic Act No. 9775 Section 3(j) of the Anti-Trafficking of Persons Act of 2003 (as amended). Pornography—refers to any representation, through publication, exhibition, cinematography, indecent shows, information technology, or by whatever means, of a person engaged in real or simulated explicit sexual activities or any representation of the sexual parts of a person for primarily sexual purposes

(continued)

Table 11.1 (continued)

Lanzarote Treaty	The Philippines
	Section 3(a) of the Anti-Child Pornography Act of 2009. For the purpose of this Act, a child shall also refer to (1) a person regardless of age who is presented, depicted or portrayed as a child as defined herein; and (2) computer-generated, digitally or manually crafted images or graphics of a person who is represented or who is made to appear to be a child as defined herein[b] Section 4 of the Anti-Child Pornography Act of 2009. Unlawful or Prohibited Acts. It shall be unlawful for any person (a) To hire, employ, use, persuade, induce or coerce a child to perform in the creation or production of any form of child pornography (b) To produce, direct, manufacture or create any form of child pornography (c) To publish, offer, transmit, sell, distribute, broadcast, advertise, promote, export or import any form of child pornography (d) To possess any form of child pornography with the intent to sell, distribute, publish or broadcast: Provided, That possession of three (3) or more articles of child pornography of the same form shall be prima facie evidence of the intent to sell, distribute, publish or broadcast (e) To knowingly, willfully and intentionally provide a venue for the commission of prohibited acts such as, but not limited to, dens, private rooms, cubicles, cinemas, houses or in establishments purporting to be a legitimate business[c] (f) For film distributors, theaters and telecommunication companies, by themselves or in cooperation with other entities, to distribute any form of child pornography (g) For a parent, legal guardian or person having custody or control of a child to knowingly permit the child to engage, participate or assist in any form of child pornography[d] (h) To engage in the luring or grooming of a child; (i) To engage in pandering of any form of child pornography (j) To willfully access any form of child pornography (k) To conspire to commit any of the prohibited acts stated in this section. Conspiracy to commit any form of child pornography shall be committed when two (2) or more persons come to an agreement concerning the commission of any of the said prohibited acts and decide to commit it; and (l) To possess any form of child pornography
Article 21. Offences concerning the participation of a child in pornographic performances	Section 9 of the Special Protection of Children Against Abuse, Exploitation and Discrimination Act. Obscene Publications and Indecent Shows. Any person who shall hire, employ, use, persuade, induce or coerce a child to perform in obscene exhibitions and indecent shows, whether live or in video, or model in obscene publications or pornographic materials or to sell or distribute the said materials shall suffer the penalty of prision mayor in its medium period If the child used as a performer, subject or seller/distributor is below twelve (12) years of age, the penalty shall be imposed in its maximum period Any ascendant, guardian, or person entrusted in any capacity with the care of a child who shall cause and/or allow such child to be employed or to participate in an obscene play, scene, act, movie or show or in any other acts[e] covered by this section shall suffer the penalty of prision mayor in its medium period
Article 22. Corruption of children	Article 340 of the Revised Penal Code. Corruption of Minors. Any person who shall habitually or with abuse of authority or confidence, promote or facilitate the prostitution or corruption of persons underage to satisfy the lust of another, shall be punished by prisión correccional in its minimum and medium periods, and if the culprit be a public officer, he shall also suffer the penalty of temporary absolute disqualification
Article 23. Solicitation of children for sexual purposes	Section 3(h) of the Anti-Child Pornography Act of 2009. "Grooming" refers to the act of preparing a child or someone who the offender believes to be a child for a sexual activity or sexual relationship by communicating any form of child pornography. It includes online enticement or enticement through any other means Section 3(i) of the Anti-Child Pornography Act of 2009. "Luring" refers to the act of communicating, by means of a computer system, with a child or someone who the offender believes to be a child for the purpose of facilitating the commission of a sexual activity or production of any form of child pornography Section 3(j) of the Anti-Child Pornography Act of 2009. "Pandering" refers to the act of offering, advertising, promoting, representing or distributing through any means any material or purported material that is intended to cause another to believe that the material or purported material contains any form of child pornography, regardless of the actual content of the material or purported material

(continued)

Table 11.1 (continued)

Lanzarote Treaty	The Philippines
Trafficking in persons	Section 3 of the Anti-Trafficking in Persons Act of 2003 (as amended). (a) Trafficking in Persons—refers to the recruitment, obtaining, hiring, providing, offering, transportation, transfer, maintaining, harboring, or receipt of persons with or without the victim's consent or knowledge, within or across national borders by means of threat, or use of force, or other forms of coercion, abduction, fraud, deception, abuse of power or of position, taking advantage of the vulnerability of the person, or, the giving or receiving of payments or benefits to achieve the consent of a person having control over another person for the purpose of exploitation which includes at a minimum, the exploitation or the prostitution of others or other forms of sexual exploitation, forced labor or services, slavery, servitude or the removal or sale of organs The recruitment, transportation, transfer, harboring, adoption or receipt of a child for the purpose of exploitation or when the adoption is induced by any form of consideration for exploitative purposes shall also be considered as 'trafficking in persons' even if it does not involve any of the means set forth in the preceding paragraph Section 4 of the Anti-Trafficking in Persons Act of 2003 (as amended). Acts of Trafficking in Persons. It shall be unlawful for any person, natural or juridical, to commit any of the following acts: (a) To recruit, obtain, hire, provide, offer, transport, transfer, maintain, harbor, or receive a person by any means, including those done under the pretext of domestic or overseas employment or training or apprenticeship, for the purpose of prostitution, pornography, or sexual exploitation;… (e) To maintain or hire a person to engage in prostitution or pornography;… (k) To recruit, transport, harbor, obtain, transfer, maintain, hire, offer, provide, adopt or receive a child for purposes of exploitation or trading them, including but not limited to, the act of buying and/or selling a child for any consideration or for barter for purposes of exploitation. Trafficking for purposes of exploitation of children shall include:… (2) The use, procuring or offering of a child for prostitution, for the production of pornography, or for pornographic performances Section 4-A of the Anti-Trafficking in Persons Act of 2003 (as amended). Attempted Trafficking in Persons. Where there are acts to initiate the commission of a trafficking offense but the offender failed to or did not execute all the elements of the crime, by accident or by reason of some cause other than voluntary desistance, such overt acts shall be deemed as an attempt to commit an act of trafficking in persons. As such, an attempt to commit any of the offenses enumerated in Section 4 of this Act shall constitute attempted trafficking in persons Section 5 of the Anti-Trafficking in Persons Act of 2003. Acts that Promote Trafficking in Persons. The following acts which promote or facilitate trafficking in persons shall be unlawful (c) To advertise, publish, print, broadcast or distribute, or cause the advertisement, publication, printing broadcasting or distribution by any means, including the use of information technology and the internet, of any brochure, flyer, or any propaganda material that promotes trafficking in persons Section 6 of the Anti-Trafficking in Persons Act of 2003. Qualified Trafficking in Persons. The following are considered as qualified trafficking (a) When the trafficked person is a child Section 11(a) of the Anti-Trafficking in Persons Act of 2003 (as amended). Use of Trafficked Persons. Any person who buys or engages the services of trafficked persons for prostitution shall be penalized.. (1) If an offense … involves sexual intercourse or lascivious conduct with a child…
Sex tourism	Section 3(g) of the Anti-Trafficking in Persons Act of 2003 (as amended). Sex Tourism—refers to a program organized by travel and tourism-related establishments and individuals which consists of tourism packages or activities, utilizing and offering escort and sexual services as enticement for tourists. This includes sexual services and practices offered during rest and recreation periods for members of the military Section 4 of the Anti-Trafficking in Persons Act of 2003 (as amended). Acts of Trafficking in Persons. It shall be unlawful for any person, natural or juridical, to commit any of the following acts (d) To undertake or organize tours and travel plans consisting of tourism packages or activities for the purpose of utilizing and offering persons for prostitution, pornography or sexual exploitation
Cybersex	Section 4(c)(1) of the Cybercrime Prevention Act of 2012. Cybersex. The willful engagement, maintenance, control, or operation, directly or indirectly, of any lascivious exhibition of sexual organs or sexual activity, with the aid of a computer system, for favor or consideration Section 5(b) of the Cybercrime Prevention Act of 2012. Attempt in the Commission of Cybercrime (in relation to cybersex). Any person who willfully attempts to commit any of the offenses enumerated in this Act shall be held liable

(continued)

Table 11.1 (continued)

Lanzarote Treaty	The Philippines
Photo or video voyeurism	Section 3 of the Anti-Photo and Video Voyeurism Act of 2009. "Photo or video voyeurism" means the act of taking photo or video coverage of a person or group of persons performing sexual act or any similar activity or of capturing an image of the private area of a person or persons without the latter's consent, under circumstances in which such person/s has/have a reasonable expectation of privacy, or the act of selling, copying, reproducing, broadcasting, sharing, showing or exhibiting the photo or video coverage or recordings of such sexual act or similar activity through VCD/DVD, internet, cellular phones and similar means or device without the written consent of the person/s involved, notwithstanding that consent to record or take photo or video coverage of same was given by such person/s Section 4 of the Anti-Photo and Video Voyeurism Act of 2009. Prohibited Acts. It is hereby prohibited and declared unlawful for any person (a) To take photo or video coverage of a person or group of persons performing sexual act or any similar activity or to capture an image of the private area of a person/s such as the naked or undergarment clad genitals, pubic area, buttocks or female breast without the consent of the person/s involved and under circumstances in which the person/s has/have a reasonable expectation of privacy; (b) To copy or reproduce, or to cause to be copied or reproduced, such photo or video or recording of sexual act or any similar activity with or without consideration (c) To sell or distribute, or cause to be sold or distributed, such photo or video or recording of sexual act, whether it be the original, copy or reproduction thereof; or (d) To publish or broadcast, or cause to be published or broadcast, whether in print or broadcast media, or show or exhibit the photo or video coverage or recordings of such sexual act or any similar activity through VCD/DVD, internet, cellular phones and other similar means or device The prohibition under paras (b), (c) and (d) shall apply notwithstanding that consent to record or take photo or video coverage of the same was given by such person/s. Any person who violates this provision shall be liable for photo or video voyeurism as defined herein Section 6 of the Anti-Photo and Video Voyeurism Act of 2009. Exemption. Nothing contained in this Act, however, shall render it unlawful or punishable for any peace officer, who is authorized by a written order of the court, to use the record or any copy thereof as evidence in any civil, criminal investigation or trial of the crime of photo or video voyeurism: Provided, That such written order shall only be issued or granted upon written application and the examination under oath or affirmation of the applicant and the witnesses he/she may produce, and upon showing that there are reasonable grounds to believe that photo or video voyeurism has been committed or is about to be committed, and that the evidence to be obtained is essential to the conviction of any person for, or to the solution or prevention of, such crime

[a]Under the Rules and Regulations on the Reporting and Investigation of Child Abuse Cases, lascivious conduct is defined as "the intentional touching, either directly or through clothing, of the genitalia, anus, groin, breast, inner thigh, or buttocks, or the introduction of any object into the genitalia, anus or mouth, of any person, whether of the same or opposite sex, with an intent to abuse, humiliate, harass, degrade, or arouse or gratify the sexual desire of any person, bestiality, masturbation, lascivious exhibition of the genitals or pubic area of a person"

[b]It appears that this provision on computer-generated representations of "child" only applies within the context the Anti-Child Pornography Act and in relation to child pornography

[c]This provision may not cover *per se* the creation or operation of websites and online chat rooms since the examples contemplate physical spaces. However, the use of these physical spaces, including internet cafes, to for purposes of webcam child sex tourism and child pornography would be covered under the law

[d]This provision may also apply to webcam child sex tourism and child pornography

[e]The term "other acts" may cover other lascivious conduct as defined under the Rules and Regulations on the Reporting and Investigation of Child Abuse Cases

Child Sexual Abuse

Crimes against persons or chastity under the Revised Penal Code like rape, acts of lasciviousness, forcible and consented abduction, and simple and qualified seduction do not apply in the online context since they require for their commission the physical presence and interaction of the victim and the offender.[15] It appears though that the crime of child abuse may be committed using online and digital means since it covers acts that cause "*psychological* and physical *abuse*, neglect, cruelty, *sexual abuse* and *emotional maltreatment*" of children.[16] Under the law, "psychological injury"

> includes any harm to a child's psychological or intellectual functioning which may be exhibited by severe anxiety, depression, withdrawal or outward aggressive behavior, or a combination of said behaviors, which may be demonstrated by a change in behavior, emotional response or cognition.[17]

Under Philippine law, sexual abuse of a child is committed through "the employment, use, persuasion, inducement, enticement or coercion of a child to engage in, or assist another person to engage in, sexual intercourse or *lascivious conduct* or the molestation, prostitution, or incest with children".[18] Based on the Rules and Regulations on the Reporting and Investigation of Child Abuse Cases, lascivious conduct is defined as

> the intentional touching, either directly or through clothing, of the genitalia, anus, groin, breast, inner thigh, or buttocks, or the introduction of any object into the genitalia, anus or mouth, of any person, whether of the same or opposite sex, with an intent to abuse, humiliate, harass, degrade, or *arouse or gratify the sexual desire of any person*, bestiality, masturbation, *lascivious exhibition of the genitals or pubic area of a person*.[19]

Willfully engaging in sexual activities with or in the presence a child or minor via a webcam or online may be considered a species of lascivious conduct that is prohibited under the law. While there is no decided case on this matter, the literal meaning of the relevant legal provisions supports this interpretation.

The Special Protection of Children Against Abuse, Exploitation and Discrimination Act further contains a catch-all provision that prohibits and penalizes "acts of neglect, abuse, cruelty or exploitation" that are "prejudicial to the child's development".[20] Section 10(a) of the Act states that

---

[15] The Revised Penal Code, Articles 226-A, 336, 337, 338, 342 and 343.

[16] Special Protection of Children Against Abuse, Exploitation and Discrimination Act, sec 3(b) (emphasis added).

[17] Rules and Regulations on the Reporting and Investigation of Child Abuse Cases, sec 2(e).

[18] Rules and Regulations on the Reporting and Investigation of Child Abuse Cases, sec 2(g) (emphasis added)

[19] Rules and Regulations on the Reporting and Investigation of Child Abuse Cases, sec 2(h) (emphasis added).

[20] Special Protection of Children Against Abuse, Exploitation and Discrimination Act, sec 10(a).

> Any person who shall commit any other acts of child abuse, cruelty or exploitation or be responsible for other conditions prejudicial to the child's development including those covered by Article 59 of Presidential Decree No. 603 [The Child and Youth Welfare Code], as amended, but not covered by the Revised Penal Code, as amended, shall suffer the penalty of *prision mayor* in its minimum period.[21]

However, the crimes involving child sex tourism provided for under Sections 10 (b), (c) and (d) of Special Protection of Children Against Abuse, Exploitation and Discrimination Act are not applicable in the case of webcam child abuse since they contemplate and require physical proximity and contact between the victim and the offender for the crimes to be committed.

Child Prostitution and Sexual Exploitation

Under Philippine law, it may be possible to commit child prostitution through digital and online means. While there can be no sexual intercourse using a webcam and other technologies, child prostitution and prostitution in general can also be committed by having "lascivious conduct" with another person.[22] As stated in the Special Protection of Children Against Abuse, Exploitation and Discrimination Act, "[t]hose who commit the act of sexual intercourse or *lascivious conduct* with a child exploited in prostitution or subjected to other sexual abuse" are subject to criminal prosecution.[23] The Supreme Court has ruled that "[t]he mere act of having sexual intercourse or committing lascivious conduct with a child who is exploited in prostitution or subjected to sexual abuse constitutes the offense. It is a *malum prohibitum*, an evil that is proscribed".[24] According to the Supreme Court, "[a] child is deemed exploited in prostitution or subjected to other sexual abuse, when the child indulges in sexual intercourse or lascivious conduct (a) for money, profit, or any other consideration; or (b) under the coercion or influence of any adult, syndicate or group".[25] The Supreme Court has also held that Section 5(b) of the Special Protection of Children Against Abuse, Exploitation and Discrimination Act "penalizes not only child prostitution, the essence of which is profit, but also other forms of sexual abuse of children.[26]

It should be noted though that, given the lack of physical presence of the parties, the crime of attempt to commit children prostitution does not apply in the context of webcam or online sexual activities since the offense requires that the offender "who,

[21] Special Protection of Children Against Abuse, Exploitation and Discrimination Act, sec 10(a).

[22] Special Protection of Children Against Abuse, Exploitation and Discrimination Act, sec 5; see also the Anti-Trafficking in Persons Act of 2003 (as amended), sec 3(c) (which defines prostitution as "any act, transaction, scheme or design involving the use of a person by another, for sexual intercourse or *lascivious conduct* in exchange for money, profit or any other consideration") (emphasis added).

[23] Special Protection of Children Against Abuse, Exploitation and Discrimination Act, sec 5(b) (emphasis added).

[24] *Malto v People*.

[25] *People v. Larin*.

[26] *People v. Larin*; see also *Olivarez v Court of Appeals*.

not being a relative of a child, is found alone with the said child inside the room or cubicle… or any other hidden or secluded area" or "is receiving services from a child in a sauna parlor or bath, massage clinic, health club and other similar establishments".[27] Following the rule of *ejusdem generis*, the phrase "other hidden and secluded areas" does not appear to cover online spaces and internet forums since the situations and examples listed in the law only cover physical locations.

In any event, persons involved in webcam child sexual abuse can be held liable for sexual exploitation under the Anti-Trafficking in Persons Act of 2003 (as amended). Under the law, sexual exploitation

> refers to *participation by a person in prostitution*, pornography or the production of pornography, in exchange for money, profit or any other consideration or where the participation is caused or facilitated by any means of intimidation or threat, use of force, or other forms of coercion, abduction, fraud, deception, debt bondage, abuse of power or of position or of legal process, taking advantage of the vulnerability of the person, or giving or receiving of payments or benefits to achieve the consent of a person having control over another person; or in sexual intercourse or *lascivious conduct* caused or facilitated by any means as provided in this Act.[28]

Procuring the services of a child prostitute and/or carrying out lewd or lascivious acts with him or her through a webcam or other means of communication can be construed as sexual exploitation and may be punishable under the law.

Child Pornography

Based on the Anti-Child Pornography Act of 2009, child pornography "refers to any representation, whether visual, audio or written combination thereof, by *electronic*, mechanical, *digital*, *optical*, magnetic or any other means, of a child engaged or involved in real or simulated explicit sexual activities".[29] Similarly, the Anti-Trafficking of Persons Act of 2003 (as amended) defines pornography as

> any representation, through publication, exhibition, cinematography, indecent shows, *information technology, or by whatever means*, of a person engaged in real or simulated explicit sexual activities or any representation of the sexual parts of a person for primarily sexual purposes.[30]

With regard to the definition of a child, Section 3(a) of the Anti-Child Pornography Act of 2009 provides that:

> "Child" refers to a person below eighteen (18) years of age or over, but is unable to fully take care of himself/herself or protect himself/herself from abuse, neglect, cruelty, exploitation or discrimination because of a physical or mental disability or condition.

For the purpose of this Act, a child shall also refer to:

[27] Special Protection of Children Against Abuse, Exploitation and Discrimination Act, sec 6.

[28] Anti-Trafficking in Persons Act of 2003 (as amended), sec 3(h).

[29] Anti-Child Pornography Act of 2009, sec 3(b) (emphasis added).

[30] Anti-Trafficking in Persons Act of 2003 (as amended), sec 3(j) (emphasis added).

(1) a person regardless of age who is presented, depicted or portrayed as a child as defined herein; and

(2) computer-generated, digitally or manually crafted images or graphics of a person who is represented or who is made to appear to be a child as defined herein.[31]

It should be noted that this provision on computer-generated representations of "child" only applies within the context the Anti-Child Pornography Act. It does not affect the general definition of a child or minor, which contemplates a real person.

The Act's definition of child pornography appears to be broad enough to cover webcam child sexual abuse since a live video feed or streaming images are undoubtedly audiovisual representations by digital or optical means of a child engaged in real or simulated explicit sexual activities.[32] Furthermore, the Cybercrime Prevention Act of 2012 expressly states that child pornography "committed through a computer system" is a criminal offense and is punishable by a penalty "one degree higher than that provided for" in the Anti-Child Pornography Act of 2009.[33] In relation to webcam child sexual abuse, the Anti-Child Pornography Act of 2009 prohibits the following acts or conduct:

(a) To hire, employ, use, persuade, induce or coerce a child to perform in the creation or production of any form of child pornography…

(c) To publish, offer, transmit, sell, distribute, broadcast, advertise, promote, export or import any form of child pornography…

(d) To possess any form of child pornography with the intent to sell, distribute, publish or broadcast: Provided, That possession of three (3) or more articles of child pornography of the same form shall be prima facie evidence of the intent to sell, distribute, publish or broadcast…

(j) To willfully access any form of child pornography…

(l) To possess any form of child pornography.[34]

Pornographic Performances and Corruption of Children

An offender who takes part in webcam child sexual abuse can also be charged for the crime of using or inducing a child to perform in obscene exhibitions and indecent shows. Under the Special Protection of Children Against Abuse, Exploitation and Discrimination Act, it is a crime to "hire, employ, use, persuade, induce or coerce a child to perform in obscene exhibitions and indecent shows, *whether live or in video*, or model in obscene publications or pornographic materials or to sell or distribute the said materials".[35] With respect to corruption of minors, while the person who "promote[s] or facilitate[s] the prostitution or

---

[31] Anti-Child Pornography Act of 2009, sec 3(a).

[32] See Anti-Child Pornography Act of 2009, sec 3(b).

[33] Cybercrime Prevention Act of 2012, sec 4(c)(2).

[34] Cybercrime Prevention Act of 2012, sec 4.

[35] Special Protection of Children Against Abuse, Exploitation and Discrimination Act, sec 9 (emphasis added).

corruption of persons underage to satisfy the lust of another" is considered the principal of the crime,[36] the person who has lascivious conduct with a minor may be charged as an accomplice under the law.[37] The sex offender may be deemed an accomplice since that person "cooperate[d] in the execution of the offense by previous or simultaneous acts".[38]

Solicitation of Children

In relation to the solicitation of children for sexual purposes, the Anti-Child Pornography Act of 2009 criminalizes the acts of "grooming" and "luring". Under the law, grooming "refers to the act of preparing a child or *someone who the offender believes to be a child* for a sexual activity or sexual relationship by communicating any form of child pornography. It includes online enticement or enticement through any other means".[39] Luring is "the act of *communicating, by means of a computer system*, with a child or *someone who the offender believes to be a child* for the purpose of facilitating the commission of a sexual activity or production of any form of child pornography".[40] These crimes are very relevant to combating webcam child sex abuse. It is worth highlighting that, with regard to the crimes of grooming and luring, the victim does not have to be an actual child and he or she can be "someone who the offender believes to be a child".[41] As of yet, there is no case law on this subject.

Trafficking in Persons

Webcam child sexual abuse may also involve the crime of trafficking in persons, which covers:

> the recruitment, obtaining, hiring, providing, offering, transportation, transfer, maintaining, harboring, or receipt of persons with or without the victim's consent or knowledge, within or across national borders by means of threat, or use of force, or other forms of coercion, abduction, fraud, deception, abuse of power or of position, taking advantage of the vulnerability of the person, or, *the giving or receiving of payments or benefits to achieve the consent of a person having control over another person for the purpose of exploitation* which includes at a minimum, *the exploitation or the prostitution of others or other forms of sexual exploitation*, forced labor or services, slavery, servitude or the removal or sale of organs.[42]

The following acts of trafficking in persons are pertinent to webcam child sexual abuse since the law does not require that the sex offender have physical contact with the child victim:

---

36 The Revised Penal Code, Article 340.

37 The Revised Penal Code, Articles 16–18.

38 The Revised Penal Code, Article 18.

39 Anti-Child Pornography Act of 2009, sec 3(h) (emphasis added).

40 Anti-Child Pornography Act of 2009, sec 3(i) (emphasis added).

41 Anti-Child Pornography Act of 2009, sec 3(h) and 3(i).

42 Anti-Trafficking in Persons Act of 2003 (as amended), sec 3 (emphasis added).

> (a) To recruit, obtain, hire, provide, offer, transport, transfer, maintain, harbor, or receive a person by any means, including those done under the pretext of domestic or overseas employment or training or apprenticeship, for the purpose of prostitution, pornography, or sexual exploitation...
>
> (e) To maintain or hire a person to engage in prostitution or pornography...
>
> (k) To recruit, transport, harbor, obtain, transfer, maintain, hire, offer, provide, adopt or receive a child for purposes of exploitation or trading them.... Trafficking for purposes of exploitation of children shall include:... The use, procuring or offering of a child for prostitution, for the production of pornography, or for pornographic performances...[43]

In addition, a person may be charged for trafficking when he or she uses trafficked persons such as when an offender "buys or engages the services of trafficked persons for prostitution" including "sexual intercourse or *lascivious conduct* with a child".[44] It should be noted that "when the trafficked person is a child", the crime committed is qualified trafficking in persons and the penalties are higher.[45] It is also possible for the sex offender to be held answerable for attempted trafficking in persons "[w]here there are acts to initiate the commission of a trafficking offense but the offender failed to or did not execute all the elements of the crime, by accident or by reason of some cause other than voluntary desistance".[46]

Child Sex Tourism and Cybersex

A crime that is closely associated with webcam child sex abuse is sex tourism, which is defined by the Anti-Trafficking in Persons Act of 2003 (as amended) as "a program organized by travel and tourism-related establishments and individuals which consists of tourism packages or activities, utilizing and offering escort and sexual services as enticement for tourists".[47] It is considered an act of trafficking in persons "[t]o undertake or organize tours and travel plans consisting of tourism packages or activities for the purpose of utilizing and offering persons for prostitution, pornography or sexual exploitation".[48] The crime of sex tourism can equally apply in an online or digital context because, as discussed earlier, prostitution (in the form of lascivious conduct) and sexual exploitation of children can also be committed online or at a distance. The Cybercrime Prevention Act of 2012 clearly states that "[a]ll crimes defined and penalized by the Revised Penal Code, as amended, and special laws, if committed by, through and with the use of information and communications technologies shall be covered by the relevant provisions of this Act".[49] According to the Supreme Court, this provision, "merely

---

[43] Anti-Trafficking in Persons Act of 2003 (as amended), sec 4.

[44] Anti-Trafficking in Persons Act of 2003 (as amended), sec 11(a) (emphasis added).

[45] Anti-Trafficking in Persons Act of 2003 (as amended), sec 6.

[46] Anti-Trafficking in Persons Act of 2003 (as amended), sec 4-A.

[47] Anti-Trafficking in Persons Act of 2003 (as amended), sec 3(g).

[48] Anti-Trafficking in Persons Act of 2003 (as amended), sec 4.

[49] Cybercrime Prevention Act of 2012, sec 6 (the penalty imposed "shall be one (1) degree higher than that provided for by the Revised Penal Code, as amended, and special laws, as the case may be".

makes commission of existing crimes through the Internet a qualifying circumstance".[50] As the Supreme Court clarifies,

> there exists a substantial distinction between crimes committed through the use of information and communications technology and similar crimes committed using other means. In using the technology in question, the offender often evades identification and is able to reach far more victims or cause greater harm. The distinction, therefore, creates a basis for higher penalties for cybercrimes.[51]

Another offense that is particularly pertinent to webcam child sex tourism is cybersex. Under Philippine law, cybersex, which is the "willful engagement, maintenance, control, or operation, directly or indirectly, of any lascivious exhibition of sexual organs or sexual activity, with the aid of a computer system, for favor or consideration", is a punishable offense.[52] The Supreme Court has ruled that the crime of cybersex only applies to "persons engaged in the business of maintaining, controlling, or operating, directly or indirectly, the lascivious exhibition of sexual organs or sexual activity with the aid of a computer system as Congress has intended".[53] As the Supreme Court further explains,

> The understanding of those who drew up the cybercrime law is that the element of 'engaging in a business' is necessary to constitute the illegal cybersex. *The Act actually seeks to punish cyber prostitution, white slave trade, and pornography for favor and consideration. This includes interactive prostitution and pornography, i.e., by webcam.*[54]

According to the implementing rules and regulations of the Cybercrime Prevention Act of 2012, "[c]ybersex involving a child shall be punished" and dealt with in the same manner as "child pornography".[55] In addition to being prosecuted under the Cybercrime Prevention Act of 2012, a person committing cybersex may be held liable for acts involving trafficking in persons.[56] There are no decided cases in this area.

Conclusion

In summary, many of the sex-related crimes in Philippine law may be used to prosecute and punish the sexual abuse and exploitation of real children and minors via webcam or other similar means or media. This is so because the elements or the conditions for the commission of these crimes do not require the physical presence, contact or interaction between the victim and the offender and can thus be committed online or at a distance. To reiterate, a person engaged or involved in the webcam child sex abuse or sex tourism may be investigated, charged and

---

50 *Disini, Jr. v. Secretary of Justice.*

51 *Disini, Jr. v. Secretary of Justice.*

52 Cybercrime Prevention Act of 2012, sec 4(c)(1).

53 *Disini, Jr. v. Secretary of Justice.*

54 *Disini, Jr. v. Secretary of Justice.*

55 Implementing Rules and Regulations of the Cybercrime Prevention Act of 2012, sec 5(2).

56 Implementing Rules and Regulations of the Cybercrime Prevention Act of 2012, sec 5(2).

prosecuted for the following crimes: child sexual abuse, child prostitution, child sexual exploitation, child pornography, pornographic performances of children, corruption of children (as accomplices), grooming, luring, qualified trafficking in persons, attempted trafficking in persons, child sex tourism, cybersex and attempted cybersex. Law enforcement authorities can conduct investigations when any of these crimes have been, are being, or are about to be committed.

Assuming the victim is an actual child or minor, the offender may still be held criminally liable even if the child victim does not actually undress because the act of undressing is not an essential element for the commission of the following crimes: child sexual abuse, child prostitution, child sexual exploitation, pornographic performances of children, corruption of children, grooming, luring, trafficking in persons, child sex tourism, and cybersex. The offender can be prosecuted based on other overt acts that he or she committed that constitute these crimes.

In relation to attempted crimes, Article 6 of the Revised Penal Code provides that "There is an attempt when the offender commences the commission of a felony directly by overt acts, and does not perform all the acts of execution which should produce the felony by reason of some cause or accident other than his own spontaneous desistance". The offender who carries out webcam or online child sexual abuse may be held liable for attempting to commit the crimes discussed above.

11.2.3 Possible Obstacles in Substantive Law Concerning Sweetie

Despite the applicability of many sex-related crimes to webcam child sex abuse or sex tourism, most of these crimes would not apply in the case of Sweetie 2.0 because they require as an essential element for their commission the involvement of a real child or minor victim. Given that Sweetie 2.0 is a computer program or system, it is not considered a "natural person"[57] under the law and, as such, has no civil or juridical capacity or legal rights and personality.[58] Because Sweetie 2.0 is not a "person", it is not protected under the law and it cannot be the victim of a crime. Further, the following crimes discussed above cannot be committed without a natural person being involved, targeted or victimized: child sexual abuse, child prostitution, child sexual exploitation, child pornography,[59] attempt to commit child pornography,[60] pornographic performances of children, corruption of children, grooming, luring, qualified trafficking in persons, attempted trafficking in persons,

[57] Civil Code of the Philippines, Article 40.

[58] Civil Code of the Philippines, Article 37.

[59] Subject to the qualification that this crime applies to computer-generated child pornography.

[60] Subject to the qualification that this crime applies to computer-generated child pornography.

and child sex tourism.[61] Some of these crimes would apply in the case of Sweetie 2.0 if the program performed "simulated explicit sexual activities" or had "any representation of the sexual parts of a person for primarily sexual purposes".[62] However, giving Sweetie 2.0 these features and making it function this way would amount to child pornography and subject Sweetie 2.0's developer and operator to potential criminal liability.

While a number of crimes may not apply to Sweetie 2.0 because it is not an actual child or minor, pursuant to Article 59 of the Revised Penal Code, the offender may still be held criminally liable in connection with an impossible crime. Article 59 states that

> When the person intending to commit an offense has already performed the acts for the execution of the same but nevertheless the crime was not produced by reason of the fact that the act intended was by its nature one of impossible accomplishment or because the means employed by such person are essentially inadequate to produce the result desired by him, the court, having in mind the social danger and the degree of criminality shown by the offender, shall impose upon him the penalty of arresto mayor or a fine ranging from 200 to 500 pesos.[63]

The penalty of arresto mayor ranges from one month and one day to six months imprisonment. According to the Supreme Court, the requisites of an impossible crime are:

> (1) that the act performed would be an offense against persons or property;
>
> (2) that the act was done with evil intent; and (3) that its accomplishment was inherently impossible, or the means employed was either inadequate or ineffectual.[64]

The Supreme Court ruled that:

> To be impossible under this clause, the act intended by the offender must be by its nature one impossible of accomplishment. There must be either (1) legal impossibility, or (2) physical impossibility of accomplishing the intended act in order to qualify the act as an impossible crime.[65]

The article on impossible crime may apply in relation to Sweetie 2.0 because: acts of webcam or online child sex abuse are offenses against persons; they are done with evil intent; and their accomplishment is inherently impossible since, unbeknownst to the offender, Sweetie 2.0 is a computer program and not an actual child.

The few sex-related activities carried out in relation to Sweetie 2.0 that may be punishable offenses and would trigger further or potential criminal investigation and/or prosecution are cybersex, attempted cybersex, grooming and luring.

[61] It should be noted that, with respect to grooming and luring, while not a child, another natural person is involved (i.e., "someone who the offender believes to be a child").

[62] Anti-Trafficking of Persons Act of 2003 (as amended), sec 3(j); Anti-Child Pornography Act of 2009, sec 3(b).

[63] The Revised Penal Code, Article 59.

[64] *Jacinto v. People*.

[65] *Jacinto v. People*.

Grooming and luring can apply in the case of Sweetie 2.0 because these offenses may involve "someone who the offender believes to be a child"[66] (e.g., an undercover police officer). While the crime of cybersex appears to contemplate two persons engaging in lascivious conduct or sexual activity using a computer for consideration or profit (e.g., in the context of prostitution or pornography), the offense may also apply in the case of Sweetie 2.0 since "any lascivious exhibition of sexual organs or sexual activity… with the aid of a computer system" is a punishable offense.[67] If an offender engages or attempts to engage in any lascivious or sexual acts with Sweetie 2.0, a crime is committed since the presence or involvement an actual victim is not necessary to be charged and prosecuted for this offense. Further, cybersex is *mala prohibita* crime and the mere carrying out of this prohibited act is already a violation of the law.

Sweetie 2.0 may also be used to investigate and prosecute crimes involving the solicitation of children (i.e., grooming and luring) since they may be committed in relation to "someone who the offender believes to be a child" and an actual child does not have to be present.[68] It should be noted though that the provisions on grooming and luring use the term "someone", which contemplates the involvement of a real person (e.g., a police officer or member of law enforcement). To get around this requirement without having to amend the law, Sweetie 2.0 can be designed to be semi-automated or not completely autonomous. Sweetie 2.0 can be programmed in such a way that, while it may generally operate or function without human intervention, it should always be subject to the control of a human operator. In this way, the offender would still be soliciting "someone who the offender believes to be a child", albeit the human operator is using a sophisticated computer system to communicate and interact with numerous suspected sex offenders around the world.

In relation to the solicitation of children, the developer or operator of Sweetie 2.0 should take care not to commit the crime of pandering. Under the Anti-Child Pornography Act of 2009, pandering is "the *act of offering*, advertising, promoting, *representing* or distributing through any means *any material or purported material that is intended to cause another to believe that the material or purported material contains any form of child pornography, regardless of the actual content of the material or purported material*".[69] In its attempt to expose, identity and gather evidence against online sex offenders using Sweetie 2.0, Sweetie 2.0's developer may end up unwittingly committing this crime. It appears though that the crime of pandering would not apply if Sweetie 2.0 was utilized as a part of a police investigation since such actions are within the remit and authority of law enforcement and thus impliedly excluded from criminal liability.[70]

---

[66] Anti-Child Pornography Act of 2009, secs 3(h) sec 3(i).

[67] Cybercrime Prevention Act of 2012, sec 4(c)(1).

[68] Anti-Child Pornography Act of 2009, sec 3(h) and 3(i).

[69] Anti-Child Pornography Act of 2009, sec 3(j).

[70] See Anti-Child Pornography Act of 2009, secs 2, 21 and 22; see Cybercrime Prevention Act of 2012, sec 2 and Chapter IV.

Finally, using Sweetie 2.0 to target and expose online sex offenders may fall under the Anti-Photo and Video Voyeurism Act of 2009. Under the law, photo or video voyeurism is a punishable offense:

> "Photo or video voyeurism" means the act of taking photo or video coverage of a person or group of persons performing sexual act or any similar activity or of capturing an image of the private area of a person or persons without the latter's consent, under circumstances in which such person/s has/have a reasonable expectation of privacy, or the act of selling, copying, reproducing, broadcasting, sharing, showing or exhibiting the photo or video coverage or recordings of such sexual act or similar activity through VCD/DVD, internet, cellular phones and similar means or device without the written consent of the person/s involved, notwithstanding that consent to record or take photo or video coverage of same was given by such person/s.[71]

The law criminalizes the following acts:

- To take photo or video coverage of a person or group of persons performing sexual act or any similar activity or to capture an image of the private area of a person/s such as the naked or undergarment clad genitals, pubic area, buttocks or female breast without the consent of the person/s involved and under circumstances in which the person/s has/have a reasonable expectation of privacy;
- To copy or reproduce, or to cause to be copied or reproduced, such photo or video or recording of sexual act or any similar activity with or without consideration;
- To sell or distribute, or cause to be sold or distributed, such photo or video or recording of sexual act, whether it be the original, copy or reproduction thereof; or
- To publish or broadcast, or cause to be published or broadcast, whether in print or broadcast media, or show or exhibit the photo or video coverage or recordings of such sexual act or any similar activity through VCD/DVD, internet, cellular phones and other similar means or device.

The prohibition under paras (b), (c) and (d) shall apply notwithstanding that consent to record or take photo or video coverage of the same was given by such person/s. Any person who violates this provision shall be liable for photo or video voyeurism as defined herein.[72]

The Act only provides one express exemption for any police or "peace officer, who is authorized by a written order of the court, to use the record or any copy thereof as evidence in any civil, criminal investigation or trial of the crime of photo or video voyeurism".[73] It should be noted that the exemption only applies if the record or copy is used as evidence in prosecuting photo or video voyeurism cases and not other crimes.[74] Save for this exemption, evidence gathered using Sweetie 2.0

---

[71] Anti-Photo and Video Voyeurism Act of 2009, sec 3.

[72] Anti-Photo and Video Voyeurism Act of 2009, sec 4.

[73] Anti-Photo and Video Voyeurism Act of 2009, sec 6.

[74] Anti-Photo and Video Voyeurism Act of 2009, sec 6.

would be inadmissible in all other cases, investigations, trials and prosecutions.[75] The Anti-Photo and Video Voyeurism Act of 2009 unequivocally states that: "Any record, photo or video, or copy thereof, obtained or secured by any person in violation of the preceding sections shall not be admissible in evidence in any judicial, quasi-judicial, legislative or administrative hearing or investigation".[76] Despite the provisions of this Act, there are legal and policy grounds to argue that this exclusionary rule does not apply to the taking of photos or recording of videos of sexual acts conducted by law enforcement authorities pursuant to a lawful court order to investigate the commission of crimes other than photo or video voyeurism. It is clearly not the purpose of a special law like the Anti-Photo and Video Voyeurism Act of 2009 to restrict or effectively bar the ability of law enforcement authorities to investigate and gather evidence on other forms of sex-related crimes found in the Revised Penal Code and other special laws. Such investigatory activities carried out by the police are clearly not "photo or video voyeurism" since they are part of lawful investigations. Furthermore, using Sweetie 2.0 may involve the Data Privacy Act of 2012, specifically the possible unauthorized processing and disclosure of sensitive personal information of suspected sex offenders.[77] The above liabilities under data privacy law do not apply if law enforcement authorities operate Sweetie 2.0 as it would be part of their lawful investigations and prosecutions.

11.3 Analysis of Criminal Procedure Law

11.3.1 General Description of Legal Framework

In the Philippines, criminal actions are instituted or commenced with the filing of a formal complaint or information. A complaint is "a sworn written statement charging a person with an offense, subscribed by the offended party, any peace officer, or other public officer charged with the enforcement of the law violated",[78] while an information is "an accusation in writing charging a person with an offense, subscribed by the prosecutor and filed with the court".[79] Save in cases where a person is lawfully arrested without a warrant or the penalty for the offense charged is less than "four (4) years, two (2) months and one (1) day [imprisonment] without regard to the fine", a preliminary investigation is normally held before a person is arraigned and tried.[80] The main aim of a preliminary investigation is "to determine whether there is sufficient ground to engender a well-founded belief that a crime has

[75] Anti-Photo and Video Voyeurism Act of 2009, sec 7.

[76] Anti-Photo and Video Voyeurism Act of 2009, sec 7.

[77] Data Privacy Act of 2012, secs 25 28 and 32.

[78] Revised Rules of Criminal Procedure, rule 110 sec 3.

[79] Revised Rules of Criminal Procedure, rule 110 sec 4.

[80] Revised Rules of Criminal Procedure, rule 112, secs 1 and 7.

been committed and the respondent is probably guilty thereof, and should be held for trial".[81] Preliminary investigations may be conducted by: "(a) Provincial or City Prosecutors and their assistants; (b) Judges of the Municipal Trial Courts and Municipal Circuit Trial Courts; (c) National and Regional State Prosecutors; and (d) Other officers as may be authorized by law".[82] The above investigating officers are responsible for determining whether to file or dismiss the complaint or information.[83] During the preliminary investigation, the relevant judge or court may issue an arrest warrant.[84] Under Philippine law, a peace officer or a private person may arrest a person without a warrant in the following circumstances:

(n) When, in his presence, the person to be arrested has committed, is actually committing, or is attempting to commit an offense;
(o) When an offense has just been committed and he has probable cause to believe based on personal knowledge of facts or circumstances that the person to be arrested has committed it; and
(p) When the person to be arrested is a prisoner who has escaped from a penal establishment or place where he is serving final judgment or is temporarily confined while his case is pending, or has escaped while being transferred from one confinement to another.[85]

Aside from arrests, criminal investigations and prosecutions often necessitate conducting searches and seizures. In the Philippines, only a judge or court may issue a search warrant.[86] As stated in the Philippine Constitution,

> no search warrant or warrant of arrest shall issue except upon probable cause to be determined personally by the judge after examination under oath or affirmation of the complainant and the witnesses he may produce, and particularly describing the place to be searched and the persons or things to be seized.[87]

A search though may also be conducted as an incident to a lawful arrest and a "person lawfully arrested may be searched for dangerous weapons or anything which may have been used or constitute proof in the commission of an offense without a search warrant".[88]

It should be noted that there are no specific laws in the Philippines on undercover policing and operations. Such activities are subject to general laws (e.g., the

[81] Revised Rules of Criminal Procedure, rule 112 sec 1.

[82] Revised Rules of Criminal Procedure, rule 112 sec 2.

[83] Revised Rules of Criminal Procedure, rule 112 sec 4.

[84] Revised Rules of Criminal Procedure, rule 112 sec 6.

[85] Revised Rules of Criminal Procedure, rule 113 sec 5.

[86] 1987 Constitution, Article III sec 2; see also Revised Rules of Criminal Procedure, rule 126 secs 1 and 2.

[87] 1987 Constitution, Article III sec 2; see also Revised Rules of Criminal Procedure, rule 126 sec 4.

[88] Revised Rules of Criminal Procedure, rule 126 sec 13.

rights and protections granted under the Philippine Constitution) and the relevant case law on entrapment.

Table 11.2 Criminal procedure rules in the Convention on Cybercrime and their Philippine equivalents

Council of Europe Convention on Cybercrime	The Philippines
Article 16. Expedited preservation of stored computer data	Section 13 of the Cybercrime Prevention Act of 2012. Preservation of Computer Data. The integrity of traffic data and subscriber information relating to communication services provided by a service provider shall be preserved for a minimum period of six (6) months from the date of the transaction. Content data shall be similarly preserved for six (6) months from the date of receipt of the order from law enforcement authorities requiring its preservation Law enforcement authorities may order a one-time extension for another six (6) months: Provided, That once computer data preserved, transmitted or stored by a service provider is used as evidence in a case, the mere furnishing to such service provider of the transmittal document to the Office of the Prosecutor shall be deemed a notification to preserve the computer data until the termination of the case The service provider ordered to preserve computer data shall keep confidential the order and its compliance
Article 17. Expedited preservation and partial disclosure of traffic data	Section 14 of the Cybercrime Prevention Act of 2012. Disclosure of Computer Data. Law enforcement, upon securing a court warrant, shall issue an order requiring any person or service provider to disclose or submit subscriber's information, traffic data or relevant data in his/its possession or control within seventy-two (72) hours from receipt of the order in relation to a valid complaint officially docketed and assigned for investigation and the disclosure is necessary and relevant for the purpose of investigation

(continued)

Table 11.2 (continued)

Council of Europe Convention on Cybercrime	The Philippines
Article 18. Production order	Section 14 of the Cybercrime Prevention Act of 2012. Disclosure of Computer Data. Law enforcement, upon securing a court warrant, shall issue an order requiring any person or service provider to disclose or submit subscriber's information, traffic data or relevant data in his/its possession or control within seventy-two (72) hours from receipt of the order in relation to a valid complaint officially docketed and assigned for investigation and the disclosure is necessary and relevant for the purpose of investigation
Article 19. Search and seizure of stored computer data	Section 15 of the Cybercrime Prevention Act of 2012. Search, Seizure and Examination of Computer Data. Where a search and seizure warrant is properly issued, the law enforcement authorities shall likewise have the following powers and duties Within the time period specified in the warrant, to conduct interception, as defined in this Act, and (a) To secure a computer system or a computer data storage medium (b) To make and retain a copy of those computer data secured (c) To maintain the integrity of the relevant stored computer data (d) To conduct forensic analysis or examination of the computer data storage medium; and (e) To render inaccessible or remove those computer data in the accessed computer or computer and communications network Pursuant thereof, the law enforcement authorities may order any person who has knowledge about the functioning of the computer system and the measures to protect and preserve the computer data therein to provide, as is reasonable, the necessary information, to enable the undertaking of the search, seizure and examination Law enforcement authorities may request for an extension of time to complete the examination of the computer data storage medium and to make a return thereon but in no case for a period longer than thirty (30) days from date of approval by the court
Article 20. Real-time collection of traffic data	There are no specific rules. The general principles and rules on searches and seizures apply
Article 21. Interception of content data	There are no specific rules. The general principles and rules on searches and seizures apply

(continued)

Table 11.2 (continued)

Council of Europe Convention on Cybercrime	The Philippines
Destruction of computer data	Section 17 of the Cybercrime Prevention Act of 2012. Destruction of Computer Data. Upon expiration of the periods as provided in Sections 13 and 15, service providers and law enforcement authorities, as the case may be, shall immediately and completely destroy the computer data subject of a preservation and examination

11.3.2 Investigatory Powers

Succinct Overview of Investigatory Powers

See Table 11.2.

Human Rights

The above investigatory powers are subject to fundamental constitutional rights and legal protections. Persons have the right against unreasonable searches and seizures. Article III Section 2 of the Philippine Constitution provides that:

> The right of the people to be secure in their persons, houses, papers, and effects against unreasonable searches and seizures of whatever nature and for any purpose shall be inviolable, and no search warrant or warrant of arrest shall issue except upon probable cause to be determined personally by the judge after examination under oath or affirmation of the complainant and the witnesses he may produce, and particularly describing the place to be searched and the persons or things to be seized.[89]

As a general rule, it is mandatory for law enforcement authorities to obtain a warrant from a judge or court in order to conduct searches and seizures. Furthermore, a person's right to privacy of communication and correspondence is protected. As stated in the Constitution: "The privacy of communication and correspondence shall be inviolable except upon lawful order of the court, or when public safety or order requires otherwise as prescribed by law".[90] A judicial warrant or court order is therefore needed to collect or intercept a person's communications or correspondence. In case any collection, interception, search or seizure does not comply with the constitutional and legal requirements, "[a]ny evidence obtained in violation of this or the preceding section shall be inadmissible for any purpose in any proceeding".[91] Criminal investigations and prosecutions must also respect the rights of the accused, including the right to due process,[92] the right of

[89] 1987 Constitution.

[90] 1987 Constitution, Article III sec 3(1).

[91] 1987 Constitution, Article III sec 3(2).

[92] 1987 Constitution, Article III sec 14(1).

presumption of innocence,[93] the right to a fair trial,[94] and the right against self-incrimination.[95]

Entrapment

Under Philippine law, there is an important distinction between entrapment and instigation. An arrest made pursuant to an entrapment is lawful, while one based on instigation or inducement is prohibited. According to the Supreme Court,

> Entrapment is the employment of such ways and means for the purpose of trapping or capturing a lawbreaker. Oftentimes it is the only effective way of apprehending a criminal in the act of the commission of the offense. In entrapment the idea to commit the crime originated from the accused. Nobody induces or prods him into committing the offense. A criminal is caught committing the act by ways and means devised by peace officers.
>
> It must be distinguished from inducement or instigation wherein the criminal intent originates in the mind of the instigator and the accused is lured into the commission of the offense charged in order to prosecute him. The instigator practically induces the would-be accused into the commission of the offense and himself becomes a co-principal. In entrapment ways and means are resorted to for the purpose of capturing the lawbreaker in *fragrante delicto*. In entrapment, the crime had already been committed while in instigation, it was not and could not have been committed were it not for the instigation by the peace officer.[96]

In another case, the Supreme Court further explains the difference between lawful entrapment as opposed to instigation:

> There is entrapment when law officers employ ruses and schemes to ensure the apprehension of the criminal while in the actual commission of the crime. There is instigation when the accused was induced to commit the crime. The difference in the nature of the two lies in the origin of the criminal intent. In entrapment, the *mens rea* originates from the mind of the criminal. The idea and the resolve to commit the crime comes from him. In instigation, the law officer conceives the commission of the crime and suggests to the accused who adopts the idea and carries it into execution.
>
> The legal effects of entrapment and instigation are also different. As already stated, entrapment does not exempt the criminal from liability. Instigation does.[97]

In light of the above court rulings, the use of Sweetie 2.0 to investigate webcam child sex abuse would be considered permissible and lawful entrapment. It is clear that the *mens rea* or criminal intent to commit sex-related offenses originated from the alleged sex offenders themselves. Since Sweetie 2.0 is only used in existing online fora and chatrooms known to be frequented by persons seeking to have online sex with children, whoever is controlling Sweetie 2.0 (whether its developer or law enforcement authorities) cannot be accused of instigating or inducing these people to commit crimes. The alleged offenders already had the intent to commit sex crimes, and the use of Sweetie 2.0 merely assisted in catching them in the act.

---

93 1987 Constitution, Article III sec 14(2).

94 1987 Constitution, Article III sec 14(2).

95 1987 Constitution, Article III sec 17.

96 *People v. Gatong-o.*

97 *Araneta v. Court of Appeals.*

11.3.3 *Succinct Overview of Investigatory Powers in an Online Context*

The general principles, rules and procedures for criminal investigations and prosecution are applicable as well in the online context. The Cybercrime Prevention Act of 2012 provides additional investigatory powers to law enforcement authorities concerning computer data and systems

Preservation and Disclosure of Computer Data
There are three types of computer data involved in an investigation: subscriber's information, traffic data, and content data. Subscriber's information is defined as:

> any information contained in the form of computer data or any other form that is held by a service provider, relating to subscribers of its services *other than traffic or content data* and by which identity can be established:
>
> The type of communication service used, the technical provisions taken thereto and the period of service;
>
> The subscriber's identity, postal or geographic address, telephone and other access numbers, any assigned network address, billing and payment information, available on the basis of the service agreement or arrangement; and
>
> Any other available information on the site of the installation of communication equipment, available on the basis of the service agreement or arrangement.[98]

On its part, traffic data or non-content data means "any computer data *other than the content of the communication* including, but not limited to, the communication's origin, destination, route, time, date, size, duration, or type of underlying service".[99] Content data "refers to the communication content of the communication, the meaning or purport of the communication, or the message or information being conveyed by the communication, *other than traffic data*".[100]

In relation to the preservation of computer data, Section 13 of the Cybercrime Prevention Act of 2012 requires that a service provider offering communications services must preserve traffic data and subscriber information for "a minimum period of six (6) months from the date of the transaction".[101] With respect to content data, it must be "preserved for six (6) months from the date of receipt of the order from law enforcement authorities requiring its preservation".[102] A court order is not required in either case. The preservation order may be extended once "for another six (6) months".[103] The law also requires the service provider "keep

98 Cybercrime Prevention Act of 2012, sec 3(o) (emphasis added).

99 Cybercrime Prevention Act of 2012, sec 3(p) and sec 12 (emphasis added).

100 Implementing Rules and Regulations of the Cybercrime Prevention Act of 2012, sec 3(m) (emphasis added).

101 Cybercrime Prevention Act of 2012, sec 13.

102 Cybercrime Prevention Act of 2012, sec 13.

103 Cybercrime Prevention Act of 2012, sec 13.

confidential the order and its compliance",[104] which rules out the possibility of the service provider notifying or informing the relevant subscriber or user about the order and the former's compliance with it.

Unlike a preservation order, a production order to disclose computer data has stricter requirements and may not be carried out on the basis of an order from law enforcement authorities. A production order requires the issuance of a "court warrant" and the order must be "in relation to a valid complaint officially docketed and assigned for investigation and the disclosure is necessary and relevant for the purpose of investigation".[105] A person or service providers who receives a valid production order must "disclose or submit subscriber's information, traffic data or relevant data in his/its possession or control within seventy-two (72) hours from receipt of the order".[106]

Search and Seizure of Computer Data

The general rule is that the search, seizure and examination of computer data must be undertaken pursuant to a court order or warrant.[107] On the basis of a validly issued warrant to search and seize computer systems and data and "[w]ithin the time period specified in the warrant", law enforcement authorities can carry out the following investigatory activities: "conduct interception"; "secure a computer system or a computer data storage medium"; "make and retain a copy of those computer data secured"; "maintain the integrity of the relevant stored computer data"; "conduct forensic analysis or examination of the computer data storage medium"; and "render inaccessible or remove those computer data in the accessed computer or computer and communications network".[108] The Supreme Court has ruled that these investigatory powers do not "appear to supersede existing search and seizure rules but merely supplements them".[109]

In addition, law enforcement authorities can require the action or assistance of a service provider or any third party. Under the law, law enforcement authorities "may order any person who has knowledge about the functioning of the computer system and the measures to protect and preserve the computer data therein to provide, as is reasonable, the necessary information, to enable the undertaking of the search, seizure and examination".[110] However, law enforcement authorities and service providers have the additional duty to "[u]pon the expiration of the periods as

104 Cybercrime Prevention Act of 2012, sec 13.
105 Cybercrime Prevention Act of 2012, sec 14.
106 Cybercrime Prevention Act of 2012, sec 14.
107 Cybercrime Prevention Act of 2012, sec 14.
108 Cybercrime Prevention Act of 2012, sec 15.
109 *Disini, Jr. v. Secretary of Justice.*
110 Cybercrime Prevention Act of 2012, sec 15.

provided... immediately and completely destroy the computer data subject of a preservation and examination" pursuant to a preservation order or a search warrant.[111]

11.3.4 Application of Relevant Investigatory Powers to the Sweetie Case

Any of the investigatory powers discussed above can be used or applied to cases of webcam child sex abuse and sex tourism involving real children and minors since they concern the investigation and prosecution of various sex-related crimes. There are grounds to undertake the search, seizure, preservation and production of computer systems and data in cases involving real children because a crime is either being, about to be, or has been committed. However, as discussed above, the only crimes applicable in the case of Sweetie 2.0 are grooming, luring, cybersex, and attempted cybersex. This means that if law enforcement authorities utilize Sweetie 2.0 to combat webcam child sex abuse, they will only be able to use these investigatory powers if these four crimes are involved. Short of this, there will be no legal basis for law enforcement authorities to search, intercept or collect data from or about suspected sex offenders since the latter are neither involved in the commission of any crime.

With regard to the admissibility of any evidence or data gathered, Sweetie 2.0's chat script may be used as evidence in criminal prosecutions for the crimes of grooming, luring, cybersex, and attempted cybersex. With respect to other sex-related crimes, any computer data gathered by law enforcement authorities could be deemed inadmissible since such investigations may be deemed invalidly undertaken since, in relation to Sweetie 2.0, those other sex crimes do not apply. Without a crime being, about to, and having been committed, there would be no legal or factual bases on the part of law enforcement to conduct an investigation. The Cybercrime Prevention Act of 2012 contains an exclusionary rule that clearly states that "[a]ny evidence procured without a valid warrant or beyond the authority of the same shall be inadmissible for any proceeding before any court or tribunal".[112]

It could be argued that any computer data collected by Sweetie 2.0's developer or operator would not be covered by the exclusionary rule since this rule only applies to unlawful searches and seizures carried out by government and not those conducted by private persons or entities.[113] The Supreme Court has ruled that, "the Bill of Rights cannot be invoked against acts of private individuals, being directed

[111] Cybercrime Prevention Act of 2012, sec 17.

[112] Cybercrime Prevention Act of 2012, sec 18.

[113] *People v. Marti.*

only against the government and its law-enforcement agencies as a limitation on official action".[114] As such, computer data gathered by a private entity like Sweetie 2.0's operator can be used as evidence to prosecute alleged sex offenders. It should be noted though that, even if the computer data collected through Sweetie 2.0 is admissible, the private person may be subject to a civil action for damages for violating the constitutional and human rights of suspected sex offenders. Under Philippine law, "[a]ny public officer or employee, *or any private individual*, who directly or indirectly obstructs, defeats, violates or in any manner impedes or impairs any of the following rights and liberties of another person shall be liable to the latter for damages" including "[t]he right to be secure in one's person, house, papers, and effects against unreasonable searches and seizures" and "[t]he privacy of communication and correspondence.[115] In its use of Sweetie 2.0, a private person may be violating the fundamental rights and liberties of people and may be liable to pay damages to the offended parties. The law does not provide any express or special exemptions or defenses to the claim for damages.

11.3.5 Relevant Aspects of Digital Forensic Evidence

The Rules on Electronic Evidence do not contain specific rules on the collection of evidence in an online context and the rules are mostly concerned with the introduction and admission of electronic documents in court proceedings.[116] There are also no formal standards or technical requirements for digital forensics.

11.4 Miscellaneous

The Anti-Child Pornography Act of 2009 imposes additional duties on internet services providers (ISPs) in relation to online child pornography. The law requires ISPs to "notify the Philippine National Police (PNP) or the National Bureau of Investigation (NBI) within seven (7) days from obtaining facts and circumstances that any form of child pornography is being committed using its server or facility".[117] ISPs must also "preserve such evidence for purposes of investigation and prosecution by relevant authorities".[118] Furthermore, they should, "upon the request of proper authorities, furnish the particulars of users who gained or attempted to gain access to an internet address which contains any form of child

[114] *People v. Domasian.*

[115] The Civil Code of the Philippines, Article 32.

[116] See Rules on Electronic Evidence.

[117] Anti-Child Pornography Act of 2009, sec 9.

[118] Anti-Child Pornography Act of 2009, sec 9.

pornography".[119] Finally, ISPs are required to "install available technology, program or software to ensure that access to or transmittal of any form of child pornography will be blocked or filtered".[120] Aside from ISPs, internet content hosts are also subject to positive duties. Internet content hosts have the legal responsibility to: "[n]ot host any form of child pornography on its internet address"; "[w]ithin seven (7) days, report the presence of any form of child pornography, as well as the particulars of the person maintaining, hosting, distributing or in any manner contributing to such internet address, to the proper authorities"; "[p]reserve such evidence for purposes of investigation and prosecution by relevant authorities"; and "upon the request of proper authorities, furnish the particulars of users who gained or attempted to gain access to an internet address that contains any form of child pornography"; and "remove any form of child pornography within forty-eight (48) hours from receiving the notice that any form of child pornography is hitting its server".[121] In relation to other business establishments, the Anti-Child Pornography Act of 2009 stipulates that "photo developers, information technology professionals, credit card companies and banks and any person who has direct knowledge of any form of child pornography activities" have a "duty to report any suspected child pornography materials or transactions to the proper authorities within seven (7) days from discovery thereof".[122] Sweetie 2.0's developer or operator can trigger the above duties of Philippine ISPs, internet content hosts and other business establishments by informing them about such facts and instances of child pornography on their systems and services that the developer or operator was able to gather through Sweetie 2.0. The ISPs, internet content hosts and other businesses would then be obligated to notify law enforcement and preserve and provide the necessary information and computer data to the authorities.

It is worth noting that child pornography is treated as a transnational crime under Philippine law and is potentially an extraditable offense.[123] In addition, the Philippines has extra-territorial jurisdiction over crimes punished under the Anti-Trafficking in Persons Act of 2003 (as amended) "even if committed outside the Philippines".[124] An act of trafficking is considered a "continuing offense",[125] which means that an offender may be arrested or searched at any time or place without need of a warrant.

[119] Anti-Child Pornography Act of 2009, sec 9.

[120] Anti-Child Pornography Act of 2009, sec 9.

[121] Anti-Child Pornography Act of 2009, sec 11.

[122] Anti-Child Pornography Act of 2009, sec 10.

[123] Anti-Child Pornography Act of 2009, secs 22 and 23.

[124] Anti-Trafficking in Persons Act of 2003 (as amended), sec 26-A.

[125] Anti-Trafficking in Persons Act of 2003 (as amended), sec 26-A.

11.5 Conclusions and Recommendations

In conclusion, a person engaged in webcam child sex abuse or sex tourism with real minors and children can be the subject of a criminal investigation and prosecution for violating a number of crimes under the Revised Penal Code and special laws, namely: child sexual abuse, child prostitution, child sexual exploitation, child pornography, pornographic performances of children, corruption of children (as accomplices), grooming, luring, qualified trafficking in persons, attempted trafficking in persons, child sex tourism, cybersex, and attempted cybersex. However, with regard to Sweetie 2.0, the possible crimes that may be committed are limited to cybersex, attempted cybersex, grooming, and luring because these are the only offenses that do not require the presence or involvement of an actual child or minor for their commission. Furthermore, since the general rule is that criminal investigations are only conducted when a crime has been, is being or is about to be committed, Sweetie 2.0 can only be used *in the first stance* to investigate these four offenses.

Nonetheless, while it may appear that the utility of Sweetie 2.0 to gather evidence about online child sexual abuse and prosecute offenders is somewhat limited, the crimes of cybersex and attempted cybersex actually cover many actions and preparatory activities that sex offenders normally undertake to commit many other sex-related crimes against children. This means that once these sex offenders perpetrate cybersex or attempted cybersex (i.e., any lascivious exhibition of sexual organs or sexual activity with the aid of a computer system), it would serve as a legal basis for law enforcement authorities to conduct further investigations not just into cybersex but also other sex-related crimes that these offenders may commit or have committed. By engaging or attempting to engage in cybersex, the offenders have committed a crime, which means that, in addition to giving grounds to arrest them, law enforcement authorities have probable cause to conduct further investigations, as well as to request a judge or court to issue the necessary orders or search warrants in connection with all of the possible crimes that these offenders may commit or have committed.

As a matter for law reform, it would be helpful in the fight against online child abuse for other countries to include cybersex in the catalogue of child sexual abuse offenses and make it a crime to commit or attempt to commit *cybersex with a child or someone who the offender believes to be a child.* With the growing use of digital and online technologies to commit child sex abuse, criminalizing cybersex with a child would improve the ability of the police and law enforcement bodies to undertake undercover operations and use advanced computer systems like Sweetie 2.0 to investigate and prosecute online sex offenders more effectively and efficiently and on a much larger scale.

With regard to procedural matters, Sweetie 2.0 can be lawfully used in the first instance to investigate sex-related crimes against children such as cybersex, attempted cybersex, grooming and luring. Since entrapment is permissible under Philippine law, Sweetie 2.0 can be used to catch sex offenders in the act of committing these offenses. In order to make to the use of Sweetie 2.0 applicable to the crimes of grooming and luring, it would be advisable to design Sweetie 2.0 to be semi-automated or not completely autonomous so that Sweetie 2.0's human operator would fall under the category of "someone who the offender believes to be a child".[126]

Finally, it is recommended that Sweetie 2.0's developer or operator work together with law enforcement authorities when using Sweetie 2.0 to investigate, identify or deter webcam child sex abuse or online sex tourism. Otherwise, the developer or operator could potentially find itself liable for the crimes of pandering and photo or video voyeurism. Moreover, by undertaking investigations on its own and without the cooperation or assistance of law enforcement authorities, Sweetie 2.0's developer or operator could be charged with violating the Data Privacy Act of 2012 for unauthorized processing and disclosure of sensitive personal data of suspected sex offenders. Furthermore, it could find itself civilly liable for damages for violating the rights and liberties of the latter.

Table of Case Law and Legislation

1987 Constitution
Access Devices Regulation Act of 1998
Anti-Child Pornography Act of 2009
Anti-Photo and Video Voyeurism Act of 2009
Anti-Trafficking in Persons Act of 2003 (as amended)
Araneta v. Court of Appeals, G.R. No. L-46638, July 9, 1986
Civil Code of the Philippines
Cybercrime Prevention Act of 2012
Data Privacy Act of 2012
Disini, Jr. v. Secretary of Justice, G.R. No. 203335, February 18, 2014
Dunlao, Sr. v. Court of Appeals, G.R. No. 111343, August 22, 1996
Electronic Commerce Act
Implementing Rules and Regulations of the Cybercrime Prevention Act of 2012
Intestate Estate of Vda. de Carungcong v. People, G.R. No. 181409, February 11, 2010
Jacinto v. People, G.R. No. 162540, July 13, 2009
Garcia v. Court of Appeals, G.R. No. 157171, March 14, 2006
Malto v. People, G.R. No. 164733, September 21, 2007
Olivarez v. Court of Appeals, G.R. No. 163866, July 29, 2005
People v. Domasian, G.R. No. 95322, March 1, 1993
People v. Gatong-o, G.R. No. 78698, December 29, 1988
People v. Larin, G.R. No. 128777, October 7, 1998
People v. Marti, G.R. No. 81561, January 18, 1991
People v. Valdez, G.R. Nos. 216007-09, December 8, 2015
Revised Rules of Criminal Procedure
Rules and Regulations on the Reporting and Investigation of Child Abuse Cases

[126] Anti-Child Pornography Act of 2009, secs 3(h) and 3(i).

Rules on Electronic Evidence
Special Protection of Children Against Abuse, Exploitation and Discrimination Act
The Revised Penal Code (as amended)
The Anti-Rape Law of 1997

Michael Anthony C. Dizon is a Lecturer in Law at the University of Waikato, New Zealand. He previously worked as an information and communications technology lawyer and researcher for institutions and organisations in the Netherlands, the United Kingdom, and the Philippines. His research mainly involves the socio-legal study of technology, creativity and innovation. He has published articles and presented papers on a wide range of law and technology related topics including hacking, open source software, computer crime, privacy and data protection, and technology regulation.

Chapter 12
Substantive and Procedural Legislation in the United States of America to Combat Webcam-Related Child Sexual Abuse

Jonathan Unikowski

Contents

Abstract This chapter studies the legality of Sweetie 2.0 under the Federal Law of the United States of America. We review both substantive and procedural criminal law, and formulate recommendations for the instigators of the project. We conclude that the design of Sweetie 2.0 that was used as a basis for this analysis comprises several flaws, both from a legal and policy standpoint. In order to address these flaws, we suggest a re-design of Sweetie as a semi-supervised bot, instead of a fully automated one.

J. Unikowski (✉)
Lexloci Inc., New York, USA
e-mail: jonathan@lexloci.com

S. van der Hof et al. (eds.), *Sweetie 2.0*, Information Technology and Law Series 31,
https://doi.org/10.1007/978-94-6265-288-0_12

Keywords Sweetie · Criminal law · Webcam · Child sexual abuse · United States of America · Automatic enforcement · Enticement · Entrapment

12.1 Introduction: Legislation in the United States of America

12.1.1 General Description of the Legal Framework

The United States of America's legal framework is characterised by two main aspects: it is a common law legal system and the constitutional structure of the country is that of a federal state.

The United States of America's legal system is a common law legal system. As a result, decisions of high courts are binding on lower courts, and can only be changed by the court that took the initial decision or a higher court. This means that particular attention should be paid to the higher jurisdictions, and especially, as far as federal law is concerned, to the decisions of the U.S. Supreme Court and the U.S. Court of Appeals.[1]

The USA is also a Federal state. As is the case in all federal states, the subject of the repartition of powers between the Federal Government and the individual states is an extremely controversial topic. In the USA, the Federal Government has enumerated powers, which are held to be drawn directly from the people, and not the individual states. Among the enumerated powers of the federal government, one that should retain our attention is the federal commerce clause. Indeed, under Article 1, Section 8, Clause 3 of the U.S. Constitution, the Federal Congress has the power to "regulate commerce with foreign nations, and among the several states, and with the Indian tribes". This simple clause is the support for numerous legislation of the Federal Government, and in particular a large set of criminal legislation that regulates the Internet. This is remarkable, especially since the Constitution does not gives a direct power to pass laws in criminal areas to the federal government.

The fact that this clause supports most of the criminal legislation applicable to crimes committed online explains why most statutes referred to in this chapter are only applicable where the "facilities or means of interstate commerce" are used, or where some transportation across state lines or in foreign commerce occurs. Since the Internet is by nature an interstate network, most of the crimes that can be

[1] The Court of Appeals are organised by circuits, which are dependent from one another. As a result, in the absence of a decision of the U.S. Supreme Court, the interpretation of a given law can vary circuit by circuit.

committed using computers connected to the Internet have be deemed to fall within the confines of the Federal Commerce Clause. As a result, it is pertinent to focus our attention on federal legislation. It shall be noted, however, that the each of the 50 states have a multitudes of other criminal statutes, which may be applicable to the crimes that Sweetie is designed to combat.

Finally, some principles of constitutional law exert significant pressures on what the legislator can actually do. First and foremost, the right of free expression enshrined in the First Amendment to the US Constitution creates considerable difficulties for the regulation of speech, which in turn accounts for complex statutes, of sometimes poor readability, in area involving speech (areas that sometimes border with certain types of "pornographic" material involving the representation of children).

12.1.2 Relevant Treaties and Cybercrime Laws

The USA has ratified some of the following relevant treaties:

(a) The USA has signed and ratified the Council of Europe's Convention on Cybercrime (entered into force on January 1, 2007). The USA emitted several reservations to this convention.[2]

(b) The USA has signed the Convention on the Rights of the Child in 1995. Remarkably, it is the only member of the UN which has not ratified this convention.[3] This lack of ratification has led to controversies both in the USA and internationally.[4]

(c) The USA has signed and ratified the Optional Protocol to the Convention on the Rights of the Child on the sale of children, child prostitution and child pornography. The USA has emitted several reservations to this Optional Protocol.[5]

(d) The USA has not signed the Council of Europe's Convention on the Protection of Children against Sexual Exploitation and Sexual Abuse.

---

[2] See the Council of Europe's website at http://www.coe.int/en/web/conventions/full-list/-/conventions/treaty/185/declarations?p_auth=Nyr2hPP6.

[3] See https://treaties.un.org/Pages/ViewDetails.aspx?src=TREATY&mtdsg_no=IV-11&chapter=4&lang=en.

[4] Rutkow and Lozman 2006, p. 161.

[5] See the UN's website, https://treaties.un.org/Pages/ViewDetails.aspx?src=TREATY&mtdsg_no=IV-11-c&chapter=4&lang=en#EndDec.

12.2 Analysis of Substantive Criminal Law

12.2.1 Introduction

Federal criminal law is codified in title 18 of the United States Code ("U.S.C."). This title contains both a substantive (part I) and procedural statutory criminal law (part II). The offences relevant for this chapter can be found in the following Chapters of part I of title 18:[6]

(e) 18 U.S.C. Chapter 71—Obscenity
(f) 18 U.S.C. Chapter 109a—Sexual Abuse
(g) 18 U.S.C. Chapter 110—Sexual Exploitation and Other Abuse of Children
(h) 18 U.S.C. Chapter 117—Transportation for Illegal Sexual Activity and Related Crimes.

Although these statutes codify the law applicable to the sexual abuse of minors, they fail to provide an unique definition of what a "minor" is. Instead, definitions can be found for each chapter in a dedicated article, or within the definition of the infractions themselves. In some instances, a "minor" is defined as a person under the age of eighteen years,[7] whereas in other cases a minor is defined as a person under the age of sixteen years.[8] Moreover, in one case, the definition of "minor" is simply not given in the statute, so that courts have had to define it through a construction of Congress's intent.[9]

12.2.2 Possibly Relevant Criminal Offences

Succinct Overview of Sexual Offences Involving Minors
Table 12.1 lists the possibly relevant provisions of criminal law in the United States of America, grouped together by the provisions of the Lanzarote Convention, which gives the most comprehensive catalogue of sexual child-abuse offences available.

Overview of Sexual Offences Related to Webcam Child Sexual Abuse
In this section, we provide a summary of the elements of each of the offences listed above, along with the case law pertaining to webcam-related child abuse.

---

[6] A large number of these statutes are extremely long. As a result, in order to keep the length of this chapter within the guidelines of the project's managers, we only included the most relevant statutes in the annex. Regardless, one can easily find all of the statutes referred to in this chapter on the Legal Information Institute's website at https://www.law.cornell.edu/uscode/text.

[7] 18 U.S.C. Chapter 110; cf. 18 U.S.C. § 2256(1).

[8] 18 U.S.C. § 2425; 18 U.S.C. § 1470

[9] 18 U.S.C. § 1466A; holding that age to be 18 years (*United States v. Handley*, 564 F. Supp. 2d 996, 1003 (S.D. Iowa 2008)).

Table 12.1 Overview of sexual offences involving minors [*Source* The author]

Lanzarote treaty	U.S. Federal law
Article 18. Sexual abuse	18 U.S.C. § 2243: sexual act consisting of penetration ("however slight") or certain touching of genitalia or anus, with a person who has attained the age of 12 years but has not attained the age of 16 years and is at least four years younger than the person performing the act 18 U.S.C. § 2241: aggravation of § 2243 where force, threat, or other means rendering the victim unable to control his or her conduct (e.g. drugs) are used; or where the victim has not attained the age of 12 18 U.S.C. § 2244: intentional touching of a person under the age of 16, either directly or through the clothing, of the genitalia, anus, groin, breast, inner thigh, or buttocks, with an intent to abuse
Article 19. Offences concerning child prostitution	18 U.S.C. § 2423(a): transportation of a person who has not attained the age of 18 years with intent that the person engage in prostitution 18 U.S.C. § 2423(b): travel to the United States, or, for U.S. citizens or permanent residents, to a foreign country, for the purposes of engaging in "any illicit sexual conduct" with a person who has not attained the age of 18 18 U.S.C. § 2423(c): U.S. citizen or permanent resident engaging in "any illicit sexual conduct" with a person who has not attained the age of 18 in a foreign country
Article 20. Offences concerning child pornography	18 U.S.C. § 2252: possession, reception, transportation, and distribution of "any visual depiction", the producing of which "involves the use of a minor engaging in sexually explicit conduct" 18 U.S.C. § 2252A: possession, transportation, and distribution of "child pornography" 18 U.S.C. § 1466A: offences regarding obscene visual representations of the sexual abuse of children
Article 21. Offences concerning the participation of a child in pornographic performances	18 U.S.C. § 2251: recruitment, use, or coercing of a person under the age of 18 years to engage in any sexually explicit conduct for the purpose of producing any visual depiction of such conduct or for the purpose of transmitting a live visual depiction of such conduct 18 U.S.C. § 2260: provisions similar to § 2251 that apply to the cases where such content is produced by a person located outside the United States and would be imported in the United States

(continued)

Table 12.1 (continued)

Lanzarote treaty	U.S. Federal law
Article 22. Corruption of children	18 U.S.C. § 1470: transfer of obscene material to an individual who has not attained the age of 16 years
Article 23. Solicitation of children for sexual purposes	18 U.S.C. § 2422: coercion or enticement of a person who has not attained the age of 18 years, using any facility or means of interstate or foreign commerce, "to engage in prostitution or any sexual activity for which any person can be charged with a criminal offence"
[other offences, not covered by the Lanzarote Convention]	18 U.S.C. § 2425: transmission of information about a person who has not attained the age of 16 years with the intent to facilitate a criminally punishable sexual activity (including the production of child pornography, per 18 U.S.C. § 2427)

Sexual Abuse

The U.S. Code contains several sexual abuse statutes, which apply in different situations, notably depending on the victim's age. If the victim has attained the age of 12 years but has not attained yet the age of 16, the applicable statute is 18 U.S.C. § 2243—in such case, the statute also requires that the victim be at least 4 years younger than the perpetrator. On the other hand, if the victim is younger than 12 years of age, such an abuse is automatically an "aggravated" sexual abuse, falling under 18 U.S.C. § 2241.[10]

It should be noted that the prosecution does not need to demonstrate that the defendant knew the age of the victim at the time of the abuse.[11] However, where the victim is older than 12 years old, the perpetrator can establish by "a preponderance of the evidence" that he or she "reasonably believed that the other person had attained the age of 16 years".[12]

In both cases, sexual abuse requires a sexual act, the definition of which is given in 18 U.S.C. § 2246(2).[13] In addition to sexual abuse, Section 2244 prohibits "sexual contacts," which are defined in Section 2246(3).[14] The penalties for "sexual contacts" vary depending on the age of the victim, in a fashion similar to that of sexual abuse charges.

As noted above, the territorial scope of these provisions are limited. In this case, they applied to crimes committed on federal land, and, in the case of aggravated

[10] This latter statute also applies to children between the age of 12 and 16 years where force, threat, or other means (such as intoxication) are used.

[11] 18 U.S.C. § 2241(d) and 18 U.S.C. § 2243(d).

[12] 18 U.S.C. § 2243(c)(1).

[13] See annex.

[14] See annex.

abuse, where the perpetrator has crossed a state line with the intent to commit the crime.

Given the definition of these offences, it is not surprising that case law dealing with webcam sex is virtually inexistent, since these offences require a physical contact between the perpetrator and the victim.[15]

Offences Concerning Child Prostitution

The offences concerning child prostitution are consolidated at 18 U.S.C. § 2423.

As noted above, U.S. criminal federal law deals with instances where a crime has an interstate or foreign dimension. As a result, the offences concerning child prostitution cited at 18 U.S.C. § 2423 involve situations where there is a transportation of either the victim or the author of the crime, either in interstate commerce or outside the USA.

When it comes to child sexual abuse through the use of webcam, it is assumed that the perpetrator will not travel (or attempt to travel) to engage in intercourse with the victim. As a result, these offences are not discussed in further detail for the purposes of this chapter.[16]

Offences Concerning Child Pornography

The U.S. Code contains several statutes with offences concerning child pornography (18 U.S.C. § 2252, 18 U.S.C. § 2252A, and 18 U.S.C. § 1466A). These statutes are poorly drafted, and extremely hard to read and understand.[17]

For the purposes of this chapter, it suffices to say that 18 U.S.C. § 2252 punishes the possession, reception, transportation, and distribution of "any visual depiction", the producing of which "involves the use of a minor engaging in sexually explicit conduct".

Conversely, 18 U.S.C. § 2252A punishes similar acts, but uses the term "child pornography" instead of the concept of "visual depiction". The definition of "child pornography" is broader than the concept of "visual depiction", since it includes not only materials produced through the abuse of real minors, but also "digital image, computer image, or computer-generated image that is, or is indistinguishable from, that of a minor engaging in sexually explicit conduct" and "visual depiction[s] [that have] been created, adapted, or modified to appear that an identifiable minor is engaging in sexually explicit conduct."[18]

A third statute completes this statutory scheme. 18 U.S.C. § 1466A prohibits the production, distribution, reception, or possession with intent to distribute of "a visual depiction of any kind, including a drawing, cartoon, sculpture, or painting,

[15] We were not able to identify cases where someone watching webcam sex abuse of a child was condemned for conspiring or aiding and abetting to the principal offence. Moreover, such a situation significantly deviates from Sweetie's use cases.

[16] See annex for a reproduction of 18 U.S.C. § 2423.

[17] Given their importance for this chapter, these statutes are reproduced at length in the annex.

[18] 18 U.S.C. § 2256(8).

that [a] depicts a minor engaging in sexually explicit conduct and is obscene; or [b] depicts an image that is, or appears to be, of a minor engaging in graphic bestiality, sadistic or masochistic abuse, or sexual intercourse, including genital-genital, oral-genital, anal-genital, or oral-anal, whether between persons of the same or opposite sex; and lacks serious literary, artistic, political, or scientific value."

In contrast with the two other statutes, 18 U.S.C. § 1466A also encompasses depictions that are distinguishable from depictions of real minors (e.g. cartoons). It should be noted that this statute does not define the term "minor", and that has a result courts have had to determine the proper interpretation of the term: the term minor should be interpreted as "a person who has not reached 18 years of age."[19]

Courts have held that a a "live video feed [transmitted] over the internet" constitutes a "visual depiction" for the purposes of 18 U.S.C. § 2256(5).[20] Although no similar cases can be found for 18 U.S.C. § 1466A, the similar wording of the statute would likely lead courts to the same result.

Although these statutes may appear to allow for the prosecution on grounds of possession or production of child pornography picturing virtual children, it is unclear whether such prosecution is likely to succeed. Indeed, a previous version of 18 U.S.C. § 2252A was challenged before the U.S. Supreme Court, leading to a landmark case, *Ashcroft v. Free Speech Coalition*.[21] In that case, the U.S. Supreme Court struck the previous definition of child pornography applicable to 18 U.S.C. § 2252A on the grounds that the prohibition of child pornography that pictures virtual children, and whose production did not involve real children, was an impermissible restriction to free speech.[22] This decision might have an impact on the renewed definitions that are now applicable. Indeed. the current definition of child pornography under 18 U.S.C. § 2256(8)(B), as well as the definition used in 18 U.S.C. § 1466A have not yet been subject to challenge before the U.S. Supreme Court,[23] and it is thus unclear whether prosecutions on the basis of these statutes are likely to succeed.

Offences Concerning the Participation of a Child in Pornographic Performances

18 U.S.C. § 2251 criminalises the production of child pornography. This statute, of poor readability, is subdivided in several sub-sections.[24] For the purposes of this chapter, it suffices to say that this statute punishes whomever "employs, uses, persuades, induces, entices, or coerces any minor to engage in, or who has a minor

[19] *United States v. Handley*, 564 F. Supp. 2d 996, 1003 (S.D. Iowa 2008).

[20] *United States v. Nichols*, 371 Fed. Appx. 546, 548 (5th Cir. 2010); *United States v. Tucker*, 305 F.3d 1193, 1204 n. 15 (10th Cir. 2002).

[21] *Ashcroft v. Free Speech Coal.*, 535 U.S. 234 (2002).

[22] See for a more thorough discussion 2 A.L.R. Fed. 2d 533 ("Validity, Construction, and Application of 18 U.S.C.A. § 2252A(a), Proscribing Certain Activities Relating to Material Constituting or Containing Child Pornography").

[23] 7 A.L.R. Fed. 2d 1.

[24] See annex for a complete reproduction of the statute.

assist any other person to engage in, any sexually explicit conduct (…) for the purpose of producing any visual depiction of such conduct." The most important distinction for the purposes of this chapter is the territorial application of the statute. 18 U.S.C. § 2251(a) addresses cases where the visual depiction is produced with goods in interstate or foreign commerce, or where the depiction is transported or transmitted "using any means or facility of interstate or foreign commerce or in or affecting interstate or foreign commerce or mailed." Conversely, 18 U.S.C. § 2251 (c) addresses cases where the depiction is produced outside of the United States and where it is the intent of the perpetrator that the depiction be transported to the United States, or where the depiction is effectively transported to the United States.

It should be noted that the same acts are punished where they take place outside of the United States. Indeed, 18 U.S.C. § 2260 punishes the production and the possession of child pornography where there is an intent to transport such material into the United States. As courts have noted, "[p]unishing the creation of child pornography outside the United States that is actually, is intended to be, or may reasonably be expected to be transported in interstate or foreign commerce is an important enforcement tool."[25]

The attempt and conspiracy to commit the crimes listed in 18 U.S.C. § 2251 is punished.[26] Defendants have been successfully prosecuted for attempt to commit the crimes listed in that statute in cases where they were interacting with undercover agents.[27]

Corruption of Children

The U.S. Code contains one statute punishing the corruption of children. Under 18 U.S.C. § 1470, the fact of "knowingly" transferring obscene matter to another individual who has not attained the age of 16 years is punished by a sentence of up to 10 years. The attempt to commit this crime is punishable.

For this provision to be applicable, the perpetrator must know that the recipient has not attained the age of 16 years. Courts have made clear that the perpetrator's belief that the person to whom obscene matter is transferred was under the age of sixteen is sufficient to convict him of attempt, "even if the recipient was actually an adult."[28]

The transfer has to occur using "any facility or means of interstate or foreign commerce". With respect to this specific disposition, courts have noted that "it is beyond debate that the Internet and email are facilities or means of interstate commerce."[29]

---

25 *United States v. Thomas*, 893 F.2d 1066, 1069 (9th Cir. 1990).

26 18 U.S.C. § 2251(e).

27 *United States v. Pierson*, 544 F.3d 933, 935 (8th Cir. 2008); *United States v. Johnson*, 376 F.3d 689, 696 (7th Cir. 2004).

28 *United States v. Spurlock*, 495 F.3d 1011, 1013 (8th Cir. 2007). This decision was based on other cases involving different statutory provisions (i.e. the statutes punishing enticement).

29 *United States v. Barlow*, 568 F.3d 215, 220 (5th Cir. 2009).

What constitutes obscene material under this statute is not clearly defined. Although the case law on the definition of "obscene" for the purposes of this statute is scarce, it seems that the criterions used in the context of speech restrictions by the U.S. Supreme Court may be applicable.[30] Phrased this way, the question would thus turn on whether the average person, applying contemporary community standards, would find that the material, taken as a whole, appeals to the prurient interest; whether the work depicts or describes, in a patently offensive way, sexual conduct or excretory functions; and whether the material, taken as a whole, lacks serious literary, artistic, political, or scientific value.[31] This rather vague definition calls for a careful case-by-case analysis. Although the U.S. Supreme Court did not lay down precisely what kind of material can be deemed "patently offensive", it nonetheless offered guidance on the type of material that would fall under that definition:

(a) Patently offensive representations or descriptions of ultimate sexual acts, normal or perverted, actual or simulated.
(b) Patently offensive representation or descriptions of masturbation, excretory functions, and lewd exhibition of the genitals.[32]

It seems beyond discussion that a video of someone masturbating would be obscene under this definition.

Solicitation of Children for Sexual Purposes
18 U.S.C. § 2422(b) criminalises the persuasion, inducement, coercion or enticement of a person who has not attained the age of 18 years, using any facility or means of interstate or foreign commerce, "to engage in prostitution or any sexual activity for which any person can be charged with a criminal offence." The attempt is punished. Under this section, the word "inducement" is to be understood in its usual meaning.[33]

Numerous cases have made clear that attempts are punishable even where no actual minor exists, or where the defendant believed mistakenly that the victim was a minor, or where the "minor" in question was actually a law enforcement officer.[34]

In at least one case, this statute has successfully been used to prosecute an individual who masturbated in front of a webcam for a recipient whom he believed

---

[30] *United States v. Rudzavice*, 586 F.3d 310, 315 (5th Cir. 2009).

[31] See *Miller v. California*, 413 U.S. 15 (1973).

[32] *Op. cit.* at 25.

[33] *Batsell v. U.S.*, 403 F.2d 395 (8th Cir. 1968)

[34] See, among a plethora of cases, *United States v. Gagliardi*, 506 F.3d 140, 145 (2d Cir. 2007); *United States v. Tykarsky*, 446 F.3d 458, 466 (3d Cir. 2006); *United States v. Spurlock*, 495 F.3d 1011, 1013 (8th Cir. 2007); *United States v. Sims*, 428 F.3d 945, 960 (10th Cir. 2005); *United States v. Farley*, 607 F.3d 1294, 1325 (11th Cir. 2010); *United States v. Cote*, 504 F.3d 682, 687 (7th Cir. 2007).

was a minor.[35] Since the recipient was actually an undercover agent, the individual was convicted of attempt to commit the crime of enticement.

Other Offences Not Covered by the Lanzarote Convention
It should be noted that one additional statute prohibits the transmission of information about an individual under the age of 16 years for the purposes of furthering the commission of an illicit sexual activity.[36] The case law analysing this statute is scarce, and of little use for the purposes of this chapter.

12.3 Conclusion

For the purposes of this section, it is assumed that the perpetrator is located in the United States.

If the perpetrator induces or forces the minor to display breasts or genitals or to perform sexual activities (e.g., masturbate) in front of the webcam, this may constitute:

(i) Production of child pornography (18 U.S.C. § 2251):

1. There must be employment, use, persuasion, inducement, enticement, or coercion of a minor under the age of 18 years
2. The act depicted must fall within the acts listed in 18 U.S.C. § 2256(2)(A)
3. The perpetrator must use the means of interstate commerce (e.g. the Internet)
4. Note: the attempt is punishable

(j) Coercion or enticement (18 U.S.C. § 2422):

5. There must be persuasion, inducement, enticement, or coercion
6. The victim must be a minor under the age of 18 years
7. The minor must be forced to engage in prostitution or "any sexual activity for which any person can be charged with a criminal offense"
8. The perpetrator must use the means of interstate commerce (e.g. the Internet)
9. Note: the attempt is punishable

---

35 *United States v. Cochran*, 534 F.3d 631 (7th Cir. 2008).

36 18 U.S.C. § 2425 prohibits the transmission, using the mail or any facility or means of interstate or foreign commerce, of "the name, address, telephone number, social security number, or electronic mail address of another individual, knowing that such other individual has not attained the age of 16 years, with the intent to entice, encourage, offer, or solicit any person to engage in any sexual activity for which any person can be charged with a criminal offense." The attempt of the same crime is also prohibited.

(k) Reception of child pornography (18 U.S.C. § 2252):

- Generally, reception suffices to enable prosecution under this statutes
- The act depicted must fall within the acts listed in 18 U.S.C. § 2256(2)(A)
- Could also lead to possession depending on technical means of transmission and storage (if only transient storage, prosecution is unlikely to be able to find evidence of possession per se)
- Note: the attempt is punishable

(l) Sexual abuse (18 U.S.C. § 2241/18 U.S.C. § 2243) or sexual contact (18 U.S. C. § 2244):

10. Hardly possible if only two parties, located in different places, are involved
11. Conceivable as inchoate offence
12. Scenario: the perpetrator is acting with a third party located in the physical vicinity of the minor, and who performs the acts on the minor
13. Note: the attempt is punishable
    If the perpetrator shows his genitals or masturbates in front of the webcam, this may constitute:

(m) Transfer of obscene material to minor (18 U.S.C. § 1470)

14. The perpetrator must have knowledge that the victim has not attained the age of 16 years
15. Masturbating in front of a camera is undoubtedly "obscene" for the purposes of 18 U.S.C. § 1470
16. Note: the attempt is punishable

(n) Coercion or enticement (18 U.S.C. § 2422):

17. There must be persuasion, inducement, enticement, or coercion
18. The victim must be a minor under the age of 18 years
19. The minor must be forced to engage in prostitution or "any sexual activity for which any person can be charged with a criminal offense"
20. The perpetrator must use the means of interstate commerce (e.g. the Internet)
21. Note: the attempt is punishable

12.3.1 Possible Obstacles in Substantive Law Concerning Sweetie

The most promising route for incriminating people interacting with Sweetie in order to obtain from her the performance of certain sexual acts would be to pursue them under the inchoate offence of attempt. There is no general crime of attempt in federal law, but as noted above (Sect. 12.2.2), all the statutes that could potentially be applicable to an individual interacting with Sweetie specifically punish attempts to commit the crimes they contain.

Defendants prosecuted for attempting to commit a crime have sometimes used defences of impossibility.[37] In the case of Sweetie, the argument would be that since Sweetie is not a child, it would be impossible to attempt to victimise her through the commission of the crimes listed above. This argument is, however, unlikely to succeed. Individuals have been successfully prosecuted for attempting to commit the crimes established the provisions listed above (Sect. 12.2.2) in cases where they believed that they were interacting with a real child, but were in fact interacting with an undercover law enforcement official. The situation of someone interaction with Sweetie is not different from that of someone interacting with an undercover agent. Indeed, Sweetie purports to be a child even though she is not—she is a computer program designed and controlled by adults. From the perpetrator's point of view, these two scenarios are indistinguishable, and indeed this distinction also seems irrelevant when it comes to the legal definition of attempt in American law.

Punishable attempt traditionally include an intent to commit the attempted crime and a "substantial step" in the direction of the commission of the crime.[38] Courts have described that substantial step as "some overt act adapted to, approximating, and which in the ordinary and likely course of things will result in, the commission of the particular crime."[39] The fact that the interlocutor of the perpetrator is a piece of software does not seem to be likely to impact the kind of "substantial steps" the perpetrator might take, or his or her intent.

12.4 Analysis of Criminal Procedure Law

12.4.1 General Description of Legal Framework

Procedural criminal law is primarily driven by the Fourth, Fifth, Sixth, and Eighth Amendments to the U.S. Constitution. These amendments are applicable equally to the federal government and the individual states. As a result, they provide a good "bird's eye" view of the general traits of U.S. criminal procedure law.

---

[37] The defence of impossibility has some complex technical ramifications which would be irrelevant for the narrow purpose of this chapter. See, for a good summary, Doyle (2015), 'Attempt: An Overview of Federal Criminal Law', available at https://www.fas.org/sgp/crs/misc/R42001.pdf; see also Rogers (2004), 'New Technology, Old Defenses: Internet Sting Operations and Attempt Liability', University of Richmond Law Review (38), 477–523. See, for application to some of the statutes listed above, United *States v. Tykarsky*, 446 F.3d 458, 466 (3d Cir. 2006); *United States v. Farner*, 251 F.3d 510, 512 (5th Cir. 2001).

[38] *Braxton v. United States*, 500 U.S. 344 (1991).

[39] *United States v. Manley*, 632 F.2d 978, 988 (2d Cir. 1980).

Most investigations start with some form of search. Searches conducted by law enforcement officers must comply with the requirement of the Fourth Amendment (see below). In most cases, law enforcement officers will have to apply for a warrant before they can conduct a search. The application for the warrant must be accompanied by a sworn statement detailing the purpose and scope of the search. The warrant can then be issued, if probable cause is found, by an independent magistrate.

When the search is followed by the arrest of an individual, a number of other constitutional rights come into play. Indeed, once a suspect is arrested, certain rights are recognised to him by the Fifth Amendment, which provides that no person "shall be compelled in any criminal case to be a witness against himself". The most typical example is the reading of "Miranda rights" (named after the Supreme Court decision that established the basic requirements for such a process), which informs the suspect that he or she has the right to remain silent and consult an attorney before speaking to the police.

The individual will then be formally indicted. The indictment is issued by a grand jury, which examines cases while they are still at the inquiry stage. The grand jury's proceedings are secret.[40]

The decision whether to prosecute can be made by the police or by the prosecutor. Indeed, in minor cases, the police officer investigating the offence can generally decide to prosecute or not (for example, by issuing a "ticket" for a traffic infraction) This does not mean, however, that the police does not have an influence of more important case. Indeed, the choice to report a case to a prosecutor is once again a discretionary decision resting on the police's shoulder. Idem victim. Moreover, at the federal level, felony cases (crimes which carry a punishment of more than one year or death) can be charged by prosecutors only with the agreement of a grand jury.[41] In such cases, at the Federal level, the pre-trial phase also include a preliminary hearing which aims at determining whether the prosecution has sufficient evidence to carry the case through the next case of the procedure. The determination is made by a magistrate.

12.4.2 *Investigatory Powers*

Succinct Overview of Investigatory Powers

See Table 12.2.

Human Rights

The fundamental rights contained in the U.S. Constitution of the United States take the form of a protection of the individual's rights against the intrusions of the

---

[40] See generally 1 Crim. Proc. Chap. 8 (4th edn.)

[41] 4 Crim. Proc. § 13.1(a).

Table 12.2 Overview of investigatory powers [*Source* The author]

Council of Europe Convention on Cybercrime	United States of America
Article 16. Expedited preservation of stored computer data	18 U.S.C. § 2704: order for requiring the creation of a backup copy of the contents of electronic communications in order to preserve them
Article 17. Expedited preservation and partial disclosure of traffic data	See supra, 18 U.S.C. § 2704
Article 18. Production order	18 U.S.C. § 2703: order for requiring disclosure of electronic communications or records
Article 19. Search and seizure of stored computer data	See supra, 18 U.S.C. § 2703; see also Fourth Amendment law, below
Article 20. Real-time collection of traffic data	18 U.S.C. Chapter 206: Pen Register and Trap And Trace Devices
Article 21. Interception of content data	18 U.S.C. Chapter 119: Wire and electronic communications interception and interception of oral communications
[Other (special) investigatory powers, not covered by the Cybercrime Convention, such as undercover operations.]	

Government. These rights are part of the "Bill of Rights," which includes a series of amendments to the U.S. Constitution adopted promptly after the Independence, in 1791. For the purposes of this chapter, we focus narrowly on the Fourth Amendment to the U.S. Constitution.

The Fourth Amendment reads as follows:

> The right of the people to be secure in their persons, houses, papers, and effects, against unreasonable searches and seizures, shall not be violated, and no Warrants shall issue, but upon probable cause, supported by Oath or affirmation, and particularly describing the place to be searched, and the persons or things to be seized.

The corollary of the Fourth Amendment is the exclusionary rule that sanction its violations. Where evidence is obtained in a manner which violates a defendant's Fourth Amendment rights, the evidence cannot be used by the prosecution at trial to establish the defendant's guilt.[42]

Modern Fourth Amendment law rests upon the notion of "reasonable expectation of privacy." This notion, pioneered by a concurring opinion in a landmark U.S. Supreme Court case, *Katz v. United States*,[43] means that the Fourth Amendment is only applicable where the thing to be searched or seized is such that the person who later opposes the search has a subjective expectation of privacy which society is

[42] *Weeks v. United States*, 232 U.S. 383 (1914).

[43] *Katz v. United States*, 389 U.S. 347 (1967).

willing to recognise as reasonable. Such is the case, for instance, of one's home. In the electronic realm, as a general rule, courts are willing to find that defendants have a reasonable expectation of privacy in the computers they own.[44] Conversely, storing information in a third-party's computer does not give rise to a reasonable expectation of privacy, since the third-party's very existence effectively destroys such an expectation.[45] In the same vein, information such as IP addresses, which are revealed, provided to, or by an internet provider, are not subject to the Fourth Amendment's expectation of privacy.[46]

Under the Fourth Amendment, law enforcement agents must generally obtain a warrant in order to conduct a search. However, a number of exceptions to that rule exist. For example, searches conducted by private persons unaffiliated with the government is not subject to the exclusionary rule, even if they are blatantly unlawful.[47] Such is the case, however, only if the private party is not an agent of the government.[48] Courts have widely varying opinions on what constitute a government agent. Generally, a vigilante acting the way law enforcement would, but without any contact with law enforcement until they reported what they found, have not been found to be government agent.[49]

Entrapment

Police participation in certain criminal activities is not by itself prohibited. Indeed, crimes committed in private by participants that have no intent or interest to report the crime to law enforcement agencies can be rather hard to uncover without some form of involvement of law enforcement agencies. These practices can, however, become problematic if they amount to entrapment, which is classically defined as "the conception and planning of an offense by an officer, and his procurement of its commission by one who would not have perpetrated it except for the trickery, persuasion, or fraud of the officer."[50]

The defence of entrapment does not have any statutory basis in Federal law. Instead, it was developed by the courts. The defence of entrapment is justified by the idea that "government agents may not originate a criminal design, implant in an innocent person's mind the disposition to commit a criminal act, and then induce commission of the crime so that the Government may prosecute."[51] Entrapment can

[44] Fishman and McKenna, Wiretapping and Eavesdropping § 22:5.

[45] See *Smith v. Maryland*, 442 U.S. 735 (1979).

[46] See *United States v. Perrine*, 518 F.3d 1196, 1204 (10th Cir. 2008).

[47] *Burdeau v. McDowell*, 256 U.S. 465 (1921). It should be noted, however, that the person conducting such a search may avail himself or herself to criminal and civil prosecution.

[48] Fishman and McKenna, Wiretapping and Eavesdropping § 23:3.

[49] In the case of Sweetie, it seems that the objective of Terre des Hommes is to hand Sweetie over to law enforcement agencies. An interesting question might come up if Sweetie was to be run by Terre des Hommes: does this rule hold true even when a vigilante is systematically pursuing policy interests that are aligned with those of the Government?

[50] *Sorrells v. United States*, 287 U.S. 435 (U.S. 1932).

[51] *Jacobson v. United States*, 503 U.S. 540, 548 (1992).

originate from a variety of sources: through an undercover agent, through a confidential informant or even through a private citizen knowingly acting under the direction of government agents.[52,53]

Courts have developed two distinct approaches to the defence of entrapment. We focus on what is commonly called the "subjective approach," which has been referred to by a majority of the Supreme Court and is adhered to by the federal courts.[54] Under this view, the defence of entrapment is composed of two elements: the government must have induced the crime, and the defendant must not have a predisposition to engage in the prosecuted criminal conduct.[55] In such a view, if the government can show that the defendant was predisposed to engage in the criminal activity for which he or she is prosecuted, the defence of entrapment will fail, regardless of whether or not the conduct was induced by the government.

12.4.3 Succinct Overview of Investigatory Powers in an Online Context

The rules applicable to investigations in an online context can either be constitutional rules derived from the Fourth Amendment, or statutory rules.

The applicability of either Fourth Amendment or statutory rules rests on the nature of the item searched. Where the investigators seek to get access to information stored in a container in which a suspect has a reasonable expectation of privacy ("inside the box"), the Fourth Amendment applies. Such is the case where the information searched is stored in the suspect's computer. On the other hand, where investigators seek access to information stored in a container in which the suspect has no such expectation ("outside the box"), the statutory rules apply. Such is the case where investigators look for information in transit (e.g. wiretapping), or where they seek to access information stored in a third-party's computer (e.g. email stored on a mail server operated by a third party).

It should be noted, however, that these lines are moving. Recent decisions of the U.S. Supreme Court have shown that the Court's doctrine may evolve, especially in

---

52 2 Crim. Proc. § 5.1(c) (4th ed.).

53 As far as private citizens are concerns, courts have held that "[f]actors in determining whether a person is a government agent include 'the nature of that person's relationship with the government, the purposes for which it was understood that person might act on behalf of the government, the instructions given to that person about the nature and extent of permissible activities, and what the government knew about those activities and permitted or used.'" *United States v. Jones*, 231 F.3d 508 (9th Cir. 2000). However, when the police is not involved at all in the scheme, the line between private entrapment and solicitation narrows greatly, and might be problematic as the private party involved could in turn be prosecuted for solicitation. See *United States v. Maddox*, 492 F.2d 104 (5th Cir. 1974).

54 2 Subst. Crim. L. § 9.8 (2d ed.).

55 *Mathews v. United States*, 485 U.S. 58 (1988).

light of the always-increasing prevalence and importance of electronic communications.[56]

Three federal statutes are applicable when dealing with investigatory powers in an online context: the Wiretap Act, the Pen Register Act, and the Stored Communication Act. The Wiretap Act and the Pen Register act deal with live interception of content and metadata, whereas the Stored Communications Act deals with access to stored data. Each of these statute has different procedures for accessing data. Remarkably, none of these statutes provide an exclusionary rule for electronic communications that have been inappropriately obtained by law enforcement officials, although they all provide for heavy criminal sanction.

The Wiretap Act

The general rule set forth in 18 U.S.C. § 2511(1)(a) is the complete prohibition of interception of "any wire, oral, or electronic communication." The definition of these concepts can be found in 18 U.S.C. § 2510, and encompasses communications made through computers connected to the internet. The disclosure of the result of an interception, or the use of the results of an interception where the user knows or has reason to know that the information was obtained through an interception, is also punished.[57] 18 U.S.C. § 2511 provides for several exceptions, among others for service providers, "whose facilities are used in the transmission of a wire or electronic communication (…) while engaged in any activity which is a necessary incident to the rendition of his service."[58] More importantly, interceptions are lawful "where [the interceptor] is a party to the communication or where one of the parties to the communication has given prior consent to such interception unless such communication is intercepted for the purpose of committing any criminal or tortious act."[59,60]

In cases where the electronic communication system is "configured so that […] electronic communications [made through it are] readily accessible to the general public," interception is also permitted (18 U.S.C. § 2511(2)(g)(i)).

Federal law enforcement authorities can apply for an order authorising them to intercept communications if they cannot justify an interception under these

[56] See *United States v. Jones*, 132 S. Ct. 945 (2012) and *Riley v. California*, 134 S. Ct. 2473 (2014).

[57] 18 U.S.C. § 2511(1)(c) and (d).

[58] 18 U.S.C. § 2511(2)(a).

[59] 18 U.S.C. § 2511(2)(d), see also 18 U.S.C. § 2511(2)(c).

[60] A number of states have "two-way consent" laws, under which both parties to a communication must consent to an interception for it to be lawful. It seems however that, as far as federal law enforcement agencies are concerned, federal courts are willing to admit in evidence contents obtained through wiretapping even if a state's two-party consent law was not observed. Fishman and McKenna, Wiretapping and Eavesdropping § 5:91. Unsurprisingly, state courts generally hold the opposite position. *Kearney v. Salomon Smith Barney, Inc.*, S124739 (Sup. Ct. Cal. July 13, 2006).

exceptions.[61] Such orders are limited to cases where the interception may provide evidence of offences listed in 18 U.S.C. § 2516(1). Most of the substantial offences listed supra are part of that list, notably Section 1591 (sex trafficking of children by force, fraud, or coercion), Section 1592 (unlawful conduct with respect to documents in furtherance of trafficking, peonage, slavery, involuntary servitude, or forced labor), Sections 2251 and 2252 (sexual exploitation of children), Section 2251A (selling or buying of children), Section 2252A (relating to material constituting or containing child pornography), Section 466A (relating to child obscenity), Section 2260 (production of sexually explicit depictions of a minor for importation into the United States), Sections 2421, 2422, 2423, and 2425 (relating to transportation for illegal sexual activity and related crimes). The formal requirement for such an order can be found at 18 U.S.C. § 2518.[62] These requirements are extremely stringent, surpassing those of a regular warrant.

The Pen Register Act

Pen register and trap and trace devices are essentially interception tools that can be used to obtain non-content data.[63] 18 U.S. Code § 3121(a) generally prohibits the use of pen register or trap and trace devices without appropriate order. The procedure for obtaining the required order is detailed in 18 U.S. Code § 3122, and is much more straightforward than that for wiretap orders and warrants, as it merely requires, "a certification by the applicant that the information likely to be obtained is relevant to an ongoing criminal investigation being conducted by that agency." There is no other limitation as to, e.g., the kind of crimes in the investigation of which these devices can be used.

The Stored Communications Act

When law enforcement agents want to access information that is not in transit, and not stored on a computer in which the suspect as a reasonable expectation of privacy, the Stored Communication Act applies. The typical case for the application of the Stored Communication Act is when law enforcement officers attempt to obtain information from internet services providers or other third party storing data for their user (e.g. a provider of an email service, with respect to the emails stored on its servers).

Except in the specific cases listed in the SCA, it is a criminal offence to intentionally access without authorisation a facility through which an electronic communication service is provided and to thereby obtain, alter, or prevent authorised access to a wire or electronic communication while it is in electronic storage.[64]

This basic rule has several exception. In certain cases, communication services providers can voluntarily disclose contents or subscriber records, while in other

[61] 18 U.S.C. § 2516(1).

[62] See appendix.

[63] See annex, 18 U.S. Code § 3127, for the legal definition of these notions.

[64] 18 U.S. Code § 2701(a).

cases, law enforcement agencies have to obtain a search warrant, or, in certain instances, emit a subpoena. The interpretation of the Stored Communication Act is made extremely complex by the fact that several, somewhat arbitrary, distinctions exist depending on the kind of service used and whether or not the service in which the information is stored is provided to the "public". Moreover, at least two federal circuits (the Sixth and Ninth Circuits) have unique interpretation of the law.

Since it is unclear at this point which type of service Sweetie will operate on, it would largely exceed the scope of this chapter to detail these rules. Instead, the reader is referred in a note to two reference articles on the topic.[65]

12.4.4 Application of Relevant Investigatory Powers to the Sweetie Case

We first examine what investigatory powers cold be applied to the Sweetie case, and then turn to the problems related to entrapment.

Investigators Powers

For the purposes of this chapter, it seems appropriate to distinguish two main phases in an investigation involving Sweetie: the "contact" between Sweetie and an individual, and the investigation that follows the initial contact.

At the stage of the contact between Sweetie and an individual, the legal provisions likely to apply are the Wiretap Act and the Pen Register Act. Indeed, it does not seem that Sweetie includes any component that would allow it to access information stored on the individual's computer.

It was indicated to the author of this chapter that Sweetie would only interact with people in "public" chatrooms. If such is the case, the conversations that Sweetie would have with third party could effectively be recorded under the exception allowing regarding public services. Moreover, if Sweetie actually engaged with third-party in the context of 1-on-1 private chats, the contents of the conversation could be recorded as Sweetie would effectively be a party to the communication who can consent to the recording.

It also seems that Sweetie will record some metadata, like IP addresses. It is for now unclear what type of mechanism will allow her to do so. If Sweetie simply copies an IP address that is publicly visible (e.g. an user's hostname as displayed on a IRC channel when the user joins or leave a channel), it seems that such an interception could fall outside the definitions of devices regulated by the Pen Register Act.[66] If however the device used is specifically designed to intercept

[65] Kerr (2004), pp. 1208–1249; Scolnik (2009), pp. 349–397.

[66] See annex, 18 U.S. Code § 3127.

non-content data that is not otherwise publicly available, an application for an order might be necessary.[67]

Once the contact between Sweetie and an individual is over, law enforcement official may want to further they investigation by collecting subscriber information from the individual's internet service provider, or by conducting a search of the individual's property (for example, by seizing his or her computer to look for further incriminating material). Such searches would be quite classically regulated by the SCA and the Fourth Amendment. It does not seem that the usage of Sweetie is likely to alter the way these provisions have been applied to searches of computers, and it is thus referred to the references cited above.

Entrapment

The application of entrapment rules to Sweetie is tricky. Under the subjective approach, defendants who have a predisposition to commit a certain crime cannot use the defence of entrapment.[68] Things become trickier, however, if the defendant does not have such a predisposition, or if the prosecution is unable to demonstrate such a predisposition. Indeed, in such cases, the question turns to whether the defendant was actually induced into committing the crime.

The question of whether one was induced to commit a crime is an eminently factual inquiry. To this day, it is unclear how Sweetie will actually interact with individuals. A number of questions come to mind: is Sweetie going to connect to chatrooms under the handle "Sweetie"? Will a picture of her be clearly visible before she even interacts with individuals? Once she starts interacting, does she actively suggests that she is willing to engage in certain act for a fee, or is she merely waiting to have her correspondent suggest certain courses of action?

In any case, it would seem most appropriate that Sweetie limits herself to a passive role, where she would merely reply to solicitations or questions with clear, unambiguous answers.

Another concern is, quite naturally, the predictability of Sweetie's answers. Indeed, even though artificial intelligence research has been making significant advances over the past few years, state-of-the-art technology is still barely able to achieve conversation results equivalent to that of a real human. Since Sweetie is supposed to represent a young, non-native English speaker, some "fuzziness" can pass muster without altering her credibility, but it nonetheless poses a problem from an entrapment perspective. Indeed, inappropriate answers to queries could effectively lead Sweetie to actually, mistakenly, induce an individual in a crime. This is likely to pose problem from a legal standpoint, but also from a policy perspective. As a general rule, it would be desirable for Sweetie to avoid inducing people into

[67] It is unclear to the author of this chapter how such a component of Sweetie would work, and requests for clarifications have not enabled him to elucidate this question.

[68] It is unclear, at this point, what contextual elements might help in demonstrating perpetrators' predisposition. The test generally applied is a "totality of the circumstances" one, under which contextual factors may help demonstrate predisposition (e.g. repeated presence in a chatroom which is unambiguously marketing itself as a hub for webcam-related child abuse).

committing crime. This would allow prosecutors to be confident that they will be able to defeat entrapment defences, but also seem like the most sensible option from a public policy standpoint (see Sect. 12.5).

12.4.5 Relevant Aspects of Digital Forensic Evidence

As a general rule, the rules applicable to digital evidence are fairly straightforward. The key aspect is to be able to guarantee that the information is authentic.[69] This can generally be done in any way, including a sworn testimony of the officer who collected the evidence. It is generally important to maintain a chain of custody, as this ensures that the evidence is not tampered with.[70] When it comes to online chat conversation, it is generally recommended that officers limit their interference with the content (e.g. preferably use a screenshot or a browser's saving function instead of simple copy-and-paste).[71]

American federal law enforcement officials have extensive experience in this domain, and have compiled an extensive manual that provides useful guidance in these areas: "Searching and Seizing Computers and Obtaining Electronic Evidence in Criminal Investigations".[72]

12.5 Miscellaneous

For the purposes of this chapter, we assumed that Sweetie would be operated by local law enforcement agencies. Where relevant, we included footnotes covering the case in which Sweetie would be functioning under the supervision of a private party.

12.6 Conclusions and Recommendations

Sweetie 2.0 was seemingly born of a desire to reproduce some of the successes of Sweetie 1.0, with an automatisation of its processes in order to significantly increase its impact while reducing its labor costs. Although one can only approve of such a goal, it also raises certain concerns. These concerns are not squarely legal concerns, but also, to a certain extent, policy concerns.

[69] 1 eDiscovery & Digital Evidence § 13:1.

[70] 1 eDiscovery & Digital Evidence § 13:2.

[71] See *United States v. Jackson*, 488 F. Supp. 2d 866 (D. Neb. 2007).

[72] Computer Crime and Intellectual Property Section of the Department of Justice (2009).

First, there is a risk that Sweetie may engage with minors and effectively commit crimes, for example, by enticing minors to perform obscene sexual acts in front of the camera, as adults perpetrators often do. Indeed, it does not seem clear at this point how Sweetie will filter out people she is not supposed to engage with. This could potentially lead to a situation where Sweetie would effectively contribute to the commission of a crime of which her interlocutor would be the victim.

An additional concern arises out of Sweetie's functioning as a chat-bot. State-of-the-art artificial intelligence can not yet produce chat-bots that interact the way a human would. As a result, and given the diversity of the exchanges Sweetie is likely to run into, it seems hard to imagine how sufficient safeguards could be built in Sweetie in order to prevent her from inadvertently inducing the commission of a crime.

In both cases, some safeguards are conceivable. However, these safeguards, to function properly (and hopefully, succeed in every case), would make Sweetie exceedingly conservative, hence reducing her efficiency.

No change to the existing legislation seems appropriate to correct these problems. Indeed, these problems are related to the design of Sweetie, and it would be highly undesirable to create legal exceptions for chat-bots such as Sweetie where such exceptions would effectively run afoul of the policy interest for certain rules (e.g. prevention of enticement, rules against criminal solicitation or entrapment). Instead, the author of this chapter suggests that the developers of Sweetie consider designing it as a semi-supervised bot, instead of a fully automated one.

Designed as a semi-supervised bot, Sweetie could work with pre-scripted sequences, that could only be triggered by a human. For instance, Sweetie could be free to talk with anyone who engage her directly and do some small talk. If the conversation takes a turn that signal that the individual talking with Sweetie might be looking to commit a crime (for instance, if the individual start asking how old Sweetie is, whether she has a webcam, etc.), an operator could be informed and decide to escalate the conversation after ensuring that Sweetie is not behaving in a way inconsistent with the rules presented in this chapter (e.g. ensure that the interlocutor is an adult, that Sweetie did not answer a question in an inappropriate way that could be likely to induce her interlocutor, etc). Such a method would effectively reduce the labor required for Sweetie to function, while maintaining some safeguards against some of the risks pointed out above.

Another consideration might also lead one to be in favour of such an approach: courts, as well as the public, are always defiant against new technologies, and more often than not have trouble understanding how they work. While people are getting increasingly used to see their behaviour regulated by machines of all sorts (from speeding radars to credit card authorisation systems to full-body scanners), the idea of enforcement by fulling automated machines is still novel, and could as such be worrisome for courts or citizens who do not fully understand how they work. Introducing some human supervision in Sweetie might not only make Sweetie more efficient and accurate, but also make its novelty more acceptable to both courts and the public.

Annex

Relevant Legal Provisions (in Original)

Note: in order to facilitate the reader's use of this annex, the statutes contained in this section are ordered sequentially by number, not by topic.

18 U.S.C. § 1466A. Obscene visual representations of the sexual abuse of children

(a) In General.—Any person who, in a circumstance described in subsection (d), knowingly produces, distributes, receives, or possesses with intent to distribute, a visual depiction of any kind, including a drawing, cartoon, sculpture, or painting, that—

(1)(A) depicts a minor engaging in sexually explicit conduct; and

(B) is obscene; or

(2)(A) depicts an image that is, or appears to be, of a minor engaging in graphic bestiality, sadistic or masochistic abuse, or sexual intercourse, including genital-genital, oral-genital, anal-genital, or oral-anal, whether between persons of the same or opposite sex; and

(B) lacks serious literary, artistic, political, or scientific value;

or attempts or conspires to do so, shall be subject to the penalties provided in Section 2252A(b)(1), including the penalties provided for cases involving a prior conviction.

(b) Additional Offenses.—Any person who, in a circumstance described in subsection (d), knowingly possesses a visual depiction of any kind, including a drawing, cartoon, sculpture, or painting, that—

(1)(A) depicts a minor engaging in sexually explicit conduct; and

(B) is obscene; or

(2)(A) depicts an image that is, or appears to be, of a minor engaging in graphic bestiality, sadistic or masochistic abuse, or sexual intercourse, including genital-genital, oral-genital, anal-genital, or oral-anal, whether between persons of the same or opposite sex; and

(B) lacks serious literary, artistic, political, or scientific value;
or attempts or conspires to do so, shall be subject to the penalties provided in Section 2252A(b)(2), including the penalties provided for cases involving a prior conviction.

(c) Nonrequired Element of Offense.—It is not a required element of any offense under this section that the minor depicted actually exist.

(d) Circumstances.—The circumstance referred to in subsections (a) and (b) is that—

(1) any communication involved in or made in furtherance of the offense is communicated or transported by the mail, or in interstate or foreign commerce by any means, including by computer, or any means or instrumentality of interstate or foreign commerce is otherwise used in committing or in furtherance of the commission of the offense;

(2) any communication involved in or made in furtherance of the offense contemplates the transmission or transportation of a visual depiction by the mail, or in interstate or foreign commerce by any means, including by computer;

(3) any person travels or is transported in interstate or foreign commerce in the course of the commission or in furtherance of the commission of the offense;

(4) any visual depiction involved in the offense has been mailed, or has been shipped or transported in interstate or foreign commerce by any means, including by computer, or was produced using materials that have been mailed, or that have been shipped or transported in interstate or foreign commerce by any means, including by computer; or

(5) the offense is committed in the special maritime and territorial jurisdiction of the United States or in any territory or possession of the United States.

(e) Affirmative Defense.—It shall be an affirmative defense to a charge of violating subsection (b) that the defendant—

(1) possessed less than 3 such visual depictions; and

(2) promptly and in good faith, and without retaining or allowing any person, other than a law enforcement agency, to access any such visual depiction—

(A) took reasonable steps to destroy each such visual depiction; or

(B) reported the matter to a law enforcement agency and afforded that agency access to each such visual depiction.

(f) Definitions.—For purposes of this section—

(1) the term "visual depiction" includes undeveloped film and videotape, and data stored on a computer disk or by electronic means which is capable of conversion into a visual image, and also includes any photograph, film, video, picture, digital image or picture, computer image or picture, or computer generated image or picture, whether made or produced by electronic, mechanical, or other means;

(2) the term "sexually explicit conduct" has the meaning given the term in Section 256(2)(A) or 2256(2)(B); and

(3) the term "graphic", when used with respect to a depiction of sexually explicit conduct, means that a viewer can observe any part of the genitals or pubic area

of any depicted person or animal during any part of the time that the sexually explicit conduct is being depicted.

18 U.S.C. § 1470. Transfer of obscene material to minors
Whoever, using the mail or any facility or means of interstate or foreign commerce, knowingly transfers obscene matter to another individual who has not attained the age of 16 years, knowing that such other individual has not attained the age of 16 years, or attempts to do so, shall be fined under this title, imprisoned not more than 10 years, or both.

18 U.S.C. § 2246. Definitions for [18 U.S.C. § 2241, § 2243 § 2244]
As used in this chapter—
[...]

(2) the term "sexual act" means—

(A) contact between the penis and the vulva or the penis and the anus, and for purposes of this subparagraph contact involving the penis occurs upon penetration, however slight;

(B) contact between the mouth and the penis, the mouth and the vulva, or the mouth and the anus;

(C) the penetration, however slight, of the anal or genital opening of another by a hand or finger or by any object, with an intent to abuse, humiliate, harass, degrade, or arouse or gratify the sexual desire of any person; or

(D) the intentional touching, not through the clothing, of the genitalia of another person who has not attained the age of 16 years with an intent to abuse, humiliate, harass, degrade, or arouse or gratify the sexual desire of any person;

(3) the term "sexual contact" means the intentional touching, either directly or through the clothing, of the genitalia, anus, groin, breast, inner thigh, or buttocks of any person with an intent to abuse, humiliate, harass, degrade, or arouse or gratify the sexual desire of any person;

[...]

18 U.S.C. § 2251. Sexual exploitation of children
(a) Any person who employs, uses, persuades, induces, entices, or coerces any minor to engage in, or who has a minor assist any other person to engage in, or who transports any minor in or affecting interstate or foreign commerce, or in any Territory or Possession of the United States, with the intent that such minor engage in, any sexually explicit conduct for the purpose of producing any visual depiction of such conduct or for the purpose of transmitting a live visual

depiction of such conduct, shall be punished as provided under subsection (e), if such person knows or has reason to know that such visual depiction will be transported or transmitted using any means or facility of interstate or foreign commerce or in or affecting interstate or foreign commerce or mailed, if that visual depiction was produced or transmitted using materials that have been mailed, shipped, or transported in or affecting interstate or foreign commerce by any means, including by computer, or if such visual depiction has actually been transported or transmitted using any means or facility of interstate or foreign commerce or in or affecting interstate or foreign commerce or mailed.

(b) Any parent, legal guardian, or person having custody or control of a minor who knowingly permits such minor to engage in, or to assist any other person to engage in, sexually explicit conduct for the purpose of producing any visual depiction of such conduct or for the purpose of transmitting a live visual depiction of such conduct shall be punished as provided under subsection (e) of this section, if such parent, legal guardian, or person knows or has reason to know that such visual depiction will be transported or transmitted using any means or facility of interstate or foreign commerce or in or affecting interstate or foreign commerce or mailed, if that visual depiction was produced or transmitted using materials that have been mailed, shipped, or transported in or affecting interstate or foreign commerce by any means, including by computer, or if such visual depiction has actually been transported or transmitted using any means or facility of interstate or foreign commerce or in or affecting interstate or foreign commerce or mailed.

(c)(1) Any person who, in a circumstance described in paragraph (2), employs, uses, persuades, induces, entices, or coerces any minor to engage in, or who has a minor assist any other person to engage in, any sexually explicit conduct outside of the United States, its territories or possessions, for the purpose of producing any visual depiction of such conduct, shall be punished as provided under subsection (e).

(2) The circumstance referred to in paragraph (1) is that—

(A) the person intends such visual depiction to be transported to the United States, its territories or possessions, by any means, including by using any means or facility of interstate or foreign commerce or mail; or

(B) the person transports such visual depiction to the United States, its territories or possessions, by any means, including by using any means or facility of interstate or foreign commerce or mail.

(d)(1) Any person who, in a circumstance described in paragraph (2), knowingly makes, prints, or publishes, or causes to be made, printed, or published, any notice or advertisement seeking or offering—

(A) to receive, exchange, buy, produce, display, distribute, or reproduce, any visual depiction, if the production of such visual depiction involves the use of a minor engaging in sexually explicit conduct and such visual depiction is of such conduct; or

(B) participation in any act of sexually explicit conduct by or with any minor for the purpose of producing a visual depiction of such conduct;
shall be punished as provided under subsection (e).

(2) The circumstance referred to in paragraph (1) is that—

(A) such person knows or has reason to know that such notice or advertisement will be transported using any means or facility of interstate or foreign commerce or in or affecting interstate or foreign commerce by any means including by computer or mailed; or

(B) such notice or advertisement is transported using any means or facility of interstate or foreign commerce or in or affecting interstate or foreign commerce by any means including by computer or mailed.

(e) Any individual who violates, or attempts or conspires to violate, this section shall be fined under this title and imprisoned not less than 15 years nor more than 30 years, but if such person has one prior conviction under this chapter, Section 1591, chapter 71, chapter 109A, or chapter 117, or under Section 920 of title 10 (Article 120 of the Uniform Code of Military Justice), or under the laws of any State relating to aggravated sexual abuse, sexual abuse, abusive sexual contact involving a minor or ward, or sex trafficking of children, or the production, possession, receipt, mailing, sale, distribution, shipment, or transportation of child pornography, such person shall be fined under this title and imprisoned for not less than 25 years nor more than 50 years, but if such person has 2 or more prior convictions under this chapter, chapter 71, chapter 109A, or chapter 117, or under Section 920 of title 10 (Article 120 of the Uniform Code of Military Justice), or under the laws of any State relating to the sexual exploitation of children, such person shall be fined under this title and imprisoned not less than 35 years nor more than life. Any organization that violates, or attempts or conspires to violate, this section shall be fined under this title. Whoever, in the course of an offense under this section, engages in conduct that results in the death of a person, shall be punished by death or imprisoned for not less than 30 years or for life.

18 U.S.C. § 2252. Certain activities relating to material involving the sexual exploitation of minors

(a) Any person who—

(1) knowingly transports or ships using any means or facility of interstate or foreign commerce or in or affecting interstate or foreign commerce by any means including by computer or mails, any visual depiction, if—

(A) the producing of such visual depiction involves the use of a minor engaging in sexually explicit conduct; and

(B) such visual depiction is of such conduct;

(2) knowingly receives, or distributes, any visual depiction using any means or facility of interstate or foreign commerce or that has been mailed, or has been shipped or transported in or affecting interstate or foreign commerce, or which contains materials which have been mailed or so shipped or transported, by any means including by computer, or knowingly reproduces any visual depiction for distribution using any means or facility of interstate or foreign commerce or in or affecting interstate or foreign commerce or through the mails, if—

(A) the producing of such visual depiction involves the use of a minor engaging in sexually explicit conduct; and

(B) such visual depiction is of such conduct;

(3) either—

(A) in the special maritime and territorial jurisdiction of the United States, or on any land or building owned by, leased to, or otherwise used by or under the control of the Government of the United States, or in the Indian country as defined in Section 1151 of this title, knowingly sells or possesses with intent to sell any visual depiction; or
(B) knowingly sells or possesses with intent to sell any visual depiction that has been mailed, shipped, or transported using any means or facility of interstate or foreign commerce, or has been shipped or transported in or affecting interstate or foreign commerce, or which was produced using materials which have been mailed or so shipped or transported using any means or facility of interstate or foreign commerce, including by computer, if—

(i) the producing of such visual depiction involves the use of a minor engaging in sexually explicit conduct; and

(ii) such visual depiction is of such conduct; or

(4) either—

(A) in the special maritime and territorial jurisdiction of the United States, or on any land or building owned by, leased to, or otherwise used by or under the control of the Government of the United States, or in the Indian country as defined in Section 1151 of this title, knowingly possesses, or knowingly accesses with intent to view, 1 or more books, magazines, periodicals, films, video tapes, or other matter which contain any visual depiction; or

(B) knowingly possesses, or knowingly accesses with intent to view, 1 or more books, magazines, periodicals, films, video tapes, or other matter which contain any visual depiction that has been mailed, or has been shipped or transported using any means or facility of interstate or foreign commerce or in or affecting interstate or foreign commerce, or which was produced using materials which have been mailed or so shipped or transported, by any means including by computer, if—

(i) the producing of such visual depiction involves the use of a minor engaging in sexually explicit conduct; and

(ii) such visual depiction is of such conduct;

shall be punished as provided in subsection (b) of this section.

(b)(1) Whoever violates, or attempts or conspires to violate, paragraph (1), (2), or (3) of subsection (a) shall be fined under this title and imprisoned not less than 5 years and not more than 20 years, but if such person has a prior conviction under this chapter, Section 1591, chapter 71, chapter 109A, or chapter 117, or under Section 920 of title 10 (Article 120 of the Uniform Code of Military Justice), or under the laws of any State relating to aggravated sexual abuse, sexual abuse, or abusive sexual conduct involving a minor or ward, or the production, possession, receipt, mailing, sale, distribution, shipment, or transportation of child pornography, or sex trafficking of children, such person shall be fined under this title and imprisoned for not less than 15 years nor more than 40 years.

(2) Whoever violates, or attempts or conspires to violate, paragraph (4) of subsection (a) shall be fined under this title or imprisoned not more than 10 years, or both, but if such person has a prior conviction under this chapter, chapter 71, chapter 109A, or chapter 117, or under Section 920 of title 10 (Article 120 of the Uniform Code of Military Justice), or under the laws of any State relating to aggravated sexual abuse, sexual abuse, or abusive sexual conduct involving a minor or ward, or the production, possession, receipt, mailing, sale, distribution, shipment, or transportation of child pornography, such person shall be fined under this title and imprisoned for not less than 10 years nor more than 20 years.

(c) Affirmative Defense.—It shall be an affirmative defense to a charge of violating paragraph (4) of subsection (a) that the defendant—

(1) possessed less than three matters containing any visual depiction proscribed by that paragraph; and

(2) promptly and in good faith, and without retaining or allowing any person, other than a law enforcement agency, to access any visual depiction or copy thereof—

(A) took reasonable steps to destroy each such visual depiction; or

(B) reported the matter to a law enforcement agency and afforded that agency access to each such visual depiction.

18 U.S. Code § 2252A—Certain activities relating to material constituting or containing child pornography

(a) Any person who—

(1) knowingly mails, or transports or ships using any means or facility of interstate or foreign commerce or in or affecting interstate or foreign commerce by any means, including by computer, any child pornography;

(2) knowingly receives or distributes—

(A) any child pornography that has been mailed, or using any means or facility of interstate or foreign commerce shipped or transported in or affecting interstate or foreign commerce by any means, including by computer; or

(B) any material that contains child pornography that has been mailed, or using any means or facility of interstate or foreign commerce shipped or transported in or affecting interstate or foreign commerce by any means, including by computer;

(3) knowingly—

(A) reproduces any child pornography for distribution through the mails, or using any means or facility of interstate or foreign commerce or in or affecting interstate or foreign commerce by any means, including by computer; or

(B) advertises, promotes, presents, distributes, or solicits through the mails, or using any means or facility of interstate or foreign commerce or in or affecting interstate or foreign commerce by any means, including by computer, any material or purported material in a manner that reflects the belief, or that is intended to cause another to believe, that the material or purported material is, or contains—

(i) an obscene visual depiction of a minor engaging in sexually explicit conduct; or

(ii) a visual depiction of an actual minor engaging in sexually explicit conduct;

(4) either—

(A) in the special maritime and territorial jurisdiction of the United States, or on any land or building owned by, leased to, or otherwise used by or under the control of the United States Government, or in the Indian country (as defined in Section 1151), knowingly sells or possesses with the intent to sell any child pornography; or

(B) knowingly sells or possesses with the intent to sell any child pornography that has been mailed, or shipped or transported using any means or facility of interstate or foreign commerce or in or affecting interstate or foreign commerce by any means, including by computer, or that was produced using materials that have been mailed, or shipped or transported in or affecting interstate or foreign commerce by any means, including by computer;

(5) either—

(A) in the special maritime and territorial jurisdiction of the United States, or on any land or building owned by, leased to, or otherwise used by or under the control of the United States Government, or in the Indian country (as defined in Section 1151), knowingly possesses, or knowingly accesses with intent to view, any book, magazine, periodical, film, videotape, computer disk, or any other material that contains an image of child pornography; or

(B) knowingly possesses, or knowingly accesses with intent to view, any book, magazine, periodical, film, videotape, computer disk, or any other material that contains an image of child pornography that has been mailed, or shipped or transported using any means or facility of interstate or foreign commerce or in or affecting interstate or foreign commerce by any means, including by computer, or that was produced using materials that have been mailed, or shipped or transported in or affecting interstate or foreign commerce by any means, including by computer; (6) knowingly distributes, offers, sends, or provides to a minor any visual depiction, including any photograph, film, video, picture, or computer generated image or picture, whether made or produced by electronic, mechanical, or other means, where such visual depiction is, or appears to be, of a minor engaging in sexually explicit conduct—

(A) that has been mailed, shipped, or transported using any means or facility of interstate or foreign commerce or in or affecting interstate or foreign commerce by any means, including by computer;

(B) that was produced using materials that have been mailed, shipped, or transported in or affecting interstate or foreign commerce by any means, including by computer; or

(C) which distribution, offer, sending, or provision is accomplished using the mails or any means or facility of interstate or foreign commerce,

for purposes of inducing or persuading a minor to participate in any activity that is illegal; or

(7) knowingly produces with intent to distribute, or distributes, by any means, including a computer, in or affecting interstate or foreign commerce, child pornography that is an adapted or modified depiction of an identifiable minor. [1]

shall be punished as provided in subsection (b).

(b)

(1) Whoever violates, or attempts or conspires to violate, paragraph (1), (2), (3), (4), or (6) of subsection (a) shall be fined under this title and imprisoned not less than 5 years and not more than 20 years, but, if such person has a prior conviction under this chapter, Section 1591, chapter 71, chapter 109A, or chapter 117, or under Section 920 of title 10 (Article 120 of the Uniform Code of Military Justice), or under the laws of any State relating to aggravated sexual abuse, sexual abuse, or abusive sexual conduct involving a minor or ward, or the production, possession, receipt, mailing, sale, distribution, shipment, or transportation of child pornography, or sex trafficking of children, such person shall be fined under this title and imprisoned for not less than 15 years nor more than 40 years.

(2) Whoever violates, or attempts or conspires to violate, subsection (a)(5) shall be fined under this title or imprisoned not more than 10 years, or both, but, if any image of child pornography involved in the offense involved a prepubescent minor or a minor who had not attained 12 years of age, such person shall be fined under this title and imprisoned for not more than 20 years, or if such person has a prior conviction under this chapter, chapter 71, chapter 109A, or chapter 117, or under Section 920 of title 10 (Article 120 of the Uniform Code of

Military Justice), or under the laws of any State relating to aggravated sexual abuse, sexual abuse, or abusive sexual conduct involving a minor or ward, or the production, possession, receipt, mailing, sale, distribution, shipment, or transportation of child pornography, such person shall be fined under this title and imprisoned for not less than 10 years nor more than 20 years.

(3) Whoever violates, or attempts or conspires to violate, subsection (a)(7) shall be fined under this title or imprisoned not more than 15 years, or both.

(c) It shall be an affirmative defense to a charge of violating paragraph (1), (2), (3)(A), (4), or (5) of subsection (a) that—

(1)

(A) the alleged child pornography was produced using an actual person or persons engaging in sexually explicit conduct; and

(B) each such person was an adult at the time the material was produced; or

(2) the alleged child pornography was not produced using any actual minor or minors.

No affirmative defense under subsection (c)(2) shall be available in any prosecution that involves child pornography as described in Section 2256(8)(C). A defendant may not assert an affirmative defense to a charge of violating paragraph (1), (2), (3)(A), (4), or (5) of subsection (a) unless, within the time provided for filing pretrial motions or at such time prior to trial as the judge may direct, but in no event later than 14 days before the commencement of the trial, the defendant provides the court and the United States with notice of the intent to assert such defense and the substance of any expert or other specialized testimony or evidence upon which the defendant intends to rely. If the defendant fails to comply with this subsection, the court shall, absent a finding of extraordinary circumstances that prevented timely compliance, prohibit the defendant from asserting such defense to a charge of violating paragraph (1), (2), (3)(A), (4), or (5) of subsection (a) or presenting any evidence for which the defendant has failed to provide proper and timely notice.

(d)Affirmative Defense.—It shall be an affirmative defense to a charge of violating subsection (a)(5) that the defendant—

(1) possessed less than three images of child pornography; and

(2) promptly and in good faith, and without retaining or allowing any person, other than a law enforcement agency, to access any image or copy thereof—

(A) took reasonable steps to destroy each such image; or

(B) reported the matter to a law enforcement agency and afforded that agency access to each such image.

(e) Admissibility of Evidence.—

On motion of the government, in any prosecution under this chapter or Section 1466A, except for good cause shown, the name, address, social security number, or other nonphysical identifying information, other than the age or approximate age, of any minor who is depicted in any child pornography shall not be admissible and may be redacted from any otherwise admissible evidence, and the jury shall be instructed, upon request of the United States, that it can draw no inference from the absence of such evidence in deciding whether the child pornography depicts an actual minor.

(f) Civil Remedies.—

(1) In general.—

Any person aggrieved by reason of the conduct prohibited under subsection (a) or (b) or Section 1466A may commence a civil action for the relief set forth in paragraph (2).

(2) Relief.—In any action commenced in accordance with paragraph (1), the court may award appropriate relief, including—

(A) temporary, preliminary, or permanent injunctive relief;

(B) compensatory and punitive damages; and

(C) the costs of the civil action and reasonable fees for attorneys and expert witnesses.

(g) Child Exploitation Enterprises.—

(1) Whoever engages in a child exploitation enterprise shall be fined under this title and imprisoned for any term of years not less than 20 or for life.

(2) A person engages in a child exploitation enterprise for the purposes of this section if the person violates Section 1591, Section 1201 if the victim is a minor, or chapter 109A (involving a minor victim), 110 (except for Sections 2257 and 2257A), or 117 (involving a minor victim), as a part of a series of felony violations constituting three or more separate incidents and involving more than one victim, and commits those offenses in concert with three or more other persons.

18 U.S.C. § 2256. Definitions for [§ 2251, § 2252 and § 2260]
For the purposes of this chapter, the term—

(1) "minor" means any person under the age of eighteen years;

(2)(A) Except as provided in subparagraph (B), "sexually explicit conduct" means actual or simulated—

(i) sexual intercourse, including genital-genital, oral-genital, anal-genital, or oral-anal, whether between persons of the same or opposite sex;

(ii) bestiality;

(iii) masturbation;

(iv) sadistic or masochistic abuse; or

(v) lascivious exhibition of the genitals or pubic area of any person;

(B) For purposes of subsection 8(B) 1 of this section, "sexually explicit conduct" means—

(i) graphic sexual intercourse, including genital-genital, oral-genital, anal-genital, or oral-anal, whether between persons of the same or opposite sex, or lascivious simulated sexual intercourse where the genitals, breast, or pubic area of any person is exhibited;

(ii) graphic or lascivious simulated;

(I) bestiality;

(II) masturbation; or

(III) sadistic or masochistic abuse; or

(iii) graphic or simulated lascivious exhibition of the genitals or pubic area of any person;

(3) "producing" means producing, directing, manufacturing, issuing, publishing, or advertising;

(4) "organization" means a person other than an individual;

(5) "visual depiction" includes undeveloped film and videotape, data stored on computer disk or by electronic means which is capable of conversion into a visual image, and data which is capable of conversion into a visual image that has been transmitted by any means, whether or not stored in a permanent format;

(6) "computer" has the meaning given that term in Section 1030 of this title;

(7) "custody or control" includes temporary supervision over or responsibility for a minor whether legally or illegally obtained;

(8) "child pornography" means any visual depiction, including any photograph, film, video, picture, or computer or computer-generated image or picture, whether made or produced by electronic, mechanical, or other means, of sexually explicit conduct, where—

(A) the production of such visual depiction involves the use of a minor engaging in sexually explicit conduct;

(B) such visual depiction is a digital image, computer image, or computer-generated image that is, or is indistinguishable from, that of a minor engaging in sexually explicit conduct; or

(C) such visual depiction has been created, adapted, or modified to appear that an identifiable minor is engaging in sexually explicit conduct.

(9) "identifiable minor"—

(A) means a person—

(i)(I) who was a minor at the time the visual depiction was created, adapted, or modified; or

(II) whose image as a minor was used in creating, adapting, or modifying the visual depiction; and

(ii) who is recognizable as an actual person by the person's face, likeness, or other distinguishing characteristic, such as a unique birthmark or other recognizable feature; and

(B) shall not be construed to require proof of the actual identity of the identifiable minor.

(10) "graphic", when used with respect to a depiction of sexually explicit conduct, means that a viewer can observe any part of the genitals or pubic area of any depicted person or animal during any part of the time that the sexually explicit conduct is being depicted; and

(11) the term "indistinguishable" used with respect to a depiction, means virtually indistinguishable, in that the depiction is such that an ordinary person viewing the depiction would conclude that the depiction is of an actual minor engaged in sexually explicit conduct. This definition does not apply to depictions that are drawings, cartoons, sculptures, or paintings depicting minors or adults.

18 U.S.C. § 2422. Coercion and enticement

(a) Whoever knowingly persuades, induces, entices, or coerces any individual to travel in interstate or foreign commerce, or in any Territory or Possession of the United States, to engage in prostitution, or in any sexual activity for which any person can be charged with a criminal offense, or attempts to do so, shall be fined under this title or imprisoned not more than 20 years, or both.

(b) Whoever, using the mail or any facility or means of interstate or foreign commerce, or within the special maritime and territorial jurisdiction of the United States knowingly persuades, induces, entices, or coerces any individual who has not attained the age of 18 years, to engage in prostitution or any sexual activity for which any person can be charged with a criminal offense, or attempts to do so, shall be fined under this title and imprisoned not less than 10 years or for life.

18 U.S.C. § 2423. Transportation of minors

(a) Transportation With Intent To Engage in Criminal Sexual Activity.—A person who knowingly transports an individual who has not attained the age of 18 years in interstate or foreign commerce, or in any commonwealth, territory or possession of the United States, with intent that the individual engage in prostitution, or in any sexual activity for which any person can be charged with a criminal offense, shall be fined under this title and imprisoned not less than 10 years or for life.

(b) Travel With Intent To Engage in Illicit Sexual Conduct.—A person who travels in interstate commerce or travels into the United States, or a United States citizen or an alien admitted for permanent residence in the United States who travels in foreign commerce, for the purpose of engaging in any illicit sexual conduct with another person shall be fined under this title or imprisoned not more than 30 years, or both.

(c) Engaging in Illicit Sexual Conduct in Foreign Places.—Any United States citizen or alien admitted for permanent residence who travels in foreign commerce, and engages in any illicit sexual conduct with another person shall be fined under this title or imprisoned not more than 30 years, or both.

(d) Ancillary Offenses.—Whoever, for the purpose of commercial advantage or private financial gain, arranges, induces, procures, or facilitates the travel of a person knowing that such a person is traveling in interstate commerce or foreign commerce for the purpose of engaging in illicit sexual conduct shall be fined under this title, imprisoned not more than 30 years, or both.

(e) Attempt and Conspiracy.—Whoever attempts or conspires to violate subsection (a), (b), (c), or (d) shall be punishable in the same manner as a completed violation of that subsection.

(f) Definition.—As used in this section, the term "illicit sexual conduct" means (1) a sexual act (as defined in Section 2246) with a person under 18 years of age that would be in violation of Chapter 109A if the sexual act occurred in the special maritime and territorial jurisdiction of the United States; or (2) any commercial sex act (as defined in Section 1591) with a person under 18 years of age.

(g) Defense.—In a prosecution under this section based on illicit sexual conduct as defined in subsection (f)(2), it is a defense, which the defendant must establish by a preponderance of the evidence, that the defendant reasonably believed that the person with whom the defendant engaged in the commercial sex act had attained the age of 18 years.

18 U.S.C. § 2425. Use of interstate facilities to transmit information about a minor

Whoever, using the mail or any facility or means of interstate or foreign commerce, or within the special maritime and territorial jurisdiction of the United States, knowingly initiates the transmission of the name, address, telephone number, social security number, or electronic mail address of another individual, knowing that such other individual has not attained the age of 16 years, with the intent to entice, encourage, offer, or solicit any person to engage in any sexual activity for which any person can be charged with a criminal offense, or attempts to do so, shall be fined under this title, imprisoned not more than 5 years, or both.

18 U.S.C. § 2427. Inclusion of offenses relating to child pornography in definition of sexual activity for which any person can be charged with a criminal offense

In this chapter, the term "sexual activity for which any person can be charged with a criminal offense" includes the production of child pornography, as defined in Section 2256(8).

18 U.S. Code § 2511—Interception and disclosure of wire, oral, or electronic communications prohibited

(1) Except as otherwise specifically provided in this chapter any person who—

(a) intentionally intercepts, endeavors to intercept, or procures any other person to intercept or endeavor to intercept, any wire, oral, or electronic communication;

(b) intentionally uses, endeavors to use, or procures any other person to use or endeavor to use any electronic, mechanical, or other device to intercept any oral communication when—

(i) such device is affixed to, or otherwise transmits a signal through, a wire, cable, or other like connection used in wire communication; or

(ii) such device transmits communications by radio, or interferes with the transmission of such communication; or

(iii) such person knows, or has reason to know, that such device or any component thereof has been sent through the mail or transported in interstate or foreign commerce; or

(iv) such use or endeavor to use (A) takes place on the premises of any business or other commercial establishment the operations of which affect interstate or foreign commerce; or (B) obtains or is for the purpose of obtaining information relating to the operations of any business or other commercial establishment the operations of which affect interstate or foreign commerce; or

(v) such person acts in the District of Columbia, the Commonwealth of Puerto Rico, or any territory or possession of the United States;

(c) intentionally discloses, or endeavors to disclose, to any other person the contents of any wire, oral, or electronic communication, knowing or having reason to know that the information was obtained through the interception of a wire, oral, or electronic communication in violation of this subsection;

(d) intentionally uses, or endeavors to use, the contents of any wire, oral, or electronic communication, knowing or having reason to know that the information was obtained through the interception of a wire, oral, or electronic communication in violation of this subsection; or

(…)

(2)

(a)

(i) It shall not be unlawful under this chapter for an operator of a switchboard, or an officer, employee, or agent of a provider of wire or electronic communication service, whose facilities are used in the transmission of a wire or electronic communication, to intercept, disclose, or use that communication in the normal course of his employment while engaged in any activity which is a necessary incident to the rendition of his service or to the protection of the rights or property of the provider of that service, except that a provider of wire communication service to the public shall not utilize service observing or random monitoring except for mechanical or service quality control checks.

(…)

(c) It shall not be unlawful under this chapter for a person acting under color of law to intercept a wire, oral, or electronic communication, where such person is a party to the communication or one of the parties to the communication has given prior consent to such interception.

(d) It shall not be unlawful under this chapter for a person not acting under color of law to intercept a wire, oral, or electronic communication where such person is a party to the communication or where one of the parties to the communication has given prior consent to such interception unless such communication is intercepted for the purpose of committing any criminal or tortious act in violation of the Constitution or laws of the United States or of any State.

(…)

(g) It shall not be unlawful under this chapter or chapter 121 of this title for any person—

(i) to intercept or access an electronic communication made through an electronic communication system that is configured so that such electronic communication is readily accessible to the general public;

(…)

(i) It shall not be unlawful under this chapter for a person acting under color of law to intercept the wire or electronic communications of a computer trespasser transmitted to, through, or from the protected computer, if—

(I) the owner or operator of the protected computer authorizes the interception of the computer trespasser's communications on the protected computer;

(II) the person acting under color of law is lawfully engaged in an investigation;

(III) the person acting under color of law has reasonable grounds to believe that the contents of the computer trespasser's communications will be relevant to the investigation; and

(IV) such interception does not acquire communications other than those transmitted to or from the computer trespasser.

(…)

18 U.S. Code § 2518—Procedure for interception of wire, oral, or electronic communications

(1) Each application for an order authorizing or approving the interception of a wire, oral, or electronic communication under this chapter shall be made in writing upon oath or affirmation to a judge of competent jurisdiction and shall state the applicant's authority to make such application. Each application shall include the following information:

(a) the identity of the investigative or law enforcement officer making the application, and the officer authorizing the application;

(b) a full and complete statement of the facts and circumstances relied upon by the applicant, to justify his belief that an order should be issued, including (i) details as to the particular offense that has been, is being, or is about to be committed, (ii) except as provided in subsection (11), a particular description of the nature and location of the facilities from which or the place where the communication is to be intercepted, (iii) a particular description of the type of communications sought to be intercepted, (iv) the identity of the person, if known, committing the offense and whose communications are to be intercepted;

(c) a full and complete statement as to whether or not other investigative procedures have been tried and failed or why they reasonably appear to be unlikely to succeed if tried or to be too dangerous;

(d) a statement of the period of time for which the interception is required to be maintained. If the nature of the investigation is such that the authorization for interception should not automatically terminate when the described type of communication has been first obtained, a particular description of facts establishing probable cause to believe that additional communications of the same type will occur thereafter;

(e) a full and complete statement of the facts concerning all previous applications known to the individual authorizing and making the application, made to any judge for authorization to intercept, or for approval of interceptions of, wire, oral, or electronic communications involving any of the same persons, facilities or places specified in the application, and the action taken by the judge on each such application; and

(f) where the application is for the extension of an order, a statement setting forth the results thus far obtained from the interception, or a reasonable explanation of the failure to obtain such results.

(2) The judge may require the applicant to furnish additional testimony or documentary evidence in support of the application.

(3) Upon such application the judge may enter an ex parte order, as requested or as modified, authorizing or approving interception of wire, oral, or electronic communications within the territorial jurisdiction of the court in which the judge is sitting (and outside that jurisdiction but within the United States in the case of a mobile interception device authorized by a Federal court within such jurisdiction), if the judge determines on the basis of the facts submitted by the applicant that—

(a) there is probable cause for belief that an individual is committing, has committed, or is about to commit a particular offense enumerated in Section 2516 of this chapter;

(b) there is probable cause for belief that particular communications concerning that offense will be obtained through such interception;

(c) normal investigative procedures have been tried and have failed or reasonably appear to be unlikely to succeed if tried or to be too dangerous;

(d) except as provided in subsection (11), there is probable cause for belief that the facilities from which, or the place where, the wire, oral, or electronic communications are to be intercepted are being used, or are about to be used, in connection with the commission of such offense, or are leased to, listed in the name of, or commonly used by such person.

(4) Each order authorizing or approving the interception of any wire, oral, or electronic communication under this chapter shall specify—

(a) the identity of the person, if known, whose communications are to be intercepted;

(b) the nature and location of the communications facilities as to which, or the place where, authority to intercept is granted;

(c) a particular description of the type of communication sought to be intercepted, and a statement of the particular offense to which it relates;

(d) the identity of the agency authorized to intercept the communications, and of the person authorizing the application; and

(e) the period of time during which such interception is authorized, including a statement as to whether or not the interception shall automatically terminate when the described communication has been first obtained.

An order authorizing the interception of a wire, oral, or electronic communication under this chapter shall, upon request of the applicant, direct that a provider of wire or electronic communication service, landlord, custodian or other person shall furnish the applicant forthwith all information, facilities, and technical assistance necessary to accomplish the interception unobtrusively and with a minimum of interference with the services that such service provider, landlord, custodian, or person is according the person whose communications are to be intercepted. Any provider of wire or electronic communication service, landlord, custodian or other person furnishing such facilities or technical assistance shall be compensated therefor by the applicant for reasonable expenses incurred in providing such facilities or assistance. Pursuant to Section 2522 of this chapter, an order may also be issued to enforce the assistance capability and capacity requirements under the Communications Assistance for Law Enforcement Act.

(5) No order entered under this section may authorize or approve the interception of any wire, oral, or electronic communication for any period longer than is necessary to achieve the objective of the authorization, nor in any event longer than thirty days. Such thirty-day period begins on the earlier of the day on which the investigative or law enforcement officer first begins to conduct an interception under the order or ten days after the order is entered. Extensions of an order may be granted, but only upon application for an extension made in accordance with subsection (1) of this section and the court making the findings required by subsection (3) of this section. The period of extension shall be no longer than the authorizing judge deems necessary to achieve the purposes for which it was granted and in no event for longer than thirty days. Every order and extension thereof shall contain a provision that the authorization to intercept shall be executed as soon as practicable, shall be conducted in such a way as to minimize the interception of communications not otherwise subject to interception under this chapter, and must terminate upon attainment of the authorized objective, or in any event in thirty days. In the event the intercepted communication is in a code or foreign language, and an expert in that foreign language or code is not reasonably available during the interception period, minimization may be accomplished as soon as practicable after such interception. An interception under this chapter may be conducted in whole or in part by Government personnel, or by an individual operating under a contract with the Government, acting under the supervision of an investigative or law enforcement officer authorized to conduct the interception.

(6) Whenever an order authorizing interception is entered pursuant to this chapter, the order may require reports to be made to the judge who issued the order showing what progress has been made toward achievement of the authorized objective and the need for continued interception. Such reports shall be made at such intervals as the judge may require.

(7) Notwithstanding any other provision of this chapter, any investigative or law enforcement officer, specially designated by the Attorney General, the Deputy Attorney General, the Associate Attorney General, or by the principal prosecuting attorney of any State or subdivision thereof acting pursuant to a statute of that State, who reasonably determines that—

(a) an emergency situation exists that involves—

(i) immediate danger of death or serious physical injury to any person,

(ii) conspiratorial activities threatening the national security interest, or

(iii) conspiratorial activities characteristic of organized crime,

that requires a wire, oral, or electronic communication to be intercepted before an order authorizing such interception can, with due diligence, be obtained, and

(b) there are grounds upon which an order could be entered under this chapter to authorize such interception,

may intercept such wire, oral, or electronic communication if an application for an order approving the interception is made in accordance with this section within forty-eight hours after the interception has occurred, or begins to occur. In the absence of an order, such interception shall immediately terminate when the communication sought is obtained or when the application for the order is denied, whichever is earlier. In the event such application for approval is denied, or in any other case where the interception is terminated without an order having been issued, the contents of any wire, oral, or electronic communication intercepted shall be treated as having been obtained in violation of this chapter, and an inventory shall be served as provided for in subsection (d) of this section on the person named in the application.

(8)

(a) The contents of any wire, oral, or electronic communication intercepted by any means authorized by this chapter shall, if possible, be recorded on tape or wire or other comparable device. The recording of the contents of any wire, oral, or electronic communication under this subsection shall be done in such a way as will protect the recording from editing or other alterations. Immediately upon the expiration of the period of the order, or extensions thereof, such recordings shall be made available to the judge issuing such order and sealed under his directions. Custody of the recordings shall be wherever the judge orders. They shall not be destroyed except upon an order of the issuing or denying judge and in any event shall be kept for ten years. Duplicate recordings may be made for use or disclosure pursuant to the provisions of subsections (1) and (2) of Section 2517 of this chapter for investigations. The presence of the seal provided for by this subsection, or a satisfactory explanation for the absence thereof, shall be a prerequisite for the use or disclosure of the contents of any wire, oral, or electronic communication or evidence derived therefrom under subsection (3) of Section 2517.

(b) Applications made and orders granted under this chapter shall be sealed by the judge. Custody of the applications and orders shall be wherever the judge directs. Such applications and orders shall be disclosed only upon a showing of good cause before a judge of competent jurisdiction and shall not be destroyed except on order of the issuing or denying judge, and in any event shall be kept for ten years.

(c) Any violation of the provisions of this subsection may be punished as contempt of the issuing or denying judge.

(d) Within a reasonable time but not later than ninety days after the filing of an application for an order of approval under Section 2518(7)(b) which is denied or the termination of the period of an order or extensions thereof, the issuing or denying judge shall cause to be served, on the persons named in the order or the application, and such other parties to intercepted communications as the judge may determine in his discretion that is in the interest of justice, an inventory which shall include notice of—

(1) the fact of the entry of the order or the application;

(2) the date of the entry and the period of authorized, approved or disapproved interception, or the denial of the application; and

(3) the fact that during the period wire, oral, or electronic communications were or were not intercepted.

The judge, upon the filing of a motion, may in his discretion make available to such person or his counsel for inspection such portions of the intercepted communications, applications and orders as the judge determines to be in the interest of justice. On an ex parte showing of good cause to a judge of competent jurisdiction the serving of the inventory required by this subsection may be postponed.

(9) The contents of any wire, oral, or electronic communication intercepted pursuant to this chapter or evidence derived therefrom shall not be received in evidence or otherwise disclosed in any trial, hearing, or other proceeding in a Federal or State court unless each party, not less than ten days before the trial, hearing, or proceeding, has been furnished with a copy of the court order, and accompanying application, under which the interception was authorized or approved. This ten-day period may be waived by the judge if he finds that it was not possible to furnish the party with the above information ten days before the trial, hearing, or proceeding and that the party will not be prejudiced by the delay in receiving such information.

(10)

(a) Any aggrieved person in any trial, hearing, or proceeding in or before any court, department, officer, agency, regulatory body, or other authority of the United States, a State, or a political subdivision thereof, may move to suppress the contents of any wire or oral communication intercepted pursuant to this chapter, or evidence derived therefrom, on the grounds that—

(i) the communication was unlawfully intercepted;

(ii) the order of authorization or approval under which it was intercepted is insufficient on its face; or

(iii) the interception was not made in conformity with the order of authorization or approval.

Such motion shall be made before the trial, hearing, or proceeding unless there was no opportunity to make such motion or the person was not aware of the grounds of the motion. If the motion is granted, the contents of the intercepted wire or oral communication, or evidence derived therefrom, shall be treated as having been obtained in violation of this chapter. The judge, upon the filing of such motion by the aggrieved person, may in his discretion make available to the aggrieved person or his counsel for inspection such portions of the intercepted communication or evidence derived therefrom as the judge determines to be in the interests of justice.

(b) In addition to any other right to appeal, the United States shall have the right to appeal from an order granting a motion to suppress made under paragraph (a) of this subsection, or the denial of an application for an order of approval, if the United States attorney shall certify to the judge or other official granting such motion or denying such application that the appeal is not taken for purposes of delay. Such appeal shall be taken within thirty days after the date the order was entered and shall be diligently prosecuted.

(c) The remedies and sanctions described in this chapter with respect to the interception of electronic communications are the only judicial remedies and sanctions for nonconstitutional violations of this chapter involving such communications.

(11) The requirements of subsections (1)(b)(ii) and (3)(d) of this section relating to the specification of the facilities from which, or the place where, the communication is to be intercepted do not apply if—

(a) in the case of an application with respect to the interception of an oral communication—

(i) the application is by a Federal investigative or law enforcement officer and is approved by the Attorney General, the Deputy Attorney General, the Associate Attorney General, an Assistant Attorney General, or an acting Assistant Attorney General;

(ii) the application contains a full and complete statement as to why such specification is not practical and identifies the person committing the offense and whose communications are to be intercepted; and

(iii) the judge finds that such specification is not practical; and

(b) in the case of an application with respect to a wire or electronic communication—

(i) the application is by a Federal investigative or law enforcement officer and is approved by the Attorney General, the Deputy Attorney General, the Associate Attorney General, an Assistant Attorney General, or an acting Assistant Attorney General;

(ii) the application identifies the person believed to be committing the offense and whose communications are to be intercepted and the applicant makes a showing that there is probable cause to believe that the person's actions could have the effect of thwarting interception from a specified facility;

(iii) the judge finds that such showing has been adequately made; and

(iv) the order authorizing or approving the interception is limited to interception only for such time as it is reasonable to presume that the person identified in the application is or was reasonably proximate to the instrument through which such communication will be or was transmitted.

(12) An interception of a communication under an order with respect to which the requirements of subsections (1)(b)(ii) and (3)(d) of this section do not apply by reason of subsection (11)(a) shall not begin until the place where the communication is to be intercepted is ascertained by the person implementing the interception order. A provider of wire or electronic communications service that has received an order as provided for in subsection (11)(b) may move the court to modify or quash the order on the ground that its assistance with respect to the interception cannot be performed in a timely or reasonable fashion. The court, upon notice to the government, shall decide such a motion expeditiously.

18 U.S. Code § 2701—Unlawful access to stored communications

(a) Offense.—Except as provided in subsection (c) of this section whoever—

(1) intentionally accesses without authorization a facility through which an electronic communication service is provided; or

(2) intentionally exceeds an authorization to access that facility;

and thereby obtains, alters, or prevents authorized access to a wire or electronic communication while it is in electronic storage in such system shall be punished as provided in subsection (b) of this section.

(…)

(c) Exceptions.—Subsection (a) of this section does not apply with respect to conduct authorized—

(1) by the person or entity providing a wire or electronic communications service;

(2) by a user of that service with respect to a communication of or intended for that user; or

(3) in Sections 2703, 2704 or 2518 of this title.

18 U.S. Code § 2711—Definitions for chapter

As used in this chapter—

(1) the terms defined in Section 2510 of this title have, respectively, the definitions given such terms in that section;

(2) the term "remote computing service" means the provision to the public of computer storage or processing services by means of an electronic communications system;

(3) the term "court of competent jurisdiction" includes—

(A) any district court of the United States (including a magistrate judge of such a court) or any United States court of appeals that—

(i) has jurisdiction over the offense being investigated;

(ii) is in or for a district in which the provider of a wire or electronic communication service is located or in which the wire or electronic communications, records, or other information are stored; or

(iii) is acting on a request for foreign assistance pursuant to Section 3512 of this title; or

(B) a court of general criminal jurisdiction of a State authorized by the law of that State to issue search warrants; and

(4) the term "governmental entity" means a department or agency of the United States or any State or political subdivision thereof.

18 U.S. Code § 3121—General prohibition on pen register and trap and trace device use; exception
(a) In General.—

Except as provided in this section, no person may install or use a pen register or a trap and trace device without first obtaining a court order under Section 3123 of this title or under the Foreign Intelligence Surveillance Act of 1978 (50 U.S.C. 1801 et seq.).

(b) Exception.—The prohibition of subsection (a) does not apply with respect to the use of a pen register or a trap and trace device by a provider of electronic or wire communication service—

(1) relating to the operation, maintenance, and testing of a wire or electronic communication service or to the protection of the rights or property of such provider, or to the protection of users of that service from abuse of service or unlawful use of service; or

(2) to record the fact that a wire or electronic communication was initiated or completed in order to protect such provider, another provider furnishing service toward the completion of the wire communication, or a user of that service, from fraudulent, unlawful or abusive use of service; or (3) where the consent of the user of that service has been obtained.

(c) Limitation.—

A government agency authorized to install and use a pen register or trap and trace device under this chapter or under State law shall use technology reasonably available to it that restricts the recording or decoding of electronic or other impulses to the dialing, routing, addressing, and signaling information utilized in the processing and transmitting of wire or electronic communications so as not to include the contents of any wire or electronic communications.

(d) Penalty.—

Whoever knowingly violates subsection (a) shall be fined under this title or imprisoned not more than one year, or both.

18 U.S. Code § 3122—Application for an order for a pen register or a trap and trace device

(a) Application.—

(1) An attorney for the Government may make application for an order or an extension of an order under Section 3123 of this title authorizing or approving the installation and use of a pen register or a trap and trace device under this chapter, in writing under oath or equivalent affirmation, to a court of competent jurisdiction.

(2) Unless prohibited by State law, a State investigative or law enforcement officer may make application for an order or an extension of an order under Section 3123 of this title authorizing or approving the installation and use of a pen register or a trap and trace device under this chapter, in writing under oath or equivalent affirmation, to a court of competent jurisdiction of such State.

(b) Contents of Application.—An application under subsection (a) of this section shall include—

(1) the identity of the attorney for the Government or the State law enforcement or investigative officer making the application and the identity of the law enforcement agency conducting the investigation; and

(2) a certification by the applicant that the information likely to be obtained is relevant to an ongoing criminal investigation being conducted by that agency.

18 U.S. Code § 3123—Issuance of an order for a pen register or a trap and trace device

(a) In General.—

(1) Attorney for the government.—

Upon an application made under Section 3122(a)(1), the court shall enter an ex parte order authorizing the installation and use of a pen register or trap and trace device anywhere within the United States, if the court finds that the attorney for the Government has certified to the court that the information likely to be obtained by such installation and use is relevant to an ongoing criminal investigation. The order, upon service of that order, shall apply to any person or entity providing wire or electronic communication service in the United States whose assistance may facilitate the execution of the order. Whenever such an order is served on any person or entity not specifically named in the order, upon request of such person or entity, the attorney for the Government or law enforcement or investigative officer that is serving the order shall provide written or electronic certification that the order applies to the person or entity being served.

(2) State investigative or law enforcement officer.—

Upon an application made under Section 3122(a)(2), the court shall enter an ex parte order authorizing the installation and use of a pen register or trap and trace device within the jurisdiction of the court, if the court finds that the State law enforcement or investigative officer has certified to the court that the information likely to be obtained by such installation and use is relevant to an ongoing criminal investigation.

(3)

(A) Where the law enforcement agency implementing an ex parte order under this subsection seeks to do so by installing and using its own pen register or trap and trace device on a packet-switched data network of a provider of electronic communication service to the public, the agency shall ensure that a record will be maintained which will identify—

(i) any officer or officers who installed the device and any officer or officers who accessed the device to obtain information from the network;

(ii) the date and time the device was installed, the date and time the device was uninstalled, and the date, time, and duration of each time the device is accessed to obtain information;

(iii) the configuration of the device at the time of its installation and any subsequent modification thereof; and

(iv) any information which has been collected by the device.

To the extent that the pen register or trap and trace device can be set automatically to record this information electronically, the record shall be maintained electronically throughout the installation and use of such device.

(B) The record maintained under subparagraph (A) shall be provided ex parte and under seal to the court which entered the ex parte order authorizing the installation and use of the device within 30 days after termination of the order (including any extensions thereof).

(b) Contents of Order.—An order issued under this section—

(1) shall specify—

(A) the identity, if known, of the person to whom is leased or in whose name is listed the telephone line or other facility to which the pen register or trap and trace device is to be attached or applied;

(B) the identity, if known, of the person who is the subject of the criminal investigation;

(C) the attributes of the communications to which the order applies, including the number or other identifier and, if known, the location of the telephone line or other facility to which the pen register or trap and trace device is to be attached or applied, and, in the case of an order authorizing installation and use of a trap and trace device under subsection (a)(2), the geographic limits of the order; and

(D) a statement of the offense to which the information likely to be obtained by the pen register or trap and trace device relates; and

(2) shall direct, upon the request of the applicant, the furnishing of information, facilities, and technical assistance necessary to accomplish the installation of the pen register or trap and trace device under Section 3124 of this title.

(c) Time Period and Extensions.—

(1) An order issued under this section shall authorize the installation and use of a pen register or a trap and trace device for a period not to exceed sixty days.

(2) Extensions of such an order may be granted, but only upon an application for an order under Section 3122 of this title and upon the judicial finding required by subsection (a) of this section. The period of extension shall be for a period not to exceed sixty days.

(d) Nondisclosure of Existence of Pen Register or a Trap and Trace Device.—An order authorizing or approving the installation and use of a pen register or a trap and trace device shall direct that—

(1) the order be sealed until otherwise ordered by the court; and

(2) the person owning or leasing the line or other facility to which the pen register or a trap and trace device is attached or applied, or who is obligated by the order to provide assistance to the applicant, not disclose the existence of the pen register or trap and trace device or the existence of the investigation to the listed subscriber, or to any other person, unless or until otherwise ordered by the court.

18 U.S. Code § 3127—Definitions for chapter

(…)

(3) the term "pen register" means a device or process which records or decodes dialing, routing, addressing, or signaling information transmitted by an instrument or facility from which a wire or electronic communication is transmitted, provided, however, that such information shall not include the contents of any

communication, but such term does not include any device or process used by a provider or customer of a wire or electronic communication service for billing, or recording as an incident to billing, for communications services provided by such provider or any device or process used by a provider or customer of a wire communication service for cost accounting or other like purposes in the ordinary course of its business;

(4) the term "trap and trace device" means a device or process which captures the incoming electronic or other impulses which identify the originating number or other dialing, routing, addressing, and signaling information reasonably likely to identify the source of a wire or electronic communication, provided, however, that such information shall not include the contents of any communication;

(…)

References

Buckman D F (2016) Validity, construction, and application of 18 U.S.C.A. § 2252A(a), Proscribing certain activities relating to material constituting or containing child pornography. American Law Reports. Thomson Reuters, New York. 533.

Computer Crime and Intellectual Property Section of the Department of Justice (2009) Searching and seizing computers and obtaining electronic evidence in criminal investigations. https://www.justice.gov/sites/default/files/criminal-ccips/legacy/2015/01/14/ssmanual2009.pdf.

Doyle C (2015) Attempt: an overview of federal criminal law. https://www.fas.org/sgp/crs/misc/R42001.pdf.

Grenig J E, Gleisner W C III (2016) eDiscovery & digital evidence. Thomson West, Eagan, Minnesota.

Kerr O S (2004) A user's guide to the Stored Communications Act, and a legislator's guide to amending it. George Washington Law Review: 1208–1249.

LaFave R V (2016) Substantive criminal law, 2nd edn. Thomson West, Eagan, Minnesota.

Rogers A (2004) New technology, old defenses: internet sting operations and attempt liability. University of Richmond Law Review: 477–523.

Scolnik A (2009) Protections for electronic communications: the Stored Communications Act and the Fourth Amendment. Fordham L Rev: 349–397.

Rutkow L, Lozman J T (2006) Suffer the children?: A call for United States Ratification of the United Nations Convention on the Rights of the Child. Harvard Human Rights Journal: 161.

Relevant Case Law

U.S. Supreme Court

Ashcroft v. Free Speech Coal., 535 U.S. 234 (2002)
Braxton v. United States, 500 U.S. 344 (1991)
Burdeau v. McDowell, 256 U.S. 465 (1921)
Jacobson v. United States, 503 U.S. 540 (1992)
United States v. Jones, 132 S. Ct. 945 (2012)
Katz v. United States, 389 U.S. 347 (1967)
Mathews v. United States, 485 U.S. 58 (1988)
Miller v. California, 413 U.S. 15 (1973)

Riley v. California, 134 S. Ct. 2473 (2014)
Smith v. Maryland, 442 U.S. 735 (1979)
Sorrells v. United States, 287 U.S. 435 (1932)
Weeks v. United States, 232 U.S. 383 (1914)

Federal Appeals Courts
Ordered by circuit and alphabetically.
United States v. Gagliardi, 506 F.3d 140 (2d Cir. 2007)
United States v. Manley, 632 F.2d 978 (2d Cir. 1980)
United States v. Tykarsky, 446 F.3d 458 (3d Cir. 2006)
United States v. Barlow, 568 F.3d 215 (5th Cir. 2009)
United States v. Farner, 251 F.3d 510 (5th Cir. 2001)
United States v. Maddox, 492 F.2d 104 (5th Cir. 1974)
United States v. Nichols, 371 Fed. Appx. 546 (5th Cir. 2010)
United States v. Rudzavice, 586 F.3d 310 (5th Cir. 2009)
United States v. Cochran, 534 F.3d 631 (7th Cir. 2008)
United States v. Cote, 504 F.3d 682 (7th Cir. 2007)
United States v. Johnson, 376 F.3d 689 (7th Cir. 2004)
Batsell v. United States, 403 F.2d 395 (8th Cir. 1968)
United States v. Pierson, 544 F.3d 933 (8th Cir. 2008)
United States v. Spurlock, 495 F.3d 1011 (8th Cir. 2007)
United States v. Jones, 231 F.3d 508 (9th Cir. 2000)
United States v. Thomas, 893 F.2d 1066 (9th Cir. 1990)
United States v. Perrine, 518 F.3d 1196 (10th Cir. 2008)
United States v. Sims, 428 F.3d 945 (10th Cir. 2005)
United States v. Tucker, 305 F.3d 1193 (10th Cir. 2002)
United States v. Farley, 607 F.3d 1294 (11th Cir. 2010)
United States v. Handley, 564 F. Supp. 2d 996 (S.D. Iowa 2008)
United States v. Jackson, 488 F. Supp. 2d 866 (D. Neb. 2007)
Kearney v. Salomon Smith Barney, Inc., S124739 (Sup. Ct. Cal. July 13, 2006)

Relevant Law

18 U.S.C. Chapter 71—Obscenity
18 U.S.C. Chapter 109a—Sexual Abuse
18 U.S.C. Chapter 110—Sexual Exploitation and Other Abuse of Children
18 U.S.C. Chapter 117—Transportation for Illegal Sexual Activity and Related Crimes
18 U.S.C. Chapter 119—Wire and Electronic Communications Interception and Interception of Oral Communications
18 U.S.C. Chapter 121—Stored Communications Act
18 U.S.C. Chapter 206—Pen Register and Trap And Trace Devices

Jonathan Unikowski is the founder of Lexloci, a legal technology startup which is the maker of a research tool that makes comparing legal documents across multiple jurisdictions fast, easy, and intuitive. Prior to founding Lexloci, Jonathan Unikowski was a Legal Fellow at the Wikimedia Foundation in San Francisco, where he advised on U.S. and International privacy law and intellectual property law for the Wikimedia Foundation's projects, and primarily Wikipedia, the free encyclopedia. Jonathan Unikowski is a graduate of the University of California, Berkeley (LL. M. '15) and the University of Brussels, Belgium (Master of Laws, Magna cum Laude, '14)

The manufacturer's authorised representative in the EU is Springer Nature Customer Service Centre GmbH, Europaplatz 3, 69115 Heidelberg, Germany. If you have any concerns regarding our products, please contact ProductSafety@springernature.com

Printed and bound by CPI Group (UK) Ltd, Croydon, CR0 4YY
30/09/2026
02225449-0017